AF386163

Progress in Mathematics
Volume 227

Series Editors
Hyman Bass
Joseph Oesterlé
Alan Weinstein

James Lepowsky

Haisheng Li

Introduction to Vertex Operator Algebras and Their Representations

Springer Science+Business Media, LLC

James Lepowsky
Department of Mathematics
Rutgers University
Hill Center, Busch Campus
Piscataway, NJ 08854
U.S.A.

Haisheng Li
Department of Mathematics
Rutgers University
Armitage Hall
Camden, NJ 08102
U.S.A.

Library of Congress Cataloging-in-Publication Data
Lepowsky, J. (James)
  Introduction to vertex operator algebras and their representations / James Lepowsky,
Haisheng Li.
    p. cm. – (Progress in mathematics ; v. 227)
  Includes bibliographical references and index.
  ISBN 978-1-4612-6480-4   ISBN 978-0-8176-8186-9 (eBook)
  DOI 10.1007/978-0-8176-8186-9
  1. Vertex operator algebras. 2. Representations of algebras. I. Li, Haisheng, 1962- II.
Title. III. Progress in mathematics (Boston, Mass.); v. 227.

QA326.L48 2004
512'.55–dc22

2003063839
CIP

AMS Subject Classifications: Primary: 17B69, 81T40; Secondary: 17B67, 17B68, 81R10

ISBN 978-1-4612-6480-4   Printed on acid-free paper.

To Lael and Mei

To Dongyuan, Joyce and Jimmy

Contents

Preface

Vertex operator algebra theory is a new area of mathematics. It has been an exciting and ever-growing subject from the beginning, starting even before R. Borcherds introduced the precise mathematical notion of "vertex algebra" in the 1980s [B1]. Having developed in conjunction with string theory in theoretical physics and with the theory of "monstrous moonshine" and infinite-dimensional Lie algebra theory in mathematics, vertex (operator) algebra theory is qualitatively different from traditional algebraic theories, reflecting the "nonclassical" nature of string theory and of monstrous moonshine. The theory has revealed new perspectives that were unavailable without it, and continues to do so.

"Monstrous moonshine" began as an astonishing set of conjectures relating the Monster finite simple group to the theory of modular functions in number theory. As is now known, vertex operator algebra theory is a foundational pillar of monstrous moonshine. With the theory available, one can formulate and try to solve new problems that have far-reaching implications in a wide range of fields that had not previously been thought of as being related.

This book systematically introduces the theory of vertex (operator) algebras from the beginning, using "formal calculus," and takes the reader through the fundamental theory to the detailed construction of examples. The axiomatic foundations of vertex operator algebras and modules are studied in detail, general construction theorems for vertex operator algebras and modules are presented, and the most basic families of vertex operator algebras are constructed and their irreducible modules are constructed and are also classified. The construction theorems for algebras and modules are based on a study of *representations* of a vertex operator algebra, as opposed to *modules* for the algebra, as developed in [Li3]. A significant feature of the theory is that in general, the construction of modules for (or representations of) a vertex operator algebra is in some sense more subtle than the construction of the algebra itself. With the body of theory presented in this book as background, the reader will be well prepared to embark on any of a vast range of directions in the theory and its applications.

In the introduction, we shall sketch the theory and the contents of this book and provide motivation. We have written this book to be self-contained. The only prerequisites are some familiarity with Lie algebras and a desire to learn vertex operator algebra theory. If the reader can follow the mathematical material in the introduction, then he or she will have no trouble following the whole book.

This book is suitable for a first graduate course on vertex operator algebra theory, and in fact, that is how we have been using drafts of this book while writing and expanding it over the years.

As we have mentioned, our treatment is axiomatic, leading to the construction of examples. By contrast, the original book [FLM6] developed examples and the theory of *vertex operators* before presenting the concept of *vertex operator algebra* as a "conceptual summary" of the fundamental properties that had been built up through an ever-deepening study of certain basic structures (structures that were needed in the construction of the "moonshine module" for the Monster finite simple group in [FLM6]). That book is not a prerequisite for this book, but the reader may wish to consult [FLM6] to get a feeling for how vertex operator algebra theory originally developed through examples and through the solution of problems.

We continue to use the historical terms "vertex algebra," which was the original term of Borcherds, and "vertex operator algebra," which was used in [FLM6] for a certain variant notion. (Both notions are discussed at great length in this book.) A vertex operator algebra satisfies a few axioms that are not part of the definition of vertex algebra, but this is a relatively minor issue; as we point out in the introduction, there are many useful variants of the definition, depending on the context that one is interested in. It is the conceptual features of the notion of vertex (operator) algebra that are emphasized in the definitions in [B1] and [FLM6] that have led us to continue to use these two pieces of terminology as they have (usually) been used up to now. These conceptual features are discussed in the introduction and in the main text of the book. For instance, the two different terms reflect the issue of whether the algebras are viewed as "algebras of operators" or as algebras with a "multiplication operation" (or rather, an infinite sequence of multiplication operations); both viewpoints are "correct," and the best thing is to keep both in mind.

Primarily expository, this book does present a number of new results developed by the authors over the years in which drafts of this book were used in courses and seminars. These results are mentioned at the end of the introduction. In addition, in many places the proofs and treatments go beyond the originals, sometimes quite a bit.

The notion of vertex (operator) *super*algebra is a slight generalization of the notion of vertex (operator) algebra, and the basic general theory of vertex (operator) superalgebras is almost identical to the theory of vertex (operator) algebras; one needs to systematically keep track of $\pm$ signs at every stage. For simplicity of exposition, in this book we treat only vertex (operator) algebras. The original paper [Li3] *was* in fact written in the generality of vertex (operator) superalgebras, and the interested reader can consult [Li3] for the necessary (minor) changes.

The work on this book began when one author (J.L.) used the Japan lecture notes [Li2] by the other author (H.L.) in lectures presenting the work [Li3] for a graduate course at Rutgers University. H.L.'s work developed a theory of "representations" (in a specific sense) of a vertex operator algebra, and used it to construct families of examples of vertex operator algebras and modules; before [Li3], there had been no *general* approach available for constructing families of examples of vertex operator algebras and modules. Over a period of years, through many graduate courses and lectures by both authors, this book grew from informal lecture notes to a detailed self-contained introduction to the theory of vertex operator algebras, starting from the elementary fundamentals and leading to [Li3], and in fact, as it turned out, to a greatly expanded treatment of [Li3].

Actually, [Li3] was the published form of half of a 1993 preprint by H.L., and the other half of this 1993 preprint was published as [Li4]. H.L. is very grateful to Weiqiang Wang, then a graduate student at M.I.T., for his early interest in this preprint. H.L. would also like to thank Masahiko Miyamoto for organizing the wonderful workshop in the summer of 1994 at the Research Institute for Mathematical Sciences, Kyoto University, in which Chongying Dong, Yi-Zhi Huang and H.L. were invited to give lectures on vertex operator algebras. We are very grateful to many students and colleagues in lectures and seminars over the years, especially Chengming Bai, Corina Calinescu, Benjamin Doyon, Liang Kong, Machiel van Frankenhuysen and Lin Zhang. Their insightful comments and questions continually led us to seek increasingly better ways of illuminating critical and subtle points, and also of enhancing the readability by means of further explanatory remarks and cross-referencing. We thank Martin Karel for reading the manuscript, and Chongying Dong, Arne Meurman, Mirko Primc, and all the reviewers for their helpful comments. For the past two years, H.L. has used the text of this book in lectures at Harbin Normal University, Harbin, China. He would like to thank Shutao Chen, Yuwen Wang, Wende Liu, Shuqing Wang and the Mathematics Department for their hospitality, and to thank all the participants, Wende Liu, Shuqing Wang, Yinhua He, Li Li, Liqin Liu, Xuemei Liu and Xiuling Wang, for their interest and enthusiasm.

The formal calculus formulation of the theory of vertex operator algebras, including the Jacobi identity axiom, comes from [FLM6] and [FHL]. The "weak commutativity" and "weak associativity" properties come from [DL3]. J.L. would like to express his thanks to his collaborators on these works: Igor Frenkel, Arne Meurman, Yi-Zhi Huang and Chongying Dong.

The treatment in [DL3] actually included the vertex (operator) *super*algebra case, and in fact, this superalgebra case was itself a very special case of the full generality of [DL3]—that of "generalized vertex algebras" and "abelian intertwining algebras." Structures similar to generalized vertex algebras were also introduced and studied in [FFR] and [Mos1]. The special case of vertex (operator) superalgebras had already been studied in [Go1] and [Ts2].

It is a pleasure to thank Ann Kostant for encouraging us to publish this book with Birkhäuser and for her and her Birkhäuser colleagues' wonderful work in making the publication process so smooth.

Both authors gratefully acknowledge the support of the National Science Foundation. In addition, H.L. thanks the National Security Agency and the Rutgers University Research Council.

*Introduction to
Vertex Operator Algebras
and Their Representations*

1

Introduction

1.1 Motivation

Vertex operator algebra theory has deep roots and deep applications in mathematics and physics. Both string theory in physics and "monstrous moonshine" in mathematics played crucial roles in the development of the theory, as we shall sketch briefly.

String theory, its initial version having been introduced in the late 1960s, is based on the premise that elementary particles manifest themselves as "vibrational modes" of fundamental *strings* moving through space, rather than as moving *points*, according to quantum field-theoretic principles (see [Wi2], [GSW], [Po], [De], [Gre] for presentations). A string sweeps out a two-dimensional "world-sheet" in space-time. It is fruitful to focus on the case in which the surface is a Riemann surface locally parametrized by a complex coordinate. The result is *two-dimensional conformal (quantum) field theory*, formalized in an algebraic spirit (on a physical level of rigor), after having been developed extensively by many physicists, by A. Belavin, A. Polyakov and A. Zamolodchikov [BPZ], who were concerned primarily with critical phenomena in two-dimensional statistical mechanics. A certain new kind of algebra of operators was emerging, called "operator algebras" in [BPZ], based on the "operator product expansion" [W1] in quantum field theory. Roughly speaking, the elements of these "algebras" are certain types of *vertex operators*. These are operators of types introduced in the early days of string theory in order to describe certain kinds of physical interactions, in the context of string theory, at a "vertex," where, for instance, two particles (or strings) enter, and as a result of the interaction, one particle (or string) exits. These "algebra elements" have certain physically motivated properties, properties that were indeed known or expected to hold, or postulated to hold, in many situations.

Meanwhile, in mathematics, the vast program to classify the finite simple groups was drawing to a close around 1980 (see [Gor] for a survey). Most of the finite simple groups—the Chevalley groups [Ch] and variants—belong to infinite families related to Lie groups. What turned out to be the largest "sporadic" (not belonging to one of the infinite families) finite simple group, the Fischer–Griess Monster M, had been predicted by B. Fischer and R. Griess and was constructed by Griess [G3] as a symmetry group (of order about 10^{54}) of a commutative, but *very highly nonassociative*, seemingly ad hoc new algebra $\mathbb{B}$ of dimension 196883. The precise structure of the Griess algebra $\mathbb{B}$ was "forced" by expected properties of the conjectured-to-exist Monster. It was proved by J. Tits that $\mathbb{M}$ is actually the *full* automorphism group of $\mathbb{B}$.

A bit earlier, J. McKay, J. Thompson, J. Conway and S. Norton had discovered astounding "numerology" culminating in the *"monstrous moonshine" conjectures* (see [Th2], [CN]; see also [O]) concerning the not-yet-proved-to-exist Monster $\mathbb{M}$, namely:

There should exist an infinite-dimensional $\mathbb{Z}$-graded module for $\mathbb{M}$ (i.e., representation of $\mathbb{M}$)

$$V = \coprod_{n \in \mathbb{Z}} V_n, \tag{1.1.1}$$

or more specifically,

$$V = \coprod_{n=-1,0,1,2,3,\dots} V_n, \tag{1.1.2}$$

such that

$$\sum_{n \geq -1} (\dim V_n) q^n = J(q), \tag{1.1.3}$$

where

$$J(q) = q^{-1} + 0 + 196884q + 21493760q^2 + \cdots \tag{1.1.4}$$

(note: 196884=196883+1—McKay's initial observation). Here $J(q)$ is the classical modular function with its constant term set to 0—the (suitably normalized) generator of the field of $SL(2, \mathbb{Z})$-modular invariant functions, with $q = e^{2\pi i \tau}$, τ in the upper half-plane. More generally, for every $g \in \mathbb{M}$ (not just $g = 1$), the generating function

$$\sum_{n \geq -1} (\mathrm{tr}\ g|_{V_n}) q^n \tag{1.1.5}$$

should be the analogous "Hauptmodul" for a suitable discrete subgroup of $SL(2, \mathbb{R})$, a subgroup having a fundamental "genus-zero" property, so that its associated field of modular-invariant functions has a single (suitably normalized) generator—the Hauptmodul.

Proving this conjecture would give a remarkable connection between classical number theory and "nonclassical" sporadic group theory. After Griess constructed $\mathbb{M}$, I. Frenkel, J. Lepowsky and A. Meurman ([FLM2], [FLM6]) proved the McKay-Thompson conjecture—that there should exist a natural infinite-dimensional $\mathbb{Z}$-graded $\mathbb{M}$-module V satisfying the condition (1.1.3). This was done by means of an explicit (and necessarily elaborate) construction of such a structure V, called the "moonshine module $V^{\natural}$" in [FLM2] and [FLM6] because of its naturality and because of the strong expectation that for *all* the elements of the Monster, the series (1.1.5) with $V = V^{\natural}$ would be the desired Hauptmoduls. (For only some, not all, elements of the Monster were the series (1.1.5) determined as part of the construction [FLM2]). A detailed guide to this construction and its relations with string theory, finite group theory, Lie algebra theory and number theory, including a brief sketch of each of these subjects in relation to their connection with the work in [FLM2] and [FLM6], can be found in the introductory material in [FLM6]; the book itself is devoted to the proof of the McKay–Thompson conjecture.

In particular, the construction [FLM2] heavily uses a network of types of *vertex operators* and their algebraic structure and relations, yielding the structure $V^\natural$ and a certain "algebra of vertex operators" acting on it, in such a way that the Monster is realized as the automorphism group of this algebra of vertex operators. The Monster, a *finite* group, was now understood in terms of a natural *infinite-dimensional* structure. At the same time, the finite-dimensional Griess algebra now gained an elegant interpretation by virtue of being embedded in a natural new algebra of vertex operators acting canonically on an *infinite-dimensional* space.

Moreover, in $V^\natural$ the 196883-dimensional algebra $\mathbb{B}$ finds itself embedded in a 196884-dimensional enlargement $\mathcal{B}$ of $\mathbb{B}$, with an identity element adjoined, and this identity element of $\mathcal{B}$ gives rise to a copy of the Virasoro Lie algebra acting on $V^\natural$. The Virasoro algebra had long been known as a fundamental "symmetry algebra" in string theory and conformal field theory. A central extension of the Lie algebra of formal vector fields in one dimension, it had been discovered in three different contexts, in [Bl], [GF] and [V]. The numerical parameter defining the action of the suitably normalized central generator of the Virasoro algebra is called the *central charge* of a given representation of the Virasoro algebra. The central charge of the moonshine module is 24. The Virasoro algebra is defined in Section 1.3 below and is discussed in much greater detail in the main text of this work.

During the "first string theory revolution," which started in the summer of 1984 [GSc], this new structure $V^\natural$ in turn came to be interpreted in retrospect as an inherently *string-theoretic* structure: the "chiral algebra" underlying the $\mathbb{Z}_2$-orbifold conformal field theory based on the 24-dimensional Leech lattice; again see the discussion in [FLM6].

Then R. Borcherds [B1] introduced the precise axiomatic notion of "vertex algebra," which in a natural way extended the relations for the vertex operators for $V^\natural$ and also many other known mathematical and physical features of vertex operators. Borcherds's axioms, which are given in Section 3.6 below, turned out in retrospect to be essentially equivalent to the Belavin–Polyakov–Zamolodchikov physical axioms for the "operator algebras," which have come to be called "chiral algebras," that underlie conformal field theory and string theory (see [MSe]).

Borcherds asserted in [B1] that $V^\natural$ admits a vertex algebra structure, generated by the algebraic structure constructed in [FLM2], on which $\mathbb{M}$ (still) acts as a symmetry group. In [FLM6] this assertion was proved, by a "lifting" of the proof of the theorem of [FLM2] to a context incorporating the desired full vertex algebra structure. Also in [FLM6], a notion of "vertex operator algebra" was introduced as a useful variant of

Borcherds's definition of vertex algebra. The definition of vertex operator algebra is previewed in Section 1.3 below and is "officially" presented in Section 3.1.

Then Borcherds, in [B6], used these concepts and results, along with new ideas, including his results on generalized Kac–Moody algebras (= Borcherds algebras), together with certain ideas from string theory, such as the "physical space" of a bosonic string along with the "no-ghost theorem" of R. Brower, P. Goddard and C. Thorn ([Br], [GT]), to prove the remaining Conway–Norton conjectures, for the structure $V^\natural$. (As we mentioned above, by the results of [FLM2] and [FLM6], what was left to prove was that the formal series $\sum (\text{tr } g|_{V_n^\natural})q^n$ for all the remaining Monster elements g were indeed the desired Hauptmoduls.) Discussions of this work can be found in [Geb], [JLW], [CG], [B8], [Go2], [Le11], [Ga2], [Ray].

So "monstrous moonshine" was much more than an astonishing relation between finite group theory and number theory; its underlying theme is, or at least includes, the new theory of vertex (operator) algebras, which is itself the foundational structure for conformal field theory, and this in turn is the foundational structure underlying string theory.

In order to motivate the precise definition of vertex (operator) algebra, we first repeat that there is a vertex operator algebra (namely, $V^\natural$) whose automorphism group is the Monster $\mathbb{M}$ and which implements the McKay–Thompson–Conway–Norton conjectures relating $\mathbb{M}$ to modular functions including $J(q)$. But in fact, this vertex operator algebra $V^\natural$ has the following simply stated properties—properties that have *nothing to do with the Monster*:

1. The vertex operator algebra $V^\natural$, which is *simple* in that it has no proper nonzero "ideals" (as proved in [FLM6]), is in fact its *only* irreducible module (proved by C. Dong [D4]). (As we shall discuss below, there is a natural notion of *module* for a vertex (operator) algebra, and as is to be expected, the theory of modules is extremely important.)
2. $\dim V_0^\natural = 0$ (corresponding to the zero constant term of $J(q)$). While the constant term of the classical modular function is essentially arbitrary and is chosen to have certain values for certain classical purposes, the constant term must be chosen to be zero for the purposes of moonshine and the moonshine module vertex operator algebra.
3. As we mentioned above, the central charge of the canonical Virasoro algebra in $V^\natural$ is 24. (The number 24 is extremely basic in number theory, modular function theory, etc., and the central charge of the Virasoro algebra in the moonshine module is this "same" 24.)

These three properties are actually "smallness" properties in the sense of conformal field theory and string theory. They allow one to say that $V^\natural$ essentially defines the smallest possible nontrivial string theory (cf. [Har], [Na] and the Introduction in [FLM6]).

Conversely, it was conjectured in [FLM6] that $V^\natural$ *is the* unique *vertex operator algebra with these three "smallness" properties*. This conjecture would be the conformal field-theoretic analogue of the uniqueness of the Leech lattice (proved in [Co2]) in sphere-packing theory and the uniqueness of the Golay code (proved in [Pl]) in error-correcting code theory. (These analogies, and the roles of the Leech lattice and the Golay code in the construction of $V^\natural$, are detailed in [FLM6].) Up to this conjecture, then, we have the following remarkable characterization of the largest sporadic finite simple group: *The Monster is the automorphism group of the smallest nontrival string theory that nature allows, or more precisely, the automorphism group of the vertex operator algebra with the three canonical "smallness" properties.* This characterization of the Monster in terms of "smallness" properties of a vertex operator algebra provides a striking motivation for the precise definition of the notion of vertex operator algebra.

So, exactly what *is* a vertex operator algebra? And what *are* vertex operators? First, with what is now understood, *vertex operators* are (or rather correspond to) *elements of vertex operator algebras*, by analogy with how (for example) *vectors* are *elements of (abstract) vector spaces*; the notion of vector space is of course in turn defined by a classical axiom system. But before abstract vector spaces had been formalized, vectors already "were" something (little arrows, etc.), and this of course helped motivate the eventual axiom system for the notion of vector space. As we mentioned above, vertex operators "already were" certain objects that described certain physical interactions in the context of string theory. Certain vertex operators also appeared in mathematics independently of string theory, as we describe next.

1.2 Example of a vertex operator

We now give a concrete example of a vertex operator, the example that shows how vertex operators first arose in mathematics. The example is easy to state in detail.

The affine Kac–Moody Lie algebra $A_1^{(1)}$, also denoted $\widehat{\mathfrak{sl}(2)}$, is the "smallest" infinite-dimensional Kac–Moody algebra (see [K1], [Mo] as well as [K]), and one of the most important. For certain mathematical reasons (see [LW1]), in the late 1970s one wanted to try to solve the "philosophical" problem of discovering some "concrete" realization of this Lie algebra in terms of some kinds of operators on some "concrete" space. The Weyl–Kac character formula ([K2]; cf. [Le2]) gave an "abstract" formula, analogous to and generalizing Weyl's character formula in the case of finite-dimensional semisimple Lie algebras, for suitable modules for $\widehat{\mathfrak{sl}(2)}$, and more generally, for any symmetrizable Kac–Moody algebra, but the "concrete structure" of the modules is not visible from this formula, just as in the case of Weyl's character formula.

We now describe the solution of this realization problem. The Lie algebra $\widehat{\mathfrak{sl}(2)}$ is the following Lie algebra:

$$\widehat{\mathfrak{sl}(2)} = \mathfrak{sl}(2) \otimes \mathbb{C}[t, t^{-1}] \oplus \mathbb{C}\mathbf{k} \tag{1.2.1}$$

(the "affinization" of $\mathfrak{sl}(2)$—the affine Lie algebra based on the Lie algebra $\mathfrak{sl}(2)$ of trace-zero 2-by-2 matrices) with Lie bracket given by

$$[a \otimes t^m, b \otimes t^n] = [a, b] \otimes t^{m+n} + \mathrm{tr}(ab)m\delta_{m+n,0}\mathbf{k} \tag{1.2.2}$$

6   1 Introduction

for $a, b \in \mathfrak{sl}(2)$, $m, n \in \mathbb{Z}$ and

$$[\mathbf{k}, \widehat{\mathfrak{sl}(2)}] = 0. \tag{1.2.3}$$

That is, the element $\mathbf{k}$ is central and the term $\mathrm{tr}(ab)m\delta_{m+n,0}\mathbf{k}$ in the commutator formula defines a central extension of the "loop algebra" $\mathfrak{sl}(2) \otimes \mathbb{C}[t, t^{-1}]$. The problem was to try to find some operator-realization of this Lie algebra.

Here is the result [LW1]. Consider the space

$$S = \mathbb{C}[y_{\frac{1}{2}}, y_{\frac{3}{2}}, y_{\frac{5}{2}}, \ldots] \tag{1.2.4}$$

of polynomials in the infinitely many formal variables y_n, $n = \frac{1}{2}, \frac{3}{2}, \ldots$; note that S is a commutative associative algebra. Form the expression

$$Y = \exp\left(\sum_{n=\frac{1}{2},\frac{3}{2},\frac{5}{2},\ldots} \frac{y_n}{n} x^n\right) \exp\left(-2 \sum_{n=\frac{1}{2},\frac{3}{2},\frac{5}{2},\ldots} \frac{\partial}{\partial y_n} x^{-n}\right), \tag{1.2.5}$$

where "exp" is the formal exponential series and x is another formal variable commuting with the y_n's. In the terminology of quantum field theory, the y_n's are "creation operators" and the $\frac{\partial}{\partial y_n}$'s are "annihilation operators" acting on the "Fock space" S. Observe that the operator Y is a well-defined formal differential operator in infinitely many formal variables. In fact, the coefficient, say Y_j, of x^j for any $j \in \frac{1}{2}\mathbb{Z}$, can be computed, and each Y_j is a well-defined linear operator from S to S—a certain formal differential operator in the form of an infinite sum of products of multiplication operators with partial differentiation operators, multiplied in this order; this infinite sum actually acts as a finite sum when applied to any given element of the space S. The explicit expression for each Y_j is of course somewhat complicated, and while one *can* write it down explicitly, one does not want to have to do this. In the original work [LW1], these explicit formal differential operators Y_j *were* in fact "directly" computed, and it was only after this that it was realized that if *all* of these complicated operators Y_j were added up and their *generating function* $\sum_{j\in\frac{1}{2}\mathbb{Z}} Y_j x^j$ was formed, all the operators Y_j could then be described by the *single product of exponentials* Y, an expression that is much simpler than any of the individual operators Y_j (again see [LW1]).

The main point of this was:

Theorem 1.2.1 [LW1] *The operators*

$$1, \quad y_n, \quad \frac{\partial}{\partial y_n} \quad \left(n = \frac{1}{2}, \frac{3}{2}, \frac{5}{2}, \ldots\right) \quad \text{and} \quad Y_j \quad \left(j \in \frac{1}{2}\mathbb{Z}\right) \tag{1.2.6}$$

(1 is the identity operator) span a Lie algebra of operators acting on S; that is, the commutators of any two of these operators are actually (finite) linear combinations of these operators. Moreover, this Lie algebra is precisely a copy of the affine Lie algebra $\widehat{\mathfrak{sl}(2)}$.

This operator Y turned out to be (a variant of) the vertex operators that arose in string theory, as H. Garland has pointed out. (But that is not how it was discovered in this mathematical context.) So it turns out that vertex operators lead to symmetry, and vice–versa.

This vertex operator realization of an affine Lie algebra was generalized in [KKLW] to all the affine Lie algebras based on the simple Lie algebras of types A_n, D_n, E_6, E_7 and E_8. With what is now known, the operator Y (1.2.5) is an example of a "twisted vertex operator," and the vertex operators in [KKLW] are also twisted vertex operators of a certain type. *Untwisted* vertex operator algebra realizations of the affine Lie algebras of types $A_n^{(1)}$, $D_n^{(1)}$ and $E_n^{(1)}$ were constructed by I. Frenkel and V. Kac [FK] and by G. Segal [Se1]. It happens that the particular vertex operator (1.2.5) and the vertex operators in [FK], [Se1] (among other much more complicated types of vertex operators) play critical roles in the construction of $V^\natural$. Such vertex operators are also crucial in a wide variety of other, very different, mathematical problems. Much of Chapter 6 below is devoted to the construction and study of vertex operator algebras and modules based on affine Lie algebras.

We have emphasized the value of forming a much simpler *generating function Y* out of a certain family of relatively complicated operators Y_j. Such formation of generating functions pervades vertex operator algebra theory, and as in the example above, one finds in general that working with the generating functions is the best and easiest way to proceed. The generating functions take on the role of fundamental objects in the theory, and results about the "components" of the generating functions (such as the operators Y_j above) turn out to be corollaries of the analysis of the generating functions.

Also, it is very important to note that the generating function $Y = \sum Y_j x^j$ is a sum over *all* the integers j—*not* just the nonnegative integers, for instance. This is a fundamental feature of the main generating functions arising everywhere in vertex operator algebra theory. Correspondingly, vertex operator algebra theory has an entirely different flavor from the classical theory of formal power series, in which the powers of the formal variables are truncated from below. One can of course multiply formal power series (when one has this truncation condition), but one *cannot* in general multiply vertex operators or other basic generating functions in vertex operator algebra theory. And yet one wants, and needs, to compose vertex operators in some appropriate way, and more generally, to multiply formal series not satisfying any truncation condition, again at least in some appropriate way. The resolution of this issue is to set up and develop a "formal calculus" in several variables, based on formal Laurent series with the powers of the formal variables allowed to be unrestricted—in particular, not truncated from below. This enables one to do everything that one wants.

At the center of this formal calculus is the "formal delta function," the formal series

$$\delta(x) = \sum_{n \in \mathbb{Z}} x^n, \tag{1.2.7}$$

the sum of all integral powers of the formal variable x. This formal series is indeed analogous to the Dirac delta function. The elegant use of the formal delta function to compute commutators of vertex operators (including the vertex operator Y above, yielding a then-new proof of Theorem 1.2.1), was demonstrated by Garland [Gar3].

Our treatment throughout this book is based on formal calculus, developed in Chapter 2, following and expanding on the treatment of formal calculus in [FLM6] and [FHL]. We also explain how the methods and results of formal calculus lead to consequences for *analytic* statements, involving complex variables and convergence. In much of the physics literature, and in some of the mathematics literature, complex variables, rather than formal variables and formal calculus, are used for formulations and proofs of basic results, but for reasons discussed throughout this book, we find it more natural, again following [FLM6] and [FHL], to use formal calculus. There is a kind of "dictionary" relating the formal calculus and analytic languages, and in Chapters 2 and 3 we explain the relation between the complex-variable and formal-calculus formulations; this relation was discussed in the Appendix of [FLM6]. From the point of view of conformal field theory as a quantum field theory, formal calculus is an efficient and straightforward way to rigorously implement the relations among operators acting on the "chiral space," the expansion coefficients of the corresponding "quantum fields," and analytic properties of correlation functions among quantum fields.

1.3 The notion of vertex operator algebra

Having provided some motivation and background, we shall give a precise definition of the notion of vertex operator algebra, as presented in [FLM6] and [FHL]; this notion, as we have been saying, is a variant of Borcherds's original definition of the notion of vertex algebra [B1]. Chapter 8 and the Appendix in [FLM6] contain further extensive motivation for the precise concept of vertex operator algebra. In the definition below we include some standard terminology from conformal field theory and some commentary on what we have discussed above. In Section 3.1, and in fact throughout Chapter 3 below, we discuss the definition in much more detail, along with the axiomatic development of the theory.

In this definition we use commuting *formal* variables x, x_1, x_2, etc. In the course of giving the definition, we include some conformal field-theoretic terms, reflecting the well-understood relation between this algebraic notion and that of chiral algebra in conformal field theory. For instance, the grading of a vertex operator algebra is by *conformal weights*. This grading is slightly "shifted" from the grading (1.1.2) that we have already been using on $V^\natural$; that grading is adapted to the modularity properties of the generating functions we have been discussing, including the J-function, while the present grading is adapted to the action of the Virasoro algebra on the vertex operator algebra, as we mention in the definition.

Definition 1.3.1 A *vertex operator algebra* consists of a $\mathbb{Z}$-graded vector space

$$V = \coprod_{n \in \mathbb{Z}} V_{(n)}; \quad \text{for } v \in V_{(n)}, \quad n = \text{wt } v, \tag{1.3.1}$$

such that two *grading restrictions* hold, namely,

$$\dim V_{(n)} < \infty \quad \text{for } n \in \mathbb{Z}, \tag{1.3.2}$$

$$V_{(n)} = 0 \quad \text{for } n \text{ sufficiently negative} \tag{1.3.3}$$

(that is, for $n << 0$), equipped with a linear map (the *"state-field correspondence"*)

$$V \to (\text{End } V)[[x, x^{-1}]]$$
$$v \mapsto Y(v, x) = \sum_{n \in \mathbb{Z}} v_n x^{-n-1}$$

(where each v_n is an element of End V, the algebra of operators on V), $Y(v, x)$ denoting the *vertex operator associated with* v and equipped also with two distinguished homogeneous vectors $\mathbf{1} \in V_{(0)}$ (the *vacuum vector*) and $\omega \in V_{(2)}$ (the *conformal vector*). *The axioms:* For $u, v \in V$,

- the *truncation condition*: $u_n v = 0$ for n sufficiently large
- $Y(\mathbf{1}, x) = $ the identity operator on V
- the *creation property*: $Y(v, x)\mathbf{1}$ has no pole in x and its constant term is the vector v (which implies that the state-field correspondence is one-to-one)
- the *Virasoro algebra relations*:

$$[L(m), L(n)] = (m - n)L(m + n) + \frac{1}{12}(m^3 - m)\delta_{m+n,0} c_V \tag{1.3.4}$$

for $m, n \in \mathbb{Z}$, where $c_V \in \mathbb{C}$ (the *central charge*) and where

$$Y(\omega, x) = \sum_{n \in \mathbb{Z}} L(n) x^{-n-2} \left(= \sum_{n \in \mathbb{Z}} \omega_n x^{-n-1} \right) \tag{1.3.5}$$

(i.e., the conformal vector generates a copy of the Virasoro algebra acting on V)
- *compatibility of $L(0)$ with the grading*:

$$L(0)v = nv = (\text{wt } v)v \quad \text{for } n \in \mathbb{Z} \text{ and } v \in V_{(n)} \tag{1.3.6}$$

- the *$L(-1)$-derivative property*:

$$Y(L(-1)v, x) = \frac{d}{dx} Y(v, x), \tag{1.3.7}$$

and finally, *the main axiom* (which carries by far most of the content of the definition of vertex operator algebra):

- the *Jacobi identity* (in Frenkel–Lepowsky–Meurman's terminology [FLM6]): Recalling the formal delta function (1.2.7), we have

$$
x_0^{-1} \delta \left(\frac{x_1 - x_2}{x_0} \right) Y(u, x_1) Y(v, x_2)
$$
$$
- x_0^{-1} \delta \left(\frac{-x_2 + x_1}{x_0} \right) Y(v, x_2) Y(u, x_1)
$$
$$
= x_2^{-1} \delta \left(\frac{x_1 - x_0}{x_2} \right) Y(Y(u, x_0)v), x_2), \tag{1.3.8}
$$

where each binomial, such as (for example) $(x_1 - x_2)^{-3}$, occurring here is understood to be expanded in nonnegative integral powers of the *second* variable; the truncation condition ensures that all the expressions are well defined.

It is important to note that in the literature there are several variant definitions of the terms "vertex algebra" and "vertex operator algebra." For instance, for some purposes one does not want to impose the grading restrictions (1.3.2) and (1.3.3). These minor variations in the definition depend on the context of interest. However, one always assumes the Jacobi identity (or axioms equivalent to it). The notion of vertex algebra in [B1] is equivalent to the notion of vertex operator algebra except that in the definition of vertex algebra no grading is assumed and no conformal vector or Virasoro algebra are assumed. Sometimes one wants these assumptions and sometimes one does not. In Chapter 3 below we present and discuss both versions of the notion. The axioms introduced in [B1] did not include the Jacobi identity, but axioms equivalent to it (see Chapter 3) instead, and the formulation of the axioms was in a different style. For reasons discussed throughout this book, we prefer to emphasize the Jacobi identity as the central axiom (as in [FLM6] and [FHL]).

Now that we have the precise definition of the notion of vertex operator algebra, we have, as we recall from above, a definition (up to a conjecture) of the Monster without reference to finite group theory: It is the symmetry group of the (conjecturally) unique vertex operator algebra (a structure satisfying the Jacobi identity and the "minor" axioms) having the three "smallness" properties mentioned above. Thus the Jacobi identity plus a few words determine the largest sporadic finite simple group.

The Jacobi identity was called by this name after it was discovered in work leading to [FLM6] and [FHL] (this identity is also implicit in [B1]; see the discussion in the Introduction of [FLM6]) because it is analogous to the classical Jacobi identity in the definition of Lie algebra: For u and v in a Lie algebra,

$$
(\mathrm{ad}\ u)(\mathrm{ad}\ v) - (\mathrm{ad}\ v)(\mathrm{ad}\ u) = \mathrm{ad}((\mathrm{ad}\ u)v), \tag{1.3.9}
$$

where $\mathrm{ad}\ u$ is the operation of left bracketing with the element u.

In fact, the notion of vertex operator algebra is indeed deeply analogous to the notion of Lie algebra, and *is actually the "one-complex-dimensional analogue" of the notion of Lie algebra* (which is the corresponding "one-real-dimensional" notion, in this sense);

this statement can be made precise using the language and viewpoint of *operads* (cf. [Hua5], [Hua13], [HL2], [HL3]).

However, the vertex operator algebra Jacobi identity is really the generating function of an *infinite list* of generally highly nontrivial identities for the component operators v_n of vertex operators, with one identity for each monomial in the three formal variables x_0, x_1 and x_2. In extremely special cases, these "component identities" include the relations defining affine Lie algebras; the Virasoro algebra; the infinite-dimensional "affinization" ([FLM2], [FLM6]) of the (modified) Griess algebra $\mathcal{B}$ (part of the algebra of vertex operators on $V^\natural$ discussed above); and a vast array of other remarkable algebraic structures. It is the *generating function form of these identities (namely, the Jacobi identity)* that is the natural analogue of the Jacobi identity in the definition of Lie algebra.

Just as the Jacobi identity (with its three formal variables) is the generating function of an infinite list of identities for the component operators v_n of vertex operators in a vertex operator algebra V, the single "x-parametrized" generating function product operation $Y(u, x)v$ can be thought of as specifying an infinite list of nonassociative product operations $u_n v$ on V, for $n \in \mathbb{Z}$, and for that matter, $Y(u, x)$ is itself the generating function of the infinite family of operators u_n acting on V, just as we saw in the concrete example of a (twisted) vertex operator above. Generating functions of otherwise very complicated objects, such as nonassociative product operations, or operators on a space, or identities among such operators, pervade vertex operator algebra theory, in the form in which we present it in this book (as in [FLM2] and [FLM6]).

But *at the same time that it is the "one-complex-dimensional analogue" of the notion of Lie algebra*, the notion of vertex operator algebra is *the "one-complex-dimensional analogue" of the notion of commutative associative algebra* (which again is the corresponding "one-real-dimensional" notion); again, operad language can be used to make this precise (see [Fr6], [Hua13], [HL2], [HL3]).

The remarkable and paradoxical-sounding fact that the notion of vertex operator algebra can be, and is, the "one-complex-dimensional analogue" of both the notions of Lie algebra and of commutative associative algebra lies behind much of the richness of the whole theory, and of string theory and conformal field theory. Of course, in string theory as well, there is also a fundamental passage from "real" to "complex," reflecting the change from point particle theory (and "world-lines") to string theory (and "world-sheets" parametrized by complex coordinates).

This analogy between vertex operator algebras and commutative associative algebras actually arose out of the fact that there are basic "commutativity" and "associativity" properties of vertex operators in a vertex operator algebra, and these two properties are, together, equivalent to the Jacobi identity axiom in the definition of vertex operator algebra (see [FLM6], [FHL]). In fact, these "commutativity" and "associativity" conditions are mathematically precise formulations of properties of vertex operators that had been well known in conformal field theory (cf. [BPZ]), and this is how one knows that the axioms in [BPZ] for the "operator algebras" that came to be called "chiral algebras" are essentially equivalent to the axioms above. These commutativity and associativity conditions are extensively discussed in Chapter 3 below.

From the comments we have just made, it is not surprising that every commutative associative algebra has a natural structure of vertex algebra, as was observed in [B1]. But it turns out that it is a highly nontrivial problem to construct essentially any other example of a vertex (operator) algebra, in particular, any of the important examples arising in mathematics and physics. Chapters 5 and 6 of this book are devoted to the theory and practice of such constructions.

1.4 Simplification of the definition

It is natural to ask, *Can the Jacobi identity axiom be simplified?* As we have already mentioned, the Jacobi identity is actually the generating function of an infinite list of generally highly nontrivial identities, and one needs many of these individual component identities in working with the theory. But is there some "simpler" condition that in fact implies the Jacobi identity (in the presence of the "minor" axioms in the definition of vertex operator algebra)?

In fact there is, and this simpler condition does indeed look much simpler than the Jacobi identity. But it turns out that if one wants to verify that one has a vertex operator algebra, *it is essentially just as difficult to verify this simpler axiom for all the elements of a proposed vertex (operator) algebra as it is to verify the Jacobi identity for all the elements.* Thus it is not conceptually useful to use this simpler condition as an "official" axiom in place of the Jacobi identity. And in the study of vertex operator algebras one needs all the information in the Jacobi identity anyway.

This replacement axiom is:

For all $u, v \in V$ (where V is a structure satisfying all the conditions in the definition of vertex operator algebra except the Jacobi identity), there exists a nonnegative integer k such that

$$(x_1 - x_2)^k [Y(u, x_1), Y(v, x_2)] = 0. \tag{1.4.1}$$

This *"weak commutivity"* condition and the theorem that it implies the Jacobi identity were discovered by Dong–Lepowsky as the vertex operator algebra special case of a much more general theorem about "generalized vertex algebras" and "abelian intertwining algebras" (see formula (1.4) in [DL3]). In [Li3], Li generalized this result to the case of vertex algebras (which are not assumed to be graded). In fact, what was proved in [DL3] and, more generally, in [Li3], was that weak commutativity implies a condition called "weak associativity" [DL3], and that together these two conditions imply the Jacobi identity. And in turn, the fact that weak commutativity implies weak associativity was a variant of an earlier result—that the "commutativity" property mentioned in the previous section implies the "associativity" property (in the presence of the "minor" axioms). This was proved in [FLM6], [FHL] and [Go1]. All these results concerning how commutativity implies associativity and hence the Jacobi identity are treated extensively in Chapter 3 below.

In the theory of *modules* for a vertex operator algebra, treated in Chapter 4 below, this weak commutativity property is *not* sufficient to imply the Jacobi identity. (This is still another reason why we have not chosen to replace the Jacobi identity by this simpler-looking weak commutativity property in the definition of vertex operator algebra.) For *modules,* it is instead the *weak associativity property* that can be used as a replacement axiom to imply the Jacobi identity, as was proved in [Li3]. (Weak commutativity *fails* as a replacement axiom in the definition of *module* for a vertex operator algebra.)

1.5 Representations and modules

The title of this book uses the word "representations" rather than "modules." What is the precise difference? In classical algebraic theories, such as the theory of associative algebras or the theory of Lie algebras, there is little "operational" difference. Let A be an associative algebra with 1. To say that a vector space M is an A-module means that there is a bilinear map $A \times M \to M$, or equivalently, a linear map

$$A \otimes M \longrightarrow M$$
$$a \otimes m \mapsto a \cdot m,$$

such that

$$1 \cdot m = m,$$
$$a \cdot (b \cdot m) = (ab) \cdot m$$

for $a, b \in A$, $m \in M$, where we denote the module action by a dot. This familiar notion of A-module amounts to the equally familiar notion of representation of A. A representation of A on the vector space M is an algebra homomorphism

$$\pi : A \longrightarrow \mathrm{End}\, M$$

from A into the associative algebra of linear endomorphisms of M, and the correspondence between the two notions is of course given by the relation

$$(\pi(a))m = a \cdot m.$$

For Lie algebras rather than asociative algebras, the situation is similar: Let A be a Lie algebra with bracket operation $[\cdot, \cdot]$ and let M be a vector space. Then M is an A-module if there is a linear map

$$A \otimes M \longrightarrow M$$
$$a \otimes m \mapsto a \cdot m,$$

such that

$$a \cdot (b \cdot m) - b \cdot (a \cdot m) = [a, b] \cdot m$$

for $a, b \in A, m \in M$, where we again denote the module action by a dot. And it is well known that this notion is equivalent to the notion of representation of A. A representation of A on the vector space M is a Lie algebra homomorphism

$$\pi : A \longrightarrow \text{End } M$$

from A into the *Lie* algebra End M, where the bracket operation is the commutator of endomorphisms; the correspondence between the two notions is again given by:

$$(\pi(a))m = a \cdot m.$$

In each of these situations, it is so easy to pass back and forth between the notion of module and the notion of representation that one often uses the words "representation" and "module" as synonyms (as we ourselves have done above).

Now, for both associative and Lie algebras, as well as for many other algebraic structures, there is a simple, general principle that "predicts" what the notion of module should be once the notion of algebra has been fixed: An A-module is a vector space M on which A acts bilinearly *such that all the axioms in the definition of algebra that make sense hold.* So for associative algebras, we use in the definition of the notion of module the left-identity property of the element 1 but not the right-identity property (because M does not have an element 1) and we use the associativity axiom, but written in a form so that the module element is on the right (because for a module action, the module element is written on the right). Similarly, for Lie algebras, we cannot use in the definition of the notion of module the skew symmetry axiom (that $[a, b] = -[b, a]$ for a, b in the algebra), but we can and do use the Jacobi identity axiom in the definition of the notion of Lie algebra, *but written, with the help of the skew symmetry axiom, in a form in which that the module element is always on the right.* That is, we use the Jacobi identity in the form

$$[a, [b, c]] - [b, [a, c]] = [[a, b], c]$$

(where $a, b, c \in A$), as in (1.3.9) above, rather than in the form

$$[a, [b, c]] + [b, [c, a]] + [c, [a, b]] = 0,$$

and we replace c by the module element m. Certainly, for an algebraic structure (such as Lie or associative algebras), when one uses this general principle to motivate the definition of the notion of module, it is automatic that an algebra is a module for itself.

The point is that in vertex operator algebra theory, where, as we know, the axiom systems are much more subtle than those for traditional algebraic structures, for each relevant variant of the notion of algebra (and as we have noted, there are a number of useful variants of the notion of algebra), this general principle leads us to the "correct" notion of module. Thus from above the reader now knows the precise definition of the notion of module for a vertex operator algebra and also for a vertex algebra (although the grading of a module for a vertex operator algebra need not be a $\mathbb{Z}$-grading). It is again automatic that an algebra is a module for itself, but more important is the fact that the (rather subtle) statements and proofs of the fundamental properties of algebras automatically carry over to the module case (that is, those among the fundamental properties that make sense when formulated for modules). See Chapter 4 below.

On the other hand, the notion of *representation* of a vertex operator algebra is a very different story. Unlike the case in classical algebraic theories, the notion of representation is in a certain sense more delicate than the notion of module, as we have suggested above, and while there is indeed an equivalence between the notion of module and the notion of representation, this equivalence is subtle. One of the problems is that for a module W for a vertex operator algebra V, End W is not a vertex operator algebra; it is not even close to being a vertex operator algebra. And yet a fruitful representation theory can indeed be developed, as we now discuss.

1.6 Construction of families of examples

As we have mentioned, in vertex operator algebra theory it is difficult to construct in full detail nontrival examples of vertex operator algebras, even examples that are much simpler than $V^\natural$, and one cannot do the theory without examples; in fact, the theory is so rich because the examples are so rich.

How can one efficiently construct families of examples of vertex operator algebras and their modules?

In classical algebraic subjects like group theory, Lie algebra theory, and so on, one of course has vast supplies of interesting examples available from the beginning, guiding the development of the general theory. In vertex operator algebra theory, *there are essentially no examples (except for commutative associative algebras, as we mentioned above), that are easy to construct and for which the axioms can be easily proved.* In "classical" mathematics, there simply *are no nontrivial examples of vertex operator algebras* "lying around waiting to be axiomatized," in contrast with vector spaces, groups, Lie algebras, topological spaces, and so on.

As we have noted, a general method for constructing vertex (operator) algebras and their modules is in fact found in [Li3]. The main idea is to formulate and study a notion of *representation of,* as opposed to *module for,* a vertex (operator) algebra, by developing the theory of a "vertex-algebra analogue," denoted by $\mathcal{E}(W)$ in this book, of the notion of the usual endomorphism algebra End W of a vector space W, and by defining a *representation of a vertex algebra V* on W to be a (suitable kind of) homomorphism from V to $\mathcal{E}(W)$. Each element of the space $\mathcal{E}(W)$ involves a formal variable, and in addition, the "multiplication" of two elements of $\mathcal{E}(W)$ involves a *second* formal variable, by analogy with how "multiplication" in a vertex (operator) algebra involves a formal variable. The "product" of two elements of $\mathcal{E}(W)$ mirrors an action of one vertex operator on another long known and used in conformal field theory. This action was also exploited in the work of B. Lian and G. Zuckerman ([LZ3], [LZ4]). The analysis in [Li3] yields sufficient conditions, that in a wide range of cases are easy to implement, for constructing families of vertex operator algebras *and* their modules. One key idea in this work was to exploit the result mentioned above that in the notion of module, weak associativity implies the Jacobi identity.

Chapter 5 below is devoted to presenting the theory of $\mathcal{E}(W)$ and representation theory (in this sense), and showing how it leads to useful general criteria for constructing vertex (operator) algebra and modules. We highlight general results of E. Frenkel–V. Kac–A. Radul–W. Wang, A. Meurman–M. Primc and X. Xu for constructing vertex (operator) algebras in this spirit. For constructing *modules*, we use the full power of the material from [Li3] treated in Chapter 5.

In Chapter 6 we shall apply the general results in Chapter 5 to construct important families of examples of vertex (operator) algebras and modules, notably, algebras and modules based on the Virasoro algebra, on affine Lie algebras, on Heisenberg (Lie) algebras and on lattices, and in Section 6.6, we treat the construction and classification of irreducible modules for the vertex operator algebras based on suitable standard (integrable highest weight) modules for affine Kac–Moody algebras. In Section 6.6 we also describe the Goddard–Kent–Olive "coset construction" ([GKO1], [GKO2]) of certain irreducible modules for the Virasoro algebra. We base our discussion of this construction on I. Frenkel–Y. Zhu's study of "commutants" (or centralizers) of subalgebras of vertex operator algebras, which we treat in detail in Section 3.11.

The treatment in Chapter 6 basically follows [Li3] with some updates, some coming from extensions of results in [Li3] presented in Chapter 5, and some involving additional applications of these methods. The original constructions of these vertex operator algebras and modules were usually quite different from the treatment here. Some of these constructions are due originally to Borcherds [B1], Frenkel–Lepowsky–Meurman [FLM6], B. Feigin–E. Frenkel [FF7], I. Frenkel–Zhu [FZ] (and its analogue for vertex operator superalgebras, [KWa]), Huang [Hua4] and Dong–Lepowsky [DL3].

Thus, the reader who wishes to see nontrivial examples of vertex operator algebras and modules, including such very important examples as those based on the Virasoro algebra, on Heisenberg algebras and on affine Lie algebras, may wish to turn to Chapter 6.

Incidentally, the construction of the moonshine module vertex operator algebra $V^\natural$ [FLM6] is not one of the examples that we handle by the methods presented in this book. The original treatment in [FLM6] actually included the construction of a number of types of vertex operator algebras and modules handled here, such as lattice vertex operator algebras and certain vertex operator algebras based on Heisenberg algebras, on the Virasoro algebra and on affine Lie algebras, by methods adapted to the needs of the theorem proved in [FLM6], but the treatement in Chapter 5 (and in [Li3]) does not serve to simplify the construction of the vertex operator algebra $V^\natural$ itself. This early mathematically precise example of a vertex operator algebra remains a singularly complex one.

1.7 Some further developments

This book presents what we view as the core of the theory of vertex operator algebras and their modules (and representations), as based on formal calculus. With this material as background, the reader should find it easy to embark on any desired direction in the vast theory or its applications. There are many beautiful further developments that we wish we had had time to include in this book. Here is a small sampling of developments that the reader will be well prepared for, and which lead to some of the frontiers of current research (note that certain other developments are also discussed in various remarks throughout the main text):

Modular invariance We have discussed the importance of the modular invariance of the "graded dimensions" of vertex operator algebras, as in (1.1.3) (for $V = V^\natural$). Analogous basic modular invariance properties hold for many important classes of vertex operator algebras and their modules. In [Z2], Zhu used vertex operator algebra theory to prove a general result establishing certain $SL(2, \mathbb{Z})$-modular invariance properties involving vertex operator algebras and their modules. This result has been generalized in a number of directions, including [DLM10], [Miy8], [Miy9] and [Hua21].

Intertwining operators and the general operator product expansion Given three modules for a vertex operator algebra, one has *intertwining operators* relating them; these are defined using the Jacobi identity, as developed in [FHL], which includes a self-contained introduction to contragredient modules as well as intertwining operators. Intertwining operators correspond to "chiral tree-level 3-point correlation functions" in conformal field theory. Intertwining operators in general involve *nonintegral* powers of the formal variables, and in composing or combining them, one encounters highly sophisticated issues of multiple-valued analytic functions and their monodromy. When such monodromy is of a certain simple type, one can combine such intertwining operators to form a "generalized vertex algebra" or "abelian intertwining algebra" [DL3]; such structure is closely related to "Z-algebras" ([LW4], [LP1]) and to "parafermionic conformal field theory" [ZF1]. The results of [Li3] have been generalized to such algebras [GaoL]. The situation is much more subtle when the monodromy of intertwining operators is not of this special type, and in this generality, one must use more analysis (along with algebra) than in the situations with simpler monodromy (such as in this book). In this general setting, Huang [Hua8] has proved that under suitable general conditions, the associativity property of vertex operators that we have discussed above (for vertex operators in a vertex operator algebra) generalizes to intertwining operators, leading to the construction of general (nonabelian) intertwining operator algebras ([Hua10], [Hua11], [Hua16]). This general theorem on the existence and associativity of the operator product expansion is a critical step in the construction [HL6] of braided tensor categories, and in fact vertex tensor categories, based on suitable modules categories for vertex operator algebras. All of this is part of a program (see especially [Hua11], the book [Hua13] and [Hua18]) to use vertex operator algebra theory to construct conformal field theories in the sense of G. Segal [Se2] and M. Kontsevich; this program was initiated by I. Frenkel in [Fr6].

Algebraic geometry and conformal field theory The viewpoint of algebraic geometry has been used for a long time to study conformal field theory. For this direction, see in particular [TK3], [TUY], [BFM], [BD], [Gai] and [NT]. See especially the book [FB], which presents the algebro-geometric approach and relates it to vertex operator algebra theory; see also [HL8].

Quantum analogues The problem of defining quantum analogues of the notions of vertex operator algebra, module and intertwining operator was formulated by I. Frenkel and N. Jing in [FJ], where certain "q-vertex operators" were introduced and used to construct certain quantum affine Lie algebras. Quantum analogues of the Knizhnik–Zamolodchikov equations [KZ] were formulated and studied by I. Frenkel and N. Reshetikhin [FR]; see the book [EFK]. Notions of "quantum vertex (operator) algebra" have been proposed and studied by E. Frenkel and N. Reshetikhin [FrR], by P. Etingof and D. Kazhdan [EK] and by Borcherds [B9].

In the bibliography, we have listed a selection of both early and more recent works relating to the theory of vertex operator algebras, including works not specifically mentioned otherwise in this book, and a number of books and monographs that present a rich variety of ideas and viewpoints, in particular, Etingof–I. Frenkel–Kirillov [EFK], Di Francesco–Mathieu–Sénéchal [FMS], E. Frenkel–Ben-Zvi [FB], Huang [Hua13], Kac [K7], Matsuo–Nagatomo [MN2] and Xu [Xu12]. We apologize to authors whose relevant work we may have neglected to list.

A few of the results in this book are new. Perhaps the main ones are the following:

As we have described, one of the basic themes treated in this book is a detailed analysis and comparison of a variety of different useful axiom systems for the notions of vertex algebra, of vertex operator algebra, and also of module (including work of Borcherds, I. Frenkel–Lepowsky–Meurman, I. Frenkel–Huang–Lepowsky, Dong–Lepowsky and Li mentioned above). In order to provide conveniently brief proofs of certain of the equivalences among the definitions, in the case when there is no grading on the structures we introduce and exploit properties that we call "formal commutativity" and "formal associativity" in Chapters 3 and 4.

Also, a convenient criterion for the extendability of a given map from a given generating subset of a vertex algebra to a homomorphism of the algebra to the space $\mathcal{E}(W)$ (mentioned above) based on a given vector space W is introduced in Chapter 5 (Theorem 5.7.6). This criterion proves to be extremely useful in constructing a wide range of important examples of modules for a vertex (operator) algebra, as demonstrated in Chapter 6.

Finally, we introduce and exploit a certain associative algebra $A(L_0)$ associated with a given even lattice L_0 (see Section 6.5). This algebra is designed so that the $A(L_0)$-modules, with suitable natural restrictions, can be proved, with the aid of the extendability theorem, to be exactly the modules for the lattice-vertex algebra V_{L_0} ([B1], [FLM6]) based on L_0. This gives a new proof of certain results in Chapter 8 of [FLM6] and of the classification of irreducible V_{L_0}-modules by Dong [D2] (see also [DLM4]). The algebra $A(L_0)$ is somewhat analogous to a Lie algebra, to a universal enveloping algebra and to Zhu's algebra [Z2], which he used in his study of modularity, but it is in fact none of these. It does, however, seem to be natural and useful.

2

Formal Calculus

In this chapter we shall present the elementary formal calculus that is basic in vertex operator algebra theory. Formal calculus has been treated in Chapters 2 and 8 of [FLM6] and in [FHL]. For our present purposes and to make this work self-contained, we present here a full but slightly different treatment.

Throughout this work we use the usual symbols $\mathbb{C}$ for the complex numbers, $\mathbb{Q}$ for the rational numbers and $\mathbb{Z}$ for the integers, and we write $\mathbb{Z}_+$ for the positive integers and $\mathbb{N}$ for the nonnegative integers.

2.1 Formal series and the formal delta function

First we introduce several basic spaces of formal series and explain some natural operations involving these spaces. We also introduce the fundamental formal delta function $\delta(x)$. Everything is based on formal Laurent series of the type $\sum_{n \in \mathbb{Z}} a_n x^n$ and analogous formal series in several variables. We emphasize that arbitrary integral powers of the formal variables will be allowed. In particular, in contrast with the situation for "truncated" formal Laurent series $\sum_{n \geq k} a_n x^n$, our "doubly infinite" formal Laurent series cannot be multiplied except under special circumstances. We shall formulate simple but crucial conditions for the multiplication of formal Laurent series to be well behaved. If we do not assume such conditions, then we encounter curious paradoxes, as we shall see.

Throughout this work, the symbols $x, y, x_0, x_1, x_2, \ldots$ will denote mutually commuting independent formal variables.

We shall work over $\mathbb{C}$, the field of complex numbers, but almost everything will be valid over any field of characteristic 0.

Even though we shall focus on formal series involving integral powers of our formal variables, many of our considerations carry over in the obvious ways to formal series involving more general powers of the formal variables, such as rational powers or even complex powers.

Let V be a vector space. We denote by $V[[x, x^{-1}]]$ the vector space of (*doubly-infinite*) *formal Laurent series* in x with coefficients in V:

$$V[[x, x^{-1}]] = \left\{ \sum_{n \in \mathbb{Z}} v_n x^n \mid v_n \in V \right\}. \tag{2.1.1}$$

Given a linear map $\pi : V \to W$ of vector spaces, we shall also write

$$\pi : V[[x, x^{-1}]] \to W[[x, x^{-1}]] \tag{2.1.2}$$

for the natural extension.

Remark 2.1.1 In vertex algebra theory, our vector space will often be of the form End V, that is, the algebra of endomorphisms of a vector space V, which in turn will be a vertex algebra (or a vertex operator algebra). Correspondingly, we will have formal series of the form $\sum_{n \in \mathbb{Z}} v_n x^n$ where $v_n \in$ End V. The most important such formal series will be those called "vertex operators." One basic property of a vertex operator in (End $V)[[x, x^{-1}]]$ is that it is parametrized by a vector v in the vertex algebra V, and the formal series expansion of this vertex operator will be written as $\sum_{n \in \mathbb{Z}} v_n x^{-n-1}$ (rather than for example as $\sum_{n \in \mathbb{Z}} v_n x^n$); so v_n will be an element of End V parametrized by the element $v \in V$ and by $n \in \mathbb{Z}$. The reason why vertex operators parametrized by elements $v \in V$ will be written as $\sum v_n x^{-n-1}$ rather than as $\sum v_n x^n$ will become clear later, when we develop the theory of vertex algebras and vertex operator algebras.

The space $V[[x, x^{-1}]]$ contains various useful subspaces:

$$V[x] = \left\{ \sum_{n \in \mathbb{N}} v_n x^n \mid v_n \in V, \ \text{all but finitely many } v_n = 0 \right\} \tag{2.1.3}$$

(space of V-*valued polynomials in* x),

$$V[x, x^{-1}] = \left\{ \sum_{n \in \mathbb{Z}} v_n x^n \mid v_n \in V, \ \text{all but finitely many } v_n = 0 \right\} \tag{2.1.4}$$

(space of *formal Laurent polynomials*),

$$V[[x]] = \left\{ \sum_{n \in \mathbb{N}} v_n x^n \mid v_n \in V \right\} \tag{2.1.5}$$

(space of *formal power series*), and

$$V((x)) = \left\{ \sum_{n \in \mathbb{Z}} v_n x^n \mid v_n \in V, \ v_n = 0 \ \text{for } n \text{ sufficiently negative} \right\} \tag{2.1.6}$$

(the space of *truncated formal Laurent series*). We have

$$V[x] = V \otimes \mathbb{C}[x], \quad V[x, x^{-1}] = V \otimes \mathbb{C}[x, x^{-1}]. \tag{2.1.7}$$

But note that $V[[x, x^{-1}]]$ is strictly larger than $V \otimes \mathbb{C}[[x, x^{-1}]]$ since each element of $V \otimes \mathbb{C}[[x, x^{-1}]]$ is a *finite* sum of elements of V times elements of $\mathbb{C}[[x, x^{-1}]]$.

Remark 2.1.2 Notice the meanings that we attach to terms like "formal power series" and "formal Laurent series"; these terms sometimes have different meanings in the literature.

Here are two simple "substitution" procedures. Let $a \in \mathbb{C}, a \neq 0$. For

$$v(x) = \sum_{n \in \mathbb{Z}} v_n x^n \in V[x, x^{-1}],$$

set

$$v(a) = \sum_{n \in \mathbb{Z}} a^n v_n, \tag{2.1.8}$$

and for

$$v(x) = \sum_{n \in \mathbb{Z}} v_n x^n \in V[[x, x^{-1}]],$$

set

$$v(ax) = \sum_{n \in \mathbb{Z}} a^n v_n x^n. \tag{2.1.9}$$

Here and later, we must be careful that in operations involving formal series, we never add more than finitely many complex numbers or vectors when we compute the coefficient of each monomial in the formal variable(s); see also the next remark.

We shall frequently need formal sums of infinitely many formal series, as well as products of finitely many formal series, of operators on some vector space. We must keep in mind the following basic, universal principle telling us when the resulting expressions are well defined.

Remark 2.1.3 A formal sum or product of formal series of operators on a vector space will be understood to exist if and only if the coefficient of any monomial (in the relevant formal variable or formal variables) in the formal sum or product acts as a *finite* sum of operators *when it is applied to any fixed, but arbitrary, vector in the space*. (Infinite sums of operators will definitely be allowed, but only under this restrictive condition.)

The following, then, are the precise definitions of some of the basic operations.

Definition 2.1.4 Let V be a vector space and let $(f_i)_{i \in I}$ be a family in End V (I an index set). We say that $(f_i)_{i \in I}$ is *summable* if for every $v \in V$, $f_i v = 0$ for all but a finite number of $i \in I$. In this case we write the corresponding "sum" operator as

$$\sum_{i \in I} f_i : V \to V$$

$$v \mapsto \sum_{i \in I} f_i v. \tag{2.1.10}$$

Of course, any finite family is summable.

Now we consider the summability, and sum, of an indexed family of *generating functions* in the formal variable x. The point here will be to focus on each fixed power of x. Let $(F_i(x))_{i \in I}$ be a family in (End V)$[[x, x^{-1}]]$ and for each $i \in I$, set $F_i(x) = \sum_{n \in \mathbb{Z}} f_i(n)x^n$. (The notation $f_i(n)$ used here for the coefficient operators of $F_i(x)$ should not be confused with the evaluation of some endomorphism f_i on some vector. Sometimes in vertex operator algebra theory, notation like $f(n)$ is used to designate an operator that might also be called f_n, as in Remark 2.1.1, or $f_{(n)}$.) We say that $\sum_{i \in I} F_i(x)$ *exists* if for every $n \in \mathbb{Z}$, the family $(f_i(n))_{i \in I}$ is summable. We then set

$$\sum_{i \in I} F_i(x) = \sum_{n \in \mathbb{Z}} \left(\sum_{i \in I} f_i(n) \right) x^n. \tag{2.1.11}$$

Note incidentally that for a *single* formal series

$$F(x) = \sum_{n \in \mathbb{Z}} f(n)x^n \in (\text{End } V)[[x, x^{-1}]],$$

the sum of the family $(f(n)x^n)_{n \in \mathbb{Z}}$ of course exists in the new sense (with the index set I now taken to be $\mathbb{Z}$) and equals $F(x)$ (since for each power of x, there is only one element to sum).

Definition 2.1.5 Let $(F_i(x))_{i=1}^r$ be a finite family in (End V)$[[x, x^{-1}]]$, with $F_i(x) = \sum_{n \in \mathbb{Z}} f_i(n)x^n$. We say that the product $F_1(x) \cdots F_r(x)$ *exists* if for every $n \in \mathbb{Z}$, the family

$$(f_1(n_1) \cdots f_r(n_r))_{\substack{n_1, \ldots, n_r \\ n_1 + \cdots + n_r = n}}$$

is summable. We then set

$$F_1(x) \cdots F_r(x) = \sum_{n \in \mathbb{Z}} \left(\sum_{n_1 + \cdots + n_r = n} f_1(n_1) \cdots f_r(n_r) \right) x^n$$

$$\in (\text{End } V)[[x, x^{-1}]]. \tag{2.1.12}$$

Suppose that $F_1(x) \cdots F_r(x)$ exists and that for a fixed q with $1 \le q < r$, $F_1(x) \cdots F_q(x)$ and $F_{q+1}(x) \cdots F_r(x)$ exist. Then their product exists and

$$F_1(x) \cdots F_r(x) = (F_1(x) \cdots F_q(x))(F_{q+1}(x) \cdots F_r(x)). \tag{2.1.13}$$

Remark 2.1.6 Note that this implies that, for example, if the three products $F_1(x)F_2(x)F_3(x)$, $F_1(x)F_2(x)$ and $F_2(x)F_3(x)$ all exist, then

$$(F_1(x)F_2(x))F_3(x) = F_1(x)(F_2(x)F_3(x)), \tag{2.1.14}$$

and the indicated "nested products" in fact exist and also equal $F_1(x)F_2(x)F_3(x)$. It is natural and important to assume that the *triple product* $F_1(x)F_2(x)F_3(x)$ (with no parentheses!) exists. Next we shall give examples to show how things can (and do) "go wrong" if multiple products (such as triple products) do not exist. This situation is analogous to the familiar situation in analysis where, for example, a natural way to ensure that different *iterated* integrals of the same function are equal is to assume that the corresponding *multiple* integral exists. In analysis, it is in general just as critical to verify the *existence* of limits as it is to determine their value. Analogously, in formal calculus it is just as important to verify the existence of such objects as multiple products as it is to manipulate them. We will be discussing many important variations on this theme throughout the theory.

An example of a nonexistent product is

$$\left(\sum_{n\geq 0} x^n\right)\left(\sum_{n\leq 0} x^n\right) \tag{2.1.15}$$

(the coefficients being viewed as scalar operators in End V or as elements of $\mathbb{C}$). Another example of a nonexistent product is

$$\left(\sum_{n\geq 0} \frac{x^n}{2^n}\right)\left(\sum_{n\leq 0} x^n\right); \tag{2.1.16}$$

even though the coefficient of each power of x in this formal product is a convergent formal sum of complex numbers, we are *not* allowing this product to exist. Such products do not arise "in real life" in vertex algebra theory, and the theory is far simpler if we do not allow analytic convergence to "rescue" algebraic operations in this way. In deeper parts of vertex operator algebra theory, convergence does indeed play a serious role, but the most natural way to proceed is to develop a rich "purely algebraic" theory and then to apply it to analytic problems at a later stage.

The following amusing and instructive paradox (which we put in quotation marks because it is a false assertion!) involves a nonexistent product of three series:

$$"\delta(x) = \left(\left(\sum_{n\geq 0} x^n\right)(1-x)\right)\delta(x) = \left(\sum_{n\geq 0} x^n\right)((1-x)\delta(x))$$
$$= \left(\sum_{n\geq 0} x^n\right)0 = 0", \tag{2.1.17}$$

where $\delta(x) = \sum_{n\in\mathbb{Z}} x^n$, as in (2.1.32) below. This paradox illustrates particularly well why we need to pay careful attention to the precise definitions and the elementary principles. Of course, the resolution of the paradox is that the triple product does not exist. The "associative law" does indeed hold, but only when it is supposed to hold— under the assumptions in (2.1.13). Another version of the same paradox is:

$$
\begin{aligned}
\text{``} \sum_{n<0} x^n &= \left(\left(\sum_{n\geq 0} x^n \right) (1-x) \right) \left(\sum_{n<0} x^n \right) \\
&= \left(\sum_{n\geq 0} x^n \right) \left((1-x) \left(\sum_{n<0} x^n \right) \right) \\
&= \left(\sum_{n\geq 0} x^n \right)(-1) = -\sum_{n\geq 0} x^n. \text{''}
\end{aligned}
\qquad (2.1.18)
$$

Here is an example of an existent product that contains a nonexistent subproduct:

$$
\left(\sum_{n\geq 0} x^n \right) \left(\sum_{n\leq 0} x^n \right) 0 = 0.
\qquad (2.1.19)
$$

We invite the reader to find a pair $f(x) \in \mathbb{C}[[x]]$, $g(x) \in \mathbb{C}[[x^{-1}]]$ such that both f and g have infinitely many nonzero coefficients and yet $f(x)g(x)$ exists. There are indeed such pairs. (Please note: This exercise is intended for amusement only; it is not needed anywhere in this book.)

The preceding notions generalize in an obvious way to the case of several commuting formal variables $x_1, x_2, \ldots, x_r$. For example,

$$
V[[x_1, x_1^{-1}, x_2, x_2^{-1}]] = \left\{ \sum_{m,n\in\mathbb{Z}} v_{mn} x_1^m x_2^n \mid v_{mn} \in V \right\},
\qquad (2.1.20)
$$

$$
V[[x_1, x_2, x_2^{-1}]] = \left\{ \sum_{m\in\mathbb{N},\, n\in\mathbb{Z}} v_{mn} x_1^m x_2^n \mid v_{mn} \in V \right\}.
\qquad (2.1.21)
$$

We now combine these ingredients to formulate a useful fact, representative of many facts concerning formal series that we shall be using regularly: Consider $\mathrm{Hom}(V, V((x)))$ as a natural subspace of $(\mathrm{End}\ V)[[x, x^{-1}]]$. Let

$$
A(x) \in (\mathrm{End}\ V)[[x, x^{-1}]]
$$

and

$$
B(x) \in \mathrm{Hom}(V, V((x))),
$$

so that for any $v \in V$, $B(x)v \in V((x))$, that is, the formal series $B(x)v$ is truncated from below. (Later, we shall take $A(x)$ and $B(x)$ to be vertex operators.) For $n \in \mathbb{Z}$, we consider $(x_1 - x_2)^n$ as an element of the space (2.1.20), where (as in Definition 2.2.1) we use the binomial series expansion in nonnegative powers of the second variable, x_2. (Of course, for $n \geq 0$, $(x_1 - x_2)^n$ is simply a polynomial.) Then

$$
\text{the product } (x_1 - x_2)^n A(x_1) B(x_2) \text{ exists for all } n \in \mathbb{Z},
\qquad (2.1.22)
$$

in particular, for $n < 0$ (the nontrivial case). In fact, when we apply this to any $v \in V$, the coefficient of x_2^k for any k is a finite sum of elements of $V[[x_1, x_1^{-1}]]$. Note that it is not necessary for us to assume that

$$B(x) \in (\text{End } V)((x))$$

(a global truncation of the series $B(x)$), which would be too restrictive for our applications.

We shall also need "formal limits." Let

$$\sum_{m,n \in \mathbb{Z}} F(m, n) x_1^m x_2^n \in (\text{End } V)[[x_1, x_1^{-1}, x_2, x_2^{-1}]].$$

We say that

$$\lim_{x_1 \to x_2} \sum_{m,n \in \mathbb{Z}} F(m, n) x_1^m x_2^n$$

exists if for every $n \in \mathbb{Z}$, the family $(F(m, n - m))_{m \in \mathbb{Z}}$ is summable. In this case we set

$$\lim_{x_1 \to x_2} \left(\sum_{m,n \in \mathbb{Z}} F(m, n) x_1^m x_2^n \right) = \sum_{n \in \mathbb{Z}} \left(\sum_{m \in \mathbb{Z}} F(m, n - m) \right) x_2^n. \qquad (2.1.23)$$

Thus the notion $\lim_{x_1 \to x_2}$ is used only as an alternative notation for $|_{x_1 = x_2}$, and we shall sometimes use forms of the latter notation.

Remark 2.1.7 The notation $\lim_{x_1 \to x_2}$ and its variants certainly suggest analysis rather than algebra, and in fact, as we have already been saying, proving the *existence* of such formal expressions as formal limits is as crucial as calculating them. Sometimes this is quite a subtle matter; we shall be discussing a number of important issues related to this in the rest of this chapter. Note in particular that the notation $\lim_{x_1 \to x_2}$ suggests that the substitution $x_1 = x_2$ *might or might not be allowed*, and that if it is allowed, this must be verified. For instance, for

$$f(x) = \sum_{n \geq -1} x^n \in \mathbb{C}((x)), \qquad (2.1.24)$$

we observe that

$$\lim_{x \to 0} f(x) \text{ does not exist}, \qquad (2.1.25)$$

but that

$$\lim_{x \to 0} x f(x) \text{ exists and equals 1}. \qquad (2.1.26)$$

Also, for any nonzero complex number c,

$$\lim_{x \to c} f(x) \text{ does not exist}. \qquad (2.1.27)$$

Here we are of course using the notation $\lim_{x \to c}$ to mean the formal substitution $|_{x = c}$.

There are natural "multiplication" bilinear maps

$$V[x, x^{-1}] \times \mathbb{C}[[x, x^{-1}]] \to V[[x, x^{-1}]] \tag{2.1.28}$$

$$\mathbb{C}[x, x^{-1}] \times V[[x, x^{-1}]] \to V[[x, x^{-1}]] \tag{2.1.29}$$

$$\mathbb{C}((x)) \times V((x)) \to V((x)), \tag{2.1.30}$$

and obvious variants, which are denoted simply by juxtaposition. Later we shall often meet products like $f(x)a(x)$ with $f(x) \in \mathbb{C}[[x, x^{-1}]]$, $a(x) \in (\text{End } V)[[x, x^{-1}]]$, and we shall have to impose conditions in order to ensure that such a product is defined. It is clear that if $f(x) \in \mathbb{C}((x))$ and $a(x) \in \text{Hom}(V, V((x)))$, then $f(x)a(x)$ exists as an element of $(\text{End } V)[[x, x^{-1}]]$. Thus we have a natural "multiplication" bilinear map

$$\mathbb{C}((x)) \times \text{Hom}(V, V((x))) \to \text{Hom}(V, V((x))) \subset (\text{End } V)[[x, x^{-1}]]. \tag{2.1.31}$$

Now we define the important formal series

$$\delta(x) = \sum_{n \in \mathbb{Z}} x^n \in \mathbb{C}[[x, x^{-1}]]. \tag{2.1.32}$$

Formally, this is the Laurent series expansion of the classical "delta function" or "Dirac delta function" distribution or generalized function at the "point" $x = 1$, a fact that motivates the following fundamental properties of $\delta(x)$.

Proposition 2.1.8 *(a) Let $f(x) \in V[x, x^{-1}]$. Then*

$$f(x)\delta(x) = f(1)\delta(x). \tag{2.1.33}$$

(b) Let

$$f(x_1, x_2) \in (\text{End } V)[[x_1, x_1^{-1}, x_2, x_2^{-1}]]$$

be such that

$$\lim_{x_1 \to x_2} f(x_1, x_2) \text{ exists.} \tag{2.1.34}$$

Then in $(\text{End } V)[[x_1, x_1^{-1}, x_2, x_2^{-1}]]$,

$$f(x_1, x_2)\delta\left(\frac{x_1}{x_2}\right) = f(x_1, x_1)\delta\left(\frac{x_1}{x_2}\right) = f(x_2, x_2)\delta\left(\frac{x_1}{x_2}\right) \tag{2.1.35}$$

and, in particular, the indicated expressions exist.

Proof. (a) This follows from the fact that $x^n\delta(x) = \delta(x)$ for any $n \in \mathbb{Z}$ and linearity.
(b) Let

$$f(x_1, x_2) = \sum_{m,n \in \mathbb{Z}} a(m, n)x_1^m x_2^n.$$

Then formally,

$$f(x_1, x_2)\delta\left(\frac{x_1}{x_2}\right) = \left(\sum_{m,n\in\mathbb{Z}} a(m,n)x_1^m x_2^n\right)\left(\sum_{k\in\mathbb{Z}} x_1^k x_2^{-k}\right)$$

$$= \sum_{m,n,k\in\mathbb{Z}} a(m,n)x_1^{m+k} x_2^{n-k}$$

$$= \sum_{m,n,k\in\mathbb{Z}} a(m,n)x_2^{m+n} x_1^{m+k} x_2^{-m-k}$$

$$= f(x_2, x_2)\delta\left(\frac{x_1}{x_2}\right), \tag{2.1.36}$$

and from this formal argument we conclude that all these expressions exist. Using the fact that $\delta(x_1/x_2) = \delta(x_2/x_1)$ and applying the equality just proved, we obtain the other equality. $\square$

Remark 2.1.9 Provided that (2.1.34) holds, when we multiply $f(x_1, x_2)$ by $\delta(x_1/x_2)$ we are allowed to formally set x_1 equal to x_2 in $f(x_1, x_2)$, and setting $x_1 = x_2$ amounts to setting x_1/x_2, the argument appearing in the formal delta function $\delta(x_1/x_2)$, equal to 1; cf. (2.1.33). The formal delta function will always have this "operational meaning," even when the argument in the delta function is more complicated; that is, $f(y)\delta(y) = f(1)\delta(y)$ even when y is a more complicated expression, as we discuss below, *but provided that suitable conditions hold.* The "suitable condition" in the case of (2.1.33) is that $f(x)$ should be a *terminating* Laurent series (that is, $f(1)$ should exist), and the "suitable condition" in (2.1.35) is that the (algebraic) limit should exist.

From Proposition 2.1.8, if $\lim_{x_1\to x_2} f(x_1, x_2)$ exists, then $f(x_1, x_2)\delta(x_1/x_2)$ exists. In fact, it is easy to see that the converse is also true. Also note that if we want the conclusion (2.1.35) to hold, then the condition (2.1.34) is certainly a necessary one, and the proposition shows that it is sufficient too.

Later we shall often see products like $(x_1 - x_2)^k\delta(x_1/x_2)$ with $k \in \mathbb{N}$, but $(x_1 - x_2)^k\delta(x_1/x_2)$ does not exist if $k < 0$. Another example of a formal series $f(x_1, x_2)$ for which $\lim_{x_1\to x_2} f(x_1, x_2)$ does not exist is $\delta(x_1/x_2)$; correspondingly, $\delta(x_1/x_2)$ cannot be squared (this is an algebraic counterpart of the analytic fact that the delta-distribution cannot be squared). However, $\delta(x_1/x_2)$ can be multiplied by any formal series in either x_1 or x_2 alone, including, say, $\delta(x_1)$.

Finally, let us record the simple and basic relations

$$x^{-1}\delta\left(\frac{y}{x}\right) = y^{-1}\delta\left(\frac{y}{x}\right) = y^{-1}\delta\left(\frac{x}{y}\right), \tag{2.1.37}$$

which will be used many times below.

2.2 Derivations and the formal Taylor Theorem

In this section we shall study the formal derivative operation on formal series and present the formal Taylor Theorem. This formula and the formal delta function play fundamental roles in vertex operator algebra theory.

As in Section 2.1, let V be a vector space. For $v(x) = \sum_{n \in \mathbb{Z}} v_n x^n \in V[[x, x^{-1}]]$, we define the formal derivative in the obvious way:

$$\frac{d}{dx} v(x) = \sum_{n \in \mathbb{Z}} n v_n x^{n-1}, \tag{2.2.1}$$

so that d/dx is an endomorphism of $V[[x, x^{-1}]]$. We shall also use the usual prime notation of calculus. For example,

$$\delta'(x) = \sum_{n \in \mathbb{Z}} n x^{n-1} \in \mathbb{C}[[x, x^{-1}]]. \tag{2.2.2}$$

In settings involving several variables $x_1, x_2, \ldots$ we shall use the usual partial derivative notation $\partial/\partial x_1, \partial/\partial x_2, \ldots$.

The operator d/dx acts as a derivation of $\mathbb{C}[x, x^{-1}]$, and in fact, more generally,

$$\frac{d}{dx}(f(x)v(x)) = \left(\frac{d}{dx} f(x)\right) v(x) + f(x) \left(\frac{d}{dx} v(x)\right) \tag{2.2.3}$$

whenever the product $f(x)v(x)$ is defined, for example, when $f(x) \in \mathbb{C}[[x, x^{-1}]]$ and $v(x) \in V[x, x^{-1}]$. Analogous formulas of course hold for partial derivatives.

We also define the following formal residue operator Res_x from $V[[x, x^{-1}]]$ to V:

$$\mathrm{Res}_x v(x) = \text{the coefficient of } x^{-1} \text{ in } v(x). \tag{2.2.4}$$

By definition, it is clear that

$$\mathrm{Res}_x v'(x) = 0 \tag{2.2.5}$$

for $v(x) \in V[[x, x^{-1}]]$. Whenever the product $f(x)v(x)$ is defined,

$$\mathrm{Res}_x(f'(x)v(x)) = -\mathrm{Res}_x(f(x)v'(x)) \tag{2.2.6}$$

since $\mathrm{Res}_x(f(x)v(x))' = 0$. The notion of residue generalizes to several variables in an obvious way and the corresponding formulas still hold. The simple relation

$$\mathrm{Res}_x y^{-1}\delta\left(\frac{x}{y}\right) = \mathrm{Res}_x x^{-1}\delta\left(\frac{y}{x}\right) = 1 \tag{2.2.7}$$

will be used frequently.

It is crucial to keep in mind the following notational device:

Definition 2.2.1 (binomial expansion convention) Throughout this work, we define $(x + y)^n$ for $n \in \mathbb{Z}$ (in particular, for $n < 0$) to be the formal series

$$(x + y)^n = \sum_{k \in \mathbb{N}} \binom{n}{k} x^{n-k} y^k, \tag{2.2.8}$$

where $\binom{n}{k} = \dfrac{n(n-1)\cdots(n-k+1)}{k!}$. That is, *binomial expressions are to be expanded in nonnegative integral powers of the second variable.* (Here x or y, but not in general both, can be replaced by a nonzero complex number.) In particular, while $(x+y)^n = (y+x)^n$ if $n \geq 0$, $(x+y)^n$ is *not* the same as $(y+x)^n$ if $n < 0$. Note that if x and y were complex numbers rather than formal variables, we would be expanding $(x+y)^n$ in the domain given by $|x| > |y|$. Also, the power n can be taken to be more general than a negative integer; it can be any complex number or any element of any associative algebra.

For $n \in \mathbb{Z}$ and $r \in \mathbb{N}$, we have the following derivative formula:

$$\frac{1}{r!}\left(\frac{\partial}{\partial x}\right)^r (x+y)^n = \binom{n}{r}(x+y)^{n-r} = \frac{1}{r!}\left(\frac{\partial}{\partial y}\right)^r (x+y)^n. \qquad (2.2.9)$$

If $v(x) = \sum_{n \in \mathbb{Z}} v_n x^n$, then by definition

$$v(x+y) = \sum_{n \in \mathbb{Z}} v_n (x+y)^n = \sum_{n \in \mathbb{Z}} \sum_{i \in \mathbb{N}} \binom{n}{i} v_n x^{n-i} y^i$$

$$\in V[[x, x^{-1}, y]], \qquad (2.2.10)$$

and, in particular, the formal series $v(x+y)$ is defined. We shall relate this to the higher derivatives of $v(x)$ by a formal Taylor Theorem.

Next we shall set up the formal exponential notation, since we want to consider all higher derivatives at once by means of expressions like $e^{y(d/dx)}$.

To formalize such expressions in general, let

$$S \in y(\text{End } V)[[y]], \qquad (2.2.11)$$

so that S has no constant term. Then

$$e^S = \sum_{n \in \mathbb{N}} \frac{1}{n!} S^n \qquad (2.2.12)$$

is a well-defined element of $(\text{End } V)[[y]]$ and hence it acts as an endomorphism of $V[[y]]$. (Each S^n is defined because the formal series S is truncated from below, and the sum (2.2.12) is defined because S involves only positive powers of y.) If S and T are commuting elements of $y(\text{End } V)[[y]]$, then

$$e^{S+T} = e^S e^T, \qquad (2.2.13)$$

as one easily checks by expanding both sides. In particular,

$$e^S e^{-S} = 1, \qquad (2.2.14)$$

so that e^S acts as a linear automorphism of $V[[y]]$.

Now since the operator d/dx is a derivation, $e^{y(d/dx)}$ ought to act formally as an automorphism. A little more generally, let $p(x) \in \mathbb{C}[x, x^{-1}]$. Then the endomorphism

$$T = T_{p(x)} = p(x)\frac{d}{dx} \tag{2.2.15}$$

of $\mathbb{C}[[x, x^{-1}]]$ acts as a derivation of $\mathbb{C}[x, x^{-1}]$, and in fact

$$T(f(x)g(x)) = (Tf(x))g(x) + f(x)(Tg(x)) \tag{2.2.16}$$

for all $f(x) \in \mathbb{C}[x, x^{-1}]$ and $g(x) \in \mathbb{C}[[x, x^{-1}]]$. By induction,

$$T^n(f(x)g(x)) = \sum_{k=0}^{n} \binom{n}{k}(T^k f(x))(T^{n-k}g(x)) \tag{2.2.17}$$

for $n \geq 0$, and so

$$e^{yT}(f(x)g(x)) = (e^{yT}f(x))(e^{yT}g(x)) \tag{2.2.18}$$

in $\mathbb{C}[[x, x^{-1}]][[y]]$. (Of course, this simple standard argument holds more generally—for instance, when $y \in w\mathbb{C}[[w]]$, w another formal variable. The case $y = \log(1 + w)$ is an especially interesting example, since e^{yT} becomes $(1 + w)^T$.)

We now have the following formal Taylor Theorem.

Proposition 2.2.2 *Let $v(x) \in V[[x, x^{-1}]]$. Then*

$$e^{y\frac{d}{dx}}v(x) = v(x + y) \tag{2.2.19}$$

(and these expressions exist).

Proof. Writing $v(x) = \sum_{n \in \mathbb{Z}} v_n x^n$, we have

$$e^{y\frac{d}{dx}}v(x) = \sum_{n \in \mathbb{Z}}\sum_{i \in \mathbb{N}} \frac{y^i}{i!}\left(\frac{d}{dx}\right)^i v_n x^n = \sum_{n \in \mathbb{Z}}\sum_{i \in \mathbb{N}} \binom{n}{i} v_n y^i x^{n-i} = v(x + y). \qquad \square$$

Remark 2.2.3 While our formal considerations are of course closely related to ordinary analysis (or calculus), the algebraic statements often deal with formal series that are far from representing any kinds of actual functions of complex variables. For instance, the formal Taylor Theorem still holds, for the same reasons as above, if $v(x)$ is as general a formal series as, say, $\sum_{n \in \mathbb{C}} v_n x^n$ ($v_n \in V$ for $n \in \mathbb{C}$ rather than for $n \in \mathbb{Z}$).

Remark 2.2.4 We may take the space V in Proposition 2.2.2 to be of the form $V[[x_2, x_2^{-1}]]$, so that $V[[x, x^{-1}]]$ becomes $V[[x, x^{-1}, x_2, x_2^{-1}]]$, and we may generalize y to

$$y \in w\mathbb{C}[x_2, x_2^{-1}][[w]], \tag{2.2.20}$$

for example. As a result, we obtain the following two-variable analogue of (2.2.19): For $v(x, x_2) \in V[[x, x^{-1}, x_2, x_2^{-1}]]$,

$$e^{y\frac{\partial}{\partial x}} v(x, x_2) = v(x + y, x_2). \tag{2.2.21}$$

This formula remains valid whenever its formal justification remains valid, for instance, for

$$y \in w\mathbb{C}((x_2))[[w]], \tag{2.2.22}$$

provided that

$$v(x, x_2) \in V((x_2))[[x, x^{-1}]]. \tag{2.2.23}$$

2.3 Expansions of zero and applications

In this section we consider what have been called "expansions of zero" (cf. [FLM6]). Briefly, an expansion of zero is a formal series which is the difference between two expansions of the *same* rational function in opposite directions (or in different domains if we think in terms of complex variables). The most important expansion of zero is the formal delta function $\delta(x)$. The use of the formal delta function in vertex operator algebra theory is fundamental. For example, the commutator of two vertex operators will be an expansion of zero in two variables expressible in terms of the delta function.

Consider the field $\mathbb{C}(x)$ of rational functions in the indeterminate x over $\mathbb{C}$, the field of fractions of the polynomial ring $\mathbb{C}[x]$. The field $\mathbb{C}(x)$ is *not* a subspace of $\mathbb{C}[[x, x^{-1}]]$; quotients of polynomials would have to be expanded before they could be viewed as formal Laurent series, and the point is that we shall be using different expansions of the same rational function.

Note that the algebra $\mathbb{C}((x))$ of truncated formal Laurent series is the field of fractions of the formal power series ring $\mathbb{C}[[x]]$. We shall often express certain useful elements of $\mathbb{C}((x))$ and $\mathbb{C}((x^{-1}))$ by means of (the Laurent expansions of) appropriate analytic functions of x and x^{-1}, respectively.

Now we formalize the process of taking opposite expansions of a rational function: There are two canonical embeddings:

$$\iota_+ : \mathbb{C}(x) \hookrightarrow \mathbb{C}((x))$$
$$\iota_- : \mathbb{C}(x) = \mathbb{C}(x^{-1}) \hookrightarrow \mathbb{C}((x^{-1})). \tag{2.3.1}$$

For $f \in \mathbb{C}(x)$, $\iota_+ f$ is the expansion of f as a formal Laurent series in x, and $\iota_- f$ is its expansion as a formal Laurent series in x^{-1}.

Next we introduce a basic linear map Θ, expressing the difference between these two expansions:

$$\Theta = \Theta_x : \mathbb{C}(x) \to \mathbb{C}[[x, x^{-1}]]$$
$$f \mapsto \iota_+ f - \iota_- f. \tag{2.3.2}$$

Viewing $\mathbb{C}[x, x^{-1}]$ as a subalgebra of $\mathbb{C}(x)$, we have

$$\text{Ker } \Theta = \mathbb{C}[x, x^{-1}]. \tag{2.3.3}$$

Motivated by the definition of Θ, we call the elements of the image Im Θ the *expansions of zero*. For $f \in \mathbb{C}[x, x^{-1}]$ and $g \in \mathbb{C}(x)$,

$$\Theta(fg) = f\Theta(g). \tag{2.3.4}$$

Thus Θ is a $\mathbb{C}[x, x^{-1}]$-module map and Im Θ is a $\mathbb{C}[x, x^{-1}]$-submodule of $\mathbb{C}[[x, x^{-1}]]$.

Remark 2.3.1 The formal delta function $\delta(x)$ (see (2.1.32)) is an expansion of zero, since

$$\iota_+((1-x)^{-1}) = \sum_{n \geq 0} x^n = (1-x)^{-1},$$

$$\iota_-((1-x)^{-1}) = \iota_-(-x^{-1}(1-x^{-1})^{-1})$$
$$= -\sum_{n < 0} x^n = -x^{-1}(1-x^{-1})^{-1} = (-x+1)^{-1}$$

and so

$$\delta(x) = \Theta((1-x)^{-1}) = (1-x)^{-1} + (x-1)^{-1}. \tag{2.3.5}$$

Note how symbols like $(1-x)^{-1}$ can have different meanings depending on whether they stand for rational functions or formal series (and note the use of the binomial expansion convention!). The context should always make it clear which meaning is intended, and the careful reader will not need different symbols to express the different objects. In any case, it is always crucial to be very alert to the exact meaning of such an expression in each context.

It is clear that the differentiation operator d/dx (defined on $\mathbb{C}(x)$ and $\mathbb{C}[[x, x^{-1}]]$) commutes with Θ:

$$\Theta \circ \frac{d}{dx} = \frac{d}{dx} \circ \Theta : \mathbb{C}(x) \to \mathbb{C}[[x, x^{-1}]]. \tag{2.3.6}$$

Combining (2.3.5) with (2.3.6) we obtain (using the standard calculus notation for the nth derivative):

Proposition 2.3.2 *For $n \in \mathbb{N}$,*

$$\frac{1}{n!}\delta^{(n)}(x) = \Theta((1-x)^{-n-1})$$
$$= (1-x)^{-n-1} - (-x+1)^{-n-1}. \quad \square \tag{2.3.7}$$

Remark 2.3.3 Note that the point here is that the nth derivative of the expression $(1-x)^{-1}$ coincides with $n!$ times its $(n+1)^{\text{st}}$ power. This classical elementary fact will play a major role for us.

The preceding notions and results of course generalize to several variables, and in fact it is the case of two or more variables that we really need in vertex algebra theory. To illustrate, we shall consider the case of two variables. Instead of the field $\mathbb{C}(x_1, x_2)$ of all rational functions in x_1 and x_2, we shall need only a certain subalgebra. Set $S = \{x_1, x_2, x_1 \pm x_2\}$. Let $\mathbb{C}[x_1, x_2]_S$ be the subalgebra of $\mathbb{C}(x_1, x_2)$ generated by $x_1^{\pm 1}$, $x_2^{\pm 1}$ and $(x_1 \pm x_2)^{-1}$. Define ι_{12} to be the linear map

$$\iota_{12} : \mathbb{C}[x_1, x_2]_S \to \mathbb{C}[[x_1, x_1^{-1}, x_2, x_2^{-1}]] \tag{2.3.8}$$

such that $\iota_{12}(f(x_1, x_2))$ is the formal Laurent series expansion of $f(x_1, x_2)$ involving only finitely many negative powers of x_2. Analogously, we define the linear map $\iota_{21}(f(x_1, x_2))$ using the opposite expansion.

Remark 2.3.4 These formal expansions yield convergent expansions in the complex domains $|x_1| > |x_2| > 0$ and $|x_2| > |x_1| > 0$, respectively. If we were to allow more general denominators, for example, denominators of the form $x_1 - cx_2$ with $c \in \mathbb{C}$ other than ± 1, then we would have different domains of convergence. But as we shall see in many ways, it is natural to use only the indicated denominators.

If our variables are labeled with subscripts other than 1 and 2 we shall use obvious analogues of the notations ι_{12} and ι_{21}.

It is clear that both ι_{12} and ι_{21} are injective. Define

$$\Theta(f) = \iota_{12}f - \iota_{21}f. \tag{2.3.9}$$

Remark 2.3.5 (cf. Remark 2.3.1) We have

$$x_2^{-1}\delta\left(\frac{x_1}{x_2}\right) = \Theta\left((x_1 - x_2)^{-1}\right) = (x_1 - x_2)^{-1} + (x_2 - x_1)^{-1} \tag{2.3.10}$$

(using the binomial expansion convention, as usual).

Since Θ commutes with $\partial/\partial x_1$ and $\partial/\partial x_2$, we also have the following analogue of Proposition 2.3.2 (also recall Remark 2.3.3):

Proposition 2.3.6 *For* $n \in \mathbb{N}$,

$$\frac{1}{n!}\left(\frac{\partial}{\partial x_2}\right)^n x_2^{-1}\delta\left(\frac{x_1}{x_2}\right) = (x_1 - x_2)^{-n-1} - (-x_2 + x_1)^{-n-1}$$

$$= \frac{(-1)^n}{n!}\left(\frac{\partial}{\partial x_1}\right)^n x_2^{-1}\delta\left(\frac{x_1}{x_2}\right). \tag{2.3.11}$$

As an immediate corollary of Proposition 2.3.6 we know how to multiply nonnegative powers of $x_1 - x_2$ by derivatives of the delta function expression (2.3.10).

Proposition 2.3.7 *For* $m, n \in \mathbb{N}$,

$$(x_1 - x_2)^m \left(\frac{\partial}{\partial x_1}\right)^n x_2^{-1}\delta\left(\frac{x_1}{x_2}\right)$$

$$= (-1)^m \frac{n!}{(n-m)!} \left(\frac{\partial}{\partial x_1}\right)^{n-m} x_2^{-1}\delta\left(\frac{x_1}{x_2}\right) \quad \text{if } m \leq n;$$

$$(2.3.12)$$

$$(x_1 - x_2)^m \left(\frac{\partial}{\partial x_1}\right)^n x_2^{-1}\delta\left(\frac{x_1}{x_2}\right) = 0 \text{ if } m > n. \quad \square \qquad (2.3.13)$$

Typically, a sequence of identities like (2.3.11) can and should be nicely written in terms of generating functions. (This will be a central theme in our treatment.) Using the formal Taylor Theorem (Proposition 2.2.2 and Remark 2.2.4) we obtain the following formal generating function for the left-hand side of (2.3.11), incorporating all the higher derivatives at once:

$$\sum_{n\in\mathbb{N}} \frac{x_0^n}{n!} \left(\frac{\partial}{\partial x_2}\right)^n x_2^{-1}\delta\left(\frac{x_1}{x_2}\right) = e^{x_0 \frac{\partial}{\partial x_2}} x_2^{-1}\delta\left(\frac{x_1}{x_2}\right) = e^{x_0 \frac{\partial}{\partial x_2}} x_1^{-1}\delta\left(\frac{x_2}{x_1}\right)$$

$$= x_1^{-1}\delta\left(\frac{x_2 + x_0}{x_1}\right). \qquad (2.3.14)$$

This generating function procedure has given us an expression that looks even simpler than the original expressions (2.3.11), *and that involves no derivatives at all!* Of course, this expression equals by definition

$$x_1^{-1}\delta\left(\frac{x_2 + x_0}{x_1}\right) = \sum_{n\in\mathbb{Z}} x_1^{-n-1}(x_2 + x_0)^n$$

$$= \sum_{n\in\mathbb{Z}}\sum_{i\in\mathbb{N}} \binom{n}{i} x_1^{-n-1} x_2^{n-i} x_0^i. \qquad (2.3.15)$$

Such three-variable expressions will be fundamental in everything that follows. We shall see again and again that generating functions such as these should be viewed as basic objects in their own right.

Similarly, the generating function for the right-hand side of (2.3.11) is

$$\sum_{n\in\mathbb{N}} \frac{x_0^n}{n!} \left(-\frac{\partial}{\partial x_1}\right)^n x_2^{-1}\delta\left(\frac{x_1}{x_2}\right) = e^{-x_0 \frac{\partial}{\partial x_1}} x_2^{-1}\delta\left(\frac{x_1}{x_2}\right)$$

$$= x_2^{-1}\delta\left(\frac{x_1 - x_0}{x_2}\right). \qquad (2.3.16)$$

We have proved the first part of the next proposition; the second part will follow simply by writing down the generating function of the middle expression in (2.3.11) (recall Remark 2.3.3). Formula (2.3.18) in fact encapsulates much of the subtlety in the definition of the notion of vertex algebra.

Proposition 2.3.8 *We have*

$$x_2^{-1}\delta\left(\frac{x_1-x_0}{x_2}\right)=x_1^{-1}\delta\left(\frac{x_2+x_0}{x_1}\right), \tag{2.3.17}$$

$$x_0^{-1}\delta\left(\frac{x_1-x_2}{x_0}\right)-x_0^{-1}\delta\left(\frac{-x_2+x_1}{x_0}\right)=x_2^{-1}\delta\left(\frac{x_1-x_0}{x_2}\right). \tag{2.3.18}$$

Proof. Since of course $(x_1-x_2)^n=(-x_2+x_1)^n$ for $n\in\mathbb{N}$, we see that the generating function of the middle expression in (2.3.11) is:

$$\sum_{n\geq 0}x_0^n\left((x_1-x_2)^{-n-1}-(-x_2+x_1)^{-n-1}\right)$$

$$=\sum_{n\in\mathbb{Z}}x_0^n\left((x_1-x_2)^{-n-1}-(-x_2+x_1)^{-n-1}\right)$$

$$=x_0^{-1}\delta\left(\frac{x_1-x_2}{x_0}\right)-x_0^{-1}\delta\left(\frac{-x_2+x_1}{x_0}\right). \tag{2.3.19}$$

Combined with the comments above, this proves the proposition. $\square$

Remark 2.3.9 Notice that the left-hand side of (2.3.18) has the obvious symmetry $(x_0,x_1,x_2)\leftrightarrow(-x_0,x_2,x_1)$ so that (2.3.17) in fact follows from (2.3.18).

Remark 2.3.10 Here is a slightly different generating function proof of (2.3.17): Since

$$\left(\frac{\partial}{\partial x_1}+\frac{\partial}{\partial x_2}\right)x_2^{-1}\delta\left(\frac{x_1}{x_2}\right)=\left(\frac{\partial}{\partial x_1}+\frac{\partial}{\partial x_2}\right)\left((x_1-x_2)^{-1}+(x_2-x_1)^{-1}\right)=0,$$

we get

$$e^{x_0\left(\frac{\partial}{\partial x_1}+\frac{\partial}{\partial x_2}\right)}x_2^{-1}\delta\left(\frac{x_1}{x_2}\right)=x_2^{-1}\delta\left(\frac{x_1}{x_2}\right).$$

Thus

$$e^{-x_0\frac{\partial}{\partial x_1}}x_2^{-1}\delta\left(\frac{x_1}{x_2}\right)=e^{x_0\frac{\partial}{\partial x_2}}x_2^{-1}\delta\left(\frac{x_1}{x_2}\right)=e^{x_0\frac{\partial}{\partial x_2}}x_1^{-1}\delta\left(\frac{x_2}{x_1}\right).$$

Combining this with the formal Taylor Theorem we obtain (2.3.17). Note that the first line of this argument holds because $x_2^{-1}\delta(x_1/x_2)$ equals a formal expression of the form $F(x_1-x_2)$, and this is one of the many related reasons why the expression $x_2^{-1}\delta(x_1/x_2)$ is more natural than the simpler-looking $\delta(x_1/x_2)$.

Remark 2.3.11 We shall also want the following relation (which is clear using (2.3.17)) among the partial derivatives of $x_0^{-1}\delta((x_1-x_2)/x_0)$:

$$\frac{\partial}{\partial x_1}x_0^{-1}\delta\left(\frac{x_1-x_2}{x_0}\right)=-\frac{\partial}{\partial x_2}x_0^{-1}\delta\left(\frac{x_1-x_2}{x_0}\right)$$

$$=-\frac{\partial}{\partial x_0}x_0^{-1}\delta\left(\frac{x_1-x_2}{x_0}\right). \tag{2.3.20}$$

Remark 2.3.12 Note that the (equal) expressions in (2.3.17) are expanded in nonnegative integral powers of x_0, while in (2.3.18) the three expressions are expanded in nonnegative powers of x_2, x_1 and x_0, respectively. Soon we shall be multiplying expressions like those in Proposition 2.3.8 by suitable formal Laurent series $f(x_0, x_1, x_2)$ and, provided that f satisfies the appropriate conditions, we will be able to formally set the argument in the delta function expression equal to 1 in $f(x_0, x_1, x_2)$, just as we discussed in Remark 2.1.9. For instance, for suitable f, we will see that

$$x_0^{-1}\delta\left(\frac{x_1 - x_2}{x_0}\right) f(x_0, x_1, x_2) = x_0^{-1}\delta\left(\frac{x_1 - x_2}{x_0}\right) f(x_1 - x_2, x_1, x_2), \quad (2.3.21)$$

where of course the binomial expansion convention is being used. One important class of formal series f for which (2.3.21) is in fact valid is the class of (terminating) Laurent polynomials in x_0, x_1 and x_2. We invite the reader to prove that (2.3.21) holds for

$$f(x_0, x_1, x_2) \in \mathbb{C}[x_0, x_0^{-1}, x_1, x_1^{-1}, x_2, x_2^{-1}];$$

this is a special case of the more general result given in (2.3.56) below. Notice that for all of the delta function expressions occurring in Proposition 2.3.8, formally setting the argument in the delta function equal to 1 always amounts to the same "virtual formula"

$$\text{``}x_0 - x_1 + x_2 = 0,\text{''}$$

which motivates the substitutions to be made. If we were to change the sign of the formal variable x_1, formulas (2.3.17) and (2.3.18) would become perfectly symmetrical.

Remark 2.3.13 In Remark 2.1.7 we gave some simple examples of formal limits in one variable that do or do not exist. The two-variable analogues of (2.1.25) and (2.1.26) are worth noting: Consider the (well-defined) formal series

$$f(x_1, x_2) = \sum_{n \geq -1} (x_1 - x_2)^n \in \mathbb{C}[[x_1, x_1^{-1}, x_2, x_2^{-1}]] \quad (2.3.22)$$

(with the binomial expansion convention of course being used). We note that

$$\lim_{x_1 \to x_2} f(x_1, x_2) \text{ does not exist} \quad (2.3.23)$$

(according, of course, to our precise definition of existence of such a formal limit, stated before Remark 2.1.7), because

$$\lim_{x_1 \to x_2} (x_1 - x_2)^{-1} \text{ does not exist.} \quad (2.3.24)$$

But

$$\lim_{x_1 \to x_2} (x_1 - x_2) f(x_1, x_2) \text{ exists and equals 1.} \quad (2.3.25)$$

Remark 2.3.14 Note that

$$\delta(x_1 - x_2) = \sum_{n \in \mathbb{Z}} (x_1 - x_2)^n \qquad (2.3.26)$$

exists; that is,

$$\lim_{x_0 \to 1} x_0^{-1} \delta\left(\frac{x_1 - x_2}{x_0}\right) \text{ exists and equals } \delta(x_1 - x_2). \qquad (2.3.27)$$

In fact, any of the three formal variables in $x_0^{-1}\delta((x_1 - x_2)/x_0)$ can be formally replaced by any nonzero complex number (and in the case of x_2, the substituted complex number is also allowed to be zero). Such formal substitutions of a nonzero complex number for one of the three formal variables in such expressions as $x_0^{-1}\delta((x_1 - x_2)/x_0)$ in fact play an important role in the tensor product theory of [HL4]–[HL7], [Hua8].

Formulas (2.3.17) and (2.3.18) and their applications below are intimately related to Cauchy's residue formula and certain contour deformations applied to rational functions of complex variables. This was discussed extensively in the Appendix of [FLM6]. We shall explain this relation in detail in the next few remarks and, in connection with the Jacobi identity axiom for vertex algebras and vertex operator algebras, in Remarks 3.1.15 and 3.1.16. At the same time, we will explain the relation between our formal delta-function expressions (and other expansions of zero) and distributions or generalized functions, in the sense of functional analysis (cf. [GS]).

Remark 2.3.15 We have already mentioned that the formal series $\delta(x)$ agrees with the Laurent series expansion of the classical "delta function" distribution at $x = 1$, and in this and the next few remarks we elaborate on the interpretation of $\delta(x)$ and other expansions of zero as generalized functions, i.e., linear functionals on suitably specified vector spaces of "test functions." In fact, the formal Laurent series $\delta(x)$ can be precisely interpreted as the linear functional

$$f(x) \mapsto f(1) \qquad (2.3.28)$$

on the space of "test functions" $f(x)$ consisting of the Laurent polynomials in x; the formula

$$\mathbb{C}[x, x^{-1}] \to \mathbb{C}$$
$$f(x) \mapsto \text{Res}_x f(x)\delta(x) \qquad (2.3.29)$$

implements this evaluation (since $\text{Res}_x f(x)\delta(x) = \text{Res}_x f(1)\delta(x) = f(1)\text{Res}_x \delta(x) = f(1)$) and also realizes this evaluation in the form of a formal integration procedure (Res_x) applied to the product of the test function and the generalized function, a traditional notation for generalized functions. We can make this integration procedure less formal using Cauchy's integral formula, by replacing x by a complex variable z and by expressing the linear functional $f(z) \mapsto f(1)$ (for $f(z) \in \mathbb{C}[z, z^{-1}]$, the space of rational functions of z with a possible pole only at $z = 0$) by

$$f(1) = \frac{1}{2\pi i} \int_C \frac{f(z)dz}{z-1}, \qquad (2.3.30)$$

where C is a counterclockwise contour around $z = 1$. Note that $f(z)$ is multiplied by $(z-1)^{-1}$ before the contour integration, and compare this to formula (2.3.5), in which $\delta(x)$ is expressed as the expansion of zero given by $(x-1)^{-1} + (1-x)^{-1}$. It is clear that $\delta(x)$ is the *unique* element of $\mathbb{C}[[x, x^{-1}]]$ such that

$$\text{Res}_x f(x)\delta(x) = f(1) \qquad (2.3.31)$$

for any $f(x) \in \mathbb{C}[x, x^{-1}]$. (If there were two such formal series, subtracting would give a formal series, say $\mu(x)$, such that $\text{Res}_x f(x)\mu(x) = 0$ for any $f(x)$, and choosing $f(x) = x^n$ for each n would show that all the coefficients of $\mu(x)$ are 0.) So $\delta(x)$ precisely amounts to the linear functional $f(x) \mapsto f(1)$. As we are about to see still further, such a distribution-theoretic interpretation of expansions of zero is very illuminating and reveals a precise connection with complex analysis. But we emphasize that one significant advantage of our formal-calculus approach to these matters is that we do not need to consider "test functions" $f(x)$ when considering expansions of zero; for instance, in place of "$f \mapsto f(1)$" or "$f \mapsto \text{Res}_x f(x)\delta(x)$" we simply write "$\delta(x)$," an element of $\mathbb{C}[[x, x^{-1}]]$.

Remark 2.3.16 It is straightforward to extend these considerations to two-variable expansions of zero such as $x_2^{-1}\delta(x_1/x_2)$. In fact, one of the ways to interpret this particular expansion of zero is as the formal substitution operation

$$f(x_1) \mapsto f(x_2) \qquad (2.3.32)$$

for a "test function" $f(x) \in \mathbb{C}[x, x^{-1}]$, since

$$\text{Res}_{x_1} f(x_1)x_2^{-1}\delta\left(\frac{x_1}{x_2}\right) = \text{Res}_{x_1} f(x_2)x_2^{-1}\delta\left(\frac{x_1}{x_2}\right) = f(x_2), \qquad (2.3.33)$$

and it is again clear that this relation (for all $f(x) \in \mathbb{C}[x, x^{-1}]$) uniquely characterizes the formal series $x_2^{-1}\delta(x_1/x_2) \in \mathbb{C}[[x_1, x_1^{-1}, x_2, x_2^{-1}]]$ (as above, take $f(x) = x^n$ for all n). The analogue of formula (2.3.30) is of course

$$f(z_2) = \frac{1}{2\pi i} \int_C \frac{f(z_1)dz_1}{z_1 - z_2}, \qquad (2.3.34)$$

where C is a counterclockwise contour around $z_1 = z_2$; recall from Remark 2.3.5 that $x_2^{-1}\delta(x_1/x_2)$ is the expansion of zero associated with $(x_1 - x_2)^{-1}$. It is worth pointing out that (2.3.33) can easily be generalized to the following analogue for a two-variable formal series in place of $f(x)$. Let V be a vector space and let

$$f(x_1, x_2) \in (\text{End } V)[[x_1, x_1^{-1}, x_2, x_2^{-1}]]$$

be such that

$$\lim_{x_1 \to x_2} f(x_1, x_2) \text{ exists,} \tag{2.3.35}$$

as in Proposition 2.1.8. Then using Proposition 2.1.8 and (2.2.7) we see that

$$\mathrm{Res}_{x_1} f(x_1, x_2) x_2^{-1} \delta\left(\frac{x_1}{x_2}\right) = f(x_2, x_2). \tag{2.3.36}$$

Remark 2.3.17 The analogous generalized function interpretation of the three-variable delta function expression $x_0^{-1}\delta((x_1 - x_2)/x_0)$ and the related expressions is more interesting. To figure out what this interpretation is, first notice that for

$$f(x_0, x_1, x_2) \in \mathbb{C}[x_0, x_0^{-1}, x_1, x_1^{-1}, x_2, x_2^{-1}], \tag{2.3.37}$$

the space of Laurent polynomials in three variables (our space of "test functions"),

$$\begin{aligned}
\mathrm{Res}_{x_0} &\left(x_0^{-1}\delta\left(\frac{x_1 - x_2}{x_0}\right) f(x_0, x_1, x_2)\right) \\
&= \mathrm{Res}_{x_0}\left(x_0^{-1}\delta\left(\frac{x_1 - x_2}{x_0}\right) f(x_1 - x_2, x_1, x_2)\right) \\
&= \mathrm{Res}_{x_0}\left(x_0^{-1}\delta\left(\frac{x_1 - x_2}{x_0}\right)\right) f(x_1 - x_2, x_1, x_2) \\
&= f(x_1 - x_2, x_1, x_2);
\end{aligned}$$

the first equality follows from Remark 2.3.12 (and (2.3.56) below). Applying the additional residue operations Res_{x_1} and Res_{x_2}, we also have

$$\begin{aligned}
\mathrm{Res}_{x_0}\mathrm{Res}_{x_1}\mathrm{Res}_{x_2} &\left(x_0^{-1}\delta\left(\frac{x_1 - x_2}{x_0}\right) f(x_0, x_1, x_2)\right) \\
&= \mathrm{Res}_{x_1}\mathrm{Res}_{x_2} f(x_1 - x_2, x_1, x_2), \tag{2.3.38}
\end{aligned}$$

and as usual, this formula (for all indicated f) clearly characterizes $x_0^{-1}\delta((x_1 - x_2)/x_0)$ as a formal Laurent series in x_0, x_1 and x_2. The right-hand side of (2.3.38) can be expressed as follows: Replace x_1 and x_2 by complex variables z_1 and z_2, respectively; expand $f(z_1 - z_2, z_1, z_2)$ (which is the most general rational function of z_1 and z_2 with possible singularities only at $z_1 = 0$, $z_2 = 0$ and $z_1 = z_2$) in the domain $|z_1| > |z_2| > 0$ (or restrict $f(z_1 - z_2, z_1, z_2)$ to this domain); and then extract the residue at $z_2 = 0$ (with z_1 held fixed) followed by the residue at $z_1 = 0$. That is,

$$\begin{aligned}
\mathrm{Res}_{x_0}\mathrm{Res}_{x_1}\mathrm{Res}_{x_2} &\left(x_0^{-1}\delta\left(\frac{x_1 - x_2}{x_0}\right) f(x_0, x_1, x_2)\right) \\
&= \mathrm{Res}_{z_1=0}\mathrm{Res}_{z_2=0} f(z_1 - z_2, z_1, z_2), \tag{2.3.39}
\end{aligned}$$

where $|z_1| > |z_2| > 0$. This procedure for assigning a complex number to a formal Laurent polynomial $f(x_0, x_1, x_2)$ or to a rational function of z_1 and z_2 with possible singularities only at $z_1 = 0$, $z_2 = 0$ and $z_1 = z_2$ thus amounts precisely to the information encoded in the formal series $x_0^{-1}\delta((x_1 - x_2)/x_0)$ (as an element of $\mathbb{C}[[x_0, x_0^{-1}, x_1, x_1^{-1}, x_2, x_2^{-1}]]$).

Remark 2.3.18 It is especially interesting to interpret the two relations (2.3.17) and (2.3.18) according to the previous remark. Of course, each of the four distinct expressions (such as $x_1^{-1}\delta((x_2+x_0)/x_1)$) appearing in (2.3.17) and (2.3.18) has its own interpretation as in Remark 2.3.17. Specifically, arguing as in Remark 2.3.17, for an arbitrary Laurent polynomial $f(x_0, x_1, x_2)$ in x_0, x_1 and x_2, we have

$$\operatorname{Res}_{x_0}\operatorname{Res}_{x_1}\operatorname{Res}_{x_2}\left(x_0^{-1}\delta\left(\frac{-x_2+x_1}{x_0}\right)f(x_0, x_1, x_2)\right)$$
$$= \operatorname{Res}_{x_1}\operatorname{Res}_{x_2} f(-x_2+x_1, x_1, x_2)$$
$$= \operatorname{Res}_{z_2=0}\operatorname{Res}_{z_1=0} f(z_1-z_2, z_1, z_2), \tag{2.3.40}$$

where $|z_2| > |z_1| > 0$;

$$\operatorname{Res}_{x_0}\operatorname{Res}_{x_1}\operatorname{Res}_{x_2}\left(x_1^{-1}\delta\left(\frac{x_2+x_0}{x_1}\right)f(x_0, x_1, x_2)\right)$$
$$= \operatorname{Res}_{x_0}\operatorname{Res}_{x_2} f(x_0, x_2+x_0, x_2)$$
$$= \operatorname{Res}_{z_2=0}\operatorname{Res}_{z_0=0} f(z_0, z_2+z_0, z_2) \quad (\text{where } |z_2| > |z_0| > 0)$$
$$= \operatorname{Res}_{z_2=0}\operatorname{Res}_{z_1=z_2} f(z_1-z_2, z_1, z_2), \tag{2.3.41}$$

where $|z_2| > |z_1-z_2| > 0$;

$$\operatorname{Res}_{x_0}\operatorname{Res}_{x_1}\operatorname{Res}_{x_2}\left(x_2^{-1}\delta\left(\frac{x_1-x_0}{x_2}\right)f(x_0, x_1, x_2)\right)$$
$$= \operatorname{Res}_{x_0}\operatorname{Res}_{x_1} f(x_0, x_1, x_1-x_0)$$
$$= \operatorname{Res}_{z_1=0}\operatorname{Res}_{z_0=0} f(z_0, z_1, z_1-z_0) \quad (\text{where } |z_1| > |z_0| > 0)$$
$$= -\operatorname{Res}_{z_1=0}\operatorname{Res}_{z_2=z_1} f(z_1-z_2, z_1, z_2), \tag{2.3.42}$$

where $|z_1| > |z_1-z_2| > 0$. We conclude that the equalities (2.3.17) and (2.3.18) amount precisely to the following assertions: For an arbitrary rational function $g(z_1, z_2)$ of z_1 and z_2 with possible singularities only at $z_1 = 0$, $z_2 = 0$ and $z_1 = z_2$,

$$\operatorname{Res}_{z_2=0}\operatorname{Res}_{z_1=z_2} g(z_1, z_2) = -\operatorname{Res}_{z_1=0}\operatorname{Res}_{z_2=z_1} g(z_1, z_2), \tag{2.3.43}$$
$$\operatorname{Res}_{z_1=0}\operatorname{Res}_{z_2=0} g(z_1, z_2) - \operatorname{Res}_{z_2=0}\operatorname{Res}_{z_1=0} g(z_1, z_2)$$
$$= -\operatorname{Res}_{z_1=0}\operatorname{Res}_{z_2=z_1} g(z_1, z_2), \tag{2.3.44}$$

where the successions of residues are evaluated in the four relevant domains. But these are simply cases of the Cauchy residue formula applied to a very special class of rational functions. That is, the formal variable identities (2.3.17) and (2.3.18) amount *precisely* to the Cauchy identities (2.3.43) and (2.3.44), respectively, for an arbitrary rational function of the indicated type. Below we will relate (2.3.44) to the Jacobi identity for vertex (operator) algebras. Again we emphasize that while the generalized function considerations lead to this interesting reformulation of such formal calculus identities as (2.3.18) in terms of complex analysis, our formal variable approach does not involve "test functions" (like $g(z_1, z_2)$ in (2.3.44)); furthermore, it is very natural to work with the three independent formal variables x_0, x_1, x_2 on an equal footing, in such relations as (2.3.18), in contrast to the two complex variables and various expansion domains in such relations as (2.3.44) (equivalent to (2.3.18)).

Remark 2.3.19 Some authors use the notation $\delta(x_1 - x_2)$ for what we write as $x_2^{-1}\delta(x_1/x_2)$, but this leaves no "room" for the third variable in the natural expression $x_2^{-1}\delta((x_1 - x_0)/x_2)$, and one loses the visible symmetry in Proposition 2.3.8, a symmetry that we will extensively exploit. Moreover, as we have discussed in Remark 2.3.14, the expression $\delta(x_1 - x_2)$ is already a well-defined formal series unequal to $x_2^{-1}\delta(x_1/x_2)$, and this formal series $\delta(x_1 - x_2)$ and its variants arise naturally in, for example, tensor product theory [HL4]–[HL7], [Hua8].

Now we shall discuss certain combinations of delta function expressions and operators that will be very important later, in particular, in the definition of the notion of vertex algebra. Let V be a vector space and let

$$A(x) \in (\text{End } V)[[x, x^{-1}]] \quad \text{and} \quad B(x) \in \text{Hom}(V, V((x))). \tag{2.3.45}$$

Notice that

$$x_0^{-1}\delta\left(\frac{x_1 - x_2}{x_0}\right) A(x_1)B(x_2) \quad \text{exists} \tag{2.3.46}$$

(as an element of $(\text{End } V)[[x_0, x_0^{-1}, x_1, x_1^{-1}, x_2, x_2^{-1}]]$) because for any $n \in \mathbb{Z}$, the coefficient of x_0^n is $(x_1 - x_2)^{-n-1}A(x_1)B(x_2)$, which in turn exists, as we observed in Section 2.1. (The coefficient of each power of x_2 gives a finite sum when we apply to an element of V.) In the definition of the notion of vertex algebra (and throughout the theory), $A(x)$ and $B(x)$ will be vertex operators. The truncation hypothesis on $B(x)$ will also be needed for $A(x)$ (and for all vertex operators) since products of pairs of vertex operators will be needed in both orders.

Remark 2.3.20 We have just illustrated the most useful strategy for verifying the existence of a somewhat complicated formal expression involving several variables. First look for a "preferred" one of the variables such that the expansion coefficients in the powers of this variable are easy to analyze. In the expression (2.3.46), the variable x_2 would also serve the purpose, because the coefficient of each power of x_2 in (2.3.46) (when applied to a vector in V) is a finite sum of various derivatives of $x_0^{-1}\delta(x_1/x_0)$ times elements of $V[[x_1, x_1^{-1}]]$. This technique of "peeling off" the variables one by one, in succession, will be used extensively in what follows, often without explicit comment. The reader will find this process to be easy and routine. But recall that one must indeed always verify that every expression that is supposed to exist does exist. Carelessness can lead to traps like (2.1.17), but in fact it is easy to develop the habit of routinely being careful about the existence of desired expressions; just as in analysis, one gets used to verifying routinely that necessary limits exist without having to write out detailed proofs.

Notice how easily the following natural generalization of Proposition 2.1.8(b), incorporating higher derivatives as well as derivations more general than ordinary differentiation, is formulated and proved using generating functions:

Proposition 2.3.21 *Let*

$$p(x_1, x_2) \in \mathbb{C}[x_1, x_1^{-1}, x_2, x_2^{-1}] \tag{2.3.47}$$

and

$$f(x_1, x_2) \in (\text{End } V)[[x_1, x_1^{-1}, x_2, x_2^{-1}]] \tag{2.3.48}$$

such that

$$\lim_{x_1 \to x_2} f(x_1, x_2) \ \text{exists.}$$

Set

$$T = p(x_1, x_2) \frac{\partial}{\partial x_1}. \tag{2.3.49}$$

Then

$$f(x_1, x_2) e^{yT} \delta\left(\frac{x_1}{x_2}\right) = (e^{-yT} f)(x_2, x_2) e^{yT} \delta\left(\frac{x_1}{x_2}\right). \tag{2.3.50}$$

Equivalently, for $n \geq 0$,

$$f(x_1, x_2) T^n \delta\left(\frac{x_1}{x_2}\right) = \sum_{k=0}^{n} (-1)^k \binom{n}{k} (T^k f)(x_2, x_2) T^{n-k} \delta\left(\frac{x_1}{x_2}\right), \tag{2.3.51}$$

all expressions existing.

Proof. First note that

$$\lim_{x_1 \to x_2} \left(\frac{\partial f}{\partial x_1}\right)(x_1, x_2) \ \text{exists}$$

and hence, by iteration, so does

$$\lim_{x_1 \to x_2} \left(\left(\frac{\partial}{\partial x_1}\right)^k f\right)(x_1, x_2)$$

for $k \geq 0$. Thus

$$\lim_{x_1 \to x_2} (T^k f)(x_1, x_2) \ \text{exists for } k \geq 0,$$

since T^k is a polynomial in $\partial/\partial x_1$ with coefficients in $\mathbb{C}[x_1, x_1^{-1}, x_2, x_2^{-1}]$. It follows that the expression on the right-hand side of (2.3.51) exists, and so the expression on the right-hand side of (2.3.50) also exists. It is equally easy to see that the left-hand side of (2.3.51), and thus of (2.3.50), exists. Using Proposition 2.1.8(b), (2.2.14) and (an obvious extension to two variables of) (2.2.18), we have

$$
\begin{aligned}
f(x_1, x_2)e^{yT}\delta\left(\frac{x_1}{x_2}\right) &= e^{yT}\left[(e^{-yT}f(x_1, x_2))\delta\left(\frac{x_1}{x_2}\right)\right] \\
&= e^{yT}\left[(e^{-yT}f)(x_2, x_2)\delta\left(\frac{x_1}{x_2}\right)\right] \\
&= (e^{-yT}f)(x_2, x_2)e^{yT}\delta\left(\frac{x_1}{x_2}\right). \quad \square
\end{aligned}
\tag{2.3.52}
$$

Remark 2.3.22 Of course, formula (2.3.51) could have been (discovered and) proved by induction on n, but the generating function approach is simpler and more elegant. While it is quite straightforward, Proposition 2.3.21 is a very "concentrated" result, in that even for small values of n, and even for the very special derivation $T = \frac{\partial}{\partial x_1}$, formula (2.3.51) contains a lot of information, crucial in vertex operator calculus. Note also that when $p(x_1, x_2)$ in (2.3.49) is any Laurent polynomial other than $p(x_1, x_2) = 1$, it would be an unpleasant task to express T^n *explicitly* as a sum of powers of $\frac{\partial}{\partial x_1}$ multiplied by Laurent polynomials, but fortunately we do not have to worry about this.

Remark 2.3.23 As we already know, the formal exponential $e^{y(d/dx)}$ can be written as an *explicit* (formal) "change-of-variable transformation," namely, formal translation by y. For general derivations such as T in (2.3.49), there is no such simple explicit change-of-variable formula for the formal transformation e^{yT} acting on formal series. In the present work, we shall need essentially only the case $T = \partial/\partial x_1$ (for which the formal Taylor Theorem gives an alternate expression for e^{yT}), but since Proposition 2.3.21 indeed holds for general T with no extra effort required, we have formulated this result in general. A deep analysis of change-of-variable transformations of the type e^{yT}, where T in (2.3.49) is replaced by various types of formal *infinite* series, plays a major role in [Hua1], [Hua2], [Hua13].

Remark 2.3.24 In vertex operator calculus we shall often need the following straightforward generalization of Proposition 2.3.21. Let

$$
f(x_1, x_2, y) \in (\text{End } V)[[x_1, x_1^{-1}, x_2, x_2^{-1}, y, y^{-1}]]
\tag{2.3.53}
$$

be such that

$$
\lim_{x_1 \to x_2} f(x_1, x_2, y) \quad \text{exists}
$$

and such that for any $v \in V$,

$$
f(x_1, x_2, y)v \in V[[x_1, x_1^{-1}, x_2, x_2^{-1}]]((y)).
$$

Then the assertions of Proposition 2.3.21 still hold, that is,

$$
f(x_1, x_2, y)e^{yT}\delta\left(\frac{x_1}{x_2}\right) = (e^{-yT}f)(x_2, x_2, y)e^{yT}\delta\left(\frac{x_1}{x_2}\right).
\tag{2.3.54}
$$

Indeed, apply Proposition 2.3.21 to $f(x_1, x_2, y_1)$ (y_1 a new formal variable) and then take $\lim_{y_1 \to y}$.

Remark 2.3.25 Here is a very important application of Remark 2.3.24. Taking $p(x_1, x_2) = 1$, i.e., $T = \partial/\partial x_1$, and then using the Taylor Theorem (and multiplying by x_2^{-1}, which, as we have already seen, is a natural thing to do in this context), we get the special case

$$x_2^{-1}\delta\left(\frac{x_1 + y}{x_2}\right) f(x_1, x_2, y) = x_2^{-1}\delta\left(\frac{x_1 + y}{x_2}\right) f(x_2 - y, x_2, y). \quad (2.3.55)$$

Then, using (2.3.17), we get an extremely useful variant of (2.3.55). Under the hypotheses in Remark 2.3.24,

$$x_1^{-1}\delta\left(\frac{x_2 - y}{x_1}\right) f(x_1, x_2, y) = x_1^{-1}\delta\left(\frac{x_2 - y}{x_1}\right) f(x_2 - y, x_2, y). \quad (2.3.56)$$

Note the important similarity between this statement and the simpler delta function substitution principle (2.1.35); in (2.3.56) we are in fact allowed to replace x_1 by $x_2 - y$, a replacement motivated by formally setting the expression $(x_2 - y)/x_1$ in the argument of the delta function equal to 1. (But note that Proposition 2.1.8 alone does not justify (2.3.56); we need the present more elaborate argument.) We shall often be using the delta function substitution formula (2.3.56), or a variant of it, without explicit comment (when the hypotheses in Remark 2.3.24 hold).

Even though the formal variable y played a special role among the three formal variables in Remark 2.3.25, in practice this variable will be on essentially equal footing with x_1 and x_2, and so we will now call it $-x_0$, as we already did in Proposition 2.3.8 and in Remarks 2.3.17 and 2.3.18. The following application of the principle in Remark 2.3.25, closely related to the discussion in Remarks 2.3.17 and 2.3.18, will be crucial in understanding the Jacobi identity for vertex operator algebras:

Proposition 2.3.26 *Consider a formal Laurent polynomial in three variables x_0, x_1, x_2, that is, a formal Laurent series of the form*

$$f(x_0, x_1, x_2) = \frac{g(x_0, x_1, x_2)}{x_0^r x_1^s x_2^t}, \quad (2.3.57)$$

where g is a polynomial and $r, s, t \in \mathbb{Z}$. Then

$$x_1^{-1}\delta\left(\frac{x_2 + x_0}{x_1}\right) \iota_{20}(f|_{x_1 = x_0 + x_2}) = x_2^{-1}\delta\left(\frac{x_1 - x_0}{x_2}\right) \iota_{10}(f|_{x_2 = x_1 - x_0}) \quad (2.3.58)$$

and

$$x_0^{-1}\delta\left(\frac{x_1 - x_2}{x_0}\right) \iota_{12}(f|_{x_0 = x_1 - x_2}) - x_0^{-1}\delta\left(\frac{x_2 - x_1}{-x_0}\right) \iota_{21}(f|_{x_0 = x_1 - x_2})$$
$$= x_2^{-1}\delta\left(\frac{x_1 - x_0}{x_2}\right) \iota_{10}(f|_{x_2 = x_1 - x_0}). \quad (2.3.59)$$

Proof. We simply multiply each of the two formulas in Proposition 2.3.8 by the Laurent polynomial (2.3.57) and observe that by (2.3.56),

$$x_1^{-1}\delta\left(\frac{x_2+x_0}{x_1}\right) f(x_0,x_1,x_2) = x_1^{-1}\delta\left(\frac{x_2+x_0}{x_1}\right)\iota_{20}(f|_{x_1=x_0+x_2}),$$

$$x_2^{-1}\delta\left(\frac{x_1-x_0}{x_2}\right) f(x_0,x_1,x_2) = x_2^{-1}\delta\left(\frac{x_1-x_0}{x_2}\right)\iota_{10}(f|_{x_2=x_1-x_0}),$$

$$x_0^{-1}\delta\left(\frac{x_1-x_2}{x_0}\right) f(x_0,x_1,x_2) = x_0^{-1}\delta\left(\frac{x_1-x_2}{x_0}\right)\iota_{12}(f|_{x_0=x_1-x_2}),$$

$$x_0^{-1}\delta\left(\frac{x_2-x_1}{-x_0}\right) f(x_0,x_1,x_2) = x_0^{-1}\delta\left(\frac{x_2-x_1}{-x_0}\right)\iota_{21}(f|_{x_0=x_1-x_2}).$$

(*Note that the four evaluations* $f|_{x_1=x_0+x_2}$, *etc., are rational functions, not formal series.*)   □

It is clear from this proof that the assertion of Proposition 2.3.26 holds more generally if

$$f(x_0,x_1,x_2) \in \mathbb{C}((x_0,x_1,x_2)),$$

that is, if f is a formal Laurent series truncated from below in all three variables. For the precise formulation of this result, it is natural for us to extend the use of the iota-maps beyond algebras of formal rational functions, as follows:

Let

$$A = \mathbb{C}((x_1,x_2))[(x_1\pm x_2)^{-1}] = \mathbb{C}[[x_1,x_2]][x_1^{-1},x_2^{-1},(x_1-x_2)^{-1},(x_1+x_2)^{-1}],$$

understood as a subalgebra of the field of fractions of the ring $\mathbb{C}[[x_1,x_2]]$ and *not* as an algebra of formal series. We have the obvious (well-defined) analogues of the iota-maps defined above; i.e., the injections

$$\iota_{12} : A \hookrightarrow \mathbb{C}[[x_1,x_1^{-1},x_2,x_2^{-1}]],$$

$$\iota_{21} : A \hookrightarrow \mathbb{C}[[x_1,x_1^{-1},x_2,x_2^{-1}]] \tag{2.3.60}$$

give the canonical expansions of an element involving only finitely many negative powers of x_2 (respectively, x_1).

In case our variables are labeled with subscripts other than 1 and 2 we shall use appropriate analogues of these notations.

Of course, we could extend the maps ι_{12} and ι_{21} to more general algebras than A, but we shall not need to.

In this setting, we have the following obvious generalization of the last result.

Proposition 2.3.27 *Let $f(x_0,x_1,x_2)$ be a formal Laurent series truncated from below in powers of x_0, x_1 and x_2, that is,*

$$f(x_0,x_1,x_2) \in \mathbb{C}((x_0,x_1,x_2)). \tag{2.3.61}$$

Then the assertions of formulas (2.3.58) and (2.3.59) hold for f.   □

Remark 2.3.28 While Proposition 2.3.26 can be interpreted by means of complex analysis and the Cauchy residue formula as in Remark 2.3.18, Proposition 2.3.27 cannot be so interpreted, because the formal series $f(x_0, x_1, x_2)$ in Proposition 2.3.27 is not in general a (convergent) expansion of a function. This is still another reason why the formal variable approach is so natural in vertex algebra theory.

3

Vertex Operator Algebras: The Axiomatic Basics

In this chapter we begin presenting the axiomatic theory of vertex operator algebras. The notion of vertex algebra was introduced in [B1] and a variant of the notion, that of vertex operator algebra, was developed in [FLM6] and [FHL]. The notion of vertex operator algebra is the mathematical counterpart of the notion of "operator algebra," or "chiral algebra," in conformal field theory, as formalized in [BPZ], and many of the algebraic considerations in this chapter were introduced and exploited in more physical language and settings in the vast physics literature on conformal field theory and string theory. Our treatment is based on [B1], [FLM6], [FHL], [DL3] and [Li3]. In Section 3.3 we include a discussion of the relations between the mathematical treatment of "associativity" presented here and operator product expansions in the sense of conformal field theory and string theory.

For the original motivations of the concepts of vertex algebra and of vertex operator algebra, and of the particular axioms used, we refer the reader to the research announcement [B1] and also the book [FLM6], especially the introductory material in [FLM6]. Many different ideas from both mathematics and physics are involved, and it would take us too far afield to review this history in any detail. Our purpose in the present work is instead to focus on guiding the reader to a fluent and flexible understanding of the axiomatic theory. Examples will be presented later, and in fact they will be very easy to present once the general theory has been discussed. The treatment in [FLM6] was quite the reverse. Special constructions were developed first, on the way to the construction of the moonshine module for the Monster; the abstract notion of vertex operator algebra, presented at the end of Chapter 8 of [FLM6], summarized the most fundamental general features appearing in the constructions. Thus we suggest that the reader use [FLM6], through at least Chapter 8, as a companion to the present work. (Chapters 9–13 of [FLM6] deal more specifically with the construction of the moonshine module.)

3.1 Definitions and some fundamental properties

Here we present and examine the definitions of the notions of vertex algebra and vertex operator algebra. We use the "Jacobi identity" [FLM6] as the main axiom for both notions. The equivalence of this definition of vertex algebra and Borcherds' definition of vertex algebra in [B1] will be shown in Section 3.6.

We shall take for granted the formal calculus that we discussed in Chapter 2, including in particular algebraic limits, the concept of the existence of products, the binomial expansion convention and the various delta function expressions and their properties.

A vertex algebra consists of a vector space equipped with a kind of "product operation" and a kind of "identity operator" satisfying a few axioms whose naturalness will become clear. (The reader who consults [FLM6] will see how the abstract notion emerges naturally out of a flow of ideas and constructions.)

Definition 3.1.1 A *vertex algebra* consists of a vector space V equipped, first, with a linear map (the *vertex operator map*) $V \otimes V \to V[[x, x^{-1}]]$, or equivalently, a linear map

$$Y(\cdot, x) : \ V \to (\text{End } V)[[x, x^{-1}]]$$
$$v \mapsto Y(v, x) = \sum_{n \in \mathbb{Z}} v_n x^{-n-1}. \tag{3.1.1}$$

(Notice that we are using the notation v_n for the element of End V which is the coefficient of x^{-n-1} in the formal series $Y(v, x)$; the notation v_n does not refer to any sort of component of the vector v.) We call $Y(v, x)$ the *vertex operator associated with* v. We also have a distinguished element $\mathbf{1}$ of V (the *vacuum vector*). The following conditions are assumed for $u, v \in V$: First, the *truncation condition*:

$$u_n v = 0 \quad \text{for } n \text{ sufficiently large}, \tag{3.1.2}$$

that is,

$$Y(u, x)v \in V((x)); \tag{3.1.3}$$

next, the following *vacuum property*:

$$Y(\mathbf{1}, x) = 1 \quad (1 \text{ on the right being the identity operator}); \tag{3.1.4}$$

also, the *creation property* (whereby $Y(v, x)$ "creates" the vector v):

$$Y(v, x)\mathbf{1} \in V[[x]] \quad \text{and} \quad \lim_{x \to 0} Y(v, x)\mathbf{1} = v \tag{3.1.5}$$

(that is, $Y(v, x)\mathbf{1}$ involves only nonnegative integral powers of x and the constant term is v); and finally, and most importantly, the *Jacobi identity*:

$$x_0^{-1}\delta\left(\frac{x_1 - x_2}{x_0}\right) Y(u, x_1)Y(v, x_2) - x_0^{-1}\delta\left(\frac{x_2 - x_1}{-x_0}\right) Y(v, x_2)Y(u, x_1)$$
$$= x_2^{-1}\delta\left(\frac{x_1 - x_0}{x_2}\right) Y(Y(u, x_0)v, x_2). \tag{3.1.6}$$

(It is straightforward—and of course necessary—to check that when each expression in (3.1.6) is applied to any element of V, the coefficient of each monomial in the three formal variables is a finite sum of vectors. On the right-hand side, the notation $Y(\cdot, x_2)$ is understood to be extended in an obvious way to $V[[x_0, x_0^{-1}]]$; this makes clear the meaning of the "iterate" $Y(Y(u, x_0)v, x_2)$ of the two vertex operators $Y(v, x_2)$ and $Y(u, x_0)$.)

We shall sometimes refer to the vertex algebra V as $(V, Y, \mathbf{1})$ if necessary.

Remark 3.1.2 Note that the vertex operator map $Y(\cdot, x)$ amounts to infinitely many product operations $u_n v$ on V parametrized by $n \in \mathbb{Z}$; this justifies the use of the notation u_n (introduced and used heavily in [B1]) for the component operator of the vertex operator $Y(u, x)$. (Even though the notation u_n looks deceptively like an indexed vector, we trust that the reader will not be confused by this notation.) A reason why the operators $v_n \in \text{End } V$ are parametrized as they are in (3.1.1) (following [B1]) rather than for example by the formula $Y(v, x) = \sum_{n \in \mathbb{Z}} v_n x^n$ is that in important special cases such as that of vertex algebras based on an affine Lie algebra associated with an underlying Lie algebra $\mathfrak{g}$ (see Section 6.2), the operators v_n for $v \in \mathfrak{g}$ are equal to operators that have traditionally been parametrized by the integer n rather than for example by $-n - 1$ (again see Section 6.2, and also Remark 3.1.25). We note also that the Jacobi identity (3.1.6) amounts to an infinite family of somewhat complicated identities parametrized by ordered triples of integers once we have computed and equated the coefficients of the monomials in x_0, x_1 and x_2; these identities will involve certain binomial coefficients. In this way, we may think of a vertex algebra as a kind of generalized nonassociative algebra, with infinitely many product operations satisfying infinitely many identities, rather than a single product operation satisfying finitely many identities. Here, in fact, is this list of identities, explicitly: For $l, m, n \in \mathbb{Z}$, equating the coefficient of $x_0^{-l-1} x_1^{-m-1} x_2^{-n-1}$ in the Jacobi identity (3.1.6) gives

$$\sum_{i \geq 0} (-1)^i \binom{l}{i} u_{m+l-i} v_{n+i} - (-1)^l \sum_{i \geq 0} (-1)^i \binom{l}{i} v_{n+l-i} u_{m+i}$$
$$= \sum_{i \geq 0} \binom{m}{i} (u_{l+i} v)_{m+n-i}; \tag{3.1.7}$$

when this identity is applied to any element of V, each sum over i becomes finite. (The sum over i on the right-hand side is already finite.) In our treatment, we shall have to use only "a few" of these individual identities; we will find it much easier and more conceptual to work directly with the Jacobi identity (3.1.6) itself in most settings. For nearly any examples of vertex operators $Y(u, x_1)$ and $Y(v, x_2)$ in nearly any vertex algebra, the identities in the infinite list (3.1.7) are highly nontrivial. The Jacobi identity in fact encodes and conceptualizes a great deal of subtle algebraic information relating the components u_n of vertex operators $Y(u, x)$. By virtue of the formal calculus that we have developed, the Jacobi identity will be quite easy to use, as we shall see. We shall be examining it in many different ways.

Remark 3.1.3 It turns out that it is conceptually more fundamental to think of the vertex operator map $Y(\cdot, x)$ as a kind of "multiplication operation with a formal parameter x" rather than as an infinite list of multiplication operations. One thinks of the vertex operator $Y(v, x)$ then as the "left multiplication" associated to v. This point of view, which we shall develop, is deeply related to geometric and physical structure, in which the formal variable x is (carefully!) replaced by a nonzero complex variable. But as we shall see extensively, this viewpoint also points to a natural way of developing the algebraic theory itself. Of course, we have here another instance of our basic reliance on generating functions—in this case, the generating function of an infinite family of products or of left-multiplication operators.

Remark 3.1.4 One can think of the axiom (3.1.4) as a "left-identity property" of $\mathbf{1}$ and (3.1.5) as a "right-identity property" of $\mathbf{1}$. The asymmetry between these two properties is one of the many reflections of the role of the parameter x. In terms of components of vertex operators, the creation property (3.1.5) amounts to the conditions that $v_n \mathbf{1} = 0$ for $n \geq 0$ and $v_{-1} \mathbf{1} = v$. We note that the vacuum vector $\mathbf{1}$ is unique, just as in classical algebra: suppose that $u \in V$ also has the vacuum property (3.1.4), i.e., $Y(u, x) = 1$. Then $u_{-1} \mathbf{1} = \mathbf{1}$. Combining this with $u_{-1} \mathbf{1} = u$ (from (3.1.5)) we get $u = \mathbf{1}$.

Remark 3.1.5 Property (2.3.18) of the delta function amounts to the special case $u = v = \mathbf{1}$ of the Jacobi identity. In particular, the existence of at least one (nonzero!) vertex algebra proves (2.3.18), but in fact this is not a good way to prove (2.3.18); when we discuss examples, we shall see that (2.3.18) is always used in the constructions. For instance, the one-dimensional space $\mathbb{C}\mathbf{1}$ provides a simple example of a vertex algebra; the reason is that (2.3.18) holds! (Recall that in Remark 2.3.18, (2.3.18) was interpreted in a precise way as a Cauchy residue identity for an arbitrary rational function of a certain type.)

Remark 3.1.6 More generally, again by (2.3.18), any commutative associative algebra A with identity element 1 clearly forms a vertex algebra, where we define $Y(a, x)b = ab$ for $a, b \in A$ and $\mathbf{1} = 1$. This observation will be discussed further in Remark 3.4.5, Example 3.4.6 and Remark 3.4.7.

Remark 3.1.7 It follows from the creation property (3.1.5) that $Y(v, x) = 0$ if and only if $v = 0$, so that the vertex operator map $Y(\cdot, x)$ is injective. Thus by transport of structure, we can put a vertex algebra structure on the vector space consisting of all the vertex operators $Y(v, x)$ ($v \in V$) (a subspace of $(\mathrm{End}\ V)[[x, x^{-1}]]$), so that the map $Y(\cdot, x)$ becomes an isomorphism between two vertex algebras. The resulting structure on this subspace of $(\mathrm{End}\ V)[[x, x^{-1}]]$ is very interesting, and we shall be discussing and using it extensively. When physicists who practice conformal theory identify "states" with "operators," they essentially have this picture in mind; we shall exploit it systematically, keeping a careful distinction between V and its image. In any case, this important picture brings out the fundamental way in which the elements v of a vertex algebra can, and often should, be thought of as the vertex operators $Y(v, x)$ themselves. Later, when we add some extra structure to a vertex algebra, we shall in fact call the resulting structures "vertex operator algebras" to emphasize this point. (However, for vertex algebras and vertex operator algebras the issue of identification of vectors with operators is the same; we use the terms "vertex algebra" and "vertex operator algebra" because these terms were used in [B1] and [FLM6], respectively.)

Remark 3.1.8 The left-hand side of the Jacobi identity clearly has the symmetry $(u, v; x_0, x_1, x_2) \leftrightarrow (v, u; -x_0, x_2, x_1)$, so that the right-hand side must also have this same symmetry (cf. Remark 2.3.9). We shall use this later.

Remark 3.1.9 The Jacobi identity clearly resembles the classical Jacobi identity in the definition of the notion of Lie algebra (see also Remark 3.1.14). Soon (using Remark 3.1.8) we shall also have a skew symmetry relation for vertex algebras (Proposition 3.1.19), an analogue of the classical skew symmetry of brackets for a Lie algebra, so that the notion of vertex algebra is indeed analogous to the classical notion of Lie algebra. This principle is deeply reflected in the tensor product theory developed in [HL1], [HL4]–[HL7] and [Hua8] for the representations of a vertex operator algebra.

Remark 3.1.10 On the other hand, we shall derive fundamental "commutativity" and "associativity" relations from the Jacobi identity, and we shall in turn recover the Jacobi identity from these relations, so that the notion of vertex algebra is *also* analogous to the classical notion of commutative associative algebra. (Remark 3.1.4 gave a further aspect of this analogy.) It seems at first paradoxical, but it is in fact true, that vertex algebras are analogues of *both* Lie algebras *and* commutative associative algebras. The commutativity and associativity relations are the focal point of the fundamental geometric interpretation of the notion of vertex algebra in terms of the "sewing" of multipunctured Riemann spheres, as first pointed out by I. Frenkel [Fr6], and this picture is deeply related to the physical formulations of conformal field theory (cf. the discussions in [Hua13]). In this book, we shall formulate and heavily use certain "weak" notions of commutativity and associativity and use these notions to develop the (algebraic) representation theory of vertex algebras.

We shall often use certain special cases of the Jacobi identity, such as the following commutator formula and "iterate formula," which we shall alternatively view as an "associator formula." First, taking Res_{x_0} of the Jacobi identity we obtain the *commutator formula*:

$$[Y(u, x_1), Y(v, x_2)] = \mathrm{Res}_{x_0} x_2^{-1} \delta\left(\frac{x_1 - x_0}{x_2}\right) Y(Y(u, x_0)v, x_2); \qquad (3.1.8)$$

notice how Res_{x_0} applied to each of the two terms on the left-hand side of (3.1.6) immediately gives the product of $Y(u, x_1)$ and $Y(v, x_2)$ in the appropriate order; recall (2.2.7). It is important to observe that the only part of $Y(u, x_0)v$ that enters into the commutator formula is its *singular part* (involving the negative powers of x_0); the *regular part* of $Y(u, x_0)v$ (the part involving the nonnegative powers of x_0) does not affect the commutator. The commutator formula (3.1.8) first appeared in [B1] in the form of the following family of identities for the commutators $[u_m, v_n]$ (which for us are the identities obtained by equating the coefficients of $x_0^{-1}x_1^{-m-1}x_2^{-n-1}$ in the Jacobi identity (3.1.6); cf. (3.1.7)): For $m, n \in \mathbb{Z}$,

$$[u_m, v_n] = \sum_{i \geq 0} \binom{m}{i} (u_i v)_{m+n-i}, \tag{3.1.9}$$

a finite sum.

Remark 3.1.11 Here is an amusing way to rewrite the commutator formula (3.1.8): Using (2.3.18) and (2.3.56) we find that

$$
\begin{aligned}
&[Y(u, x_1), Y(v, x_2)] \\
&= \mathrm{Res}_{x_0} Y\left(\left(x_0^{-1}\delta\left(\frac{x_1 - x_2}{x_0}\right) Y(u, x_0) - x_0^{-1}\delta\left(\frac{-x_2 + x_1}{x_0}\right) Y(u, x_0)\right) v, x_2\right) \\
&= \mathrm{Res}_{x_0} Y\left(\left(x_0^{-1}\delta\left(\frac{x_1 - x_2}{x_0}\right) Y(u, x_1 - x_2)\right.\right. \\
&\qquad\qquad \left.\left. -x_0^{-1}\delta\left(\frac{-x_2 + x_1}{x_0}\right) Y(u, -x_2 + x_1)\right) v, x_2\right) \\
&= Y((Y(u, x_1 - x_2) - Y(u, -x_2 + x_1))v, x_2). \tag{3.1.10}
\end{aligned}
$$

Note that we are *not* allowed to use "linearity" to write the right-hand side as a difference of $Y(Y(u, x_1 - x_2)v, x_2)$ and $Y(Y(u, -x_2 + x_1)v, x_2)$; neither of these expressions in general exists! In the expression

$$(Y(u, x_1 - x_2) - Y(u, -x_2 + x_1))v,$$

the regular part of $Y(u, x)$ cancels out and the singular part becomes an expansion of zero, as in (2.3.11).

Now taking Res_{x_1} instead of Res_{x_0} of the Jacobi identity, we obtain, using (2.3.17) to rewrite the delta function expression on the right-hand side of (3.1.6), the *iterate formula* (the formula for the iterate of the vertex operators $Y(v, x_2)$ and $Y(u, x_0)$):

$$
\begin{aligned}
Y(Y(u, x_0)v, x_2) = \mathrm{Res}_{x_1}\Big(&x_0^{-1}\delta\left(\frac{x_1 - x_2}{x_0}\right) Y(u, x_1)Y(v, x_2) \\
&-x_0^{-1}\delta\left(\frac{x_2 - x_1}{-x_0}\right) Y(v, x_2)Y(u, x_1)\Big) \tag{3.1.11}
\end{aligned}
$$

which appeared, in component form, in [B1] (cf. Definition 3.6.5): For $m, n \in \mathbb{Z}$,

$$(u_m v)_n = \sum_{i \geq 0} (-1)^i \binom{m}{i} \left(u_{m-i}v_{n+i} - (-1)^m v_{m+n-i}u_i\right) \tag{3.1.12}$$

(a finite sum over i when applied to any element of V); this identity comes from equating the coefficients of $x_0^{-m-1}x_1^{-1}x_2^{-n-1}$ in the Jacobi identity (3.1.6); cf. (3.1.7).

We can easily simplify the first term on the right-hand side of (3.1.11) by using (2.3.17), (2.3.56) and then (2.2.7), so that (3.1.11) turns into the following *associator formula* (which should be compared with (3.1.8)):

$$Y(Y(u, x_0)v, x_2) - Y(u, x_0 + x_2)Y(v, x_2)$$
$$= -\mathrm{Res}_{x_1} x_0^{-1} \delta \left(\frac{x_2 - x_1}{-x_0} \right) Y(v, x_2)Y(u, x_1). \tag{3.1.13}$$

The term "associator formula" is motivated by the analogy with nonassociative algebra theory, where the associator of elements a, b and c is $(ab)c - a(bc)$; recall that $Y(v, x)$ should be thought as a left-multiplication operator. (Strictly speaking, we should apply both sides of (3.1.13) to a vector $w \in V$ before allowing ourselves to call (3.1.13) the "associator formula," but we shall use the term anyway, the element w being implicit.)

Remark 3.1.12 Just as in Remark 3.1.11, we can rewrite the associator formula as follows (using (2.3.17)):

$$Y(Y(u, x_0)v, x_2) - Y(u, x_0 + x_2)Y(v, x_2)$$
$$= \mathrm{Res}_{x_1} Y(v, x_2) \left(x_2^{-1} \delta \left(\frac{x_1 - x_0}{x_2} \right) Y(u, x_1) - x_0^{-1} \delta \left(\frac{x_1 - x_2}{x_0} \right) Y(u, x_1) \right)$$
$$= \mathrm{Res}_{x_1} Y(v, x_2) \left(x_1^{-1} \delta \left(\frac{x_2 + x_0}{x_1} \right) Y(u, x_1) - x_1^{-1} \delta \left(\frac{x_0 + x_2}{x_1} \right) Y(u, x_1) \right)$$
$$= Y(v, x_2)(Y(u, x_2 + x_0) - Y(u, x_0 + x_2)) \tag{3.1.14}$$

and just as in Remark 3.1.11, we are not allowed to write this as the difference of $Y(v, x_2)Y(u, x_2 + x_0)$ and $Y(v, x_2)Y(u, x_0 + x_2)$ since these expressions do not in general exist. Notice again that the regular part of $Y(u, x)$ cancels out and the singular part becomes an expansion of zero.

Remark 3.1.13 The formulas (3.1.8) and (3.1.13) in fact describe the failure of the vertex operators to commute and to "associate." By the following simple operation, we can in fact make the right-hand sides of (3.1.8) and (3.1.13) vanish. By multiplying both sides of (3.1.8) by the polynomial $(x_1 - x_2)^k$, where k is a sufficiently large nonnegative integer, we actually get

$$(x_1 - x_2)^k [Y(u, x_1), Y(v, x_2)] = 0, \tag{3.1.15}$$

which looks like an assertion of commutativity. (The reader is invited to prove (3.1.15) as an exercise; we shall carry this out in Section 3.2.) But of course we are not allowed to divide by $(x_1 - x_2)^k$, and vertex operators do not in general commute. We call (3.1.15) the *weak commutativity* relation. Analogously, we can make the right-hand side of (3.1.13) vanish. First, apply (3.1.13) to a third vector w. Then by multiplying both sides of (3.1.13) by $(x_0 + x_2)^l$ for a sufficiently large nonnegative integer l, we can make the right-hand side of (3.1.13) vanish (as we shall again see in detail in Section 3.3), so that we get

$$(x_0 + x_2)^l Y(Y(u, x_0)v, x_2)w = (x_0 + x_2)^l Y(u, x_0 + x_2)Y(v, x_2)w. \quad (3.1.16)$$

We call this the *weak associativity* relation; if we were allowed to divide by $(x_0 + x_2)^l$ (which we are not), we would have an associativity relation.

Remark 3.1.14 The commutator formula (3.1.8) resembles the classical Jacobi identity for a Lie algebra, expressed using the adjoint representation ad:

$$\mathrm{ad}(u)\mathrm{ad}(v) - \mathrm{ad}(v)\mathrm{ad}(u) = \mathrm{ad}(\mathrm{ad}(u)v).$$

The iterate formula (3.1.11) *also* looks like the (same) classical Jacobi identity, written as

$$\mathrm{ad}(\mathrm{ad}(u)v) = \mathrm{ad}(u)\mathrm{ad}(v) - \mathrm{ad}(v)\mathrm{ad}(u).$$

The commutator and the iterate formulas for vertex algebras will both play important roles.

Remark 3.1.15 But the Jacobi identity (3.1.6) is really a "better" analogue of the classical Jacobi identity for a Lie algebra (also recall Remark 3.1.9), since it contains "all" the information and since the three terms in the Jacobi identity resemble one another; this point will be elaborated in Section 3.7. The Jacobi identity for vertex algebras is formally a combination of the classical Jacobi identity for Lie algebras and the three-term delta function relation (2.3.18), which in turn can be naturally and precisely interpreted using complex analysis, as explained in Remark 2.3.18. For this reason, the Jacobi identity was alternatively called the *Jacobi–Cauchy identity* in [FHL]. In the next remark we explain this in detail.

Remark 3.1.16 The Jacobi identity (3.1.6) can in fact be interpreted in a precise way, in the spirit of Remark 2.3.18 (see also Remark 2.3.28), as an equality of "operator-valued formal distributions (or generalized functions)" as follows. As in Remarks 2.3.17 and 2.3.18, consider an arbitrary "test function"

$$f(x_0, x_1, x_2) \in \mathbb{C}[x_0, x_0^{-1}, x_1, x_1^{-1}, x_2, x_2^{-1}],$$

i.e., a Laurent polynomial in three variables. Exactly as in these earlier remarks, we see that if we multiply each of the three terms in the Jacobi identity by f and then extract $\mathrm{Res}_{x_0}\mathrm{Res}_{x_1}\mathrm{Res}_{x_2}$, we obtain an operator-valued linear function on the space $\mathbb{C}[x_0, x_0^{-1}, x_1, x_1^{-1}, x_2, x_2^{-1}]$, a linear function that is equivalent to the content of the term in the Jacobi identity itself. For instance, the first term on the left-hand-side of (3.1.6) (an operator-valued formal series in x_0, x_1 and x_2) yields and is in turn characterized by the linear function

$$\mathbb{C}[x_0, x_0^{-1}, x_1, x_1^{-1}, x_2, x_2^{-1}] \to \mathrm{End}\ V$$

given by

$$\mathrm{Res}_{x_0}\mathrm{Res}_{x_1}\mathrm{Res}_{x_2}\left(x_0^{-1}\delta\left(\frac{x_1 - x_2}{x_0}\right) Y(u, x_1)Y(v, x_2)f(x_0, x_1, x_2)\right). \quad (3.1.17)$$

But just as in Remark 2.3.17, this expression equals

$$\mathrm{Res}_{x_1}\mathrm{Res}_{x_2}Y(u,x_1)Y(v,x_2)f(x_1-x_2,x_1,x_2), \tag{3.1.18}$$

and as in Remark 2.3.18 we find that the Jacobi identity (3.1.6) is equivalent, then, to the following relation for an arbitrary f:

$$\mathrm{Res}_{x_1}\mathrm{Res}_{x_2}Y(u,x_1)Y(v,x_2)f(x_1-x_2,x_1,x_2)$$
$$-\mathrm{Res}_{x_1}\mathrm{Res}_{x_2}Y(v,x_2)Y(u,x_1)f(-x_2+x_1,x_1,x_2)$$
$$=\mathrm{Res}_{x_0}\mathrm{Res}_{x_2}Y(Y(u,x_0)v,x_2)f(x_0,x_0+x_2,x_2), \tag{3.1.19}$$

using the binomial expansion convention and the first equality in (2.3.41), and one might be tempted to abuse notation and write the right-hand side as

$$\mathrm{Res}_{x_1-x_2}\mathrm{Res}_{x_2}Y(Y(u,x_1-x_2)v,x_2)f(x_1-x_2,x_1,x_2), \tag{3.1.20}$$

although this expression is not well defined. Because the operator-valued formal series appearing here, like $Y(u,x_1)Y(v,x_2)$, are not expansions of rational functions, we cannot rewrite this Jacobi identity further using complex analysis, as we did in (2.3.44). Neverthless, the formal similarity between (2.3.44) (combined with (2.3.43)) and (3.1.19) is clear. It is important to note, however, that for vertex *operator* algebras, which we define and discuss below (rather than for vertex algebras in general), we *can* create genuine rational functions from $Y(u,x_1)Y(v,x_2)$ and the other two operator-valued formal series appearing in (3.1.19), as follows. Let w be an arbitrary element of the vertex operator algebra V and let w' be an arbitrary element of the restricted dual V' (defined in (3.2.3) below) of V. Then as we shall see in Proposition 3.2.7, Proposition 3.3.5 and Proposition 3.3.8 below, the three formal series

$$\langle w', Y(u,x_1)Y(v,x_2)w\rangle,$$
$$\langle w', Y(v,x_2)Y(u,x_1)w\rangle,$$
$$\langle w', Y(Y(u,x_0)v,x_2)w\rangle \tag{3.1.21}$$

are in fact the expansions of *a common rational function* in three different domains, and therefore the Jacobi identity for a vertex *operator* algebra can be equivalently expressed using complex variables, in a direct extension of (2.3.44) (combined with (2.3.43)), as the following assertion: For $g(z_1,z_2)$ as in (2.3.43) and (2.3.44),

$$\mathrm{Res}_{z_1=0}\mathrm{Res}_{z_2=0}\langle w', Y(u,z_1)Y(v,z_2)w\rangle g(z_1,z_2)$$
$$-\mathrm{Res}_{z_2=0}\mathrm{Res}_{z_1=0}\langle w', Y(v,z_2)Y(u,z_1)w\rangle g(z_1,z_2)$$
$$=\mathrm{Res}_{z_2=0}\mathrm{Res}_{z_1=z_2}\langle w', Y(Y(u,z_1-z_2)v,z_2)w\rangle g(z_1,z_2). \tag{3.1.22}$$

(In contrast with (3.1.20), the expression on the right-hand side of (3.1.22) is indeed well defined.) This formula (originally derived and explained in the Appendix of [FLM6]) justifies the alternative term "Jacobi–Cauchy identity" [FHL] for the Jacobi identity (see also Remark 3.1.17), and interprets this identity in a precise way as a relation among operator-valued generalized functions. But for the reasons that we have been presenting, it is more natural for us to express the Jacobi identity in the form (3.1.6).

Remark 3.1.17 As in the Appendix of [FLM6], we may write (3.1.22) using contour-integral notation, as follows:

$$\frac{1}{(2\pi i)^2} \int_{C_1^R} \int_{C_2^\rho} \langle w', Y(u, z_1)Y(v, z_2)w\rangle g(z_1, z_2)dz_1 dz_2$$

$$-\frac{1}{(2\pi i)^2} \int_{C_2^\rho} \int_{C_1^r} \langle w', Y(v, z_2)Y(u, z_1)w\rangle g(z_1, z_2)dz_1 dz_2$$

$$= \frac{1}{(2\pi i)^2} \int_{C_2^\rho} \int_{C_1^\epsilon(z_2)} \langle w', Y(Y(u, z_1 - z_2)v, z_2)w\rangle g(z_1, z_2)dz_1 dz_2, \qquad (3.1.23)$$

where $C_i^r(z)$ denotes the circular contour, with counterclockwise orientation, of radius r around the point $z \in \mathbb{C}$ in the variable z_i; $C_i^r = C_i^r(0)$; and r, R, ρ and ϵ are radii such that $R > \rho > r$ and ϵ is sufficiently small.

Since a vertex operator $Y(v, x)$ depends on the formal variable x, one naturally wants to know what happens if the formal derivative operator d/dx is applied to $Y(v, x)$. In the following proposition we show that the formal derivative of a vertex operator $Y(v, x)$ is always (with no further hypotheses needed) also a vertex operator, so that we have a natural endomorphism $\mathcal{D}$ of V because $Y(\cdot, x)$ is injective (recall Remark 3.1.7). This map $\mathcal{D}$ is in fact given by $v \mapsto v_{-2}\mathbf{1}$ (recall the notation (3.1.1)), that is, $\mathcal{D}(v)$ is the coefficient of x^1 in the expression $Y(v, x)\mathbf{1}$.

Proposition 3.1.18 *Let V be a vertex algebra and let $\mathcal{D}$ be the endomorphism of V defined by*

$$\mathcal{D}(v) = v_{-2}\mathbf{1} \ \text{for} \ v \in V. \qquad (3.1.24)$$

Then

$$Y(\mathcal{D}v, x) = \frac{d}{dx}Y(v, x). \qquad (3.1.25)$$

Proof. Using the iterate formula (3.1.11) (actually, the coefficient of x_0^1 in (3.1.11)), $Y(\mathbf{1}, x) = 1$, (2.3.11), (2.2.6) and (2.1.35) in succession, we get:

$$Y(\mathcal{D}v, x_2) = Y(v_{-2}\mathbf{1}, x_2)$$
$$= \mathrm{Res}_{x_1}\left((x_1 - x_2)^{-2}Y(v, x_1) - (-x_2 + x_1)^{-2}Y(v, x_1)\right)$$
$$= \mathrm{Res}_{x_1}\left((x_1 - x_2)^{-2} - (-x_2 + x_1)^{-2}\right)Y(v, x_1)$$
$$= -\mathrm{Res}_{x_1}\left(\frac{\partial}{\partial x_1}x_2^{-1}\delta\left(\frac{x_1}{x_2}\right)\right)Y(v, x_1)$$
$$= \mathrm{Res}_{x_1}x_2^{-1}\delta\left(\frac{x_1}{x_2}\right)\frac{d}{dx_1}Y(v, x_1)$$
$$= \mathrm{Res}_{x_1}x_2^{-1}\delta\left(\frac{x_1}{x_2}\right)\frac{d}{dx_2}Y(v, x_2)$$
$$= \frac{d}{dx_2}Y(v, x_2). \qquad \square$$

Note that

$$\mathcal{D}(\mathbf{1}) = 0, \tag{3.1.26}$$

from either (3.1.24) or (3.1.25).

We easily obtain the following exponential ("global") form of the "$\mathcal{D}$-derivative property" (3.1.25):

$$Y(e^{x_0 \mathcal{D}} v, x) = e^{x_0 \frac{d}{dx}} Y(v, x). \tag{3.1.27}$$

(To prove this, simply take the nth iterate of (3.1.25), multiply by $x_0^n / n!$ and sum over n.) Of course, the global formula (3.1.27) implies (3.1.25). By the formal Taylor Theorem we get the alternative form

$$Y(e^{x_0 \mathcal{D}} v, x) = Y(v, x + x_0). \tag{3.1.28}$$

Now we apply (3.1.28) to the vacuum vector $\mathbf{1}$. Since $Y(v, x + x_0)\mathbf{1}$ involves only nonnegative powers of $x + x_0$, we may specialize x to 0. Using the creation property (and then changing x_0 to x) we obtain the relation

$$Y(v, x)\mathbf{1} = e^{x \mathcal{D}} v \quad \text{for } v \in V. \tag{3.1.29}$$

This is actually an enhanced form of the creation property, which in turn is an immediate consequence.

For a nonassociative algebra, one is interested in knowing whether the multiplication is, for instance, commutative, anticommutative or associative, and as we have mentioned, we shall be studying similar properties for vertex algebras. We have the following analogue for vertex algebras of the classical skew symmetry of brackets for a Lie algebra; this formula, which generalizes (3.1.29), appeared in [B1] (cf. Definition 3.6.5 below).

Proposition 3.1.19 (skew symmetry) *Let V be a vertex algebra and recall the operator $\mathcal{D}$ defined in Proposition 3.1.18. Then*

$$Y(u, x)v = e^{x \mathcal{D}} Y(v, -x)u \tag{3.1.30}$$

for $u, v \in V$.

Proof. Using the symmetry $(u, v; x_0, x_1, x_2) \leftrightarrow (v, u; -x_0, x_2, x_1)$ of the left-hand side of the Jacobi identity (recall Remark 3.1.8) and also using the basic principle (2.3.56) and (3.1.28), we find that

$$x_2^{-1}\delta\left(\frac{x_1 - x_0}{x_2}\right) Y(Y(u, x_0)v, x_2) = x_1^{-1}\delta\left(\frac{x_2 + x_0}{x_1}\right) Y(Y(v, -x_0)u, x_1)$$

$$= x_1^{-1}\delta\left(\frac{x_2 + x_0}{x_1}\right) Y(Y(v, -x_0)u, x_2 + x_0)$$

$$= x_1^{-1}\delta\left(\frac{x_2 + x_0}{x_1}\right) Y(e^{x_0 \mathcal{D}} Y(v, -x_0)u, x_2).$$

Taking Res_{x_1} and using (2.3.17) we get

$$Y(Y(u, x_0)v, x_2) = Y(e^{x_0 \mathcal{D}} Y(v, -x_0)u, x_2). \tag{3.1.31}$$

Now we invoke the injectivity of the vertex operator map $Y(\cdot, x)$ (recall Remark 3.1.7). $\qquad\square$

Remark 3.1.20 This proof is exactly analogous to the following argument in the classical setting of Lie algebras. Suppose a nonassociative algebra satisfies the condition

$$\mathrm{ad}(u)\mathrm{ad}(v) - \mathrm{ad}(v)\mathrm{ad}(u) = \mathrm{ad}(\mathrm{ad}(u)v)$$

for all u, v, where $\mathrm{ad}(u)$ designates left multiplication by u (cf. Remark 3.1.14). Since the left-hand side changes sign when u and v are reversed, so does the right-hand side. If ad is an injective map, we conclude that $\mathrm{ad}(u)v = -\mathrm{ad}(v)u$, which is the skew symmetry of brackets for a Lie algebra.

We also have the following "$\mathcal{D}$-bracket formulas" for a vertex algebra.

Proposition 3.1.21 *Let V be a vertex algebra and let $\mathcal{D}$ be as in Proposition 3.1.18. Then*

$$[\mathcal{D}, Y(v, x)] = \frac{d}{dx} Y(v, x) \tag{3.1.32}$$

$$[\mathcal{D}, Y(v, x)] = Y(\mathcal{D}v, x) \tag{3.1.33}$$

for $v \in V$.

Proof. For $u, v \in V$, using (3.1.30) and (3.1.25) we obtain

$$\begin{aligned}
\frac{d}{dx} Y(u, x)v &= \mathcal{D}e^{x\mathcal{D}} Y(v, -x)u + e^{x\mathcal{D}} \frac{d}{dx} Y(v, -x)u \\
&= \mathcal{D}Y(u, x)v - e^{x\mathcal{D}} Y(\mathcal{D}v, -x)u \\
&= \mathcal{D}Y(u, x)v - Y(u, x)\mathcal{D}v.
\end{aligned} \tag{3.1.34}$$

Thus $\frac{d}{dx} Y(u, x) = [\mathcal{D}, Y(u, x)]$ for $u \in V$, and (3.1.33) follows from Proposition 3.1.18. $\quad\square$

Combining Proposition 3.1.21 with the formal Taylor Theorem we get the following conjugation formula (cf. (3.1.28)):

$$e^{x_0 \mathcal{D}} Y(v, x) e^{-x_0 \mathcal{D}} = Y(v, x + x_0) \quad \text{for } v \in V. \tag{3.1.35}$$

To verify this, we write (3.1.32) in the form

$$(L_{\mathcal{D}} - R_{\mathcal{D}})Y(v, x) = \frac{d}{dx} Y(v, x), \tag{3.1.36}$$

where $L_{\mathcal{D}}$ and $R_{\mathcal{D}}$ denote the operations of left and right multiplication by $\mathcal{D}$, respectively. We then iterate and sum as in the proof of (3.1.27), and since the operators $L_{\mathcal{D}}$ and $R_{\mathcal{D}}$ commute, we simply use (2.2.13).

In Lie algebra theory, most of the more interesting algebras are either finite dimensional or, more generally, $\mathbb{Z}$-graded with finite-dimensional homogeneous subspaces. Correspondingly, we are often interested in vertex algebras that satisfy certain natural conditions, especially those vertex algebras called "vertex operator algebras" in [FLM6] and [FHL]. As we mentioned above, we refer the reader to [B1] and [FLM6] for extensive motivations for this concept.

Definition 3.1.22 A *vertex operator algebra* is a $\mathbb{Z}$-graded vector space (graded by *weights*)

$$V = \coprod_{n \in \mathbb{Z}} V_{(n)} \quad \text{for } v \in V_{(n)}, \quad n = \text{wt } v, \tag{3.1.37}$$

such that

$$\dim V_{(n)} < \infty \quad \text{for } n \in \mathbb{Z} \tag{3.1.38}$$

$$V_{(n)} = 0 \quad \text{for } n \text{ sufficiently negative} \tag{3.1.39}$$

(*grading restrictions*), equipped with a vertex algebra structure $(V, Y, \mathbf{1})$ and a distinguished homogeneous vector ω (the *conformal vector*) of weight 2 ($\omega \in V_{(2)}$), satisfying the following conditions:

$$[L(m), L(n)] = (m - n)L(m + n) + \frac{1}{12}(m^3 - m)\delta_{m+n,0}c_V \tag{3.1.40}$$

for $m, n \in \mathbb{Z}$ (*Virasoro algebra relations*), where

$$Y(\omega, x) = \sum_{n \in \mathbb{Z}} L(n)x^{-n-2} \ (= \sum_{n \in \mathbb{Z}} \omega_n x^{-n-1}), \tag{3.1.41}$$

that is,

$$L(n) = \omega_{n+1} \quad \text{for } n \in \mathbb{Z} \tag{3.1.42}$$

(using the notation (3.1.1)), and

$$c_V \in \mathbb{C} \tag{3.1.43}$$

(*central charge* or *rank* of V); we also have

$$L(0)v = nv = (\text{wt } v)v \quad \text{for } n \in \mathbb{Z} \text{ and } v \in V_{(n)} \tag{3.1.44}$$

(that is, the $L(0)$-eigenspace decomposition of V coincides with the grading of V); and

$$Y(L(-1)v, x) = \frac{d}{dx}Y(v, x) \tag{3.1.45}$$

($L(-1)$-*derivative property*).

We shall sometimes refer to the vertex operator algebra V as $(V, Y, \mathbf{1}, \omega)$ if necessary.

Remark 3.1.23 The *Virasoro algebra* is defined as the Lie algebra $\mathcal{L}$ with basis $\{L_n \mid n \in \mathbb{Z}\} \cup \{\mathbf{c}\}$ equipped with the bracket relations

$$[L_m, L_n] = (m - n)L_{m+n} + \frac{1}{12}(m^3 - m)\delta_{m+n,0}\mathbf{c} \qquad (3.1.46)$$

together with the condition that $\mathbf{c}$ is a central element of $\mathcal{L}$. These relations indeed define a Lie algebra, and formula (3.1.40) asserts that the correspondence given by $L_n \mapsto L(n)$ for $n \in \mathbb{Z}$ and $\mathbf{c} \mapsto c_V$ (the scalar operator) defines a representation of the Virasoro algebra on the space V (or equivalently, equips V with the structure of a module for the Virasoro algebra) with "central charge" equal to the number c_V. This is why c_V is called the central charge of V; the reason why c_V is also called the rank of V (as in [FLM6] and [FHL]) is that for the important case of vertex operator algebras based on suitable lattices, c_V equals the rank of the lattice, as we discuss in Sections 6.4 and 6.5. Axioms (3.1.44) and (3.1.45) relate the components $L(0)$ and $L(-1)$ of $Y(\omega, x)$ to the grading of V and to the operator $\mathcal{D}$, respectively.

Remark 3.1.24 Here is why the vector $\omega \in V$ giving rise to the Virasoro algebra relations (3.1.40) is called "the conformal vector": Recall from (2.2.15) the derivations

$$T_{p(x)} = p(x)\frac{d}{dx} \qquad (3.1.47)$$

of $\mathbb{C}[x, x^{-1}]$, where $p(x) \in \mathbb{C}[x, x^{-1}]$. These derivations form a Lie algebra with the bracket relations

$$[T_{p(x)}, T_{q(x)}] = T_{p(x)q'(x)-q(x)p'(x)} \qquad (3.1.48)$$

for $p(x), q(x) \in \mathbb{C}[x, x^{-1}]$. In fact, *all* derivations of $\mathbb{C}[x, x^{-1}]$ are of the form $T_{p(x)}$. (Proof: Let T be a derivation and set $p(x) = T(x)$. Then $T(1) = 0$ and $0 = T(xx^{-1}) = T(x)x^{-1} + xT(x^{-1})$, so that $T(x^{-1}) = -x^{-2}T(x)$. It follows that T and $T_{p(x)}$ agree on all powers of x.) Write $\mathrm{Der}\,\mathbb{C}[x, x^{-1}]$ for this Lie algebra. With respect to the basis

$$d_n = -x^{n+1}\frac{d}{dx}, \quad n \in \mathbb{Z}, \qquad (3.1.49)$$

of $\mathrm{Der}\,\mathbb{C}[x, x^{-1}]$, the commutators have the form

$$[d_m, d_n] = (m - n)d_{m+n} \quad \text{for } m, n \in \mathbb{Z}. \qquad (3.1.50)$$

Comparing (3.1.50) with (3.1.46) we see that the Virasoro algebra $\mathcal{L}$ is a central extension of $\mathrm{Der}\,\mathbb{C}[x, x^{-1}]$:

$$0 \to \mathbb{C}\mathbf{c} \to \mathcal{L} \to \mathrm{Der}\,\mathbb{C}[x, x^{-1}] \to 0,$$
$$L_n \mapsto -x^{n+1}\frac{d}{dx}, \qquad (3.1.51)$$

that is, Der $\mathbb{C}[x, x^{-1}]$ is the image of $\mathcal{L}$ when the central element $\mathbf{c}$ is specialized to 0. Also, $\mathbb{C}[x, x^{-1}]$ is an $\mathcal{L}$-module with central charge 0. On the other hand, Der $\mathbb{C}[x, x^{-1}]$ can be interpreted as a Lie algebra of formal infinitesimal conformal transformations of a complex variable. An example of this phenomenon is the formal Taylor Theorem, Proposition 2.2.2, in which the derivation d/dx appears as a formal infinitesimal translation, and in fact the analogous formal exponentials of all the basis elements $d_n = -x^{n+1}\frac{d}{dx}$ can also be explicitly computed as conformal transformations of a complex variable, in a formal sense (cf. Section 8.3 of [FLM6]). This is why the vector ω is termed the "conformal vector."

Notice that $L(-1) = \omega_0$ and $L(0) = \omega_1$. Applying Res_{x_1} and the operation $\mathrm{Res}_{x_1} x_1$ to the commutator formula (3.1.8) with $u = \omega$ and using (2.3.17), we obtain

$$[L(-1), Y(v, x)] = Y(L(-1)v, x) \quad \left(= \frac{d}{dx}Y(v, x)\right) \tag{3.1.52}$$

$$[L(0), Y(v, x)] = xY(L(-1)v, x) + Y(L(0)v, x). \tag{3.1.53}$$

Furthermore, (3.1.53), (3.1.44) and (3.1.45) imply that if v is homogeneous, then

$$\mathrm{wt}\ v_n = \mathrm{wt}\ v - n - 1, \tag{3.1.54}$$

that is, v_n is a homogeneous operator that maps $V_{(m)}$ to $V_{(m+\mathrm{wt}\ v-n-1)}$ for all $m, n \in \mathbb{Z}$. (What we actually verify here is that $[L(0), v_n] = (\mathrm{wt} v - n - 1)v_n$.) Incidentally, notice that the truncation condition (3.1.2), (3.1.3) in the definition of vertex operator algebra formally follows from (3.1.54) together with the condition (3.1.39) that the grading is truncated.

Remark 3.1.25 Sometimes one wants to use the following variants $X(v, x)$ of the vertex operators $Y(v, x)$ in a vertex operator algebra V. First, for a *homogeneous* element $v \in V$, set

$$X(v, x) = x^{\mathrm{wt}\ v}Y(v, x) \tag{3.1.55}$$

$$= \sum_{n\in\mathbb{Z}} v_n x^{\mathrm{wt}\ v-n-1} \tag{3.1.56}$$

$$= \sum_{n\in\mathbb{Z}} v_{n+\mathrm{wt}\ v-1} x^{-n} \tag{3.1.57}$$

and for $n \in \mathbb{Z}$ write

$$v_{[n]} = v_{n+\mathrm{wt}\ v-1}, \tag{3.1.58}$$

so that

$$X(v, x) = \sum_{n\in\mathbb{Z}} v_{[n]} x^{-n} \tag{3.1.59}$$

and, from (3.1.54),

$$\mathrm{wt}\ v_{[n]} = -n, \tag{3.1.60}$$

a "simpler" formula than (3.1.54). We also have

$$Y(v, x) = x^{-\mathrm{wt}\ v} X(v, x) = \sum_{n \in \mathbb{Z}} v_{[n]} x^{-n-\mathrm{wt}\ v}. \tag{3.1.61}$$

Note that

$$v_{[n]} = v_n \quad \text{in the case wt}\ v = 1, \tag{3.1.62}$$

but this is not true otherwise. We can easily extend the correspondence $v \mapsto X(v, x)$ from *homogeneous* elements v to *arbitrary* elements $v \in V$ by linearity, giving the uniform formula

$$V \to (\mathrm{End}\ V)[[x, x^{-1}]]$$
$$v \mapsto X(v, x) = Y(x^{L(0)}v, x) \tag{3.1.63}$$

for $X(v, x)$. In view of (3.1.60) the operator $X(v, x)$ is sometimes called the *homogeneous vertex operator* associated with v. These homogeneous vertex operators frequently arise in examples, as in [FLM6], for instance, since (3.1.59) and (3.1.60) look "simpler" than (3.1.1) and (3.1.54). One such example has already arisen above. For the conformal vector ω, we have

$$X(\omega, x) = \sum_{n \in \mathbb{Z}} L(n)x^{-n}, \tag{3.1.64}$$

since wt $\omega = 2$; that is $\omega_{[n]} = L(n)$ for $n \in \mathbb{Z}$. However, as we shall see throughout this work, the vertex operators $Y(v, x)$ are really the natural ones; even the fundamental Jacobi identity (3.1.6) itself does not have any obvious reformulation in terms of the homogeneous vertex operators (cf. Section 8.6 of [FLM6], where this issue arose to some extent). On the other hand, a general and conceptual "Jacobi identity" was in fact discovered for the homogeneous vertex operators in [Le12] and [Le13]; this is related to a fundamental formal "change-of-variables" theorem of Zhu ([Z1], [Z2]).

We have

$$L(-1)\mathbf{1} = L(0)\mathbf{1} = L(n)\mathbf{1} = 0 \tag{3.1.65}$$

for $n > 0$ from the creation property (3.1.5), so that the vacuum vector is homogeneous of weight 0:

$$\mathrm{wt}\ \mathbf{1} = 0. \tag{3.1.66}$$

Note that we are not assuming that $V_{(n)} = 0$ for $n < 0$, although most of the useful vertex operator algebras have this property.

Notice that (3.1.45) and Proposition 3.1.18 imply that for a vertex operator algebra, the operator $\mathcal{D}$ coincides with $L(-1)$:

$$\mathcal{D} = L(-1), \tag{3.1.67}$$

and in particular,

$$L(-1)v = v_{-2}\mathbf{1} \quad \text{for } v \in V. \tag{3.1.68}$$

By the creation property (3.1.5), we have

$$\omega = L(-2)\mathbf{1} \ (= \omega_{-1}\mathbf{1}), \tag{3.1.69}$$

and using the Virasoro algebra relations (3.1.40) we obtain

$$\omega_0\omega = L(-1)\omega, \tag{3.1.70}$$

$$\omega_1\omega = L(0)L(-2)\mathbf{1} = 2L(-2)\mathbf{1} = 2\omega, \tag{3.1.71}$$

$$\omega_2\omega = L(1)L(-2)\mathbf{1} = 0, \tag{3.1.72}$$

$$\omega_3\omega = L(2)L(-2)\mathbf{1} = \frac{1}{2}c_V\mathbf{1}, \tag{3.1.73}$$

$$\omega_n\omega = 0 \quad \text{for } n \geq 4. \tag{3.1.74}$$

Conversely, it is easy to see that the relations (3.1.70)–(3.1.74) imply the Virasoro algebra relations (3.1.40), where we of course use the commutator formula (3.1.8) and the $L(-1)$-derivative property (3.1.45). (The relations (3.1.70)–(3.1.74) describe the singular part of $Y(\omega, x)\omega$, which as we saw earlier is the only part of $Y(\omega, x)\omega$ that enters into the commutator formula.)

Remark 3.1.26 The one-dimensional space $\mathbb{C}\mathbf{1}$ carries a natural vertex operator algebra structure of central charge 0 and weight 0, with $\omega = 0$. We also view the space 0 as a (degenerate) vertex operator algebra of all central charges simultaneously; note that in the Virasoro algebra commutation relations (3.1.40), c_V really refers to a multiple of the identity operator on V.

3.2 Commutativity properties

Here we systematically discuss the commutativity properties of a vertex algebra, and in the next section we continue with the associativity properties. These features are fundamental. They are central to the subtle analogy between vertex algebras and commutative associative algebras, to the geometric and conformal field-theoretic meaning of vertex operator algebra theory, and to the algebraic theory itself.

Throughout this section we fix a vertex algebra $(V, Y, \mathbf{1})$. While two given vertex operators of course do not in general commute, we have the following basic general fact (recall Remark 3.1.13).

Proposition 3.2.1 (weak commutativity) *For $u, v \in V$, there exists a nonnegative integer k such that*

$$(x_1 - x_2)^k [Y(u, x_1), Y(v, x_2)] = 0. \tag{3.2.1}$$

Proof. We want to make the expression on the right-hand side of the Jacobi identity (3.1.6) vanish. Let $k \geq 0$. Multiplying the Jacobi identity by x_0^k and then taking Res_{x_0} gives

$$(x_1 - x_2)^k Y(u, x_1) Y(v, x_2) - (x_1 - x_2)^k Y(v, x_2) Y(u, x_1)$$
$$= \mathrm{Res}_{x_0} x_2^{-1} \delta \left(\frac{x_1 - x_0}{x_2} \right) x_0^k Y(Y(u, x_0)v, x_2). \tag{3.2.2}$$

(Note that we are indeed using the assumption that $k \geq 0$.) Now simply choose k such that $u_n v = 0$ for $n \geq k$. $\square$

Remark 3.2.2 The multiplication by $(x_1 - x_2)^k$ amounts to the clearing of a (formal) pole, and this phenomenon will pervade the discussions of all the variants of commutativity (and of associativity). The proof shows that we may take $k \geq 0$ such that $u_n v = 0$ for $n \geq k$. Note that the assertion of (3.2.1) for some fixed $k \geq 0$ implies the assertion for all larger k (since we can multiply the relation by a polynomial), and if we know the smallest k for which it holds, then we have the maximum amount of information. As we shall see, though, the qualitative statement of (3.2.1) (for *some* $k \geq 0$) turns out to be very powerful, and will essentially enable us to reconstruct all the information, including the full Jacobi identity. Note also that (3.2.1) can be viewed as the generating function of an infinite list of nontrivial relations (involving binomial coefficients) among the operators u_m and v_n, obtained by extracting the coefficient of each monomial in x_1 and x_2.

Remark 3.2.3 As we shall discuss in detail later when we develop examples, the integer k in (3.2.1) can often be taken to be "very small" when u and v range through convenient generating subspaces of many important vertex operator algebras. For instance, for vertex operator algebras based on affine Lie algebras, $k = 2$ is sufficient when u and v range through the standard generating space $V_{(1)}$ of weight one. For the moonshine module $V^\natural$, whose automorphism group is the Monster, $k = 4$ gives zero when u and v range through the 196,884-dimensional generating subspace $V_{(2)}^\natural$ of weight two. When the value of k is suitably reduced below this critical value on the left-hand side of (3.2.1), we can get important "product operations" on the generating subspaces. For instance, $k = 0$ of course gives the commutator of the two vertex operators (recall (3.1.8)), which in the case of a vertex operator algebra based on an affine Lie algebra includes the information of the bracket operation on the underlying finite-dimensional Lie algebra in a natural way; this bracket is in fact given simply by $[u, v] = u_0 v$ for $u, v \in V_{(1)}$. The "product operation" obtained by taking $k = 1$ on the left-hand-side of (3.2.1) is called the *cross-bracket* of the vertex operators $Y(u, x_1)$ and $Y(v, x_2)$, and in the case of $V^\natural$, when u and v lie in the 196,884-dimensional subspace $V_{(2)}^\natural$, this cross-bracket exactly yields the (highly nontrivial) commutative nonassociative product operation on the Griess algebra in a natural way (see [FLM6]); in fact, the operation is given by $u_1 v$ for $u, v \in V_{(2)}^{\natural}$.

The weak commutativity relation should be compared with the commutator formula (3.1.8), which amounts to Res_{x_0} of the Jacobi identity. The term "weak commutativity" of course refers to the presence of the factor $(x_1 - x_2)^k$ in what would otherwise be a commutativity relation. From Remark 3.2.2 or the commutator formula (3.1.8), we know when two vertex operators actually commute in the strict sense:

Remark 3.2.4 For $u, v \in V$, consider the assertion that $[Y(u, x_1), Y(v, x_2)] = 0$. This means that $[u_m, v_n] = 0$ for all $m, n \in \mathbb{Z}$. We observe that $[Y(u, x_1), Y(v, x_2)] = 0$ if and only if $u_n v = 0$ for $n \geq 0$, or equivalently, if and only if $Y(u, x)v$ involves only nonnegative powers of x. In fact, the "if" part is clear and the "only if" holds because for $n \geq 0$,

$$u_n v = u_n v_{-1} \mathbf{1} = v_{-1} u_n \mathbf{1} = 0$$

(in view of (3.1.5)). Of course, $Y(u, x)v \in V[[x]]$ if and only if $Y(v, x)u \in V[[x]]$.

Remark 3.2.5 One might be (briefly!) tempted to multiply the relation (3.2.1) by either the formal series $(x_1 - x_2)^{-k}$ or the formal series $(-x_2 + x_1)^{-k}$ (each of course expanded according to the binomial expansion convention). This multiplication operation is indeed rigorous, but from this we of course *cannot* conclude that $[Y(u, x_1), Y(v, x_2)] = 0$, because the associative law fails to hold in this case. Notice that the situation here is similar to that in the paradox (2.1.17); the formal product

$$(x_1 - x_2)^{-k}(x_1 - x_2)^k [Y(u, x_1), Y(v, x_2)]$$

of three formal series in two variables fails to exist. The similarity with (2.1.17) is actually very close, since the commutator $[Y(u, x_1), Y(v, x_2)]$ involves the delta function. See also Remark 2.1.6.

Remark 3.2.6 Notice that the weak commutativity relation (3.2.1) is an analogue of the commutativity *of the left multiplication operations* associated to two elements of a commutative associative algebra (it is not directly analogous to the commutativity of multiplication itself), while the skew symmetry relation (3.1.30) is an analogue of the anti-commutativity of multiplication in a Lie algebra.

Next, we shall formulate the notion of what was called "commutativity" in [FLM6] and [FHL]. Since this notion uses the "matrix coefficients" of products of vertex operators, it is appropriate to assume now that V is a vertex operator algebra rather than just a vertex algebra. (Actually, we do not need all of the properties.)

For a $\mathbb{Z}$-graded vector space such as V, set

$$V' = \coprod_{n \in \mathbb{Z}} V_{(n)}^* \tag{3.2.3}$$

(the *restricted dual*), the direct sum of the dual spaces of the homogeneous subspaces $V_{(n)}$ of V—the space of linear functionals on V vanishing on all but finitely many $V_{(n)}$. (Of course, this notion and notation apply to vector spaces graded with respect to any set; later we shall consider the restricted duals of $\mathbb{C}$-graded modules.) Denote by $\langle \cdot, \cdot \rangle$ the natural pairing between V' and V. Note that for $u, v \in V$ and $v' \in V'$,

$$\langle v', Y(u, x)v \rangle \in \mathbb{C}[x, x^{-1}] \tag{3.2.4}$$

from (3.1.54). (To verify this, we may assume without loss of generality that the vectors are homogeneous vectors. Here we are definitely using the restricted dual, together with (3.1.54); if v' were an unrestricted element of the full dual space V^*, the formal series $\langle v', Y(u, x)v \rangle$ would not in general be truncated from above in powers of x.)

This phenomenon generalizes to the fundamental "rationality" properties of products and iterates of vertex operators, leading to the commutativity property, and later, the associativity property. In the following statement of "rationality of products" and "commutativity," we use the algebra $\mathbb{C}[x_1, x_2]_S$ and the iota-maps introduced in Section 2.3.

Proposition 3.2.7 *Let V be a vertex operator algebra and let $u, v, w \in V$ and $w' \in V'$ be arbitrary. We have:*
(a) **(rationality of products)** *The formal series*

$$\langle w', Y(u, x_1)Y(v, x_2)w \rangle \quad \left(= \sum_{m,n \in \mathbb{Z}} \langle w', u_m v_n w \rangle x_1^{-m-1} x_2^{-n-1} \right)$$

lies in the image of the map ι_{12}:

$$\langle w', Y(u, x_1)Y(v, x_2)w \rangle = \iota_{12} f(x_1, x_2), \tag{3.2.5}$$

where the (uniquely determined) element $f \in \mathbb{C}[x_1, x_2]_S$ is of the form

$$f(x_1, x_2) = \frac{g(x_1, x_2)}{(x_1 - x_2)^k x_1^l x_2^m} \tag{3.2.6}$$

for some $g \in \mathbb{C}[x_1, x_2]$ and $k, l, m \in \mathbb{Z}$, where k depends only on u and v; it is independent of w and w'.
(b) **(commutativity)** *We also have*

$$\langle w', Y(v, x_2)Y(u, x_1)w \rangle = \iota_{21} f(x_1, x_2), \tag{3.2.7}$$

that is, in informal language,

$$\text{"}Y(u, x_1)Y(v, x_2) \quad \text{agrees with} \quad Y(v, x_2)Y(u, x_1) \tag{3.2.8}$$

as operator-valued rational functions."

Proof. Let k be a nonnegative integer such that the weak commutativity relation (3.2.1) holds for u and v. Then

$$(x_1 - x_2)^k \langle w', Y(u, x_1)Y(v, x_2)w \rangle = (x_1 - x_2)^k \langle w', Y(v, x_2)Y(u, x_1)w \rangle.$$

$$(3.2.9)$$

The left-hand side of (3.2.9) involves only finitely many negative powers of x_2, by (3.1.3), and only finitely many positive powers of x_1, by (3.1.54). Similarly, the right-hand side of (3.2.9) involves only finitely many negative powers of x_1 and only finitely many positive powers of x_2. Thus both sides of (3.2.9) involve only finitely many powers of both x_1 and x_2, so that each side of (3.2.9) must be a Laurent polynomial, that is, a formal Laurent series of the form $h(x_1, x_2) \in \mathbb{C}[x_1, x_1^{-1}, x_2, x_2^{-1}]$. Then the rational function

$$f(x_1, x_2) = \frac{h(x_1, x_2)}{(x_1 - x_2)^k}$$

satisfies the desired conditions. In fact, the left-hand side of (3.2.5) involves only finitely many negative powers of x_2 and so can be multiplied by $(x_1 - x_2)^{-k}$, and analogously, the left-hand side of (3.2.7) involves only finitely many negative powers of x_1 and so can be multiplied by $(-x_2 + x_1)^{-k}$. Here we are using a natural two-variable analogue of (2.1.13), or more specifically, of (2.1.14); see Remarks 2.1.6 and 3.2.5. $\square$

Remark 3.2.8 If one tried to cancel $(x_1 - x_2)^k$ from both sides of (3.2.9) by multiplying both sides by the formal series $(x_1 - x_2)^{-k}$, one would of course fail for the reason pointed out in Remark 3.2.5.

Remark 3.2.9 The proof of Proposition 3.2.7 in fact shows that the weak commutativity relation (3.2.1) implies (rationality and) commutativity (i.e., the assertions of Proposition 3.2.7). More precisely, if we assume all the axioms for a vertex operator algebra except for the Jacobi identity, and if we also assume (3.1.53), and therefore (3.1.54), then weak commutativity implies commutativity. Conversely, commutativity implies weak commutativity, since

$$(x_1 - x_2)^k \langle w', Y(u, x_1)Y(v, x_2)w \rangle - (x_1 - x_2)^k \langle w', Y(v, x_2)Y(u, x_1)w \rangle$$

$$= \iota_{12}\left(\frac{g(x_1, x_2)}{x_1^l x_2^m}\right) - \iota_{21}\left(\frac{g(x_1, x_2)}{x_1^l x_2^m}\right)$$

$$= 0.$$

$$(3.2.10)$$

We emphasize that our (purely algebraic) commutativity results immediately imply the following results involving convergent series in suitable domains; we shall replace the formal variables x_0, x_1 and x_2 by the complex variables z_0, z_1 and z_2. (For the first time, we now use special properties of the ground field $\mathbb{C}$, besides the fact that it has characteristic 0.)

Corollary 3.2.10 *The two formal series*

$$\langle w', Y(u, z_1)Y(v, z_2)w\rangle \quad \left(= \sum_{m,n\in\mathbb{Z}} \langle w', u_m v_n w\rangle z_1^{-m-1} z_2^{-n-1}\right) \qquad (3.2.11)$$

and

$$\langle w', Y(v, z_2)Y(u, z_1)w\rangle \quad \left(= \sum_{m,n\in\mathbb{Z}} \langle w', v_n u_m w\rangle z_1^{-m-1} z_2^{-n-1}\right) \qquad (3.2.12)$$

in two complex variables are absolutely convergent to a common rational function $f(z_1, z_2)$ *(recall (3.2.6)) in the (disjoint) domains*

$$|z_1| > |z_2| > 0 \quad and \quad |z_2| > |z_1| > 0, \qquad (3.2.13)$$

respectively. $\square$

Remark 3.2.11 Notice that the convergence is essentially nothing but the convergence of the binomial series $(x_1 - x_2)^{-k}$ (recall Proposition 3.2.7). The "rationality" of products means more specifically that the matrix coefficients (3.2.5) are formal expansions, in the direction of nonnegative powers of x_2 rather than of x_1, of formal rational functions *of the very special type* (3.2.6)—that is, with formal poles only at $x_1 = 0$, $x_2 = 0$ or $x_1 = x_2$. When the x_i are specialized to complex variables z_i, we thus get convergence in the domains (3.2.13).

The informal statement (3.2.8) or some form of Corollary 3.2.10 is called the "mutual locality" of two vertex operators in two-dimensional conformal field theory ("locality" of "quantum fields" is a fundamental concept in quantum field theory; cf. [Wi] and [Go1]), and Proposition 3.2.7 precisely interprets this in an algebraic sense. The rationality of products and commutativity, together with the rationality of iterates and associativity, which will be discussed in the next section, are basic to the geometric interpretation of the notion of vertex operator algebra.

It turns out that for a vertex algebra (without grading, and so on), there is actually a close and useful analogue of Proposition 3.2.7, "formal commutativity," with a similar (easy) proof. Instead of rational functions we have to be content with formal Laurent series that are not necessarily convergent in any sense. Also, instead of the restricted dual of a vertex operator algebra (recall (3.2.3)) we have available only the dual space V^* of a given vector space V without grading. We shall continue to use the notation $\langle \cdot, \cdot \rangle$ for the pairing between the dual of a vector space and the vector space. We shall use the iota-maps defined in the general setting before Proposition 2.3.27.

The analogue of Proposition 3.2.7 states (in very general form, which we shall use later for the action of a vertex algebra V on a module W and other applications):

Proposition 3.2.12 (formal commutativity) *Let V and W be vector spaces (possibly the same) and suppose that we have a linear map*

$$Y_W(\cdot, x) : \quad V \to (\text{End } W)[[x, x^{-1}]]$$
$$v \mapsto Y_W(v, x). \tag{3.2.14}$$

*Let $u, v \in V$, $w \in W$ and $w^* \in W^*$. Assume that the following truncation conditions
hold: $Y_W(u, x)w \in W((x))$ and $Y_W(v, x)w \in W((x))$. Assume also that the assertion
of weak commutativity holds for the pair (u, v) acting on the vector w, i.e., there exists
$k \in \mathbb{N}$ (depending only on u and v) such that*

$$(x_1 - x_2)^k [Y_W(u, x_1), Y_W(v, x_2)]w = 0. \tag{3.2.15}$$

Then the formal series

$$\langle w^*, Y_W(u, x_1)Y_W(v, x_2)w \rangle \tag{3.2.16}$$

lies in the image of the map ι_{12}:

$$\langle w^*, Y_W(u, x_1)Y_W(v, x_2)w \rangle = \iota_{12} f(x_1, x_2), \tag{3.2.17}$$

where the (uniquely determined) element

$$f \in \mathbb{C}[[x_1, x_2]][x_1^{-1}, x_2^{-1}, (x_1 - x_2)^{-1}]$$

is of the form

$$f(x_1, x_2) = \frac{g(x_1, x_2)}{(x_1 - x_2)^k x_1^l x_2^m} \tag{3.2.18}$$

*for some $g \in \mathbb{C}[[x_1, x_2]]$ and $k, l, m \in \mathbb{Z}$, where k depends only on u and v (and not on
w or w^*). Moreover,*

$$\langle w^*, Y_W(v, x_2)Y_W(u, x_1)w \rangle = \iota_{21} f(x_1, x_2). \tag{3.2.19}$$

Proof. The proof is exactly parallel to that of Proposition 3.2.7 except that the formal
series

$$(x_1 - x_2)^k \langle w^*, Y_W(u, x_1)Y_W(v, x_2)w \rangle = (x_1 - x_2)^k \langle w^*, Y_W(v, x_2)Y_W(u, x_1)w \rangle \tag{3.2.20}$$

lies in $\mathbb{C}((x_1, x_2))$ rather than in $\mathbb{C}[x_1, x_1^{-1}, x_2, x_2^{-1}]$. $\quad\square$

Remark 3.2.13 Just as in Remark 3.2.9, formal commutativity clearly also implies
weak commutativity (in the generality of any vector spaces V and W together with a
linear map

$$Y_W(\cdot, x) : V \to (\text{End } W)[[x, x^{-1}]]). \tag{3.2.21}$$

3.3 Associativity properties

Next we discuss the associativity properties of a vertex algebra. As we shall see, the associativity properties are sometimes similar to the commutativity properties, but there are also significant differences. The reader should carefully compare the situations in this section and the previous section and should notice the extent to which the comments in Section 3.2 carry over to analogous comments concerning associativity.

Throughout this section we fix a vertex algebra $(V, Y, \mathbf{1})$. The "weak associativity" relation is stated as follows:

Proposition 3.3.1 (weak associativity) *For $u, w \in V$, there exists a nonnegative integer l (depending only on u and w) such that for any $v \in V$,*

$$(x_0 + x_2)^l Y(Y(u, x_0)v, x_2)w = (x_0 + x_2)^l Y(u, x_0 + x_2)Y(v, x_2)w. \qquad (3.3.1)$$

Proof. By analogy with the proof of Proposition 3.2.1, we want to force the second term on the left-hand side of the Jacobi identity to vanish by clearing a suitable pole. First, by using the delta function property (2.3.17) we obtain

$$x_1^{-1}\delta\left(\frac{x_0 + x_2}{x_1}\right) Y(u, x_1)Y(v, x_2) - x_1^{-1}\delta\left(\frac{x_2 + x_0}{x_1}\right) Y(Y(u, x_0)v, x_2)$$

$$= x_0^{-1}\delta\left(\frac{x_2 - x_1}{-x_0}\right) Y(v, x_2)Y(u, x_1). \qquad (3.3.2)$$

In the first term on the left-hand side, we can replace $Y(u, x_1)$ by $Y(u, x_0 + x_2)$ (by (2.3.56)). Let $w \in V$ and let $l \geq 0$. Multiplying the Jacobi identity (3.3.2) by x_1^l and then taking Res_{x_1} we obtain

$$(x_0 + x_2)^l \left(Y(u, x_0 + x_2)Y(v, x_2)w - Y(Y(u, x_0)v, x_2)w\right)$$

$$= \text{Res}_{x_1} x_1^l x_0^{-1}\delta\left(\frac{x_2 - x_1}{-x_0}\right) Y(v, x_2)Y(u, x_1)w. \qquad (3.3.3)$$

Then simply choose l such that $u_n w = 0$ for $n \geq l$. $\square$

Remark 3.3.2 The proof shows that we may take $l \geq 0$ to be such that $u_n w = 0$ for $n \geq l$. Furthermore, the assertion (3.3.1) for some fixed $l \geq 0$ implies the assertion for all larger l, and knowing the smallest l for which it holds gives us the maximum amount of information (cf. Remark 3.2.2). Unlike the weak commutativity relation, the weak associativity relation does not contain enough information for us to be able to reconstruct the Jacobi identity from it alone. (This is not surprising, since weak associativity says nothing about what happens when the order of two vectors is reversed.) However, as we shall see, the weak associativity relation and the skew symmetry relation (3.1.30) are together strong enough to yield the Jacobi identity.

Remark 3.3.3 The designation "weak associativity" is of course motivated by the analogy with associativity for an algebra.

Remark 3.3.4 Notice that on the right-hand side of (3.3.1) we have the formal translation $x_0 \mapsto x_0 + x_2$, and this formal translation is an echo of formal translations that we have already seen—in the Jacobi identity and in formula (3.1.28), for instance. All of this is a manifestation of the fundamental geometric and conformal field-theoretic nature of vertex operator algebra theory (cf. [Hua13]).

Now we can formulate the notion of "associativity" (cf. [FLM6] and [FHL]), as opposed to weak associativity. As was the case for the notion of commutativity, this (parallel) notion uses "matrix coefficients." So we assume now that V is a vertex operator algebra rather than just a vertex algebra. (As was the case with commutativity, we do not need all of the properties.) Using the notion V' (recall (3.2.3)) and the algebras of rational functions and iota-maps introduced in Section 2.3, we first have the following "rationality of iterates" and preliminary version of associativity:

Proposition 3.3.5 *Let V be a vertex operator algebra and let $u, v, w \in V$ and $w' \in V'$ be arbitrary. We have:*

(a) (**rationality of iterates**) *The formal series*

$$\langle w', Y(Y(u, x_0)v, x_2)w \rangle \quad \left(= \sum_{m,n \in \mathbb{Z}} \langle w', (u_m v)_n w \rangle x_0^{-m-1} x_2^{-n-1} \right)$$

lies in the image of the map ι_{20}:

$$\langle w', Y(Y(u, x_0)v, x_2)w \rangle = \iota_{20} p(x_0, x_2), \tag{3.3.4}$$

where the (uniquely determined) element $p \in \mathbb{C}[x_0, x_2]_S$ is of the form

$$p(x_0, x_2) = \frac{q(x_0, x_2)}{x_0^k (x_0 + x_2)^l x_2^m} \tag{3.3.5}$$

for some $q \in \mathbb{C}[x_0, x_2]$ and $k, l, m \in \mathbb{Z}$, where l depends only on u and w; it is independent of v and w'.

(b) The series

$$\langle w', Y(u, x_0 + x_2)Y(v, x_2)w \rangle$$

$$\left(= \sum_{m,n \in \mathbb{Z}} \sum_{i \geq 0} \binom{-m-1}{i} \langle w', u_m v_n w \rangle x_0^{-m-1-i} x_2^{i-n-1} \right)$$

lies in the image of ι_{02}, and in fact

$$\langle w', Y(u, x_0 + x_2)Y(v, x_2)w \rangle = \iota_{02} p(x_0, x_2). \tag{3.3.6}$$

That is, in informal language,

$$\text{``} Y(Y(u, x_0)v, x_2) \quad \text{agrees with} \quad Y(u, x_0 + x_2)Y(v, x_2) \tag{3.3.7}$$

as operator-valued rational functions.''

Proof. Let l be a nonnegative integer such that the weak associativity relation (3.3.1) holds for u, v and w. Then

$$(x_0 + x_2)^l \langle w', Y(u, x_0 + x_2)Y(v, x_2)w \rangle = (x_0 + x_2)^l \langle w', Y(Y(u, x_0)v, x_2)w \rangle.$$
$$(3.3.8)$$

The left-hand side of (3.3.8) involves only finitely many negative powers of x_2, by (3.1.3), and only finitely many positive powers of x_0, by (3.1.54). Similarly, the right-hand side of (3.3.8) involves only finitely many negative powers of x_0 and only finitely many positive powers of x_2. Thus the common series is a finite series and thus a formal Laurent series of the form $r(x_0, x_2) \in \mathbb{C}[x_0, x_0^{-1}, x_2, x_2^{-1}]$. Then the rational function

$$p(x_0, x_2) = \frac{r(x_0, x_2)}{(x_0 + x_2)^l}$$

satisfies the desired conditions (as in the end of the proof of Proposition 3.2.7). $\square$

Remark 3.3.6 Notice that the proof of Proposition 3.3.5 shows that weak associativity implies the assertions of Proposition 3.3.5 in the presence of (3.1.54) (which follows from the $L(0)$-bracket formula (3.1.53)). Conversely, the assertions of Proposition 3.3.5 clearly imply weak associativity.

The "strong" version of associativity, which relates $\langle w', Y(u, z_1)Y(v, z_2)w \rangle$ with a suitable iterate, requires another step. We see that for the rational function $p(x_0, x_2)$ having the form indicated in (3.3.5),

$$\iota_{12}p(x_1 - x_2, x_2) = (\iota_{02}p(x_0, x_2))|_{x_0 = x_1 - x_2}. \tag{3.3.9}$$

Here we are comparing the substitutions $x_0 = x_1 - x_2$ before and after expanding the rational function, and it is sufficient to verify (3.3.9) for $p(x_0, x_2)$ of the following special forms: a polynomial in x_0 and x_2; x_0^{-1}; x_2^{-1}; and $(x_0 + x_2)^{-1}$. The most interesting cases are x_0^{-1} and $(x_0 + x_2)^{-1}$. For this last case, what we need to verify is that the "nested binomial expansion" $((x_1 - x_2) + x_2)^{-1}$ equals x_1^{-1}, which is true.

Now consider the formal series on the left-hand side of (3.3.6). We may perform the substitution $x_0 = x_1 - x_2$ in this series. Each negative power $(x_0 + x_2)^n$ of $x_0 + x_2$ is then replaced by the nested binomial expansion $((x_1 - x_2) + x_2)^n$, which equals x_1^n (an example of the kind of substitution that has arisen in Remark 2.3.25, for instance), and we have

$$\langle w', Y(u, x_0 + x_2)Y(v, x_2)w \rangle|_{x_0 = x_1 - x_2} = \langle w', Y(u, x_1)Y(v, x_2)w \rangle. \tag{3.3.10}$$

Thus by (3.3.6) and (3.3.9) we get

$$\langle w', Y(u, x_1)Y(v, x_2)w \rangle = \iota_{12}p(x_1 - x_2, x_2), \tag{3.3.11}$$

and in particular, the left-hand side, which is the same as the left-hand side of (3.2.5), is the ι_{12} expansion of the rational function

$$p(x_1 - x_2, x_2) = \frac{q(x_1 - x_2, x_2)}{(x_1 - x_2)^k x_1^l x_2^m}. \tag{3.3.12}$$

Remark 3.3.7 Notice that we have given a second proof, using weak associativity instead of weak commutativity, of a version of the rationality of products (Proposition 3.2.7(a)). However, in place of the exact assertion of Proposition 3.2.7(a), we have the existence of a (unique) $f(x_1, x_2)$ as in (3.2.6) such that l depends only on u and w; the present argument does not include the information that k depends only on u and v.

In any case, using Proposition 3.2.7 we see that

$$f(x_1, x_2) = p(x_1 - x_2, x_2), \tag{3.3.13}$$

or equivalently,

$$p(x_0, x_2) = f(x_0 + x_2, x_2). \tag{3.3.14}$$

Formula (3.3.13) gives:

Proposition 3.3.8 (associativity) *We have*

$$\iota_{12}^{-1}\langle w', Y(u, x_1)Y(v, x_2)w\rangle = \left(\iota_{20}^{-1}\langle w', Y(Y(u, x_0)v, x_2)w\rangle\right)\big|_{x_0 = x_1 - x_2}. \tag{3.3.15}$$

That is, in informal language,

$$\text{``}Y(u, x_1)Y(v, x_2) \quad \text{agrees with} \quad Y(Y(u, x_1 - x_2)v, x_2) \tag{3.3.16}$$

as operator-valued rational functions, where the right-hand expression is to be expanded as a Laurent series in $x_1 - x_2$." $\square$

Remark 3.3.9 We emphasize that the expression

$$Y(Y(u, x_1 - x_2)v, x_2)$$

does not in general exist as a formal series in x_1 and x_2. The formal translation $x_0 \mapsto x_1 - x_2$ must be performed at the level of rational functions, not formal series. This subtle point is reflected in the expansion domains arising in the next corollary.

As with commutativity, our algebraic associativity results immediately imply the following result involving convergent series in suitable domains:

Corollary 3.3.10 *The formal series of complex numbers $\langle w', Y(u, z_1)Y(v, z_2)w\rangle$ and $\langle w', Y(Y(u, z_1 - z_2)v, z_2)w\rangle$ are absolutely convergent to the common rational function $f(z_1, z_2)$ (recall (3.2.6)) in the domains*

$$|z_1| > |z_2| > 0 \quad \text{and} \quad |z_2| > |z_1 - z_2| > 0, \tag{3.3.17}$$

respectively, and in particular in the common domain

$$|z_1| > |z_2| > |z_1 - z_2| > 0. \tag{3.3.18}$$

Remark 3.3.11 As in Remark 3.2.11, we point out that the "rationality" of iterates refers to a particular expansion of a *very special type of rational function* (3.3.5), with formal poles only at $x_0 = 0$, $x_2 = 0$ or $x_0 = -x_2$. When we specialize the formal variables x_i to complex variables z_i, and when we use only the two variables z_1 and z_2, we get the domains indicated in Corollary 3.3.10.

Remark 3.3.12 The assertion (3.3.16), or better, its precise version, Corollary 3.3.10, is an important case of the "associativity of the operator product expansion" in two-dimensional conformal field theory, and Proposition 3.3.8 (and Remark 3.3.9) precisely interpret this algebraically; note that the expansion of $Y(Y(u, z_1 - z_2)v, z_2)$ in powers of $z_1 - z_2$ represents the "operator product" $Y(u, z_1)Y(v, z_2)$. As in the case of commutativity, the convergence involved is nothing but the convergence of binomial series, when, of course, "matrix coefficients" involving the arbitrary elements $w \in W$ and $w' \in W'$ are used. As we have just explained, the product $Y(u, z_1)Y(v, z_2)$ can be expressed as an (infinite) sum of the form

$$Y(u, z_1)Y(v, z_2) = \sum_{n \leq N} Y(w_{(n)}, z_2)(z_1 - z_2)^{-n-1}, \qquad (3.3.19)$$

where the $w_{(n)}$, $n \leq N$, are vectors in V (namely, $w_{(n)} = u_n v$; N is chosen so that $u_n v = 0$ for $n > N$) and where it is understood that it is actually the "matrix coefficients" of the two sides of (3.3.19) that are being equated; in this sense, the operator product expansion (3.3.19) indeed converges to the product of the two operators. Not only do we have the *existence* of the operator product expansion for the vertex operators $Y(u, z_1)$ and $Y(v, z_2)$, but we also have the *associativity* property, which amounts to the assertion that the vectors $w_{(n)}$ on the right-hand side of (3.3.19) are of the form $u_n v$.

Remark 3.3.13 When we discussed the commutator formula (3.1.8) above, we emphasized that only the *singular part* of $Y(u, x_0)v$ enters into the commutator. For this reason, the operator product expansion (3.3.19) is often written in the literature on conformal field theory as

$$Y(u, z_1)Y(v, z_2) \sim \sum_{0 \leq n \leq N} Y(w_{(n)}, z_2)(z_1 - z_2)^{-n-1}, \qquad (3.3.20)$$

where the right-hand side of (3.3.20), a finite sum, now reflects only the *singular* part (with respect to x_0) of $Y(Y(u, x_0)v, x_2)$; the symbol "$\sim$" indicates that the *regular* part of the expansion is dropped. Note the precise correspondence between (3.3.20) and the formula (3.1.10); recall the discussion in Remark 3.1.11. All the information needed to compute the commutator of the vertex operators $Y(u, z_1)$ and $Y(v, z_2)$ is contained in the (finitely many) vertex operators $Y(w_{(n)}, z_2)$ appearing in (3.3.20). (But the reader should keep in mind that the commutator formula is only one "slice" of the Jacobi identity, which gives "all" the information.) In fact, the procedure for extracting the commutator from the singular part of the operator product expansion has already been described in Remark 3.1.16: In formula (3.1.22), one simply takes the "test functions" $g(z_1, z_2)$ to be of the form

$$g(z_1, z_2) = z_1^m z_2^n \tag{3.3.21}$$

for integers m and n; that is, we do not allow $g(z_1, z_2)$ to have any pole at $z_1 = z_2$. Then the left-hand side of (3.1.22) becomes the commutator $[u_m, v_n]$ of the (arbitrary) components u_m and v_n of the vertex operators $Y(u, z_1)$ and $Y(v, z_2)$ (or rather, it becomes the "matrix coefficient" $\langle w', [u_m, v_n]w \rangle$), and the right-hand side of (3.1.22) gives the desired commutator; the answer is of course identical to the component form (3.1.9) of the commutator formula (3.1.8). Indeed, formula (3.1.22) specializes to

$$
\begin{aligned}
&\mathrm{Res}_{z_1=0}\mathrm{Res}_{z_2=0}\langle w', (Y(u, z_1)z_1^m)(Y(v, z_2)z_2^n)w \rangle \\
&\quad -\mathrm{Res}_{z_2=0}\mathrm{Res}_{z_1=0}\langle w', (Y(v, z_2)z_2^n)(Y(u, z_1)z_1^m)w \rangle \\
&= \mathrm{Res}_{z_2=0}\mathrm{Res}_{z_0=0}\langle w', Y(Y(u, z_0)v, z_2)w \rangle (z_2 + z_0)^m z_2^n,
\end{aligned}
\tag{3.3.22}
$$

where on the right-hand side we write z_1 as $z_2 + z_0$; this can of course also be written in the contour-integral form (3.1.23). Note that only the singular part of $Y(u, z_0)v$ enters into the right-hand side. The operator product expansion notation (3.3.20) thus suggests the use of the contour-deformation technique described in Remarks 3.1.16 and 3.1.17, and in fact in the conformal field theory literature, this contour-deformation technique is indeed often used in conjunction with the operator product expansion in the form (3.3.20), for computing commutators of vertex operators. This approach, along with its connection with the Jacobi identity, is discussed in detail in the Appendix of [FLM6]. The reader may wish to do the (easy) exercise of writing down the alternative versions using (3.3.20), (3.1.22) and (3.1.23) of the calculations of commutators of the "concrete" vertex operators that arise in many places in this book, especially in the construction of families of vertex operator algebras in Chapter 6; in Remark 3.3.14, we give two significant examples. In the present work, however, as in much of the theory of vertex operator algebras, the formal-calculus approach (reflected for instance in the formulation (3.1.8) of the commutator), rather than the contour-deformation technique for computing commutators, allows us to gain insights, and indeed, to formulate and prove the results we want. For instance, as we illustrate in this work, the very statement of the Jacobi identity (3.1.6), with its three formal variables, enables us to use the powerful techniques of formal calculus, which are not readily available if the identity is written either using two complex variables (as in (3.1.22)) or using binomial coefficients and no formal variables (as in (3.1.7)).

Remark 3.3.14 Here are a couple of important "concrete" examples of (3.3.20). For the generating function

$$L(z) = \sum_{n \in \mathbb{Z}} L(n)z^{-n-2} \tag{3.3.23}$$

(using the notation "z" rather than "x") of the Virasoro algebra operators $L(n)$ in the definition of the notion of vertex operator algebra (recall (3.1.41)), (3.3.20) becomes

$$L(z_1)L(z_2) \sim \frac{L'(z_2)}{z_1 - z_2} + \frac{2L(z_2)}{(z_1 - z_2)^2} + \frac{c_V/2}{(z_1 - z_2)^4}, \tag{3.3.24}$$

which follows immediately from (3.1.70)–(3.1.74), and by what we have just explained, this yields the commutator relations (3.1.40) for the $L(n)$; cf. the discussion after (3.1.70)–(3.1.74). The reader should compare this with Example 5.5.6, where the Virasoro algebra bracket relations (3.1.40) are "translated" into generating function form; the right-hand side of (5.5.5) involves the formal delta function and its first and third derivatives in a way that agrees with the commutator formula (3.1.8) when we apply the formal Taylor Theorem to the right-hand side of (3.1.8). Note the correspondence between the three terms on the right-hand side of (3.3.24) and of (5.5.5). Analogously, as we shall see in Section 6.2 (see Remark 6.2.17), for an affine Lie algebra in place of the Virasoro algebra, we have the following analogue of (3.3.24), using the notation (6.2.1), (6.2.2) and (6.2.7):

$$a(z_1)b(z_2) \sim \frac{[a,b](z_2)}{z_1 - z_2} + \frac{\langle a, b \rangle \mathbf{k}}{(z_1 - z_2)^2}; \tag{3.3.25}$$

this is the counterpart of the affine Lie algebra bracket relations written in the generating function form (6.2.8). (The Lie algebra element $\mathbf{k}$ is typically replaced by a scalar.)

Remark 3.3.15 All of the discussion in Remarks 3.3.12–3.3.14 concerning the operator product expansion and the calculation of commutators for vertex operator algebras applies equally well to *modules* for a vertex operator algebra; see Chapter 4.

Remark 3.3.16 We mentioned at the beginning of Remark 3.3.12 that (3.3.16) is "an important case of" the associativity of the operator product expansion. The *general* case of the operator product expansion in vertex operator algebra theory (and conformal field theory) is much deeper than what we have just been discussing. As we will see in detail in the next chapter, the Jacobi identity, along with the appropriate "minor axioms," defines the notion of *module* for a vertex operator algebra, and from the definition one gets many properties. Given three modules (not necessarily distinct) for a vertex operator algebra, one can further use the Jacobi identity to define a natural notion of *intertwining operator* among the three modules. This notion of intertwining operator is essentially the vertex algebra analogue of the Lie algebraic notion of module map (module homomorphism) from the tensor product of two modules for a given Lie algebra to a third module. This notion was described in a conformal field-theoretic context in [MSe] and was defined in the setting of formal calculus in [FHL]; intertwining operators correspond to "chiral tree-level 3-point correlation functions" in conformal field theory. The *general* operator product expansion in vertex operator algebra theory is an assertion analogous to (3.3.19), but where the two vertex operators on the left-hand side are generalized to *intertwining operators*. In this general setting, the existence, together with the associativity, of the operator product expansion is indeed true under suitable, natural, hypotheses, but in this generality it is a deep theorem; it does *not* follow easily from the axioms, and even in special cases it is not easy. The general theorem is due to Y.-Z. Huang [Hua8]; the formulation and proof of this theorem involve the theory developed in [HL1] and [HL4]–[HL7] of tensor products of modules for a vertex operator algebra. In a number of places in the literature, such existence and associativity of the operator product expansion for intertwining operators is assumed without proof in general, such as in [MSe], where this major assumption is clearly identified as an assumption in general rather than as a theorem. The hypotheses of Huang's general theorem in [Hua8] have been established in many contexts of interest; see for instance [FZ], [HL9], [Wa1], [Hua10]. More generally, in [Hua19], these hypotheses have been proved for *all* vertex operator algebras satisfying certain conditions of finiteness and complete reducibility of suitably generalized modules.

Just as we did at the end of the last section, we can easily generalize the associativity considerations to vertex algebras (without grading). Using the algebra

$$\mathbb{C}[[x_0, x_2]][x_0^{-1}, x_2^{-1}, (x_0 + x_2)^{-1}] = \mathbb{C}((x_0, x_2, x_0 + x_2)) \tag{3.3.26}$$

and the corresponding maps ι_{02} and ι_{20}, we have the following precise analogue of Proposition 3.3.5, with the obvious proof:

Proposition 3.3.17 *Let V and W be vector spaces equipped with a linear map*

$$Y(\cdot, x) : V \to (\mathrm{End}\ V)[[x, x^{-1}]] \tag{3.3.27}$$

and also a linear map

$$Y_W(\cdot, x) : V \to (\mathrm{End}\ W)[[x, x^{-1}]]. \tag{3.3.28}$$

Let $u, v \in V$, $w \in W$ and $w^* \in W^*$. Assume that the truncation conditions $Y(u, x)v \in V((x))$ and $Y_W(v, x)w \in W((x))$ hold and assume also that the assertion of weak associativity holds for the (ordered) triple (u, v, w), that is, there exists a nonnegative integer l depending only on u and w such that

$$(x_0 + x_2)^l Y_W(Y(u, x_0)v, x_2)w = (x_0 + x_2)^l Y_W(u, x_0 + x_2)Y_W(v, x_2)w.$$

$$\tag{3.3.29}$$

Then the formal series

$$\langle w^*, Y_W(Y(u, x_0)v, x_2)w \rangle \tag{3.3.30}$$

lies in the image of the map ι_{20}:

$$\langle w^*, Y_W(Y(u, x_0)v, x_2)w \rangle = \iota_{20} p(x_0, x_2), \tag{3.3.31}$$

where the (uniquely determined) element

$$p \in \mathbb{C}[[x_0, x_2]][x_0^{-1}, x_2^{-1}, (x_0 + x_2)^{-1}]$$

is of the form

$$p(x_0, x_2) = \frac{q(x_0, x_2)}{x_0^k (x_0 + x_2)^l x_2^m} \tag{3.3.32}$$

for some $q \in \mathbb{C}[[x_0, x_2]]$ and $k, l, m \in \mathbb{Z}$, where l depends only on u and w (and not on v or w^*). Moreover,

$$\langle w^*, Y_W(u, x_0 + x_2)Y_W(v, x_2)w \rangle = \iota_{02} p(x_0, x_2). \quad \square \tag{3.3.33}$$

Remark 3.3.18 As the proof shows, if we remove the assumption that l depends only on u and w (so that l might also depend on v), then the result is still valid, except that we cannot conclude that l is independent of v.

Now we need to imitate the proof of Proposition 3.3.8 (associativity) in order to obtain the analogous, precisely formulated, result in the present generality:

Proposition 3.3.19 (formal associativity) *Under the hypotheses of Proposition 3.3.17, with $u, v \in V$, $w \in W$ and $w^* \in W^*$, the formal series*

$$\langle w^*, Y_W(u, x_1)Y_W(v, x_2)w \rangle$$

lies in the image of the map ι_{12} defined in (2.3.60):

$$\langle w^*, Y_W(u, x_1)Y_W(v, x_2)w \rangle = \iota_{12}f(x_1, x_2), \tag{3.3.34}$$

where the (uniquely determined) element

$$f \in \mathbb{C}[[x_1, x_2]][x_1^{-1}, x_2^{-1}, (x_1 - x_2)^{-1}]$$

is of the form

$$f(x_1, x_2) = \frac{g(x_1, x_2)}{(x_1 - x_2)^k x_1^l x_2^m} \tag{3.3.35}$$

for some $g \in \mathbb{C}[[x_1, x_2]]$ and $k, l, m \in \mathbb{Z}$, where l depends only on u and w (and not on v or w^*). Moreover,

$$f(x_1, x_2) = p(x_1 - x_2, x_2), \tag{3.3.36}$$

or equivalently,

$$\iota_{12}^{-1}\langle w^*, Y_W(u, x_1)Y_W(v, x_2)w \rangle = \left(\iota_{20}^{-1}\langle w^*, Y_W(Y(u, x_0)v, x_2)w \rangle\right)\big|_{x_0 = x_1 - x_2} \tag{3.3.37}$$

(as elements of $\mathbb{C}[[x_1, x_2]][x_1^{-1}, x_2^{-1}, (x_1 - x_2)^{-1}]$).

Proof. We argue just as in (3.3.9)-(3.3.12) and Remark 3.3.7. We have a (well-defined) canonical map

$$\mathbb{C}[[x_1, x_2]][x_1^{-1}, x_2^{-1}, (x_1 - x_2)^{-1}] \to \mathbb{C}[[x_0, x_2]][x_0^{-1}, x_2^{-1}, (x_0 + x_2)^{-1}]$$

$$s(x_1, x_2) \mapsto s(x_1, x_2)\big|_{x_1 = x_0 + x_2} = s(x_0 + x_2, x_2), \tag{3.3.38}$$

and the reverse map

$$t(x_0, x_2) \mapsto t(x_0, x_2)\big|_{x_0 = x_1 - x_2} = t(x_1 - x_2, x_2) \tag{3.3.39}$$

is the inverse map. Moreover, for such $t(x_0, x_2)$, we have

$$\iota_{12}t(x_1 - x_2, x_2) = \left(\iota_{02}t(x_0, x_2)\right)\big|_{x_0 = x_1 - x_2}. \tag{3.3.40}$$

The rest is clear. $\square$

Remark 3.3.20 As in Remark 3.3.18, we observe that if in the hypotheses of Proposition 3.3.17 we omit the condition that l is independent of v, then Proposition 3.3.19 still holds, except that l can depend on v.

3.4 The Jacobi identity from commutativity and associativity

In Sections 3.2 and 3.3 we have seen that the Jacobi identity implies commutativity and associativity properties. Here we shall prove that conversely, in two different settings, the Jacobi identity also follows from such properties.

First we use Proposition 2.3.26 to obtain:

Proposition 3.4.1 *The Jacobi identity for a vertex operator algebra follows from the rationality of products and iterates, and commutativity and associativity (the assertions of Proposition 3.2.7(a),(b), Proposition 3.3.5(a) and Proposition 3.3.8). In particular, in the definition of the notion of vertex operator algebra, the Jacobi identity can be replaced by these properties.*

Proof. For the formal Laurent polynomial $f(x_0, x_1, x_2)$ in Proposition 2.3.26 we simply take $g(x_0, x_1)/x_0^k x_1^l x_2^m$ in the notation of Proposition 3.2.7. Then by Proposition 3.2.7,

$$\iota_{12}\left(f|_{x_0=x_1-x_2}\right) = \langle w', Y(u, x_1)Y(v, x_2)w\rangle \tag{3.4.1}$$

and

$$\iota_{21}\left(f|_{x_0=x_1-x_2}\right) = \langle w', Y(v, x_2)Y(u, x_1)w\rangle, \tag{3.4.2}$$

and by Proposition 3.3.5(a), Proposition 3.3.8 and (3.4.1),

$$\begin{aligned}
\iota_{20}\left(f|_{x_1=x_0+x_2}\right) &= \iota_{20}\left(f(x_1 - x_2, x_1, x_2)|_{x_1=x_0+x_2}\right) \\
&= \iota_{20}\left(\left(\iota_{12}^{-1}\langle w', Y(u, x_1)Y(v, x_2)w\rangle\right)|_{x_1=x_0+x_2}\right) \\
&= \langle w', Y(Y(u, x_0)v, x_2)w\rangle.
\end{aligned} \tag{3.4.3}$$

The Jacobi identity follows. $\square$

Remark 3.4.2 As the proof shows, the dependence of k and l on only certain combinations of u, v, w and w' (see Proposition 3.2.7(a) and Proposition 3.3.5(a)) is *not* needed; thus Proposition 3.4.1 remains valid if we assume only the corresponding *milder* versions of rationality, commutativity and associativity, with k and l depending on u, v, w and w'.

We have seen that commutativity, formal commutativity and weak commutativity are essentially equivalent, in the appropriate settings, and similarly for associativity. So it is natural to ask: For a vertex algebra (without grading), even though we do not have rational functions available, does the Jacobi identity follow from weak commutativity together with weak associativity? The answer is indeed yes, under the assumption of only the truncation condition, and the reason is that the proof of Proposition 3.4.1 carries over to this situation. We formulate the result in the following precise way, so that we can later apply it to modules as well as algebras (and also in order to emphasize that the proof works "element-wise"; as in the case of several other results, hypotheses concerning particular vectors imply conclusions about those particular vectors):

Proposition 3.4.3 *Let V, W, Y and Y_W be as in Propositions 3.2.12 and 3.3.19, let $u, v \in V$ and $w \in W$, and assume that the formal series $Y(u, x)v$, $Y_W(u, x)w$ and $Y_W(v, x)w$ are all truncated, as in Propositions 3.2.12 and 3.3.19. Then the Jacobi identity for the ordered pair (u, v), applied to the vector w, i.e., the assertion*

$$x_0^{-1}\delta\left(\frac{x_1 - x_2}{x_0}\right) Y_W(u, x_1)Y_W(v, x_2)w$$

$$-x_0^{-1}\delta\left(\frac{x_2 - x_1}{-x_0}\right) Y_W(v, x_2)Y_W(u, x_1)w$$

$$= x_2^{-1}\delta\left(\frac{x_1 - x_0}{x_2}\right) Y_W(Y(u, x_0)v, x_2)w, \tag{3.4.4}$$

is equivalent to weak commutativity for the pair (u, v) acting on the vector w (recall (3.2.15)) together with weak associativity for the ordered triple (u, v, w) (recall (3.3.29)). In particular, in the definition of vertex algebra, the Jacobi identity can be replaced by weak commutativity (the assertion of Proposition 3.2.1) and weak associativity (the assertion of Proposition 3.3.1).

Proof. The Jacobi identity for (u, v, w) implies weak commutativity and weak associativity for (u, v, w) by the proofs of Propositions 3.2.1 and 3.3.1. Conversely, using Proposition 3.2.12 (formal commutativity) and Proposition 3.3.19 (formal associativity), we see that in view of Proposition 2.3.27, the proof of Proposition 3.4.1 still works with obvious small changes. $\square$

Remark 3.4.4 Of course, just as in Remark 3.4.2, when we assume weak commutativity for (u, v) acting on w, we do not need to assume that k depends only on u and v, and when we assume weak associativity for (u, v, w), we do not need to assume that l depends only on u and w.

Remark 3.4.5 We now know that the Jacobi identity is (among other things!) a device by which commutativity and associativity can be combined into a single axiom. As a trivial case of Proposition 3.4.3, we observe (as we had already seen in Remark 3.1.6) that any commutative associative algebra A with identity 1 becomes a vertex algebra if we define $Y(a, x)b = ab$ for $a, b \in A$ (so that $Y(a, x)b = ab$ involves no nonzero powers of x) and if we take the vacuum vector $\mathbf{1}$ to be 1. In the same spirit, note that in ordinary classical algebra theory, one could combine the commutativity and associativity axioms for a nonassociative algebra with 1 into the single axiom

$$x_0^{-1}\delta\left(\frac{x_1-x_2}{x_0}\right)a(bc) - x_0^{-1}\delta\left(\frac{x_2-x_1}{-x_0}\right)b(ac) = x_2^{-1}\delta\left(\frac{x_1-x_0}{x_2}\right)(ab)c.$$

$$(3.4.5)$$

In particular, a commutative associative algebra (with identity 1) amounts precisely to a vertex algebra A for which $Y(a, x)$ is constant (i.e., involves no nonzero powers of x) for all $a \in A$. (These are also exactly those vertex algebras for which $\mathcal{D} = 0$, by Proposition 3.1.18.)

Example 3.4.6 Here we give Borcherds' construction [B1] of a vertex algebra from a commutative associative algebra with derivation (cf. Remark 3.4.5). Let A be a commutative associative algebra with identity together with a derivation $\mathcal{D}$. (The previous Remark concerns the case $\mathcal{D} = 0$.) For $a \in A$, we define $Y(a, x) \in (\text{End } A)[[x]]$ by

$$Y(a, x)b = (e^{x\mathcal{D}}a)b = \sum_{n \in \mathbb{N}} \frac{1}{n!}x^n(\mathcal{D}^n a)b \qquad (3.4.6)$$

for $b \in A$. That is,

$$a_n b = 0, \quad a_{-n-1}b = \frac{1}{n!}(\mathcal{D}^n a)b \qquad (3.4.7)$$

for $a, b \in A$, $n \in \mathbb{N}$. Then $Y(a, x)1 = e^{x\mathcal{D}}a$ for $a \in A$, and since $\mathcal{D}(1) = 0$, we have $Y(1, x)b = b$ for $b \in A$. This gives the truncation condition, the creation property and the vacuum property, and also the fact that $\mathcal{D}$ as defined in (3.1.24) agrees with our derivation $\mathcal{D}$. Since A is commutative and associative, we have

$$Y(a, x_1)Y(b, x_2) = Y(b, x_2)Y(a, x_1) \qquad \text{for } a, b \in A,$$

giving weak commutativity. Extend $\mathcal{D}$ to a derivation of the commutative associative algebra $A[[x_0, x_2]]$ with $\mathcal{D}(x_0) = \mathcal{D}(x_2) = 0$. Then $e^{x_2\mathcal{D}}$ is an automorphism of $A[[x_0, x_2]]$, and

$$\begin{aligned}
Y(a, x_0 + x_2)Y(b, x_2)c &= \left(e^{(x_0+x_2)\mathcal{D}}a\right)\left(e^{x_2\mathcal{D}}b\right)c \\
&= \left(e^{x_2\mathcal{D}}\left(\left(e^{x_0\mathcal{D}}a\right)b\right)\right)c \\
&= Y(Y(a, x_0)b, x_2)c, \qquad (3.4.8)
\end{aligned}$$

so that weak associativity also holds. Thus by Proposition 3.4.3, $(A, Y, 1)$ is a vertex algebra. In particular, by taking $A = \mathbb{C}[t, t^{-1}]$, $\mathcal{D} = d/dt$ we obtain a vertex algebra $\mathbb{C}[t, t^{-1}]$, where

$$Y(f(t), x)g(t) = \left(e^{x\frac{d}{dt}}f(t)\right)g(t) = f(t + x)g(t) \qquad (3.4.9)$$

for $f(t), g(t) \in A$.

Remark 3.4.7 Returning to the considerations of Remark 3.4.5, suppose that V is a vertex operator algebra with conformal vector ω equal to 0. Then $L(-1) = 0$, and since $L(-1) = \mathcal{D}$ by (3.1.67), V is simply a commutative associative algebra, and is also finite-dimensional since the eigenspace for $L(0) = 0$ must be finite-dimensional. Conversely, a finite-dimensional commutative associative algebra (with identity) clearly carries vertex operator algebra structure when we take $\omega = 0$. Thus we see that the vertex operator algebras with $\omega = 0$ amount precisely to the finite-dimensional commutative associative algebras.

3.5 The Jacobi identity from commutativity

As we noticed in Remark 3.2.6, the weak commutativity relation (3.2.1) is really an analogue of the commutativity of left multiplication operations in a classical algebra, not the commutativity of multiplication itself. A closer analogue of classical commutativity of multiplication is skew symmetry (3.1.30) (even though this is also analogous to the skew symmetry of Lie bracket in Lie algebra theory!). Let A be a nonassociative algebra with a right identity 1 and denote by L_a the operation of left multiplication by $a \in A$. Suppose that $L_a L_b = L_b L_a$ for any $a, b \in A$, or equivalently, $a(bc) = b(ac)$ for any $a, b, c \in A$. Setting $c = 1$, we obtain the commutativity relation $ab = ba$. We then obtain associativity since

$$a(bc) = a(cb) = c(ab) = (ab)c \quad \text{for any } a, b, c \in A, \tag{3.5.1}$$

so that A is a commutative associative algebra with identity.

This classical fact seems to suggest that weak commutativity together with the creation property (an analogue of the right identity property; recall Remark 3.1.4) might imply weak associativity, hence the Jacobi identity. This is essentially the case.

Theorem 3.5.1 *The Jacobi identity for a vertex algebra follows from weak commutativity (the assertion of Proposition 3.2.1) in the presence of the other axioms together with the $\mathcal{D}$-bracket-derivative formula (3.1.32). In particular, in the definition of the notion of vertex algebra, the Jacobi identity can be replaced by these properties.*

Proof. We shall imitate the classical argument, first proving skew symmetry and then weak associativity, and the Jacobi identity will then follow from Proposition 3.4.3 (which in turn used formal commutativity—Proposition 3.2.12—and formal associativity—Proposition 3.3.19).

We first prove (3.1.29), a special case of skew symmetry. From the vacuum property we have $\mathcal{D}(1) = 1_{-2}1 = 0$ (cf. (3.1.26)). Recall that the $\mathcal{D}$-bracket-derivative formula (3.1.32) implies the conjugation formula (3.1.35), which gives

$$Y(v, x_0 + x)\mathbf{1} = e^{x\mathcal{D}}Y(v, x_0)e^{-x\mathcal{D}}\mathbf{1} = e^{x\mathcal{D}}Y(v, x_0)\mathbf{1}. \tag{3.5.2}$$

In view of the creation property we can set $x_0 = 0$ in (3.5.2), and we find that $Y(v, x)\mathbf{1} = e^{x\mathcal{D}}v$ (cf. (3.1.29)).

To prove skew symmetry in general, let $u, v \in V$ and let $k \in \mathbb{N}$ be such that $x^k Y(v, x)u$ involves only nonnegative powers of x and such that the weak commutativity relation (3.2.1) for (u, v) holds. Then using (3.1.29) and (3.1.35) we get

$$
\begin{aligned}
(x_1 - x_2)^k Y(u, x_1) Y(v, x_2) \mathbf{1} &= (x_1 - x_2)^k Y(v, x_2) Y(u, x_1) \mathbf{1} \\
&= (x_1 - x_2)^k Y(v, x_2) e^{x_1 \mathcal{D}} u \\
&= (x_1 - x_2)^k e^{x_1 \mathcal{D}} Y(v, x_2 - x_1)u.
\end{aligned}
\tag{3.5.3}
$$

In the last expression we may set $x_2 = 0$ because $(x_1 - x_2)^k Y(v, x_2 - x_1)u$ involves only nonnegative powers of $x_2 - x_1$. Setting $x_2 = 0$ and using the creation property we get

$$
x_1^k Y(u, x_1)v = x_1^k e^{x_1 \mathcal{D}} Y(v, -x_1)u.
\tag{3.5.4}
$$

Multiplying both sides by x_1^{-k} we obtain the skew symmetry relation for (u, v).

Now we prove weak associativity, using (3.5.1) as a guide. For any $u, v, w \in V$, let $k \in \mathbb{N}$ be such that the weak commutativity relation (3.2.1) holds for u and w. Then by using skew symmetry (for the pairs (v, w) and $(Y(u, x_0)v, w)$) and the conjugation formula (3.1.35), derived above, we obtain weak associativity:

$$
\begin{aligned}
(x_0 + x_2)^k Y(u, x_0 + x_2) Y(v, x_2)w \\
= (x_0 + x_2)^k Y(u, x_0 + x_2) e^{x_2 \mathcal{D}} Y(w, -x_2)v \\
= e^{x_2 \mathcal{D}} (x_0 + x_2)^k Y(u, x_0) Y(w, -x_2)v \\
= e^{x_2 \mathcal{D}} (x_0 + x_2)^k Y(w, -x_2) Y(u, x_0)v \\
= (x_0 + x_2)^k Y(Y(u, x_0)v, x_2)w. \qquad \square
\end{aligned}
\tag{3.5.5}
$$

Remark 3.5.2 The proof of Theorem 3.5.1 shows that skew symmetry follows from weak commutativity and the formula (3.1.32) in the presence of all the axioms except the Jacobi identity. This gives a second proof of skew symmetry (recall Proposition 3.1.19).

Remark 3.5.3 Notice that in the argument for the classical analogue, we only used the right identity and commutativity properties. However, in Theorem 3.5.1, formula (3.1.32) needs to be assumed, as the following counterexample shows. Let A be a commutative associative algebra with identity of dimension greater than one, so that we can write $A = \mathbb{C} \oplus U$ for some nonzero subspace U. Define

$$
Y(\lambda + u, x)a = \lambda a + (1 + x)ua \quad \text{for } \lambda \in \mathbb{C}, \ u \in U, \ a \in A.
\tag{3.5.6}
$$

Then the truncation condition, the vacuum property (with 1 being the vacuum vector) and the creation property clearly hold. Furthermore, one can easily verify that $Y(u, x_1) Y(v, x_2) = Y(v, x_2) Y(u, x_1)$ for any $u, v \in U \cup \{1\}$, so that by linearity weak commutativity holds. However, the formula (3.1.32) fails. (Of course, $(A, Y, 1)$ is not a vertex algebra.) First, since $Y(1, x)1 = 1$ and $Y(u, x)1 = (1 + x)u$ for $u \in U$, we have $\mathcal{D}(1) = 0$ and $\mathcal{D}(u) \ (= u_{-2}1) = u$. Then $\frac{d}{dx} Y(u, x)1 = u$ and

$$[\mathcal{D}, Y(u, x)]\mathbf{1} = \mathcal{D}Y(u, x)\mathbf{1} = (1 + x)\mathcal{D}(u) = (1 + x)u,$$

so that $[\mathcal{D}, Y(u, x)] \neq \frac{d}{dx}Y(u, x)$ for $0 \neq u \in U$.

Now commutativity for a vertex operator algebra implies weak commutativity (recall Remark 3.2.9), and $L(-1) = \mathcal{D}$ (recall (3.1.67)). Now let us *not* assume that we know that $L(-1) = \mathcal{D}$. The proof of Theorem 3.5.1 holds for $L(-1)$ in place of $\mathcal{D}$, and we have:

Theorem 3.5.4 *In the definition of the notion of vertex operator algebra, the Jacobi identity can be replaced by commutativity and the $L(-1)$-bracket-derivative formula (the outer equality in (3.1.52)).* $\square$

3.6 The Jacobi identity from skew symmetry and associativity

Here we shall prove that the Jacobi identity follows from skew symmetry and associativity in the presence of other axioms. As a consequence, we show that Definition 3.1.1 is equivalent to Borcherds' original definition of the notion of vertex algebra [B1]. This section is based on [Li3].

As we noted before, for a nonassociative algebra A with identity the commutativity of left multiplications amounts to the commutativity and associativity of multiplication in A. We also noted that skew symmetry in vertex algebra theory is an analogue of the commutativity of multiplication in a commutative algebra. By this analogy, skew symmetry and associativity properties ought to imply commutativity properties and hence the Jacobi identity, and this is indeed the case.

Theorem 3.6.1 *The Jacobi identity for a vertex algebra follows from skew symmetry (the assertion of Proposition 3.1.19) and weak associativity (with the dependence of l on only u and w not needed; see Proposition 3.3.1) in the presence of the other axioms. In particular, in the definition of the notion of vertex algebra the Jacobi identity can be replaced by these properties.*

Proof. This argument will be very similar to the proof of Theorem 3.5.1. The classical "guide," formula (3.6.4) below, is similar to the guide (3.5.1) for Theorem 3.5.1, except that the middle equality in (3.6.4) is the outer equality in (3.5.1) and vice–versa; the arguments (3.5.5) and (3.6.6) are analogously related.

We shall first derive the conjugation formula (3.1.35), the global form of the $\mathcal{D}$-bracket-derivative formula. Note that the vacuum property (3.1.4) and skew symmetry (3.1.30) immediately imply (3.1.29), that is, $Y(v, x)\mathbf{1} = e^{x\mathcal{D}}v$ for $v \in V$. For $u, v \in V$, from weak associativity there is a nonnegative integer l such that

$$(x_0 + x_2)^l Y(u, x_0 + x_2)Y(v, x_2)\mathbf{1} = (x_0 + x_2)^l Y(Y(u, x_0)v, x_2)\mathbf{1}, \qquad (3.6.1)$$

which together with (3.1.29) gives

$$(x_0 + x_2)^l Y(u, x_0 + x_2)e^{x_2 \mathcal{D}} v = (x_0 + x_2)^l e^{x_2 \mathcal{D}} Y(u, x_0)v. \qquad (3.6.2)$$

Since both sides of (3.6.2) involve only nonnegative powers of x_2, we can multiply by $(x_0 + x_2)^{-l}$ to get an equivalent form of (3.1.35):

$$Y(u, x_0 + x_2)e^{x_2 \mathcal{D}} v = e^{x_2 \mathcal{D}} Y(u, x_0)v. \qquad (3.6.3)$$

Now we prove weak commutativity, using the following argument in the classical analogue as a guide:

$$a(bc) = (bc)a = b(ca) = b(ac). \qquad (3.6.4)$$

For $u, v, w \in V$, let l be a nonnegative integer such that the weak associativity relation (3.3.1) for (v, w, u) holds (where we do not need to assume that l is independent of w), so that

$$(x_2 - x_1)^l Y(Y(v, x_2)w, -x_1)u = (x_2 - x_1)^l Y(v, x_2 - x_1)Y(w, -x_1)u. \qquad (3.6.5)$$

Then using skew symmetry for $(u, Y(v, x_2)w)$ and for (u, w) and using (3.6.3) we obtain weak commutativity in the following form:

$$\begin{aligned}
(x_2 - x_1)^l & Y(u, x_1)Y(v, x_2)w \\
&= (x_2 - x_1)^l e^{x_1 \mathcal{D}} Y(Y(v, x_2)w, -x_1)u \\
&= (x_2 - x_1)^l e^{x_1 \mathcal{D}} Y(v, x_2 - x_1)Y(w, -x_1)u \\
&= (x_2 - x_1)^l Y(v, x_2)e^{x_1 \mathcal{D}} Y(w, -x_1)u \\
&= (x_2 - x_1)^l Y(v, x_2)Y(u, x_1)w. \qquad (3.6.6)
\end{aligned}$$

Note that l is allowed to depend on w as well as on u and v; the desired result follows from Proposition 3.4.3 together with Remark 3.4.4. $\square$

Remark 3.6.2 What the proofs of Theorems 3.5.1 and 3.6.1 actually show is that skew symmetry and weak associativity are together equivalent to the $\mathcal{D}$-bracket-derivative formula and weak commutativity.

When we study and construct modules for vertex algebras, it turns out that we shall need a result similar to Theorem 3.6.1, but stronger. We note that there is another path besides (3.6.4) to proving the commutativity of left-multiplication operations for a commutative associative algebra, namely,

$$a(bc) = (ab)c = (ba)c = b(ac), \qquad (3.6.7)$$

and it is important that (3.6.7) has the following obvious generalization:

$$a(bu) = (ab)u = (ba)u = b(au), \qquad (3.6.8)$$

where a and b are elements of the algebra and u is an element of *an arbitrary module* for the algebra. (This exhibits the trivial fact that for a module for a commutative associative algebra, the actions of algebra elements commute.) The path given in (3.6.4) does not have this generality because the element c in (3.6.4) does not always remain on the right. Just as (3.6.4) did, the path (3.6.7) will lead to a proof of Theorem 3.6.1, a proof which, most importantly, is general enough to show (as we shall see later) that a suitable space W has the structure of a module for a vertex algebra V. Here is the more general result, stated (as in Propositions 3.2.12, 3.3.17 and 3.3.19) in terms of spaces V and W:

Theorem 3.6.3 *Let $(V, Y, \mathbf{1})$ be a triple that satisfies all the axioms in the definition of the notion of vertex algebra except for the Jacobi identity, and in addition has the skew symmetry property (3.1.30). Let W be a vector space and let $Y_W(\cdot, x)$ be a linear map from V to $(\mathrm{End}\ W)[[x, x^{-1}]]$ such that $Y_W(\mathbf{1}, x) = 1$ and $Y_W(v, x)w \in W((x))$ for $v \in V$ and $w \in W$. Assume too that weak associativity holds for any $u, v \in V$ and $w \in W$, in the sense that there exists $l \in \mathbb{N}$ (depending on u, v and w) such that*

$$(x_0 + x_2)^l Y_W(Y(u, x_0)v, x_2)w = (x_0 + x_2)^l Y_W(u, x_0 + x_2)Y_W(v, x_2)w. \quad (3.6.9)$$

Then the Jacobi identity (3.4.4) holds for any $u, v \in V$ and $w \in W$.

Proof. First we prove the $\mathcal{D}$-derivative formula (3.1.25) for Y_W. For $u \in V$ and $w \in W$, by the weak associativity assumption there exists $l \in \mathbb{N}$ such that

$$(x_0 + x_2)^l Y_W(Y(u, x_0)\mathbf{1}, x_2)w = (x_0 + x_2)^l Y_W(u, x_0 + x_2)Y_W(\mathbf{1}, x_2)w.$$

Using the vacuum property for $\mathbf{1}$ acting on W and the formula $Y(u, x)\mathbf{1} = e^{x\mathcal{D}}u$, which follows from the vacuum property for V and skew symmetry (as at the beginning of the proof of Theorem 3.6.1), we obtain

$$(x_0 + x_2)^l Y_W(e^{x_0\mathcal{D}}u, x_2)w = (x_0 + x_2)^l Y_W(u, x_0 + x_2)w. \quad (3.6.10)$$

From the assumed truncation condition, we choose l larger if necessary so that $x^l Y_W(u, x)w \in W[[x]]$, and we have

$$(x_0 + x_2)^l Y_W(e^{x_0\mathcal{D}}u, x_2)w = (x_0 + x_2)^l Y_W(u, x_2 + x_0)w, \quad (3.6.11)$$

since we can now change $x_0 + x_2$ to $x_2 + x_0$ on the right-hand side. Since all the factors now involve only nonnegative powers of x_0, we can multiply by $(x_2 + x_0)^{-l}$ to get

$$Y_W(e^{x_0\mathcal{D}}u, x_2)w = Y_W(u, x_2 + x_0)w, \quad (3.6.12)$$

which is the global form (3.1.28) of the $\mathcal{D}$-derivative formula (3.1.25) for Y_W.

To prove the Jacobi identity, we recall from Propositions 3.3.17 and 3.3.19 and Remarks 3.3.18 and 3.3.20 that formal associativity holds for Y_W (with the integer l in (3.3.35) possibly depending on v). Hence from the proof of Proposition 3.4.3, it suffices to prove that the following version of formal commutativity holds for Y_W: For $u, v \in V$, $w \in W$ and $w^* \in W^*$, there exists

$$f(x_1, x_2) \in \mathbb{C}[[x_1, x_2]][x_1^{-1}, x_2^{-1}, (x_1 - x_2)^{-1}] \quad (3.6.13)$$

such that

$$\langle w^*, Y_W(u, x_1)Y_W(v, x_2)w \rangle = \iota_{12} f(x_1, x_2), \qquad (3.6.14)$$

$$\langle w^*, Y_W(v, x_2)Y_W(u, x_1)w \rangle = \iota_{21} f(x_1, x_2). \qquad (3.6.15)$$

(It is *not* necessary to know that the power k of $x_1 - x_2$ in the denominator of f (recall (3.2.18)) depends only on u and v.)

In view of Propositions 3.3.17 and 3.3.19, there exists a (uniquely determined) element f as in (3.6.13) such that (3.6.14) holds and such that

$$\langle w^*, Y_W(Y(u, x_0)v, x_2)w \rangle = \iota_{20} f(x_0 + x_2, x_2). \qquad (3.6.16)$$

Similarly, if we simply reverse the roles of u and v, of x_1 and x_2, and of x_0 and $-x_0$, we obtain a unique element

$$f'(x_1, x_2) \in \mathbb{C}[[x_1, x_2]][x_1^{-1}, x_2^{-1}, (x_1 - x_2)^{-1}] \qquad (3.6.17)$$

such that (3.6.15) holds with f' in place of f and such that

$$\langle w^*, Y_W(Y(v, -x_0)u, x_1)w \rangle = \iota_{10} f'(x_1, x_1 - x_0). \qquad (3.6.18)$$

It suffices to prove that $f = f'$.

But by the assumed skew symmetry, the $\mathcal{D}$-derivative formula for Y_W and (3.3.40) (with the roles of the variables permuted),

$$\begin{aligned}
\langle w^*, &Y_W(Y(u, x_0)v, x_2)w \rangle \\
&= \langle w^*, Y_W(e^{x_0 \mathcal{D}} Y(v, -x_0)u, x_2)w \rangle \\
&= \langle w^*, Y_W(Y(v, -x_0)u, x_2 + x_0)w \rangle \\
&= \left(\iota_{10} f'(x_1, x_1 - x_0) \right)|_{x_1 = x_2 + x_0} \\
&= \iota_{20} f'(x_0 + x_2, x_2),
\end{aligned} \qquad (3.6.19)$$

and by (3.6.16) we see that $f = f'$. $\quad\square$

Remark 3.6.4 Notice that this argument rounds out our picture in a number of ways. For instance, earlier we had extensively considered the two products $Y(u, x_1)Y(v, x_2)$ and $Y(v, x_2)Y(u, x_1)$ and the one iterate $Y(Y(u, x_0)v, x_2)$. Now the "missing" iterate $Y(Y(v, -x_0)u, x_1)$ comes to the forefront. Also, in the statement of commutativity or formal commutativity, in the formal function $f(x_1, x_2)$ of (3.2.18), we initially knew that the power k of $x_1 - x_2$ in the denominator depended only on u and v. Now we also know that the power l of x_1 depends only on u and w and that the power m of x_2 depends only on v and w. We also comment that the formal analogue of analytic continuation in the proof of Theorem 3.6.3 corresponds to the complex domains

$$|z_1| > |z_2| > 0 \quad \text{and} \quad |z_2| > |z_1 - z_2| > 0 \qquad (3.6.20)$$

when we consider $Y_W(u, x_1)Y_W(v, x_2)$ and $Y_W(Y(u, x_0)v, x_2)$, respectively, and to the domains

$$|z_2| > |z_1| > 0 \quad \text{and} \quad |z_1| > |z_1 - z_2| > 0 \qquad (3.6.21)$$

when we consider $Y_W(v, x_2)Y_W(u, x_1)$ and $Y_W(Y(v, -x_0)u, x_1)$, respectively. We are (formally) "analytically continuing" from $|z_1| > |z_2| > 0$ to $|z_2| > |z_1| > 0$ via such a chain of domains.

Now we use the fact that weak associativity and skew symmetry imply the Jacobi identity to show that Borcherds' original set of axioms for his notion of vertex algebra and Definition 3.1.1 are equivalent.

Definition 3.6.5 In [B1], Borcherds gave the following definition: A *vertex algebra* is a vector space V (actually, this definition works over $\mathbb{Z}$) equipped with an element $\mathbf{1}$, linear operators $\mathcal{D}^{(i)}$ on V for $i \in \mathbb{Z}$ and bilinear operations $(u, v) \mapsto u_n(v)$ from $V \times V$ to V, for $n \in \mathbb{Z}$, satisfying the following relations (i)–(v) for $u, v, w \in V$ and $m, n \in \mathbb{Z}$:

(i) $u_n(w) = 0$ for n sufficiently large (depending on u and w).
(ii) $\mathbf{1}_n(w) = 0$ if $n \neq -1$, w if $n = -1$.
(iii) $u_n(\mathbf{1}) = \mathcal{D}^{(-n-1)}(u)$.
(iv) $u_n(v) = \sum_{i \geq 0}(-1)^{i+n+1}\mathcal{D}^{(i)}(v_{n+i}(u))$.
(v) $(u_m(v))_n(w) = \sum_{i \geq 0}(-1)^i \binom{m}{i}(u_{m-i}(v_{n+i}(w)) - (-1)^m v_{m+n-i}(u_i(w)))$.

We shall show that this definition is equivalent to Definition 3.1.1. First note that (i), (ii) and (v) are the truncation condition (3.1.3), the vacuum property (3.1.4) and the iterate formula (3.1.11), respectively, expressed in terms of components using the vertex operator component notation of (3.1.1) (recall the comments in Section 3.1 on the component-forms of such identities as the iterate formula (3.1.11)). Next, from (iii), $u_{-2}(\mathbf{1}) = \mathcal{D}^{(1)}(u)$, so that $\mathcal{D}^{(1)}$ is the operator $\mathcal{D}$ defined in (3.1.24). If we can show that $\mathcal{D}^{(n)} = \frac{1}{n!}(\mathcal{D}^{(1)})^n$ for $n \geq 0$ and 0 for $n < 0$, then we will have that (iii) is the relation (3.1.29) and (iv) is skew symmetry (3.1.30) (but note that at this point we do not even know that $\mathcal{D}^{(0)}$ is the identity operator, which is part of what we have to prove).

To verify this, first take $v = \mathbf{1}$ in (iv) and use (ii) to get $u_n(\mathbf{1}) = 0$ for $n \geq 0$ and $u_n(\mathbf{1}) = \mathcal{D}^{(-n-1)}(u)$ for $n < 0$. Then by (iii), $\mathcal{D}^{(-n-1)}(u) = u_n(\mathbf{1}) = 0$ for $n \geq 0$, so that $\mathcal{D}^{(i)} = 0$ for $i < 0$. (Notice that (iii) is not redundant because it gives information about $\mathcal{D}^{(i)}$ for $i < 0$.)

Now we derive a formula for multiplying the operators $\mathcal{D}^{(i)}$ using (v). Let $m, n \geq 0$. By taking $v = w = \mathbf{1}$ in (v), replacing m, n by $-m - 1, -n - 1$, respectively, and using (ii) we obtain

$$(u_{-m-1}(\mathbf{1}))_{-n-1}(\mathbf{1}) = (-1)^n \binom{-m-1}{n} u_{-m-n-1}(\mathbf{1}) = \binom{m+n}{n} u_{-m-n-1}(\mathbf{1}),$$

which says that

$$\mathcal{D}^{(n)}\mathcal{D}^{(m)}(u) = \binom{m+n}{n}\mathcal{D}^{(m+n)}(u).$$

Thus

$$\mathcal{D}^{(n)}\mathcal{D}^{(m)} = \binom{m+n}{n}\mathcal{D}^{(m+n)} \qquad (3.6.22)$$

and from this we get

$$\mathcal{D}^{(n)}\mathcal{D}^{(1)} = (n+1)\mathcal{D}^{(n+1)} \qquad \text{for } n \geq 0,$$

so that

$$n!\mathcal{D}^{(n)} = (\mathcal{D}^{(1)})^n \qquad \text{for } n \geq 1. \qquad (3.6.23)$$

We still need to prove that $\mathcal{D}^{(0)}u = u$ for $u \in V$. Set $v = \mathcal{D}^{(0)}(u) - u$. Since $\mathcal{D}^{(0)}\mathcal{D}^{(0)} = \mathcal{D}^{(0)}$, we have $\mathcal{D}^{(0)}(v) = 0$. Using (ii), (iv) (with $u = \mathbf{1}$) and (iii) we have

$$v = \mathbf{1}_{-1}(v) = \sum_{i \geq 0}(-1)^i \mathcal{D}^{(i)}(v_{i-1}(\mathbf{1})) = \sum_{i \geq 0}(-1)^i \mathcal{D}^{(i)}\mathcal{D}^{(-i)}v = \mathcal{D}^{(0)}\mathcal{D}^{(0)}(v) = 0,$$

so that $\mathcal{D}^{(0)}$ is indeed the identity operator on V. We have also just established the creation property (3.1.5).

To finish proving that Definitions 3.1.1 and 3.6.5 are equivalent, given a vertex algebra in the sense of Definition 3.1.1, we define the operators $\mathcal{D}^{(i)}$ as indicated above, and we get the conditions of Definition 3.6.5 by properties we have established. Conversely, given a vertex algebra in the sense of Definition 3.6.5, it remains to prove the Jacobi identity. We have the associator formula, i.e., the variant (3.1.13) of the iterate formula (3.1.11). Applying to w and multiplying by $(x_0 + x_2)^l$ we obtain (3.3.3) and hence weak associativity. Now we just apply Theorem 3.6.1 (or Theorem 3.6.3). We have proved:

Proposition 3.6.6 *Definitions 3.1.1 and 3.6.5 of the notion of vertex algebra are equivalent.* $\square$

At the end of this section we discuss a certain redundancy in the axioms for a vertex algebra.

Proposition 3.6.7 *The vacuum property $Y(\mathbf{1}, x) = 1$ follows from the other axioms in Definition 3.1.1.*

Proof. Assume all the axioms except for the vacuum property in Definition 3.1.1. For any $v \in V$, since $Y(v, x)\mathbf{1} \in V[[x]]$, we have

$$[Y(v, x_1), Y(\mathbf{1}, x_2)] = 0 \qquad (3.6.24)$$

by Remark 3.2.4, and together with the creation property (3.1.5) this implies

$$Y(\mathbf{1}, x)v = Y(\mathbf{1}, x)v_{-1}\mathbf{1} = v_{-1}Y(\mathbf{1}, x)\mathbf{1} \in V[[x]] \qquad (3.6.25)$$

and in addition,

$$\lim_{x \to 0} Y(\mathbf{1}, x)v = \lim_{x \to 0} v_{-1} Y(\mathbf{1}, x)\mathbf{1} = v_{-1}\mathbf{1} = v. \tag{3.6.26}$$

Thus $Y(\mathbf{1}, x) \in (\text{End } V)[[x]]$ and its constant term is the identity operator on V. Let A be the subalgebra of End V generated by the operators $\mathbf{1}_n$ for $n \in \mathbb{Z}$. Then A is commutative and $Y(\mathbf{1}, x) \in A[[x]]$. Since the constant term of $Y(\mathbf{1}, x)$ is 1, $Y(\mathbf{1}, x)$ is an invertible element of $A[[x]]$.

Set $u = v = \mathbf{1}$ in the Jacobi identity (3.1.6). Since (by (3.6.24))

$$Y(\mathbf{1}, x_1)Y(\mathbf{1}, x_2) = Y(\mathbf{1}, x_2)Y(\mathbf{1}, x_1),$$

the Jacobi identity simplifies to

$$x_2^{-1}\delta\left(\frac{x_1 - x_0}{x_2}\right) Y(\mathbf{1}, x_1)Y(\mathbf{1}, x_2) = x_2^{-1}\delta\left(\frac{x_1 - x_0}{x_2}\right) Y(Y(\mathbf{1}, x_0)\mathbf{1}, x_2). \tag{3.6.27}$$

Taking $\text{Res}_{x_1}\text{Res}_{x_0}x_0^{-1}$, by the creation property again we obtain

$$Y(\mathbf{1}, x_2)Y(\mathbf{1}, x_2) = Y(\mathbf{1}, x_2). \tag{3.6.28}$$

Since $Y(\mathbf{1}, x)$ is invertible, $Y(\mathbf{1}, x) = 1$. $\square$

Remark 3.6.8 Recall from Remark 3.1.4 that the vacuum property (3.1.4) and the creation property (3.1.5) can be considered as left-identity and right-identity properties of $\mathbf{1}$, respectively. However, the following example shows that the creation property (3.1.5) does not follow from the other axioms (cf. Proposition 3.6.7): Let U be any nonzero vector space and set $A = \mathbb{C} \oplus U$. Define $Y(\lambda + u, x) = \lambda$ for $\lambda \in \mathbb{C}$, $u \in U$. The truncation condition and the vacuum property, with $\mathbf{1}$ the vacuum vector, clearly hold. One can also easily verify the Jacobi identity for the pair (u, v) by considering $u \in \mathbb{C} \cup U$. But clearly the creation property fails.

3.7 $\mathcal{S}_3$-symmetry of the Jacobi identity

We now discuss a symmetry property of the Jacobi identity under the symmetric group $\mathcal{S}_3$, derived in [FHL]. We have already seen a great deal of "hidden symmetry" among the properties of a vertex algebra, for example, in the proof of skew symmetry (Proposition 3.1.19) using a symmetry of the Jacobi identity and a permutation of vectors and variables, and in the "symmetry" between commutativity and associativity. Using arguments that are familiar by now, we make explicit the $\mathcal{S}_3$-symmetry of the Jacobi identity. This is analogous to the simple statement that in classical Lie theory, the Jacobi identity for an ordered triple of Lie algebra elements implies the Jacobi identity for any permutation of the triple, under the assumption of skew symmetry for all pairs of elements.

Let us retain the axioms for a vertex algebra except for the Jacobi identity, and let us call the assertion that (3.1.6) holds when applied to a vector w "the Jacobi identity for the ordered triple (u, v, w)." Under natural assumptions, we shall prove the Jacobi identity for the ordered triple (v, u, w) and then for the ordered triple (u, w, v), and this will give the full $\mathcal{S}_3$-symmetry.

By skew symmetry (3.1.30) for the pair (u, v) and the $\mathcal{D}$-derivative property in the form (3.1.28) for the element $Y(v, -x_0)u$, we have

$$Y(Y(u, x_0)v, x_2)w = Y(e^{x_0 \mathcal{D}} Y(v, -x_0)u, x_2)w$$
$$= Y(Y(v, -x_0)u, x_2 + x_0)w. \tag{3.7.1}$$

Thus from the general delta function properties (2.3.17) and (2.3.56), the Jacobi identity for (u, v, w) gives

$$(-x_0)^{-1}\delta\left(\frac{x_2 - x_1}{-x_0}\right) Y(v, x_2)Y(u, x_1)w$$
$$-(-x_0)^{-1}\delta\left(\frac{x_1 - x_2}{-(-x_0)}\right) Y(u, x_1)Y(v, x_2)w$$
$$= x_1^{-1}\delta\left(\frac{x_2 - (-x_0)}{x_1}\right) Y(Y(v, -x_0)u, x_1)w, \tag{3.7.2}$$

which is the Jacobi identity for (v, u, w) (with (x_0, x_1, x_2) replaced by $(-x_0, x_2, x_1)$). (Cf. proof of Proposition 3.1.19.)

Next, multiplying both sides of the Jacobi identity for (u, v, w) by $e^{-x_2 \mathcal{D}}$ and using skew symmetry for the pairs (v, w), $(v, Y(u, x_1)w)$ and $(Y(u, x_0)v, w)$ and the form (3.1.35) of the $\mathcal{D}$-bracket-derivative formula (3.1.32) for the vector u, we obtain

$$x_0^{-1}\delta\left(\frac{x_1 - x_2}{x_0}\right) Y(u, x_1 - x_2)Y(w, -x_2)v$$
$$-x_0^{-1}\delta\left(\frac{x_2 - x_1}{-x_0}\right) Y(Y(u, x_1)w, -x_2)v$$
$$= x_2^{-1}\delta\left(\frac{x_1 - x_0}{x_2}\right) Y(w, -x_2)Y(u, x_0)v. \tag{3.7.3}$$

Then (2.3.17) and (2.3.56) give

$$x_1^{-1}\delta\left(\frac{x_0 + x_2}{x_1}\right) Y(u, x_0)Y(w, -x_2)v$$
$$+x_2^{-1}\delta\left(\frac{x_0 - x_1}{-x_2}\right) Y(Y(u, x_1)w, -x_2)v$$
$$= x_1^{-1}\delta\left(\frac{-x_2 - x_0}{-x_1}\right) Y(w, -x_2)Y(u, x_0)v, \tag{3.7.4}$$

that is,

$$x_1^{-1}\delta\left(\frac{x_0-(-x_2)}{x_1}\right)Y(u,x_0)Y(w,-x_2)v$$

$$-x_1^{-1}\delta\left(\frac{(-x_2)-x_0}{-x_1}\right)Y(w,-x_2)Y(u,x_0)v$$

$$=(-x_2)^{-1}\delta\left(\frac{x_0-x_1}{-x_2}\right)Y(Y(u,x_1)w,-x_2)v, \tag{3.7.5}$$

the Jacobi identity for (u,w,v) (and $(x_1,x_0,-x_2)$ in place of (x_0,x_1,x_2)). We conclude:

Proposition 3.7.1 *Assume the axioms for a vertex algebra except for the Jacobi identity, and in addition assume skew symmetry (3.1.30), the $\mathcal{D}$-derivative property (3.1.25) and $\mathcal{D}$-bracket-derivative property (3.1.32). Then the Jacobi identity for an ordered triple implies the identity for any permutation of this triple.* $\square$

Remark 3.7.2 The proof of Proposition 3.1.21 shows that if we assume the axioms for a vertex algebra except for the Jacobi identity and if we also assume skew symmetry (3.1.30), then the $\mathcal{D}$-derivative property (3.1.25) is in fact equivalent to the $\mathcal{D}$-bracket-derivative property (3.1.32). Thus in Proposition 3.7.1 we can in fact remove a hypothesis.

Remark 3.7.3 The second part of the argument proving Proposition 3.7.1 also shows that under the assumptions of skew symmetry and the $\mathcal{D}$-bracket-derivative formula (or equivalently, the $\mathcal{D}$-derivative formula), the commutator formula (3.1.8) is equivalent to the iterate formula (3.1.11) (or equivalently, the associator formula (3.1.13)); to see this, simply take Res_{x_0}. This observation should of course be compared with earlier arguments showing that commutativity and associativity are equivalent in various forms.

3.8 The iterate formula and normal-ordered products

We already know that the commutator of vertex operators $Y(u,x_1)$ and $Y(v,x_2)$ depends only on the singular part, with respect to the variable x_0, of $Y(Y(u,x_0)v,x_2)$; recall the commutator formula (3.1.8) and Remark 3.1.11. It is natural to ask: What is the significance of the regular part, with respect to x_0, of $Y(Y(u,x_0)v,x_2)$? The answer, which is quite straightforward, involves, and in fact motivates, a general, abstract notion of "normal ordering," as we now explain.

The iterate formula (3.1.11) and its variant (3.1.14) express $Y(Y(u,x_0)v,x_2)$ for $u,v\in V$ in terms of the products of the vertex operators $Y(u,x_1)$ and $Y(v,x_2)$ in both orders. We shall obtain canonical expressions for both the regular part and the singular part (with respect to x_0) of $Y(Y(u,x_0)v,x_2)$, in terms of $Y(v,x_2)$ and the regular and singular parts of $Y(u,x_1)$. By reading off components, we shall in particular have useful formulas for $Y(u_nv,x)$.

For $u\in V$, we write

$$Y(u,x)=Y^+(u,x)+Y^-(u,x) \tag{3.8.1}$$

where $Y^+(u, x) = \sum_{n<0} u_n x^{-n-1}$ and $Y^-(u, x) = \sum_{n\geq 0} u_n x^{-n-1}$ are the regular and singular parts of $Y(u, x)$, respectively. We shall compute $Y(Y^\pm(u, x_0)v, x_2)$.

Using obvious notation for the regular and singular parts with respect to particular variables, we observe the following simple facts:

$$x_0-\mathrm{reg}(Y(u, x_0 + x_2)) = Y^+(u, x_0 + x_2) = Y^+(u, x_2 + x_0),$$
$$x_0-\mathrm{sing}(Y(u, x_0 + x_2)) = Y^-(u, x_0 + x_2),$$
$$x_0-\mathrm{reg}(Y(u, x_2 + x_0)) = Y(u, x_2 + x_0),$$
$$x_0-\mathrm{sing}(Y(u, x_2 + x_0)) = 0 \tag{3.8.2}$$

(we are of course using the binomial expansion convention as always).

Now we examine the formula (3.1.14). Extracting the regular part with respect to x_0 we immediately obtain:

$$
\begin{aligned}
Y&(Y^+(u, x_0)v, x_2) \\
&= Y^+(u, x_2 + x_0)Y(v, x_2) + Y(v, x_2)Y^-(u, x_2 + x_0) \\
&= {}^\circ_\circ Y(u, x_2 + x_0)Y(v, x_2){}^\circ_\circ,
\end{aligned}
\tag{3.8.3}
$$

where we define the *normal-ordering operation* (or *normal-ordered product*) ${}^\circ_\circ \,\cdot\, {}^\circ_\circ$ by

$$
{}^\circ_\circ u_m v_n {}^\circ_\circ = \begin{cases} u_m v_n & \text{if } m < 0 \\ v_n u_m & \text{if } m \geq 0 \end{cases}
\tag{3.8.4}
$$

and we extend this definition to vertex operators correspondingly:

$$
{}^\circ_\circ Y(u, x_1)Y(v, x_2){}^\circ_\circ = Y^+(u, x_1)Y(v, x_2) + Y(v, x_2)Y^-(u, x_1).
\tag{3.8.5}
$$

Remark 3.8.1 It is important to note that the normal-ordered product ${}^\circ_\circ u_m v_n {}^\circ_\circ$ depends not only on the *product* $u_m v_n$ but in fact on the individual vectors u and v and also on the indices m and n; it is actually a function of four variables. Likewise, the normal-ordered product ${}^\circ_\circ Y(u, x_1)Y(v, x_2){}^\circ_\circ$ depends on u and v (and on x_1 and x_2), and not just on the product $Y(u, x_1)Y(v, x_2)$. To this extent, the notations (3.8.4) and (3.8.5) are misleading, but we will use them anyway.

Note that we *are* allowed to set $x_1 = x_2$ in this normal-ordered product (in contrast with the situation for an ordinary product of vertex operators):

$$
{}^\circ_\circ Y(u, x)Y(v, x){}^\circ_\circ = Y^+(u, x)Y(v, x) + Y(v, x)Y^-(u, x).
\tag{3.8.6}
$$

(When this expression is applied to a vector, the first term is a product of two lower-truncated formal Laurent series and the second is a product of an unrestricted formal Laurent series and a Laurent polynomial.)

Note also that this normal-ordered product is *not in general commutative*, that is, ${}^\circ_\circ u_m v_n {}^\circ_\circ \neq {}^\circ_\circ v_n u_m {}^\circ_\circ$ in general.

Remark 3.8.2 This notion of normal-ordered product canonically generalizes certain (commutative) "normal-ordered products" of certain kinds of operators in quantum field theory and conformal field theory. See for example [FLM6], and Chapter 6, for discussions and applications of normal-ordering operations for operators associated with Heisenberg algebras and with vertex operator algebras based on lattices; in these special cases, normal ordering is based on "annihilation and creation operators." Normal-ordered products are typically summed over infinite families of operators (as in the present very general vertex operator context), and one important purpose of the normal-ordering operation is to eliminate "infinities" (again, as in the present situation). We have just seen that this general normal-ordering procedure is actually dictated by natural considerations, not merely the elimination of infinities.

Extracting instead the *singular* part with respect to x_0 of (3.1.14) and again using (3.8.2), we also find that

$$Y(Y^-(u, x_0)v, x_2)$$
$$= Y^-(u, x_0 + x_2)Y(v, x_2) - Y(v, x_2)Y^-(u, x_0 + x_2)$$
$$= [Y^-(u, x_0 + x_2), Y(v, x_2)]. \tag{3.8.7}$$

Formula (3.8.3) can be combined with the formal Taylor Theorem and the $\mathcal{D}$-derivative property to give

$$Y(u_{-m}v, x) = \frac{1}{(m-1)!} {}^{\circ}_{\circ} \left(\left(\frac{d}{dx} \right)^{m-1} Y(u, x) \right) Y(v, x) {}^{\circ}_{\circ} \tag{3.8.8}$$

$$= \frac{1}{(m-1)!} {}^{\circ}_{\circ} Y(\mathcal{D}^{m-1}u, x) Y(v, x) {}^{\circ}_{\circ} \tag{3.8.9}$$

for all $m \geq 1$. The right-hand side of (3.8.8) is defined by the obvious slight generalization of (3.8.5) or (3.8.6). Note that the regular part of $\left(\frac{d}{dx} \right)^{m-1} Y(u, x)$ agrees with $\left(\frac{d}{dx} \right)^{m-1} Y^+(u, x)$.

The important special case $m = 1$ expresses the normal-ordered product of $Y(u, x)$ and $Y(v, x)$ as $Y(u_{-1}v, x)$:

$$Y(u_{-1}v, x) = {}^{\circ}_{\circ} Y(u, x) Y(v, x) {}^{\circ}_{\circ}; \tag{3.8.10}$$

this case can of course be proved by a much shorter version of the general argument (we extract the constant term in x_0).

Note that from (3.8.9) and (3.8.10), we have

$$Y(u_{-m}v, x) = \frac{1}{(m-1)!} Y((\mathcal{D}^{m-1}u)_{-1}v, x) \tag{3.8.11}$$

for $m \geq 1$.

In case $[Y(u, x_1), Y(v, x_2)] = 0$, which is equivalent to the condition that $u_n v = 0$ for $n \geq 0$ in view of Remark 3.2.4, we can drop the normal ordering and we have

$$Y(u_{-m}v, x) = \frac{1}{(m-1)!} \left(\left(\frac{d}{dx} \right)^{m-1} Y(u, x) \right) Y(v, x) \qquad (3.8.12)$$

for all $m \geq 1$ and in particular,

$$Y(u_{-1}v, x) = Y(u, x)Y(v, x); \qquad (3.8.13)$$

these expressions are indeed defined because they equal the corresponding normal-ordered expressions.

Remark 3.8.3 Notice that the case $m = 2$ and $v = 1$ of (3.8.12) is exactly the $\mathcal{D}$-derivative formula (3.1.25).

Likewise, formula (3.8.7) can be written in component form as follows: For $n \geq 0$,

$$Y(u_n v, x) = \mathrm{Res}_{x_1}(x_1 - x)^n[Y(u, x_1), Y(v, x)], \qquad (3.8.14)$$

as we see by multiplying (3.8.7) by $x_1^{-1}\delta((x_0 + x_2)/x_1)$ (which is in fact a permissible operation) and then extracting Res_{x_1} and the coefficient of x_0^{-n-1}, using (2.3.17) on the right-hand side; note that $Y^-(u, x_1)$ can indeed be replaced by $Y(u, x_1)$ in the last step of this argument. This relation should be compared with weak commutativity and the discussion in Remark 3.2.3. The case $n = 0$ gives

$$Y(u_0 v, x) = [u_0, Y(v, x)], \qquad (3.8.15)$$

the familiar formula obtained by applying $\mathrm{Res}_{x_0}\mathrm{Res}_{x_1}$ to the Jacobi identity.

Actually, formula (3.8.14) can easily be generalized to all $n \in \mathbb{Z}$, giving an expression somewhat different from (3.8.8) in the case $n < 0$, namely,

$$Y(u_n v, x) = \mathrm{Res}_{x_1}((x_1 - x)^n Y(u, x_1)Y(v, x) - (-x + x_1)^n Y(v, x)Y(u, x_1));$$
$$(3.8.16)$$

this follows immediately from the iterate formula (3.1.11). This formula will be very significant in Chapter 5.

Remark 3.8.4 Later we shall discuss a notion of "weak nilpotence," and for this we shall need normal ordered products of the type $\,{}^{\circ}_{\circ}Y(v, x)^r\,{}^{\circ}_{\circ}$. The normal ordered product $\,{}^{\circ}_{\circ}Y(v, x)^2\,{}^{\circ}_{\circ}$ has been defined. More generally, we define the normal ordered product $\,{}^{\circ}_{\circ}Y(v^{(1)}, x_1) \cdots Y(v^{(r)}, x_r)\,{}^{\circ}_{\circ}$ for $v^{(1)}, \ldots, v^{(r)} \in V$ with $r \geq 1$. Of course, we should define

$$\,{}^{\circ}_{\circ}Y(v, x)\,{}^{\circ}_{\circ} = Y(v, x). \qquad (3.8.17)$$

For $r \geq 2$, we recursively define

$$\,{}^{\circ}_{\circ}Y(v^{(1)}, x_1) \cdots Y(v^{(r)}, x_r)\,{}^{\circ}_{\circ} = \,{}^{\circ}_{\circ}Y(v^{(1)}, x_1)(\,{}^{\circ}_{\circ}Y(v^{(2)}, x_2) \cdots Y(v^{(r)}, x_r)\,{}^{\circ}_{\circ})\,{}^{\circ}_{\circ},$$
$$(3.8.18)$$

where it is understood that we use the same prescription as in (3.8.4)–(3.8.5). Different candidate definitions of the normal ordered products of more than two vertex operators might give different objects, but this is the natural one, as we shall see in Proposition 3.10.2.

3.9 Further elementary notions

In this section we shall continue discussing basic notions that are natural analogues of those in classical associative or Lie algebra theory but at the same time often have non-classical features, as usual. Specifically, we shall introduce the notions of homomorphism, subalgebra, subalgebra generated by a subset and ideal.

First we discuss the simple notion of homomorphism and related notions. Let $(V_1, Y, \mathbf{1})$ and $(V_2, Y, \mathbf{1})$ be vertex algebras. A *homomorphism* $f : V_1 \to V_2$ is defined to be a linear map such that

$$f(Y(u, x)v) = Y(f(u), x)f(v) \quad \text{for } u, v \in V_1, \tag{3.9.1}$$

or equivalently,

$$f(u_n v) = f(u)_n f(v) \quad \text{for } u, v \in V_1, \ n \in \mathbb{Z}, \tag{3.9.2}$$

and such that

$$f(\mathbf{1}) = \mathbf{1}. \tag{3.9.3}$$

(We are of course extending f canonically to a map on $V_1[[x, x^{-1}]]$.)

If V_1 and V_2 are vertex operator algebras, a homomorphism f from V_1 to V_2 is required in addition to satisfy the condition

$$f(\omega^1) = \omega^2, \tag{3.9.4}$$

where ω^1 and ω^2 are the conformal vectors for V_1 and V_2, respectively. Given such a homomorphism,

$$f L^1(n) = L^2(n) f \quad \text{for } n \in \mathbb{Z} \tag{3.9.5}$$

(using obvious notation); so f is automatically grading-preserving (since $f L^1(0) = L^2(0) f$). Furthermore, since by (3.1.73) we have

$$L^i(2) L^i(-2)\mathbf{1} = \frac{1}{2} c_{V_i} \mathbf{1} \quad \text{for } i = 1, 2, \tag{3.9.6}$$

we see using (3.9.5) that V_1 and V_2 must have the same central charge. (This is actually true even if V_1 or V_2 is the zero algebra; recall Remark 3.1.26.)

The notions of isomorphism, endomorphism and automorphism are defined in the obvious ways.

Definition 3.9.1 A *vertex subalgebra* of a vertex algebra $(V, Y, \mathbf{1})$ is a vector subspace U of V such that $\mathbf{1} \in U$ and such that $(U, Y, \mathbf{1})$ is itself a vertex algebra.

This amounts to saying that $\mathbf{1} \in U$ and that $Y(u, x)v \in U((x))$ for $u, v \in U$, i.e., $u_n v \in U$ for $u, v \in U, n \in \mathbb{Z}$. The smallest subalgebra of V is $\mathbb{C}\mathbf{1}$.

Definition 3.9.2 Suppose that $(V, Y, \mathbf{1}, \omega)$ is a vertex operator algebra. A *vertex operator subalgebra* of V is a vector subspace U of V such that $\mathbf{1} \in U$ and $\omega \in U$ and such that $(U, Y, \mathbf{1}, \omega)$ is itself a vertex operator algebra. Equivalently, a vertex operator subalgebra is a vertex subalgebra that contains ω; such a substructure is automatically a graded subspace because it is stable under $L(0)$, whose eigenspaces give the grading of V.

It sometimes happens that a vertex subalgebra U of a vertex operator algebra V together with a *new* conformal vector ω', possibly different from the conformal vector of V, is a vertex operator algebra. In such a case, we abuse terminology and call U a *vertex operator subalgebra with a possibly different conformal vector*; if the context makes the situation clear, we simply say "vertex operator subalgebra." We shall study this concept in Section 3.11.

As in classical algebraic settings, the notion of "subalgebra generated by a subset" is very convenient. For a subset S of the vertex algebra V we define $\langle S \rangle$ to be the smallest vertex subalgebra containing S and we call $\langle S \rangle$ the *vertex subalgebra generated by S*. Then $\langle S \rangle$ is the intersection of all vertex subalgebras of V containing S.

For the example of the null set, we have $\langle \emptyset \rangle = \mathbb{C}\mathbf{1}$.

To construct $\langle S \rangle$ in general, we in principle need to use products and iterates of vertex operators repeatedly, starting from elements of S. This process is greatly simplified using the following result:

Proposition 3.9.3 *For a subset S of the vertex algebra V,*

$$\langle S \rangle = \mathrm{span}\left\{ u_{n_1}^{(1)} \cdots u_{n_r}^{(r)} \mathbf{1} \mid r \in \mathbb{N}, \; u^{(1)}, \ldots, u^{(r)} \in S, \; n_1, \ldots, n_r \in \mathbb{Z} \right\}. \quad (3.9.7)$$

Proof. Let U be the subspace on the right-hand side of (3.9.7). Clearly, $S \subset U$ and any vertex subalgebra that contains S contains U, so it suffices to prove that U is a vertex subalgebra of V. Since $\mathbf{1} \in U$, it remains to prove that $Y(a, x)b \in U((x))$ for $a, b \in U$. Set

$$K = \{a \in U \mid Y(a, x)U \subset U((x))\}. \quad (3.9.8)$$

We need to show that $K = U$. From the definitions, $\{\mathbf{1}\} \cup S \subset K$. Let $a \in K, u \in S$. Then it follows from (3.1.11) with $v = a$ and the assumption on a that $Y(Y(u, x_0)a, x_2)U \subset U[[x_0, x_0^{-1}, x_2, x_2^{-1}]]$, so that $Y(u, x_0)a$, which lies in $U((x_0))$, also lies in $K((x_0))$. By induction, $U \subset K$ and so $K = U$. $\quad\square$

Remark 3.9.4 This is analogous to the statement that for a subset S of a given Lie algebra, the Lie subalgebra generated by S is the span of the products of the operators $\mathrm{ad}(s_i)$ $(s_i \in S)$ applied to elements of S.

Definition 3.9.5 Let S be a subset of a vertex operator algebra $(V, Y, \mathbf{1}, \omega)$. The *vertex operator subalgebra generated by S* is the smallest vertex operator subalgebra containing S. It is just the vertex subalgebra $\langle S \cup \{\omega\} \rangle$.

Remark 3.9.6 The smallest vertex operator subalgebra of a vertex operator algebra V is the vertex subalgebra $\langle \omega \rangle$, which by Proposition 3.9.3 is just the submodule of V for the Virasoro algebra generated by $\mathbf{1}$:

$$\text{span} \left\{ L(n_1) \cdots L(n_r)\mathbf{1} \mid n_j \in \mathbb{Z} \right\} \tag{3.9.9}$$

(recall Remark 3.1.23 and also that $\omega = L(-2)\mathbf{1}$). Notice that since $L(-1)\mathbf{1} = 0$ (recall (3.1.65)), there is always a certain "degeneracy" in this submodule.

For a subset S of the vertex algebra V, if $\langle S \rangle = V$ we say that S *generates* V. Minimal generating subspaces or subsets of certain types are of interest, and later we shall study important families of examples and in fact construct vertex algebras from objects that will turn out to be generating subsets. In Lie algebra theory it is sometimes interesting to study minimal generating sets for a Lie algebra, but it is a general and fundamental fact that any Lie algebra $\mathfrak{g}$ generates its own universal enveloping (associative) algebra $U(\mathfrak{g})$ (one knows that $\mathfrak{g} \subset U(\mathfrak{g})$ by the Poincaré–Birkhoff–Witt Theorem). The Lie algebra as a generating subspace of $U(\mathfrak{g})$ certainly plays an important role even though it is usually not a minimal generating subspace. Generating spaces of an analogous type for vertex operator algebras should play an important role; for a general study of such matters, we refer interested readers to [KLi].

In classical algebraic theories, the notion of ideal is very important. We define such a notion for vertex algebras.

Definition 3.9.7 An *ideal* of the vertex algebra V is a subspace I such that for all $v \in V$ and $w \in I$,

$$Y(v, x)w \in I((x)), \quad Y(w, x)v \in I((x)), \tag{3.9.10}$$

i.e., $v_n w \in I$ and $w_n v \in I$ for all $v \in V$, $w \in I$ and $n \in \mathbb{Z}$.

Clearly, 0 and V are ideals of V. If $V \neq 0$ and V is the only nonzero ideal we say that V is *simple*. If I is an ideal, we have $\mathcal{D}(w) = w_{-2}\mathbf{1} \in I$ for $w \in I$, so that $\mathcal{D}(I) \subset I$.

Remark 3.9.8 The notion of ideal resembles the notion of (two-sided) ideal for an associative algebra (or for a commutative associative algebra) as well as the notion of ideal for a Lie algebra. In view of skew symmetry (3.1.30), under the condition that $\mathcal{D}I \subset I$ (which is necessary), the *left-ideal condition*, $Y(v, x)w \in I((x))$ for $v \in V$, $w \in I$, is equivalent to the *right-ideal condition*: $Y(w, x)v \in I$ for $v \in V$, $w \in I$. Also, if V is a vertex operator algebra, then either the left-ideal condition or the right-ideal condition implies that $\mathcal{D}I \subset I$ because $\mathcal{D} = L(-1) = \omega_0$ and $L(-1)v = v_{-2}\mathbf{1}$. Thus the left-ideal condition is equivalent to the right-ideal condition for a vertex operator algebra.

Remark 3.9.9 Let I be an ideal of the vertex algebra V. Then just as in any classical algebraic theory, we have a natural quotient vertex algebra V/I together with the canonical homomorphism $V \to V/I$, where $\mathbf{1} + I$ is the vacuum vector and

$$(u + I)_n(v + I) = u_n v + I \quad \text{for } u, v \in V, \ n \in \mathbb{Z}. \tag{3.9.11}$$

(The ideal I is certainly allowed to equal V; recall that 0 is a vertex algebra.) We of course have standard facts. For example, if we have a homomorphism between vertex algebras, then its kernel is an ideal and the induced quotient map is an isomorphism onto a vertex subalgebra.

Remark 3.9.10 Suppose that V is a vertex operator algebra and I is an ideal. Since $L(0) = \omega_1$, we have $L(0)I \subset I$, so that I is a graded subspace of V and V/I is in fact a vertex operator algebra of the same central charge as V; we have the canonical map $V \to V/I$ of vertex operator algebras. (Again, I can equal V; recall Remark 3.1.26.)

Remark 3.9.11 Let V be a vertex operator algebra having the special property $V_{(0)} = \mathbb{C}1 \neq 0$. Any ideal is graded and the weight-zero subspace of any proper ideal must be zero since any ideal containing 1 must equal V. Thus the sum I of all proper ideals of V is the (unique) largest proper ideal, so that V/I is a simple vertex operator algebra.

3.10 Weak nilpotence and nilpotence

In classical associative algebra theory, the notion of nilpotent element plays an important role. Here we shall introduce the notions of weakly nilpotent element and nilpotent element and their basic properties.

Let V be a vertex algebra. We would like to consider the powers of a vertex operator, which, as we know, is the analogue of a left multiplication operator. But recall that for $v \in V$ and a positive integer r, $Y(v, x)^r$ does not in general exist and that a remedy for this nonexistence is to consider the normal-ordered product ${}^{\circ}_{\circ} Y(v, x)^r {}^{\circ}_{\circ}$ (see Remark 3.8.4). Now we define a notion of what we call "weakly nilpotent element," the first of two natural notions of nilpotence:

Definition 3.10.1 An element v of a vertex algebra is said to be *weakly nilpotent* if ${}^{\circ}_{\circ} Y(v, x)^r {}^{\circ}_{\circ} = 0$ for some positive integer r.

The following simple result [DL3] relates the weak nilpotency of an element v to powers of the operator v_{-1}.

Proposition 3.10.2 *Let $v \in V$. Then for $r > 0$,*

$$Y((v_{-1})^r \mathbf{1}, x) = {}^{\circ}_{\circ} Y(v, x)^r {}^{\circ}_{\circ}. \tag{3.10.1}$$

In particular, v is weakly nilpotent if and only if $(v_{-1})^r \mathbf{1} = 0$ for some $r > 0$.

Proof. Recall from Remark 3.8.4 that ${}^{\circ}_{\circ} Y(v, x)^{r+1} {}^{\circ}_{\circ} = {}^{\circ}_{\circ} Y(v, x)({}^{\circ}_{\circ} Y(v, x)^r {}^{\circ}_{\circ}) {}^{\circ}_{\circ}$ for $r > 0$. Then (3.10.1) follows immediately from (3.8.10), induction and the fact that $v_{-1}\mathbf{1} = v$. $\square$

Remark 3.10.3 We use the terminology "weakly nilpotent" rather than "nilpotent" because in commutative associative algebra theory, all nilpotent elements form an ideal (since if $a^r = 0 = b^r$, then $(a+b)^{2r-1} = 0$, and if $a^r = 0$, then $(ab)^r = a^r b^r = 0$), and in view of the analogy between the notion of vertex algebra and that of commutative associative algebra, "nilpotent elements" of a vertex algebra should also form an ideal; hence simple vertex algebras should not have nonzero "nilpotent elements." But many simple vertex operator algebras contain nonzero weakly nilpotent elements, as we shall see later, in particular, in Section 6.6.

Just as notions of nilpotent element play important roles in the structure and representation theory of Lie algebras and associative algebras, the notion of weakly nilpotent element and Proposition 3.10.2 will be important in vertex operator algebra theory, as we shall discuss in Section 6.6 (cf. [DL3], [Li3]). In applications, we often consider weakly nilpotent elements v with the property that $[Y(v, x_1), Y(v, x_2)] = 0$, or equivalently, $v_n v = 0$ for $n \geq 0$ (in view of Remark 3.2.4). The following result follows immediately from Proposition 3.10.2, or from (3.8.13) and induction:

Corollary 3.10.4 *Let v be an element of a vertex algebra such that $v_n v = 0$ for $n \geq 0$. Then*

$$Y((v_{-1})^r \mathbf{1}, x) = Y(v, x)^r \quad \text{for } r > 0. \quad \Box \tag{3.10.2}$$

On the other hand, the product $Y(v, x_1) \cdots Y(v, x_r)$ always exists (for mutually commuting independent formal variables $x_1, \ldots, x_r$). Using this we define the notion of "nilpotent element":

Definition 3.10.5 Let V be a vertex algebra. An element v of V is said to be *nilpotent* if there exists a positive integer r such that

$$Y(v, x_1) \cdots Y(v, x_r) = 0. \tag{3.10.3}$$

Clearly, nilpotence implies weak nilpotence. In terms of components, nilpotence amounts to saying that

$$v_{n_1} \cdots v_{n_r} = 0 \tag{3.10.4}$$

for arbitrary $n_1, \ldots, n_r \in \mathbb{Z}$. In contrast with the situation for weakly nilpotent elements, we have the following result, whose proof provides a good illustration of the subtleties of formal calculus:

Proposition 3.10.6 *Let V be a vertex algebra. Then the nilpotent elements of V form an ideal.*

Proof. The two parts of this proof are directly motivated by the two parts of the trivial proof that the nilpotent elements in a commutative associative algebra form an ideal (recall Remark 3.10.3).

We first prove that the nilpotent elements form a subspace of V. If u is nilpotent, it is clear that any multiple of u is nilpotent. If u and v are nilpotent, there exists a positive integer r such that

$$Y(u, x_1) \cdots Y(u, x_r) = 0 = Y(v, x_1) \cdots Y(v, x_r). \qquad (3.10.5)$$

It follows from weak commutativity (recall Proposition 3.2.1) that there exists $k \geq 0$ such that

$$(x_i - x_j)^k Y(u, x_i) Y(v, x_j) = (x_i - x_j)^k Y(v, x_j) Y(u, x_i) \qquad (3.10.6)$$

for $1 \leq i, j \leq r$ with $i \neq j$. Then (using formal variables $y_1, \ldots, y_{2r-1}$)

$$\left(\prod_{1 \leq p < q \leq 2r-1} (y_p - y_q)^k \right) Y(u + v, y_1) \cdots Y(u + v, y_{2r-1}) = 0 \qquad (3.10.7)$$

because by (3.10.6) the left-hand side is the sum of 2^{2r-1} terms of the form

$$\left(\prod_{1 \leq p < q \leq 2r-1} (y_p - y_q)^k \right) Y(u, y_{i_1}) \cdots Y(u, y_{i_t}) Y(v, y_{j_1}) \cdots Y(v, y_{j_{2r-1-i_t}}),$$

$$(3.10.8)$$

where $i_1 < \cdots < i_t$, $j_1 < \cdots < j_{2r-1-i_t}$ and

$$\{i_1, \ldots, i_t\} \cup \{j_1, \ldots, j_{2r-1-i_t}\} = \{1, 2, \ldots, 2r - 1\};$$

since either t or $2r - 1 - t \geq r$, the expression (3.10.8) is zero. It is permissible to multiply (3.10.7) by

$$\prod_{1 \leq p < q \leq 2r-1} (y_p - y_q)^{-k}$$

(to see this, first "peel off" the variable y_{2r-1}, in the sense of Remark 2.3.20, then the variable y_{2r-2}, and so on) and it follows that $u + v$ is nilpotent. Thus the nilpotent elements form a subspace.

Let *either u or v* be a nilpotent element of V, and choose $r > 0$ such that either $Y(u, x_1) \cdots Y(u, x_r) = 0$ or $Y(v, x_1) \cdots Y(v, x_r) = 0$. Then we have (using variables $x_{0i}, x_{1i}, x_{2i}, i = 1, \ldots, r$)

$$x_{2i}^{-1} \delta \left(\frac{x_{1i} - x_{0i}}{x_{2i}} \right) Y(Y(u, x_{0i})v, x_{2i})$$

$$= x_{0i}^{-1} \delta \left(\frac{x_{1i} - x_{2i}}{x_{0i}} \right) Y(u, x_{1i}) Y(v, x_{2i})$$

$$- x_{0i}^{-1} \delta \left(\frac{x_{2i} - x_{1i}}{-x_{0i}} \right) Y(v, x_{2i}) Y(u, x_{1i}). \qquad (3.10.9)$$

Let $k \geq 0$ be such that the weak commutativity relation (3.2.1) holds for (u, v). Then

$$x_{2i}^{-1}\delta\left(\frac{x_{1i} - x_{0i}}{x_{2i}}\right) x_{0i}^{k} Y(Y(u, x_{0i})v, x_{2i})$$

$$= x_{2i}^{-1}\delta\left(\frac{x_{1i} - x_{0i}}{x_{2i}}\right) \left[(x_{1i} - x_{2i})^{k} Y(u, x_{1i})Y(v, x_{2i})\right], \qquad (3.10.10)$$

where of course the brackets on the right-hand side are necessary. Now we can use a device similar to the one used above and find that

$$\left(\prod_{1 \leq p < q \leq r} (x_{2p} - x_{1q})^{k}\right) \prod_{i=1}^{r} x_{2i}^{-1}\delta\left(\frac{x_{1i} - x_{0i}}{x_{2i}}\right) x_{0i}^{k} Y(Y(u, x_{0i})v, x_{2i}) = 0$$

$$(3.10.11)$$

(ordered product of operators); note that this holds when *either u or v* is assumed nilpotent. Next we need to justify the cancellation of $\prod(x_{2p} - x_{1q})^{k}$ from this equation. For this we apply (2.3.17) to $x_{2i}^{-1}\delta((x_{1i}-x_{0i})/x_{2i})$ and then use delta function substitution to rewrite (3.10.11) as

$$\left(\prod_{1 \leq p < q \leq r} (x_{2p} - x_{2q} - x_{0q})^{k}\right) \prod_{i=1}^{r} x_{1i}^{-1}\delta\left(\frac{x_{2i} + x_{0i}}{x_{1i}}\right) x_{0i}^{k} Y(Y(u, x_{0i})v, x_{2i}) = 0.$$

$$(3.10.12)$$

By multiplying by $\prod_{i=1}^{r} x_{0i}^{-k}$ and then performing the operation $\prod_{i=1}^{r} \mathrm{Res}_{x_{1i}}$ we now get

$$\left(\prod_{1 \leq p < q \leq r} (x_{2p} - x_{2q} - x_{0q})^{k}\right) \prod_{i=1}^{r} Y(Y(u, x_{0i})v, x_{2i}) = 0, \qquad (3.10.13)$$

and we find by our usual analysis that we can multiply this by $\prod(x_{2p} - x_{2q} - x_{0q})^{-k}$ to get

$$\prod_{i=1}^{r} Y(Y(u, x_{0i})v, x_{2i}) = 0. \qquad (3.10.14)$$

Consequently, for any $n \in \mathbb{Z}$, $u_n v$ is nilpotent; we even have more generally that

$$Y(u_{n_1}v, x_1) \cdots Y(u_{n_r}v, x_r) = 0 \qquad (3.10.15)$$

for any $n_i \in \mathbb{Z}$, $i = 1, \ldots, r$. Thus the nilpotent elements form an ideal of V. $\square$

3.11 Centralizers and the center

In this section we shall introduce the notions of centralizer and center, and we shall present a theorem of I. Frenkel–Zhu [FZ] concerning the centralizers of a pair of vertex operator subalgebras with possibly different conformal vectors from that of the large vertex operator algebra. This theorem generalizes the "coset construction" [GKO1], [GKO2] for certain vertex operator algebras, as we shall explain in Remark 6.6.24 below.

Let V be a vertex algebra. Given a subset S of V, we define

$$C_V(S) = \{v \in V \mid [Y(v, x_1), Y(s, x_2)] = 0 \quad \text{for all } s \in S\} \qquad (3.11.1)$$

and we call $C_V(S)$ the *centralizer* (or *commutant*, as in [FZ]) of S in V. In the case $S = V$ we call $C_V(V)$ the *center* of V and write

$$C(V) = C_V(V). \qquad (3.11.2)$$

In view of Remark 3.2.4,

$$C_V(S) = \{v \in V \mid v_n s = 0 \quad \text{for all } s \in S, \ n \geq 0\} \qquad (3.11.3)$$
$$= \{v \in V \mid s_n v = 0 \quad \text{for all } s \in S, \ n \geq 0\} \qquad (3.11.4)$$

(cf. comments preceding Corollary 3.10.4). Clearly, $\mathbf{1} \in C_V(S)$. Furthermore, for $u, v \in C_V(S)$, $s \in S$ and $n \in \mathbb{Z}$,

$$s_i(u_n v) = u_n s_i v = 0$$

for all $i \geq 0$, so that $u_n v \in C_V(S)$. Thus $C_V(S)$ is a vertex subalgebra. As in classical theories, we have $C_V(S) = C_V(\langle S \rangle)$, which simply follows from the fact that for $u \in C_V(S)$, we have that $C_V(\{u\})$, being a subalgebra, contains $\langle S \rangle$ (recall the notation $\langle S \rangle$ from Section 3.9). Clearly, $S \subset C_V(C_V(S))$. In the following we consider the case when $S = U$ is a vertex operator subalgebra of a vertex operator algebra V.

Remark 3.11.1 We are about to consider the kernel Ker $L(-1)$ of the operator $L(-1)$, for a vertex operator algebra V. By the $L(-1)$-derivative property (3.1.45) and the injectivity of the vertex operator map $Y(\cdot, x)$, we see that for $v \in V$, $L(-1)v = 0$ if and only if $Y(v, x)$ involves no nonzero powers of x, that is, $Y(v, x) = v_{-1}$. By considering centralizers, we shall see next that in fact Ker $L(-1)$ always lies in Ker $L(0)$, that is, Ker $L(-1) \subset V_{(0)}$. Moreover, Ker $L(-1)$ coincides with the centralizer of *any* vertex operator subalgebra of V and is a commutative associative algebra in a natural way.

Proposition 3.11.2 *Let V be a vertex operator algebra and let U be any vertex operator subalgebra (with the same conformal vector), for example, $U = \langle \omega \rangle$. Then the vertex subalgebra $C_V(U)$ is independent of U and*

$$C_V(U) = \text{Ker } L(-1) \subset V_{(0)} \ (= \text{Ker } L(0)). \qquad (3.11.5)$$

In particular,

$$C(V) = C_V(U) = C_V(\langle \omega \rangle) = \text{Ker } L(-1) \subset V_{(0)}. \qquad (3.11.6)$$

Furthermore, $Y(v, x) = v_{-1}$ for $v \in C(V)$ (that is, $Y(v, x)$ involves no nonzero powers of x) and $C(V)$ is a commutative associative algebra with $\mathbf{1}$ as its identity element, under the multiplication $u \cdot v = u_{-1}v \; (= Y(u, x)v)$ for $u, v \in C(V)$.

Proof. If $v \in C_V(U)$, for $n \geq -1$ we have (recall (3.11.4))

$$L(n)v = \omega_{n+1}v = 0.$$

In particular, $L(-1)v = 0 = L(0)v$, so that $v \in \text{Ker } L(-1)$ and $v \in V_{(0)}$. Conversely, if $L(-1)v = 0$, then $\frac{d}{dx}Y(v, x) = Y(L(-1)v, x) = 0$, and so $Y(v, x) = v_{-1}$, that is, $v_n = 0$ for $n \neq -1$. By Remark 3.2.4, $v \in C(V) \subset C_V(U)$. Thus $C_V(U) = \text{Ker } L(-1) \subset V_{(0)}$. Clearly, the vacuum vector $\mathbf{1}$ is the (two-sided) identity for the nonassociative algebra $C(V)$ with multiplication defined by $u \cdot v = u_{-1}v = Y(u, x)v$. Since the left-multiplication operators for $C(V)$ commute with one another, $C(V)$ is a commutative associative algebra (recall the comments at the beginning of Section 3.5). $\qquad \square$

Remark 3.11.3 It follows from Proposition 3.11.2 that $C_V(C_V(U)) = C_V(C(V)) = V$. In view of this, the double centralizer $C_V(C_V(U))$ is strictly larger than U if $U \neq V$.

The center of a simple vertex operator algebra is one-dimensional:

Proposition 3.11.4 *Suppose that V is a simple vertex operator algebra and that U is any vertex operator subalgebra (with the same conformal vector). Then*

$$C_V(U) = \mathbb{C}\mathbf{1}. \qquad (3.11.7)$$

In particular,

$$\text{Ker } L(-1) = C(V) = C_V(\langle \omega \rangle) = \mathbb{C}\mathbf{1}. \qquad (3.11.8)$$

Proof. In view of Proposition 3.11.2, it suffices to prove the Proposition for $U = V$. Let $v \in C(V)$. We shall show that the operator v_{-1}, which is the constant term of $Y(v, x)$, is a scalar multiplication operator. Since $[v_{-1}, L(0)] = [v_{-1}, \omega_1] = 0$, $v_{-1}V_{(n)} \subset V_{(n)}$ for $n \in \mathbb{Z}$. Let λ be an eigenvalue of v_{-1} on $V_{(0)}$. (We are using the finite dimensionality of $V_{(0)}$ and the fact that our field is $\mathbb{C}$.) Since $[v_{-1}, w_m] = 0$ for $w \in V$ and $m \in \mathbb{Z}$, we find that $\text{Ker}(v_{-1} - \lambda)$ is an ideal of V (recall Remark 3.9.8). Since V is simple and $\text{Ker}(v_{-1} - \lambda) \neq 0$, we have $\text{Ker}(v_{-1} - \lambda) = V$, that is, v_{-1} acts as λ on V. Thus $v = v_{-1}\mathbf{1} = \lambda\mathbf{1} \in \mathbb{C}\mathbf{1}$. $\quad \square$

Remark 3.11.5 This argument of course includes a variant of the classical justification for Schur's Lemma. In fact, the simplicity of V amounts to the irreducibility of the adjoint V-module (defined in Chapter 4) and a Schur's Lemma for any irreducible V-module is easily obtained (see Lemma 4.5.5 below), as in [FHL].

Now we consider a general situation where U is a vertex operator subalgebra of a vertex operator algebra V with a possibly different conformal vector (a notion mentioned in Section 3.9).

Definition 3.11.6 Let $(V, Y, \mathbf{1}, \omega)$ be a vertex operator algebra. A *vertex operator subalgebra with a possibly different conformal vector* is a vertex subalgebra $(U, Y, \mathbf{1})$ of V together with an element ω' of U such that $(U, Y, \mathbf{1}, \omega')$ is a vertex operator algebra (with a grading possibly different from that of V). By abuse of terminology we shall sometimes call $(U, Y, \mathbf{1}, \omega')$ a *vertex operator subalgebra* of V if the context makes the situation clear.

Remark 3.11.7 Let $(U, Y, \mathbf{1}, \omega')$ be a vertex operator subalgebra of a vertex operator algebra $(V, Y, \mathbf{1}, \omega)$, with a possibly different conformal vector. We shall write

$$Y(\omega', x) = \sum_{n \in \mathbb{Z}} L'(n) x^{-n-2}, \tag{3.11.9}$$

where we view the operators $L'(n)$ as acting on V, not just on U. Since $(U, Y, \mathbf{1}, \omega')$ is a vertex operator algebra, the operators $L'(n)$ acting on U give rise to a representation of the Virasoro algebra on U with the central element acting as the scalar c_U (the central charge of U) on U, and the $L'(-1)$-derivative and $L'(-1)$-bracket formulas hold on U. (In the case $U = V$ but $\omega' \neq \omega$, when we write "c_U" we shall mean the central charge of U with respect to ω' rather than ω.) We shall show next that all these conditions also hold when the $L'(n)$ are considered as operators on the whole space V. We start with the following result, expressed in a general form so that we can later apply it to modules as well as algebras:

Proposition 3.11.8 *Let V be a vertex operator algebra and let W be a vector space equipped with a linear map*

$$Y_W(\cdot, x) : V \to (\text{End } W)[[x, x^{-1}]] \tag{3.11.10}$$

satisfying the truncation condition $Y_W(v, x)w \in W((x))$ for $v \in V$, $w \in W$, the vacuum property $Y_W(\mathbf{1}, x) = 1$ and the Jacobi identity (3.4.4) for all $u, v \in V$ and $w \in W$. Then for $v \in V$,

$$[L(-1), Y_W(v, x)] = Y_W(L(-1)v, x) = \frac{d}{dx} Y_W(v, x), \tag{3.11.11}$$

where the first operator $L(-1)$ acts on W and the second operator $L(-1)$ acts on V.

Proof. From the proof of Proposition 3.1.18, we have $Y_W(\mathcal{D}v, x) = \frac{d}{dx} Y_W(v, x)$ for $v \in V$. (In view of Remark 3.8.3, this $\mathcal{D}$-derivative property also follows from (3.8.12).) Since $L(-1) = \mathcal{D}$ (acting on V) and $L(-1) = \omega_0$, the rest is clear (as in (3.1.52). $\quad\square$

As an immediate consequence of Proposition 3.11.8 we have:

Corollary 3.11.9 *Let $(U, Y, \mathbf{1}, \omega')$ be a vertex operator subalgebra of a vertex operator algebra $(V, Y, \mathbf{1}, \omega)$. Then for $u \in U$,*

$$[L'(-1), Y(u, x)] = Y(L'(-1)u, x) = \frac{d}{dx}Y(u, x) \qquad (3.11.12)$$

acting on V, where $Y(\omega', x) = \sum_{n \in \mathbb{Z}} L'(n)x^{-n-2}$. $\square$

In the setting of this Corollary, from the Virasoro algebra relations we have (cf. (3.1.70)–(3.1.74)):

$$\omega'_0\omega' = L'(-1)\omega' = \mathcal{D}(\omega') = L(-1)\omega', \qquad (3.11.13)$$

$$\omega'_1\omega' = 2\omega', \qquad (3.11.14)$$

$$\omega'_2\omega' = 0, \qquad (3.11.15)$$

$$\omega'_3\omega' = \frac{1}{2}c_U\mathbf{1}, \qquad (3.11.16)$$

$$\omega'_n\omega' = 0 \quad \text{for } n \geq 4. \qquad (3.11.17)$$

(Actually, Corollary 3.11.9 does not use the element ω of V or the operator $L(-1)$, but formula (3.11.13) does use $L(-1)$.) It follows from Corollary 3.11.9 and the argument at the end of Section 3.1 that the operators $L'(m)$ $(m \in \mathbb{Z})$ acting on the *whole space V* give rise to a representation of the Virasoro algebra with the central element acting as the scalar c_U. That is:

Corollary 3.11.10 *In the setting of Corollary 3.11.9, for $m, n \in \mathbb{Z}$ we have*

$$[L'(m), L'(n)] = (m - n)L'(m + n) + \frac{1}{12}(m^3 - m)\delta_{m+n,0}c_U \qquad (3.11.18)$$

acting on V. $\square$

The following result generalizes part of Proposition 3.11.2:

Corollary 3.11.11 *In the setting of Corollary 3.11.9,*

$$C_V(U) = \mathrm{Ker}_V L'(-1), \qquad (3.11.19)$$

the kernel of $L'(-1)$ acting on V (cf. (3.11.5)).

Proof. The inclusion $C_V(U) \subset \mathrm{Ker}_V L'(-1)$ being clear (since $L'(-1) = \omega'_0$), we need only prove the other inclusion. Let $v \in \mathrm{Ker}_V L'(-1)$ and let $u \in U$. We shall show that $Y(u, x)v$ involves no negative powers of x. Using Corollary 3.11.9 we get

$$e^{x_1 L'(-1)}Y(u, x)v = e^{x_1 L'(-1)}Y(u, x)e^{-x_1 L'(-1)}v = e^{x_1 \frac{d}{dx}}Y(u, x)v, \qquad (3.11.20)$$

so that for $n \geq 0$,

$$L'(-1)^n Y(u, x)v = \left(\frac{d}{dx}\right)^n Y(u, x)v. \qquad (3.11.21)$$

Since the powers of x on the left-hand side of (3.11.21) are universally truncated from below, independently of $n \geq 0$, we see from the right-hand side that $Y(u, x)v$ must contain only nonnegative powers of x. Thus $v \in C_V(U)$. $\quad\square$

The following result, due to I. Frenkel–Zhu [FZ], states that under suitable conditions the centralizer of a vertex operator subalgebra U of a vertex operator algebra V is also a vertex operator subalgebra.

Theorem 3.11.12 *Let* $(V, Y, \mathbf{1}, \omega)$ *be a nonzero vertex operator algebra such that* $V_{(n)} = 0$ *for* $n < 0$ *and* $V_{(0)} = \mathbb{C}\mathbf{1}$. *Let* $(U, Y, \mathbf{1}, \omega')$ *be a vertex operator subalgebra of* V *and assume that*

$$\omega' \in U \cap V_{(2)} \tag{3.11.22}$$

and that

$$L(1)\omega' = 0. \tag{3.11.23}$$

Set

$$Y(\omega', x) = \sum_{n \in \mathbb{Z}} L'(n) x^{-n-2}, \tag{3.11.24}$$

acting on all of V *(as in (3.11.9)). Then the gradings of* V *and* U *are compatible (i.e.,* $L(0) = L'(0)$ *on* U*), and more generally,*

$$L(n) = L'(n) \ \text{ on } U \text{ for all } n \geq -1. \tag{3.11.25}$$

Set

$$\omega'' = \omega - \omega'. \tag{3.11.26}$$

Then $\omega'' \in C_V(U)$*, and* $(C_V(U), Y, \mathbf{1}, \omega'')$ *is a vertex operator subalgebra of* V *of central charge equal to* $c_V - c_U$*, and we have*

$$\omega'' \in C_V(U) \cap V_{(2)} \tag{3.11.27}$$

and

$$L(1)\omega'' = 0. \tag{3.11.28}$$

Proof. Set

$$Y(\omega'', x) = \sum_{n \in \mathbb{Z}} L''(n) x^{-n-2} \tag{3.11.29}$$

(acting on V), so that $L''(n) = L(n) - L'(n)$ for $n \in \mathbb{Z}$. In view of Corollary 3.11.10 we have the Virasoro relations for $m, n \in \mathbb{Z}$,

$$[L'(m), L'(n)] = (m - n)L'(m + n) + \frac{1}{12}(m^3 - m)\delta_{m+n,0}c_U, \quad (3.11.30)$$

acting on V. Since $V_{(n)} = 0$ for $n < 0$, $V_{(0)} = \mathbb{C}\mathbf{1}$ and $\omega' \in V_{(2)}$, we have $L(n)\omega' = 0$ for $n \geq 3$ and $L(2)\omega' = \frac{1}{2}\alpha\mathbf{1}$ for some $\alpha \in \mathbb{C}$. Since $L(1)\omega' = 0$ as well, it follows from the commutator formula (3.1.8), with $u = \omega$, $v = \omega'$, and the $L(-1)$-derivative property that

$$[L(m), L'(n)] = (m - n)L'(m + n) + \frac{1}{12}(m^3 - m)\delta_{m+n,0}\alpha \quad (3.11.31)$$

for $m, n \in \mathbb{Z}$, just as at the end of Section 3.1. Combining (3.11.30) with (3.11.31) we obtain

$$[L''(m), L'(n)] = 0 \quad \text{for } m \neq -n. \quad (3.11.32)$$

Thus

$$L'(-1)\omega'' = L'(-1)L''(-2)\mathbf{1} = L''(-2)L'(-1)\mathbf{1} = 0, \quad (3.11.33)$$

so that $\omega'' \in C_V(U)$ by Corollary 3.11.11 (and in particular, (3.11.32) holds even when $m = n$). Thus $L''(n)u = \omega''_{n+1}u = 0$ for $n \geq -1$, $u \in U$, and so for $n \geq -1$,

$$L(n) = L'(n) \quad \text{on } U. \quad (3.11.34)$$

The fact that $\omega'' \in C_V(U)$ also implies that $L'(n)\omega'' = 0$ and $L''(n)\omega' = 0$ for $n \geq -1$. Consequently,

$$0 = L''(2)\omega' = (L(2) - L'(2))\omega' = \frac{1}{2}(\alpha - c_U)\mathbf{1}$$

(recall (3.1.73)), which implies that $\alpha = c_U$. With all these properties we have

$$\omega''_0\omega'' = L(-1)\omega'' - L'(-1)\omega'' = L(-1)\omega'', \quad (3.11.35)$$

$$\omega''_1\omega'' = L(0)\omega'' - L'(0)\omega'' = 2\omega'', \quad (3.11.36)$$

$$\omega''_2\omega'' = L(1)\omega'' = 0, \quad (3.11.37)$$

$$\omega''_3\omega'' = L(2)\omega'' = \frac{1}{2}(c_V - c_U)\mathbf{1}, \quad (3.11.38)$$

$$\omega''_n\omega'' = 0 \quad \text{for } n \geq 4, \quad (3.11.39)$$

and again, just as at the end of Section 3.1 we see that the operators $L''(n)$ satisfy the Virasoro algebra relations on V with central charge $c_V - c_U$. Furthermore, for $v \in C_V(U)$ and $n \geq -1$,

$$L''(n)v = L(n)v - L'(n)v = L(n)v \quad (3.11.40)$$

because $L'(n)v = 0$. In particular,

$$L''(-1)v = L(-1)v, \quad L''(0)v = L(0)v. \tag{3.11.41}$$

Consequently,

$$Y(L''(-1)v, x) = \frac{d}{dx} Y(v, x) \tag{3.11.42}$$

for $v \in C_V(U)$ and $C_V(U)$ is $L(0)$-stable and hence graded (by the grading of V), with the grading given by the $L''(0)$-eigenvalues. We conclude that $(C_V(U), Y, \mathbf{1}, \omega'')$ is a vertex operator algebra of central charge equal to $c_V - c_U$. $\quad\square$

Remark 3.11.13 Note that $C_V(U)$ satisfies the same conditions as does U, so the theorem can be applied to $C_V(U)$.

The following corollary gives a necessary and sufficient condition that $C_V(C_V(U)) = U$ (cf. [FZ] and Remark 3.11.3):

Corollary 3.11.14 *In the setting of Theorem 3.11.12, $C_V(C_V(U)) = U$ if and only if $(U, Y, \mathbf{1}, \omega')$ is maximal in the sense that if $(T, Y, \mathbf{1}, \omega')$ is a vertex operator subalgebra of V, then $T \subset U$.*

Proof. By Theorem 3.11.12 applied to $C_V(U)$, we have that $(C_V(C_V(U)), Y, \mathbf{1}, \omega')$ is a vertex operator subalgebra of V, and by Corollary 3.11.11 applied to $C_V(U)$,

$$C_V(C_V(U)) = \mathrm{Ker}_V L''(-1) = \mathrm{Ker}_V(L(-1) - L'(-1))$$
$$= \{v \in V \mid L'(-1)v = L(-1)v = v_{-2}\mathbf{1}\}. \tag{3.11.43}$$

Thus if U is maximal, then $C_V(C_V(U)) = U$, and conversely, if $C_V(C_V(U) = U$ and $(T, Y, \mathbf{1}, \omega')$ is a vertex operator subalgebra of V, then $v \in T$ implies $L'(-1)v = v_{-2}\mathbf{1}$, and hence $v \in C_V(C_V(U))$ by (3.11.43); thus $v \in U$, so that $T \subset U$. $\quad\square$

3.12 Direct product and tensor product vertex algebras

In classical associative algebra theory, the direct product (sometimes called direct sum) and the tensor product of finitely many associative algebras carry natural associative algebra structures. This is also true for vertex operator algebras, as we now explain. First we discuss the simpler of the two notions.

Let $(V_1, Y, \mathbf{1}), \ldots, (V_r, Y, \mathbf{1})$ be vertex algebras. The *direct product* vertex algebra

$$V = V_1 \times \cdots \times V_r \tag{3.12.1}$$

is constructed on the direct sum $V_1 \oplus \cdots \oplus V_r$ of the corresponding vector spaces, where the linear map $Y(\cdot, x)$ from V to $(\mathrm{End}\ V)[[x, x^{-1}]]$ is defined by

$$Y((v^{(1)}, \ldots, v^{(r)}), x) = (Y(v^{(1)}, x), \ldots, Y(v^{(r)}, x)) \tag{3.12.2}$$

for $v^{(i)} \in V_i$, $1 \leq i \leq r$, and the vacuum vector is

$$\mathbf{1} = (\mathbf{1}, \ldots, \mathbf{1}).\tag{3.12.3}$$

Indeed, the truncation condition, the vacuum property and the creation property clearly hold. The Jacobi identity holds for any triple of vectors in $V_1 \cup \cdots \cup V_r$, where we view each V_i as a subspace of V in the obvious way, and so by linearity the Jacobi identity holds for any triple of vectors in V; we are using the terminology "Jacobi identity for a triple" of Section 3.7. Clearly, each V_i is an ideal of V, and so is the direct product of any subfamily of the V_i. The projection of V onto the direct product of any subfamily of the V_i is a vertex algebra homomorphism whose kernel is the direct product of the complementary V_j. Summarizing, we have:

Proposition 3.12.1 *Let* $(V_1, Y, \mathbf{1}), \ldots, (V_r, Y, \mathbf{1})$ *be vertex algebras. Then the triple* $(V, Y, \mathbf{1})$ *defined in (3.12.1)–(3.12.3) carries the structure of a vertex algebra, and each subproduct is an ideal and a quotient algebra as well.* $\square$

Remark 3.12.2 As we expect, the following universal property holds. Given a vertex algebra U and vertex algebra homomorphisms $f_i : U \to V_i$ for $1 \leq i \leq r$, there exists a unique vertex algebra homomorphism $f : U \to V$ such that $f_i = p_i f$ for all i, where $p_i : V \to V_i$ is the canonical projection. The proof is straightforward.

Remark 3.12.3 Notice that the embedding of V_i into V is not a vertex algebra homomorphism unless $V_j = 0$ for all $j \neq i$, since this embedding does not map the vacuum vector $\mathbf{1}$ of V_j to the vacuum vector $(\mathbf{1}, \ldots, \mathbf{1})$ of V. In particular, V_i is not a vertex subalgebra of V. (This situation is the same as in classical associative algebra theory, where a factor A_i of a direct product $A_1 \times \cdots \times A_r$ of associative algebras is not a subalgebra because the identity elements do not match.)

Suppose now that the V_i are vertex operator algebras of various central charges with conformal vectors $\omega^{(i)}$, respectively. Endow V with the natural $\mathbb{Z}$-grading $V = \coprod_{n \in \mathbb{Z}} V_{(n)}$, where

$$V_{(n)} = (V_1)_{(n)} \oplus \cdots \oplus (V_r)_{(n)}.\tag{3.12.4}$$

Set

$$\omega = (\omega^{(1)}, \ldots, \omega^{(r)})\tag{3.12.5}$$

and $Y(\omega, x) = \sum_{n \in \mathbb{Z}} L(n)x^{-n-2}$ as usual. Then for $n \in \mathbb{Z}$,

$$L(n) = (L^{(1)}(n), \ldots, L^{(r)}(n)),\tag{3.12.6}$$

where $Y(\omega^{(i)}, x) = \sum_{n \in \mathbb{Z}} L^{(i)}(n)x^{-n-2}$. It is clear that $L(0)$ gives rise to the grading on V and that $L(-1)$ satisfies the derivative property (3.1.45). To prove that $(V, Y, \mathbf{1}, \omega)$ is a vertex operator algebra, we must show that the operators $L(n)$ $(n \in \mathbb{Z})$ on V give rise to a representation of the Virasoro algebra (recall Remark 3.1.23) with the central element acting as *a scalar*. It is easy to see that the operators $L(n)$ give rise to a representation of the Virasoro algebra on V with the central element $\mathbf{c}$ acting on V as the r-tuple $(c_{V_1}, \ldots, c_{V_r})$, which is a scalar multiple of the vacuum vector if and only if these central charges are all equal. Thus we have:

Proposition 3.12.4 *Let $(V_1, Y, 1, \omega^{(1)}), \ldots, (V_r, Y, 1, \omega^{(r)})$ be vertex operator algebras of the same central charge. Then the direct product vertex algebra $V = V_1 \times \cdots \times V_r$ is a vertex operator algebra of the same central charge, with the conformal vector ω defined as in (3.12.5).* $\square$

Next we consider the tensor product vertex algebra, a more sophisticated notion. The *tensor product* of vertex algebras $V_1, \ldots, V_r$ is constructed on the tensor product vector space

$$V = V_1 \otimes \cdots \otimes V_r, \tag{3.12.7}$$

where the linear map Y is (uniquely) defined by

$$Y(v^{(1)} \otimes \cdots \otimes v^{(r)}, x) = Y(v^{(1)}, x) \otimes \cdots \otimes Y(v^{(r)}, x) \tag{3.12.8}$$

for $v^{(i)} \in V_i$ and the vacuum vector is

$$1 = 1 \otimes \cdots \otimes 1. \tag{3.12.9}$$

(Here we use the notation 1 for the vacuum vectors of V and each V_i.)

Proposition 3.12.5 *Let $V_1, \ldots, V_r$ be vertex algebras. Then the triple $(V, Y, 1)$ defined in (3.12.7)–(3.12.9) carries the structure of a vertex algebra.*

Proof. The truncation condition, the vacuum property and the creation property clearly hold. The simplest way to prove the Jacobi identity is to use weak commutativity and associativity. Writing formula (3.2.15) (weak commutativity) in the form

$$(x_1 - x_2)^k Y(u, x_1) Y(v, x_2) w = (x_1 - x_2)^k Y(v, x_2) Y(u, x_1) w, \tag{3.12.10}$$

we see that weak commutativity (3.2.15) and weak associativity (3.3.29) hold for elements of the form $v^{(1)} \otimes \cdots \otimes v^{(r)}$. Thus the Jacobi identity follows from Proposition 3.4.3 and linearity. $\square$

Remark 3.12.6 This natural use of weak commutativity and weak associativity of course exploits the analogy between vertex algebras and commutative associative algebras rather than the analogy with Lie algebras.

Before extending the notion of tensor product from vertex algebras to vertex operator algebras, we discuss a universal property of the tensor product vertex algebra V. Let e_i be the natural map from V_i to V, sending $v \in V_i$ to $1 \otimes \cdots \otimes v \otimes \cdots \otimes 1$. Notice that

$$Y(e_i v, x) = Y(1 \otimes \cdots \otimes v \otimes \cdots \otimes 1, x) = 1 \otimes \cdots \otimes Y(v, x) \otimes \cdots \otimes 1 \tag{3.12.11}$$

for $v \in V_i$. It is clear that e_i is a vertex algebra homomorphism. (However, e_i will not in general be a vertex operator algebra homomorphism when all the V_j are vertex operator algebras, because of the conformal vectors; see Remark 3.12.9.) From (3.12.11), for $j \neq i$,

$$e_i(V_i) \subset C_V(e_j(V_j)) \tag{3.12.12}$$

(recall (3.11.1)). Furthermore,

$$Y(v^{(1)} \otimes \cdots \otimes v^{(r)}, x) = Y(e_1 v^{(1)}, x) \cdots Y(e_r v^{(r)}, x). \tag{3.12.13}$$

Notice that the product on the right hand side can be written in any order. Now we have the following universal property for the tensor product vertex algebra:

Proposition 3.12.7 *Let $V_1, \ldots, V_r$ be vertex algebras and let $V = V_1 \otimes \cdots \otimes V_r$ be the tensor product vertex algebra. Let U be a vertex algebra and let g_i be vertex algebra homomorphisms from V_i to U for $i = 1, 2, \ldots, r$ such that for $j \neq i$,*

$$g_i(V_i) \subset C_U(g_j(V_j)). \tag{3.12.14}$$

Then there exists a unique vertex algebra homomorphism g from V to U such that $g_i = ge_i$ for all i.

Proof. By (3.12.13) and the creation property we have

$$v^{(1)} \otimes \cdots \otimes v^{(r)} = \lim_{x \to 0} (Y(e_1(v^{(1)}), x) \cdots Y(e_r(v^{(r)}), x))\mathbf{1} \tag{3.12.15}$$

for $v^{(i)} \in V_i$, where $\mathbf{1}$ is the vacuum vector of V. If g is a vertex algebra homomorphism from V to U such that $ge_i = g_i$ for all i, then

$$g(v^{(1)} \otimes \cdots \otimes v^{(r)}) = \lim_{x \to 0} g(Y(e_1(v^{(1)}), x) \cdots Y(e_r(v^{(r)}), x)\mathbf{1})$$

$$= \lim_{x \to 0} (Y(ge_1(v^{(1)}), x) \cdots Y(ge_r(v^{(r)}), x))g(\mathbf{1})$$

$$= \lim_{x \to 0} (Y(g_1(v^{(1)}), x) \cdots Y(g_r(v^{(r)}), x))\mathbf{1}, \tag{3.12.16}$$

and so g is unique if it exists. Now we define g to be the linear map from V to U such that

$$g(v^{(1)} \otimes \cdots \otimes v^{(r)}) = \lim_{x \to 0} Y(g_1(v^{(1)}), x) \cdots Y(g_r(v^{(r)}), x)\mathbf{1}$$

$$= (g_1 v^{(1)})_{-1} \cdots (g_r v^{(r)})_{-1}\mathbf{1} \tag{3.12.17}$$

for $v^{(i)} \in V_i$, where the $\mathbf{1}$ is the vacuum vector of U. Clearly, $ge_i = g_i$ for all i. It remains to prove that g is a vertex algebra homomorphism.

It is clear that g sends the vacuum vector of V to the vacuum vector of U. From Remark 3.3.2, if $u_n w = 0$ for $n \geq 0$, then in the weak associativity relation we can take the integer l to be zero, so that

$$Y(u, x_0 + x_2)Y(v, x_2)w = Y(Y(u, x_0)v, x_2)w \tag{3.12.18}$$

for $v \in V$. Notice that if $[Y(a^{(i)}, x_1), Y(a^{(j)}, x_2)] = 0$ for all $i \neq j$, then using (3.8.13) we have

$$Y((a^{(1)})_{-1} \cdots (a^{(r)})_{-1}\mathbf{1}, x) = Y(a^{(1)}, x) \cdots Y(a^{(r)}, x). \tag{3.12.19}$$

Now we shall use these facts to finish the proof. Let $u^{(i)}, v^{(i)} \in V_i$ $(1 \le i \le r)$. Then

$$\begin{aligned}
g(Y(u^{(1)} &\otimes \cdots \otimes u^{(r)}, x_1)(v^{(1)} \otimes \cdots \otimes v^{(r)})) \\
&= g(Y(u^{(1)}, x_1)v^{(1)} \otimes \cdots \otimes Y(u^{(r)}, x_1)v^{(r)}) \\
&= \lim_{x \to 0} Y(g_1 Y(u^{(1)}, x_1)v^{(1)}, x) \cdots Y(g_r Y(u^{(r)}, x_1)v^{(r)}, x)\mathbf{1} \\
&= \lim_{x \to 0} Y(Y(g_1 u^{(1)}, x_1)g_1 v^{(1)}, x) \cdots Y(Y(g_r u^{(r)}, x_1)g_r v^{(r)}, x)\mathbf{1} \\
&= \lim_{x \to 0} Y(g_1 u^{(1)}, x_1 + x)Y(g_1 v^{(1)}, x) \cdots Y(g_r u^{(r)}, x_1 + x)Y(g_r v^{(r)}, x)\mathbf{1} \\
&= \lim_{x \to 0} Y(g_1 u^{(1)}, x_1 + x) \cdots Y(g_r u^{(r)}, x_1 + x)Y(g_1 v^{(1)}, x) \cdots Y(g_r v^{(r)}, x)\mathbf{1} \\
&= Y(g_1 u^{(1)}, x_1) \cdots Y(g_r u^{(r)}, x_1)g(v^{(1)} \otimes \cdots \otimes v^{(r)}) \\
&= Y((g_1 u^{(1)})_{-1} \cdots (g_r u^{(r)})_{-1}\mathbf{1}, x_1)g(v^{(1)} \otimes \cdots \otimes v^{(r)}) \\
&= Y(g(u^{(1)} \otimes \cdots \otimes u^{(r)}), x_1)g(v^{(1)} \otimes \cdots \otimes v^{(r)}). \tag{3.12.20}
\end{aligned}$$

This completes the proof. $\square$

Now suppose that each V_i is a vertex operator algebra with conformal vector ω^i for $i = 1, \ldots, r$. Then $V = V_1 \otimes \cdots \otimes V_r$ is naturally $\mathbb{Z}$-graded as $V = \coprod_{n \in \mathbb{Z}} V_{(n)}$, where

$$V_{(n)} = \sum_{m_1 + \cdots + m_r = n} (V_1)_{(m_1)} \otimes \cdots \otimes (V_r)_{(m_r)}. \tag{3.12.21}$$

The two grading restrictions in the definition of the notion of vertex operator algebra clearly hold.

Set

$$\omega = \omega \otimes \cdots \otimes \mathbf{1} + \cdots + \mathbf{1} \otimes \cdots \otimes \omega. \tag{3.12.22}$$

Then

$$L(n) = L(n) \otimes \cdots \otimes 1 + \cdots + 1 \otimes \cdots \otimes L(n) \tag{3.12.23}$$

for $n \in \mathbb{Z}$. (Here we use the notation $L(n)$ for different operators.) If M_i, for $i = 1, \ldots, r$, is a module for the Virasoro algebra of central charge c_i, then the tensor product $\otimes_{i=1}^r M_i$ is again a module, of central charge $c_1 + \cdots + c_r$. It follows from (3.12.23) that the $\mathbb{Z}$-grading on V is given by $L(0)$. It is also clear that $L(-1)$ satisfies the derivative property (3.1.45). Thus we have:

Proposition 3.12.8 *The tensor product vertex algebra of finitely many vertex operator algebras is a vertex operator algebra whose central charge is the sum of the central charges of the tensor factors.* $\square$

Remark 3.12.9 Notice that if the V_i are vertex operator algebras, the embedding e_i is not in general a vertex operator algebra homomorphism because e_i sends the conformal vector ω of V_i to

$$\mathbf{1} \otimes \cdots \otimes \omega \otimes \cdots \otimes \mathbf{1},$$

which is not the conformal vector of V in (3.12.22) unless the conformal vector ω of V_j is zero for each $j \neq i$. But this would imply that each such V_j is a finite-dimensional commutative associative algebra by Remark 3.4.7 (or by Proposition 3.11.2, which shows that $V_j = \operatorname{Ker} L(-1) = C(V_j)$ is a commutative associative algebra).

Remark 3.12.10 With respect to the notion of tensor product of vertex algebras, vertex algebras are like commutative associative algebras but are *not* like Lie algebras, since the tensor product of Lie algebras, viewed as a nonassociative algebra, is not a Lie algebra.

4

Modules

In this chapter we discuss the notion of module for a vertex (operator) algebra V. A V-module is defined, as expected, to be a vector space W equipped with a linear map

$$Y_W : V \to (\text{End } W)[[x, x^{-1}]] \tag{4.0.1}$$

such that *all the defining properties of a vertex algebra that make sense hold.* (Actually, this is the typical principle for defining the notion of module for categories of algebras in general—Lie algebras, associative algebras, etc. A module is a vector space equipped with a linear action of the algebra such that all the algebra axioms that make sense hold.) Specifically, these defining properties are the truncation condition, the vacuum property and the Jacobi identity. (The creation property, for instance, would not make sense, so it will not be an axiom.) Accordingly, almost all of the assertions in Chapter 3 that make sense also hold and a large amount of material in Chapter 3 carries over in the obvious ways, often without change. In this chapter, we carry out many of these analogues, and we also discuss some additional concepts.

One part of Chapter 3 that will not carry over to modules is Section 3.5; for modules, commutativity does not imply the Jacobi identity. Here is why. As we have discussed extensively, the Jacobi identity is a natural combination of commutativity and associativity, and the notion of vertex algebra is analogous to the notion of commutative associative algebra. The notion of module for a commutative associative algebra is defined by using the associativity of the action ($a \cdot (b \cdot m) = (ab) \cdot m$ for a and b in the algebra and m in the module), and the commutativity of the action of algebra elements on the module is an immediate consequence ($a \cdot (b \cdot m) = b \cdot (a \cdot m)$). But this commutativity of action does not imply associativity; it does not imply that we in fact have a module. We certainly expect the same phenomenon in vertex algebra theory, and in fact, since commutative associative algebras with identity are examples of vertex algebras, this is indeed the case in vertex algebra theory; commutativity does not imply associativity for module action. On the other hand, this classical analogue suggests that *associativity* of action should imply *commutativity* of action. This is in fact true in the setting of vertex (operator) algebras and is proved in this chapter (Theorem 4.4.5) by invoking the second of our two proofs of Theorem 3.6.1 (recall Theorem 3.6.3 and the discussion preceding it).

In addition to duality (commutativity and associativity) properties we also introduce certain notions analogous to those in classical associative algebra theory.

4.1 Definition and some consequences

Let us fix a vertex algebra $V = (V, Y, \mathbf{1})$ throughout this chapter.

Definition 4.1.1 A V-*module* is a vector space W equipped with a linear map $V \otimes W \to W[[x, x^{-1}]]$, which can be equivalently expressed as a linear map

$$Y_W(\cdot, x) : \ V \to (\text{End } W)[[x, x^{-1}]]$$
$$v \mapsto Y_W(v, x) = \sum_{n \in \mathbb{Z}} v_n x^{-n-1}, \tag{4.1.1}$$

such that "all the defining properties of a vertex algebra that make sense hold." That is, for $u, v \in V$ and $w \in W$, the endomorphisms u_n of W satisfy the condition

$$u_n w = 0 \quad \text{for } n \ \text{sufficiently large,} \tag{4.1.2}$$

or equivalently,

$$Y_W(u, x)w \in W((x)) \tag{4.1.3}$$

(*truncation condition*);

$$Y_W(\mathbf{1}, x) = 1 \quad (1 \text{ on the right being the identity operator on } W) \tag{4.1.4}$$

(*vacuum property*); and

$$x_0^{-1} \delta \left(\frac{x_1 - x_2}{x_0} \right) Y_W(u, x_1) Y_W(v, x_2) - x_0^{-1} \delta \left(\frac{x_2 - x_1}{-x_0} \right) Y_W(v, x_2) Y_W(u, x_1)$$
$$= x_2^{-1} \delta \left(\frac{x_1 - x_0}{x_2} \right) Y_W(Y(u, x_0)v, x_2) \tag{4.1.5}$$

(*Jacobi identity*); note the roles of both Y_W and Y (on the right-hand side, $Y(u, x_0)$ acts on V, not W).

We sometimes refer to the V-module W as (W, Y_W) if necessary. Later we shall also freely use the notation Y for Y_W.

Remark 4.1.2 We shall sometimes call $Y_W(v, x)$ the *vertex operator on W associated with $v \in V$*. Later, in Chapter 5, we shall intensively investigate the extent to which formal series of endomorphisms of a given vector space W, of the form (4.1.1), can indeed be viewed as vertex operators on W associated with elements of some vertex algebra acting on W.

The vertex algebra V is itself clearly a V-module (since a module is a space on which V acts such that all the defining properties of a vertex algebra that make sense hold!). As such, it is called the *adjoint module*, as in Lie algebra theory.

We of course have a collection of comments on and consequences of the definition just as in Section 3.1, to which we refer the reader again. Notably, the commutator formula (3.1.8) (the residue with respect to x_0 of the Jacobi identity) and its variant (3.1.10), the iterate formula (3.1.11) (the residue with respect to x_1 of the Jacobi identity) and the associator formula (3.1.13), or (3.1.14) (reformulations of (3.1.11)) hold on a module W. Rather than repeating these comments and consequences here, we shall take the liberty of freely quoting from Chapter 3 those properties of vertex algebras that carry over without change to modules. But note that of course certain comments in Chapter 3 certainly do *not* apply to modules. For instance, in contrast to Remark 3.1.7, the map $Y_W(\cdot, x)$ is not in general injective, and so a module does not in general provide an identification of V with a space of vertex operators.

The proof of Proposition 3.1.18 immediately gives:

Proposition 4.1.3 *Let (W, Y_W) be a V-module. Then*

$$Y_W(\mathcal{D}v, x) = \frac{d}{dx} Y_W(v, x) \tag{4.1.6}$$

for $v \in V$. $\square$

The "global" forms (3.1.27) and (3.1.28) of this $\mathcal{D}$-derivative property also hold.

Remark 4.1.4 While the $\mathcal{D}$-operator is a canonical operator on the vertex algebra V (recall that $\mathcal{D}v = v_{-2}\mathbf{1}$ for $v \in V$), $\mathcal{D}$ does not by definition act on a V-module, so that we do not in general have an analogue of the $\mathcal{D}$-bracket formulas (3.1.32) and (3.1.33). But such formulas are important enough so that we shall introduce the term "V-module (W, Y_W, d)" to refer to a V-module (W, Y_W) equipped with an endomorphism d of W such that for $v \in V$,

$$[d, Y_W(v, x)] = Y_W(\mathcal{D}v, x) \left(= \frac{d}{dx} Y_W(v, x) \right). \tag{4.1.7}$$

When we have this structure, we also have the conjugation formula

$$e^{x_0 d} Y_W(v, x) e^{-x_0 d} = Y_W(e^{x_0 \mathcal{D}} v, x) = Y_W(v, x + x_0), \tag{4.1.8}$$

as in (3.1.35).

Now suppose that V is a vertex operator algebra with conformal vector ω and let (W, Y_W) be a module for V viewed as a vertex algebra. Since $\mathcal{D} = L(-1)$ on V (recall (3.1.67)), it follows from Proposition 4.1.3 that

$$Y_W(L(-1)v, x) = \frac{d}{dx} Y_W(v, x) \tag{4.1.9}$$

(cf. (3.1.45)) for $v \in V$.

We next consider the action of the operators $L(n)$, $n \in \mathbb{Z}$, on W defined by

$$Y_W(\omega, x) = \sum_{n \in \mathbb{Z}} L(n) x^{-n-2} = \sum_{n \in \mathbb{Z}} \omega_n x^{-n-1} \qquad (4.1.10)$$

(cf. (3.1.41)); we are using the notation $L(n)$ and ω_n for operators on both V and W. Because $L(-1) = \omega_0$ on W, as in (3.1.52) we have

$$[L(-1), Y_W(v, x)] = Y_W(L(-1)v, x) \qquad (4.1.11)$$

for $v \in V$. Such properties are also given in Proposition 3.11.8. In addition, the Virasoro algebra relations (3.1.40) on W follow exactly as in Section 3.1 from the commutator formula (3.1.8), the $L(-1)$-derivative property (4.1.9) and (3.1.70)–(3.1.74). Summarizing, we have:

Proposition 4.1.5 *Let V be a vertex operator algebra and let (W, Y_W) be a module for V viewed as a vertex algebra. Then the following relations hold on W:*

$$[L(-1), Y_W(v, x)] = Y_W(L(-1)v, x) = \frac{d}{dx} Y_W(v, x) \quad \text{for } v \in V, \quad (4.1.12)$$

so that $(W, Y_W, L(-1))$ is a V-module in the sense of Remark 4.1.4, and

$$[L(m), L(n)] = (m - n)L(m + n) + \frac{1}{12}(m^3 - m)\delta_{m+n,0} c_V \qquad (4.1.13)$$

for $m, n \in \mathbb{Z}$. $\square$

Now we define the notion of module for a vertex operator algebra.

Definition 4.1.6 Let V be a vertex operator algebra. A V-*module* is a module W for V viewed as a vertex algebra such that

$$W = \coprod_{h \in \mathbb{C}} W_{(h)}, \qquad (4.1.14)$$

where

$$W_{(h)} = \{w \in W \mid L(0)w = hw\}, \qquad (4.1.15)$$

the subspace of W of vectors of *weight* h, and such that the following grading restriction conditions hold:

$$\dim W_{(h)} < \infty \quad \text{for } h \in \mathbb{C}, \qquad (4.1.16)$$

$$W_{(h)} = 0 \quad \text{for } h \text{ whose real part is sufficiently negative.} \qquad (4.1.17)$$

In this definition, we have (in view of Proposition 4.1.5) kept to the principle that "all the defining properties of a vertex operator algebra that make sense hold," except that a module is generally $\mathbb{C}$-graded instead of $\mathbb{Z}$-graded. This grading relaxation is natural and appropriate in general, although in many applications modules are graded by the rational numbers $\mathbb{Q}$, and sometimes this is assumed in the definition of the notion of module. It would be too restrictive to insist that modules be $\mathbb{Z}$-graded, as the reader will see from the examples in Chapter 6.

Again, we have the usual consequences as in Section 3.1. As in (3.1.53) we have

$$[L(0), Y_W(v, x)] = x Y_W(L(-1)v, x) + Y_W(L(0)v, x) \tag{4.1.18}$$

for $v \in V$. If v is homogeneous, (4.1.9), (4.1.15) and (4.1.18) imply

$$\text{wt } v_n = \text{wt } v - n - 1, \tag{4.1.19}$$

that is, v_n maps $W_{(h)}$ to $W_{(h+\text{wt } v-n-1)}$ for $h \in \mathbb{C}$, $n \in \mathbb{Z}$. It follows that a V-module W decomposes into submodules corresponding to the congruence classes mod $\mathbb{Z}$: For $\alpha \in \mathbb{C}/\mathbb{Z}$, let

$$W_{[\alpha]} = \coprod_{h+\mathbb{Z}=\alpha} W_{(h)}. \tag{4.1.20}$$

Then

$$W = \coprod_{\alpha \in \mathbb{C}/\mathbb{Z}} W_{[\alpha]}. \tag{4.1.21}$$

In particular, if W is irreducible, then

$$W = W_{[\alpha]} \tag{4.1.22}$$

for some α.

4.2 Commutativity properties

In this section we carry out analogues for V-modules of the weak commutativity, commutativity and formal commutativity properties discussed in Section 3.2. These analogues are straightforward and the earlier statements and proofs, with obvious changes, work fine.

We shall present those analogues in order. First, we observe that the proof of Proposition 3.2.1 immediately gives:

Proposition 4.2.1 (weak commutativity) *Let* (W, Y_W) *be a* V-*module. Then for* $u, v \in V$, *there exists a nonnegative integer* k *such that*

$$(x_1 - x_2)^k [Y_W(u, x_1), Y_W(v, x_2)] = 0. \quad \square \tag{4.2.1}$$

Remark 4.2.2 As in Remark 3.2.2, we may take $k \geq 0$ such that $u_n v = 0$ for $n \geq k$, that is, $x^k Y(u, x)v \in V[[x]]$. Note that this k works for *all* V-modules.

Remark 4.2.3 Recall from Remark 3.2.4 that for $u, v \in V$, $[Y(u, x_1), Y(v, x_2)] = 0$ (acting on V) if and only if $u_n v = 0$ for $n \geq 0$. Then in view of Remark 4.2.2, $[Y(u, x_1), Y(v, x_2)] = 0$ (on V) implies $[Y_W(u, x_1), Y_W(v, x_2)] = 0$ for any V-module (W, Y_W).

Next, we discuss the analogue of the notion of commutativity, as discussed in Section 3.2. Let V be a vertex operator algebra and let $W = \coprod_{h \in \mathbb{C}} W_{(h)}$ be a V-module. Set

$$W' = \coprod_{h \in \mathbb{C}} W_{(h)}^* \tag{4.2.2}$$

(the *restricted dual*) (cf. (3.2.3)). We shall also use the notation $\langle \cdot, \cdot \rangle$ for the natural pairing between W' and W. As in (3.2.4), we have

$$\langle w', Y_W(u, x)w \rangle \in \mathbb{C}[x, x^{-1}] \tag{4.2.3}$$

for $u \in V$, $w \in W$ and $w' \in W'$ by (4.1.19). The proof of Proposition 3.2.7 immediately gives the following result on the rationality of products and commutativity:

Proposition 4.2.4 *Suppose that V is a vertex operator algebra, let (W, Y_W) be a V-module and let $u, v \in V$, $w \in W$ and $w' \in W'$ be arbitrary. We have:*
(a) (rationality of products) The formal series

$$\langle w', Y_W(u, x_1)Y_W(v, x_2)w \rangle \left(= \sum_{m,n \in \mathbb{Z}} \langle w', u_m v_n w \rangle x_1^{-m-1} x_2^{-n-1} \right)$$

lies in the image of the map ι_{12}:

$$\langle w', Y_W(u, x_1)Y_W(v, x_2)w \rangle = \iota_{12} f(x_1, x_2), \tag{4.2.4}$$

where the (uniquely determined) element $f \in \mathbb{C}[x_1, x_2]_S$ is of the form

$$f(x_1, x_2) = \frac{g(x_1, x_2)}{(x_1 - x_2)^k x_1^l x_2^m} \tag{4.2.5}$$

for some $g \in \mathbb{C}[x_1, x_2]$ and $k, l, m \in \mathbb{Z}$, where k depends only on u and v; it is independent of w and w'.
(b) (commutativity) We also have

$$\langle w', Y_W(v, x_2)Y_W(u, x_1)w \rangle = \iota_{21} f(x_1, x_2), \tag{4.2.6}$$

that is, in informal language,

$$\text{``}Y_W(u, x_1)Y_W(v, x_2) \quad \text{agrees with} \quad Y_W(v, x_2)Y_W(u, x_1) \tag{4.2.7}$$

as operator-valued rational functions.'' $\square$

Remark 4.2.5 Commutativity and weak commutativity are equivalent in exactly the same sense as in Remark 3.2.9.

As in Section 3.2, the algebraic commutativity results immediately imply the following results involving convergent series in suitable domains; we replace the formal variables x_0, x_1 and x_2 by the complex variables z_0, z_1 and z_2:

Corollary 4.2.6 *The two formal series*

$$\langle w', Y_W(u, z_1) Y_W(v, z_2) w \rangle \quad \left(= \sum_{m,n \in \mathbb{Z}} \langle w', u_m v_n w \rangle z_1^{-m-1} z_2^{-n-1} \right) \tag{4.2.8}$$

and

$$\langle w', Y_W(v, z_2) Y_W(u, z_1) w \rangle \quad \left(= \sum_{m,n \in \mathbb{Z}} \langle w', v_n u_m w \rangle z_1^{-m-1} z_2^{-n-1} \right) \tag{4.2.9}$$

in two complex variables are absolutely convergent to a common rational function $f(z_1, z_2)$ *(recall (4.2.5)) in the (disjoint) domains*

$$|z_1| > |z_2| > 0 \quad and \quad |z_2| > |z_1| > 0, \tag{4.2.10}$$

respectively. □

Finally, we discuss the analogue of formal commutativity, which, as we recall from Section 3.2, is a natural analogue of commutativity but which works more generally for an arbitrary vertex algebra. We mention again that we continue to use the notation $\langle \cdot, \cdot \rangle$ for the pairing between the dual of a vector space and the vector space, and we use the iota-maps defined in the general setting before Proposition 2.3.27.

Combining Propositions 3.2.12 (which was formulated so as to apply in the generality of modules) and 4.2.1 we immediately have:

Proposition 4.2.7 (formal commutativity) *Let* (W, Y_W) *be a* V*-module. Let* $u, v \in V$, $w \in W$ *and* $w^* \in W^*$. *Then the formal series*

$$\langle w^*, Y_W(u, x_1) Y_W(v, x_2) w \rangle \tag{4.2.11}$$

lies in the image of the map ι_{12}*:*

$$\langle w^*, Y_W(u, x_1) Y_W(v, x_2) w \rangle = \iota_{12} f(x_1, x_2), \tag{4.2.12}$$

where the (uniquely determined) element

$$f \in \mathbb{C}[[x_1, x_2]][x_1^{-1}, x_2^{-1}, (x_1 - x_2)^{-1}]$$

is of the form

$$f(x_1, x_2) = \frac{g(x_1, x_2)}{(x_1 - x_2)^k x_1^l x_2^m} \tag{4.2.13}$$

for some $g \in \mathbb{C}[[x_1, x_2]]$ and $k, l, m \in \mathbb{Z}$, where k depends only on u and v (and not on w or w^*). Moreover,

$$\langle w^*, Y_W(v, x_2)Y_W(u, x_1)w \rangle = \iota_{21} f(x_1, x_2). \quad \square \tag{4.2.14}$$

Remark 4.2.8 Weak commutativity follows from formal commutativity, as in Remark 3.2.13.

4.3 Associativity properties

Next we discuss the analogues for V-modules of the associativity properties presented in Section 3.3. Just as with the analogues of the commutativity properties discussed in Section 4.2, these analogues are straightforward and follow directly from the earlier proofs.

The first analogue is the following form of weak associativity, which follows immediately from the proof of Proposition 3.3.1.

Proposition 4.3.1 (weak associativity) *Let (W, Y_W) be a V-module. Then for $u \in V$, $w \in W$, there exists a nonnegative integer l (depending only on u and w) such that for any $v \in V$,*

$$(x_0 + x_2)^l Y_W(Y(u, x_0)v, x_2)w = (x_0 + x_2)^l Y_W(u, x_0 + x_2)Y_W(v, x_2)w. \quad \square \tag{4.3.1}$$

Remark 4.3.2 As in Remark 3.3.2, we may take $l \geq 0$ to be such that $u_n w = 0$ for $n \geq l$, or equivalently, $x^l Y_W(u, x)w \in W[[x]]$. In particular, if $u_n w = 0$ for $n \geq 0$, we have

$$Y_W(u, x_0 + x_2)Y_W(v, x_2)w = Y_W(Y(u, x_0)v, x_2)w. \tag{4.3.2}$$

Unlike k in the weak commutativity relation (4.2.1), l depends on u and w, though it does not depend on v. In contrast to the fact that the weak commutativity relation is a *global* identity on all modules (recall Remark 4.2.2), the weak associativity relation is only a *local* identity on a fixed element of a module.

In the case of vertex operator algebras, we have the following pair of results giving the analogue of associativity, which follows immediately from the proofs of Propositions 3.3.5 and 3.3.8:

Proposition 4.3.3 *Assume that V is a vertex operator algebra, let (W, Y_W) be a V-module and let $u, v \in V$, $w \in W$ and $w' \in W'$ be arbitrary. We have:*
   (a) **(rationality of iterates)** *The formal series*

$$\langle w', Y_W(Y(u, x_0)v, x_2)w\rangle \left(= \sum_{m,n\in\mathbb{Z}} \langle w', (u_m v)_n w\rangle x_0^{-m-1} x_2^{-n-1} \right)$$

lies in the image of the map ι_{20}:

$$\langle w', Y_W(Y(u, x_0)v, x_2)w\rangle = \iota_{20} p(x_0, x_2), \tag{4.3.3}$$

where the (uniquely determined) element $p \in \mathbb{C}[x_0, x_2]_S$ is of the form

$$p(x_0, x_2) = \frac{q(x_0, x_2)}{x_0^k (x_0 + x_2)^l x_2^m} \tag{4.3.4}$$

for some $q \in \mathbb{C}[x_0, x_2]$ and $k, l, m \in \mathbb{Z}$, where l depends only on u and w; it is independent of v and w'.

(b) The series

$$\langle w', Y_W(u, x_0 + x_2) Y_W(v, x_2)w\rangle$$

$$\left(= \sum_{m,n\in\mathbb{Z}} \sum_{i\geq 0} \binom{-m-1}{i} \langle w', u_m v_n w\rangle x_0^{-m-1-i} x_2^{i-n-1} \right)$$

lies in the image of ι_{02}, and in fact

$$\langle w', Y_W(u, x_0 + x_2) Y_W(v, x_2)w\rangle = \iota_{02} p(x_0, x_2). \tag{4.3.5}$$

That is, in informal language,

$$\text{``}Y_W(Y(u, x_0)v, x_2) \quad \text{agrees with} \quad Y_W(u, x_0 + x_2) Y_W(v, x_2) \tag{4.3.6}$$

as operator-valued rational functions.'' $\square$

Proposition 4.3.4 (associativity) *We have*

$$\iota_{12}^{-1} \langle w', Y_W(u, x_1) Y_W(v, x_2)w\rangle = \left(\iota_{20}^{-1} \langle w', Y_W(Y(u, x_0)v, x_2)w\rangle \right) |_{x_0 = x_1 - x_2}. \tag{4.3.7}$$

That is, in informal language,

$$\text{``}Y_W(u, x_1) Y_W(v, x_2) \quad \text{agrees with} \quad Y_W(Y(u, x_1 - x_2)v, x_2) \tag{4.3.8}$$

as operator-valued rational functions, where the right-hand expression is to be expanded as a Laurent series in $x_1 - x_2$.'' $\square$

As in Section 3.3, our algebraic associativity results immediately imply the following result involving convergent series in suitable domains:

Corollary 4.3.5 *The formal series of complex numbers* $\langle w', Y_W(u, z_1)Y_W(v, z_2)w \rangle$ *and* $\langle w', Y_W(Y(u, z_1 - z_2)v, z_2)w \rangle$ *are absolutely convergent to the common rational function* $f(z_1, z_2)$ *(recall (4.2.5)) in the domains*

$$|z_1| > |z_2| > 0 \quad and \quad |z_2| > |z_1 - z_2| > 0, \tag{4.3.9}$$

respectively, and in particular in the common domain

$$|z_1| > |z_2| > |z_1 - z_2| > 0. \quad \square \tag{4.3.10}$$

Remark 4.3.6 All the considerations concerning the operator product expansion and commutators discussed in Remarks 3.3.12 and 3.3.13 of course apply in the context of modules for a vertex operator algebra; recall Remarks 3.3.15 and 3.3.16.

For the analogue of formal associativity presented at the end of Section 3.3, combining Proposition 3.3.17 (which already applies in the generality of modules) with our current results for modules, the arguments in Section 3.3 immediately give:

Proposition 4.3.7 (formal associativity) *Let* (W, Y_W) *be a module for the vertex algebra* V *and let* $u, v \in V$, $w \in W$ *and* $w^* \in W^*$. *Then the formal series*

$$\langle w^*, Y_W(u, x_1)Y_W(v, x_2)w \rangle$$

lies in the image of the map ι_{12} *defined in (2.3.60):*

$$\langle w^*, Y_W(u, x_1)Y_W(v, x_2)w \rangle = \iota_{12}f(x_1, x_2), \tag{4.3.11}$$

where the (uniquely determined) element

$$f \in \mathbb{C}[[x_1, x_2]][x_1^{-1}, x_2^{-1}, (x_1 - x_2)^{-1}]$$

is of the form

$$f(x_1, x_2) = \frac{g(x_1, x_2)}{(x_1 - x_2)^k x_1^l x_2^m} \tag{4.3.12}$$

for some $g \in \mathbb{C}[[x_1, x_2]]$ *and* $k, l, m \in \mathbb{Z}$, *where* l *depends only on* u *and* w *(and not on* v *or* w^**). Moreover,*

$$f(x_1, x_2) = p(x_1 - x_2, x_2), \tag{4.3.13}$$

or equivalently,

$$\iota_{12}^{-1} \langle w^*, Y_W(u, x_1)Y_W(v, x_2)w \rangle = \left(\iota_{20}^{-1} \langle w^*, Y_W(Y(u, x_0)v, x_2)w \rangle \right)|_{x_0 = x_1 - x_2} \tag{4.3.14}$$

(as elements of $\mathbb{C}[[x_1, x_2]][x_1^{-1}, x_2^{-1}, (x_1 - x_2)^{-1}]$). $\quad \square$

Remark 4.3.8 Just as in Section 3.3, weak associativity, associativity and formal associativity are essentially equivalent to one another, under the relevant hypotheses.

4.4 The Jacobi identity as a consequence of associativity and commutativity properties

In this section we show that the Jacobi identity follows from commutativity and associativity properties, and also from associativity properties alone. Since we already formulated the relevant results in sufficient generality in Sections 3.4–3.6, we need only state the appropriate analogous results in our module setting.

First, in view of Proposition 3.4.3 we immediately have:

Proposition 4.4.1 *In the definition of the notion of V-module, the Jacobi identity can be replaced by weak commutativity (the assertion of Proposition 4.2.1) and weak associativity (the assertion of Proposition 4.3.1).* $\square$

Remark 4.4.2 As we noted in Remark 3.4.4, when we assume weak commutativity and weak associativity for a triple (u, v, w) (with $u, v \in V$ and $w \in W$), we can assume that k and l depend on all three of u, v and w.

Next, in the setting of vertex operator algebras, in terms of commutativity and associativity, the proof of Proposition 3.4.1 immediately gives:

Proposition 4.4.3 *Let V be a vertex operator algebra. Then the Jacobi identity for a V-module follows from the rationality of products and iterates, and commutativity and associativity (the assertions of Proposition 4.2.4(a),(b), Proposition 4.3.3(a) and Proposition 4.3.4). In particular, in the definition of the notion of V-module, the Jacobi identity can be replaced by these properties.* $\square$

Remark 4.4.4 Again (as in Remark 3.4.2), only the milder versions of rationality, commutativity and associativity have to be assumed, with k and l depending on all four elements u, v, w and w'.

We have seen that for vertex algebras and vertex operator algebras, (weak) commutativity essentially implies (weak) associativity, and hence the Jacobi identity. However, in the notion of module for a given vertex algebra, it is instead the converse that is true, as we discussed at the beginning of this chapter. Recalling this point, let us consider a commutative associative algebra A. An A-module is defined to be a vector space U equipped with a linear action of A on U such that

$$a(bu) = (ab)u \quad \text{for } a, b \in A, \ u \in U \tag{4.4.1}$$

(associativity). As a consequence, the commutativity of the action of algebra elements, $a(bu) = b(au)$, holds for $a, b \in A$, $u \in U$, since

$$a(bu) = (ab)u = (ba)u = b(au). \tag{4.4.2}$$

Theorem 3.6.3 is exactly in this spirit and it immediately implies:

Theorem 4.4.5 *Let V be a vertex algebra, W a vector space and Y_W a linear map from V to $(\mathrm{End}\ W)[[x, x^{-1}]]$ such that $Y_W(\mathbf{1}, x) = 1$, $Y_W(v, x)w \in W((x))$ for $v \in V$, $w \in W$, and such that weak associativity for Y_W holds (with l in (4.3.1) depending on u, v and w). Then Y_W satisfies the Jacobi identity and (W, Y_W) is a V-module. In particular, in the definition of the notion of V-module, the Jacobi identity may be replaced by weak associativity for Y_W.* $\square$

Remark 4.4.6 Because commutative associative algebras with identity are vertex algebras, it is easy to find an example to show that in the notion of V-module, associativity properties are genuinely stronger than commutativity properties. Here, in fact, is a simple example where commutativity of the action holds but associativity fails. Take V to be the commutative associative algebra with a basis $\{1, b\}$ such that $b^2 = 1$ and W to be a 1-dimensional space on which 1 acts as the identity operator and b as 0. Thus in the definition of the notion of V-module, commutativity properties alone *do not* contain enough information to recover the Jacobi identity.

In Chapter 5 below, the iterate formula (cf. (3.1.11))

$$Y_W(Y(u, x_0)v, x_2) = \mathrm{Res}_{x_1}\left(x_0^{-1}\delta\left(\frac{x_1 - x_2}{x_0}\right) Y_W(u, x_1)Y_W(v, x_2) \right.$$
$$\left. - x_0^{-1}\delta\left(\frac{x_2 - x_1}{-x_0}\right) Y_W(v, x_2)Y_W(u, x_1) \right) \quad (4.4.3)$$

for an action of V on a space W will follow from certain considerations, and we will want to know that we consequently have a V-module. For this, for $w \in W$ let l be a nonnegative integer such that $x^l Y_W(u, x)w \in W[[x]]$. Applying (4.4.3) to w and then multiplying by $(x_0 + x_2)^l$ we obtain the weak associativity relation (4.3.1). Thus in view of Theorem 4.4.5 we have:

Corollary 4.4.7 *Let V be a vertex algebra, W a vector space and Y_W a linear map from V to $(\mathrm{End}\ W)[[x, x^{-1}]]$ such that $Y_W(\mathbf{1}, x) = 1$, $Y_W(v, x)w \in W((x))$ for $v \in V$, $w \in W$, and such that the iterate formula (4.4.3) holds for $u, v \in V$. Then Y_W satisfies the Jacobi identity and (W, Y_W) is a V-module. In particular, in the definition of the notion of V-module, the Jacobi identity may be replaced by the iterate formula (4.4.3).* $\square$

Remark 4.4.8 Since we have just discussed the iterate formula, it is natural to comment here that everything in Section 3.8 (except Remark 3.8.3), concerning the notion of normal-ordered product and its relation with the iterate formula, and so on, carries over immediately to module actions.

4.5 Further elementary notions

Here we introduce elementary notions such as the notions of homomorphism, submodule, submodule generated by a subset, annihilating ideal and annihilating submodule. Two of the results (Propositions 4.5.7 and 4.5.8) give new associativity- and commutativity-type properties. We also consider the weak nilpotency on modules of the vertex operator associated to a weakly nilpotent element. Quite a number of properties discussed in this section tend to reinforce the analogy between vertex algebras and commutative associative algebras.

Starting in this section we shall often drop the symbol W from the notation Y_W for a module action.

First we discuss the notion of module homomorphism and related notions. Let W_1 and W_2 be V-modules. A *V-homomorphism* (or *module map*) from W_1 to W_2 is a linear map ψ such that

$$\psi(Y(v, x)w) = Y(v, x)\psi(w) \quad \text{for } v \in V, \ w \in W_1, \tag{4.5.1}$$

or equivalently,

$$\psi(v_n w) = v_n \psi(w) \quad \text{for } v \in V, \ w \in W_1, \ n \in \mathbb{Z}. \tag{4.5.2}$$

The space of module maps is denoted $\mathrm{Hom}_V(W_1, W_2)$.

If V is a vertex operator algebra and W_1 and W_2 are V-modules, then a V-homomorphism ψ from W_1 to W_2 is compatible with the gradings:

$$\psi((W_1)_{(h)}) \subset (W_2)_{(h)} \quad \text{for } h \in \mathbb{C}, \tag{4.5.3}$$

because ψ commutes with $L(0) \ (= \omega_1)$ (by (4.5.2)).

To determine that a given linear map is a homomorphism of V-modules, one need only check (4.5.1) for a generating set of our vertex algebra V:

Proposition 4.5.1 *Let W_1 and W_2 be V-modules and let $\psi \in \mathrm{Hom}_{\mathbb{C}}(W_1, W_2)$. Suppose that*

$$Y(a, x)\psi = \psi Y(a, x) \quad \text{for } a \in S, \tag{4.5.4}$$

where S is a given generating set of V. Then ψ is a V-homomorphism.

Proof. Set

$$C_V(\psi) = \{v \in V \mid Y(v, x)\psi = \psi Y(v, x)\}. \tag{4.5.5}$$

We must prove that $C_V(\psi) = V$. Since $S \subset C_V(\psi)$ and $\langle S \rangle = V$ by assumption, it suffices to show $C_V(\psi)$ is a vertex subalgebra of V. Clearly, $C_V(\psi)$ is a vector subspace and contains $\mathbf{1}$. For $u, v \in C_V(\psi)$ and $n \in \mathbb{Z}$, it follows from (3.8.16), on W_1 and on W_2, that $u_n v \in C_V(\psi)$, and so $C_V(\psi)$ is a vertex subalgebra of V. $\quad\square$

The notions of isomorphism, endomorphism and automorphism are defined in the obvious ways. The algebra of V-module endomorphisms of a module W is denoted $\mathrm{End}_V(W)$.

Definition 4.5.2 Let W be a V-module. A *submodule* of W is a subspace U such that (U, Y) is itself a V-module.

This amounts to saying that $Y(v, x)w \in U((x))$ for $v \in V$, $w \in U$, i.e., $v_n w \in U$ for $v \in V$, $n \in \mathbb{Z}$, $w \in U$.

Remark 4.5.3 Let V be a vertex operator algebra, and W a V-module. Then any submodule of W for V as a vertex algebra is automatically a module for V as a vertex operator algebra because the grading is given by the eigenspaces of $L(0)$ $(= \omega_1)$.

Remark 4.5.4 In view of Remark 3.9.8, a submodule of the adjoint module V amounts to a left ideal, while a $\mathcal{D}$-stable submodule of the adjoint module V amounts to an ideal. If V is a vertex operator algebra, then since $\mathcal{D} = L(-1) = \omega_0$, a submodule of the adjoint module is automatically $\mathcal{D}$-stable, and so a submodule of the adjoint module amounts to an ideal.

The notion of irreducible module and completely reducible (or semisimple) module are defined in the obvious ways. We have the following Schur's Lemma (cf. [FHL] and Proposition 3.11.4 and Remark 3.11.5):

Proposition 4.5.5 *Let V be a vertex operator algebra, and W an irreducible V-module. Then*

$$\mathrm{End}_V(W) = \mathbb{C}. \tag{4.5.6}$$

Proof. We must show that each V-endomorphism ψ of W is a scalar. For $\lambda \in \mathbb{C}$, denote by W_λ^ψ the λ-eigenspace of ψ. What we must show is that $W = W_\lambda^\psi$ for some $\lambda \in \mathbb{C}$. Since ψ commutes with all the operators v_n for $v \in V$ and $n \in \mathbb{Z}$, each W_λ^ψ is a V-submodule of W, and so from the irreducibility of W we have $W = W_\lambda^\psi$ if $W_\lambda^\psi \neq 0$. Thus we are reduced to showing that $W_\lambda^\psi \neq 0$ for some $\lambda \in \mathbb{C}$.

Choose any $h \in \mathbb{C}$ such that $W_{(h)} \neq 0$. In view of (4.5.3), ψ preserves $W_{(h)}$. Since $\dim W_{(h)} < \infty$ (recall (4.1.16)) and we are working over $\mathbb{C}$, ψ has an eigenvector in $W_{(h)}$, and so $W_\lambda^\psi \neq 0$ for some $\lambda \in \mathbb{C}$. $\square$

Now we discuss the notion of submodule generated by a subset of a module. For a subset T of a V-module W we define $\langle T \rangle$ to be the smallest submodule containing T and we call $\langle T \rangle$ the *submodule generated by T*, so that $\langle T \rangle$ is the intersection of all submodules of W containing T, and also

$$\langle T \rangle = \mathrm{span}\ \{v_{n_1}^{(1)} \cdots v_{n_r}^{(r)} w \mid r \in \mathbb{N},\ v^{(1)}, \ldots, v^{(r)} \in V,\ n_1, \ldots, n_r \in \mathbb{Z},\ w \in T\}. \tag{4.5.7}$$

Of course, for an associative algebra A and an A-module M, the submodule generated by a subset T of M is simply AT, the linear span of the elements at for $a \in A$, $t \in T$. It turns out that we have the following nontrivial analogue ([DM5], [Li5]) of this fact, which is related to the weak associativity property in vertex algebra theory, and which strengthens (4.5.7).

Proposition 4.5.6 *Let W be a V-module and let T be a subset of W. Then*

$$\langle T \rangle = \mathrm{span}\ \{v_n w \mid v \in V,\ n \in \mathbb{Z},\ w \in T\}. \tag{4.5.8}$$

This proposition will follow immediately from the following result ([DLM8], [Li15]), which applies to vertex algebras as well as modules.

Proposition 4.5.7 *Let W be a V-module (possibly equal to the vertex algebra V) and let $u, v \in V$, $p, q \in \mathbb{Z}$ and $w \in W$. Then $u_p v_q w$ can be expressed as a linear combination of elements of the form $t_r w$, with $t \in V$ and $r \in \mathbb{Z}$, and the elements t can be chosen to be of the form $u_s v$ for $s \in \mathbb{Z}$. A precise formula is as follows: Let l be a nonnegative integer such that*

$$u_n w = 0 \quad \text{for } n \geq l \tag{4.5.9}$$

and let m be a nonnegative integer such that

$$v_n w = 0 \quad \text{for } n > m + q. \tag{4.5.10}$$

Then

$$u_p v_q w = \sum_{i=0}^{m} \sum_{j=0}^{l} \binom{p-l}{i} \binom{l}{j} (u_{p-l-i+j} v)_{q+l+i-j} w. \tag{4.5.11}$$

Proof. As one expects, this will follow from the weak associativity formula. But the argument will necessarily be subtle.

Since $u_n w = 0$ for $n \geq l$, from Remark 4.3.2 we have

$$(x_0 + x_2)^l Y(u, x_0 + x_2) Y(v, x_2) w = (x_0 + x_2)^l Y(Y(u, x_0) v, x_2) w. \tag{4.5.12}$$

In order to extract $u_p v_q w$ from the left-hand side, we first note that

$$
\begin{aligned}
u_p v_q w &= \mathrm{Res}_{x_1} \mathrm{Res}_{x_2} x_1^p x_2^q Y(u, x_1) Y(v, x_2) w \\
&= \mathrm{Res}_{x_0} \mathrm{Res}_{x_1} \mathrm{Res}_{x_2} x_0^{-1} \delta\left(\frac{x_1 - x_2}{x_0}\right) x_1^p x_2^q Y(u, x_1) Y(v, x_2) w \\
&= \mathrm{Res}_{x_0} \mathrm{Res}_{x_1} \mathrm{Res}_{x_2} x_1^{-1} \delta\left(\frac{x_0 + x_2}{x_1}\right) x_1^p x_2^q Y(u, x_1) Y(v, x_2) w \\
&= \mathrm{Res}_{x_0} \mathrm{Res}_{x_1} \mathrm{Res}_{x_2} x_1^{-1} \delta\left(\frac{x_0 + x_2}{x_1}\right) (x_0 + x_2)^p x_2^q Y(u, x_0 + x_2) Y(v, x_2) w \\
&= \mathrm{Res}_{x_0} \mathrm{Res}_{x_2} (x_0 + x_2)^p x_2^q Y(u, x_0 + x_2) Y(v, x_2) w. \tag{4.5.13}
\end{aligned}
$$

In fact, this is clearly a special case of the general fact (proved the same way)

$$\mathrm{Res}_{x_1} \mathrm{Res}_{x_2} F(x_1, x_2) = \mathrm{Res}_{x_0} \mathrm{Res}_{x_2} F(x_0 + x_2, x_2) \tag{4.5.14}$$

for any doubly-infinite formal Laurent series $F(x_1, x_2)$ truncated from below in x_2.

It now follows from (4.5.12) and (4.5.13) that

$$
\begin{aligned}
u_p v_q w &= \mathrm{Res}_{x_0} \mathrm{Res}_{x_2} (x_0 + x_2)^{p-l} x_2^q \left[(x_0 + x_2)^l Y(u, x_0 + x_2) Y(v, x_2) w \right] \\
&= \mathrm{Res}_{x_0} \mathrm{Res}_{x_2} (x_0 + x_2)^{p-l} x_2^q \left[(x_0 + x_2)^l Y(Y(u, x_0) v, x_2) w \right], \tag{4.5.15}
\end{aligned}
$$

and it is crucial to note that in this last expression, we of course cannot cancel the expression $(x_0 + x_2)^l$; the brackets are necessary. Of course, the reason (4.5.15) is well defined is that in the expression in brackets, the powers of x_2 are truncated from below (recall (4.5.12)); so only a polynomial truncation of the formal series $(x_0 + x_2)^{p-l}$ contributes to the residue with respect to x_2 in (4.5.15). It is now clear that $u_p v_q w$ equals some linear combination of elements of the form $(u_s v)_r w$, giving the first assertion of the Proposition.

To give a formula for the expression, from (4.5.12) and (4.5.10) we see that we may take the polynomial truncation of $(x_0 + x_2)^{p-l}$ to be

$$p(x_0, x_2) = \sum_{i=0}^{m} \binom{p-l}{i} x_0^{p-l-i} x_2^i \qquad (4.5.16)$$

(with highest exponent of x_2 equal to m). Thus we have our desired result:

$$u_p v_q w = \mathrm{Res}_{x_0} \mathrm{Res}_{x_2} p(x_0, x_2) x_2^q (x_0 + x_2)^l Y(Y(u, x_0)v, x_2)w, \qquad (4.5.17)$$

which immediately gives (4.5.11). $\square$

As one might expect, Proposition 4.5.7 has a natural analogue based on the weak commutativity relation in place of the weak associativity relation (cf. Lemma 3.10 of [Li14] or Lemma 2.1 of [Li15]). This analogue, which follows, of course applies to algebras as well as modules.

Proposition 4.5.8 *Let W be a V-module (possibly equal to V) and let $u, v \in V$, $p, q \in \mathbb{Z}$ and $w \in W$. Then $u_p v_q w$ can be expressed as a linear combination of elements of the form $v_r u_s w$ for $r, s \in \mathbb{Z}$. A precise formula is as follows: Let k be a nonnegative integer such that*

$$u_n v = 0 \quad for\ n \geq k \qquad (4.5.18)$$

and let m be a nonnegative integer such that

$$v_n w = 0 \quad for\ n > m + q. \qquad (4.5.19)$$

Then

$$u_p v_q w = \sum_{i=0}^{m} \sum_{j=0}^{k} (-1)^{i+j} \binom{-k}{i} \binom{k}{j} v_{q+i+j} u_{p-i-j} w. \qquad (4.5.20)$$

Proof. We use weak commutativity. Since $u_n v = 0$ for $n \geq k$, we have

$$(x_1 - x_2)^k Y(u, x_1)Y(v, x_2) = (x_1 - x_2)^k Y(v, x_2)Y(u, x_1). \qquad (4.5.21)$$

Proceeding somewhat as in the previous proof, we have

$$
\begin{aligned}
u_p v_q w &= \operatorname{Res}_{x_1}\operatorname{Res}_{x_2} x_1^p x_2^q Y(u, x_1)Y(v, x_2)w \\
&= \operatorname{Res}_{x_1}\operatorname{Res}_{x_2} x_1^p x_2^q (x_1 - x_2)^{-k}\left[(x_1 - x_2)^k Y(u, x_1)Y(v, x_2)w\right] \\
&= \operatorname{Res}_{x_1}\operatorname{Res}_{x_2} x_1^p x_2^q (x_1 - x_2)^{-k}\left[(x_1 - x_2)^k Y(v, x_2)Y(u, x_1)w\right],
\end{aligned}
\qquad (4.5.22)
$$

and, as in the previous proof, we of course cannot remove the brackets in this last expression; the reason (4.5.22) is well defined is that in this bracketed expression, the powers of x_2 are truncated from below (because (4.5.21) is applied to w), so that only a polynomial truncation of the formal series $(x_1 - x_2)^{-k}$ contributes to the residue with respect to x_2. Thus $u_p v_q w$ clearly equals some linear combination of elements of the form $v_r u_s w$, justifying the first assertion of the Proposition.

As for a precise formula, we see that we may take the polynomial truncation of $(x_1 - x_2)^{-k}$ to be

$$
q(x_1, x_2) = \sum_{i=0}^{m} (-1)^i \binom{-k}{i} x_1^{-k-i} x_2^i,
\qquad (4.5.23)
$$

and we have the desired formula:

$$
u_p v_q w = \operatorname{Res}_{x_1}\operatorname{Res}_{x_2} x_1^p x_2^q q(x_1, x_2)(x_1 - x_2)^k Y(v, x_2)Y(u, x_1)w,
\qquad (4.5.24)
$$

giving (4.5.20). $\square$

Remark 4.5.9 Since the polynomial $q(x_1, x_2)$ in (4.5.23) and (4.5.24) is a "good approximation" of $(x_1 - x_2)^{-k}$, the factor $q(x_1, x_2)(x_1 - x_2)^k$ in (4.5.24) is "approximately" 1, and to this extent, the right-hand side of (4.5.24) is "approximately"

$$
\operatorname{Res}_{x_1}\operatorname{Res}_{x_2} x_1^p x_2^q Y(v, x_2)Y(u, x_1)w,
\qquad (4.5.25)
$$

which simply equals $v_q u_p w$. That is, in a suitable sense,

$$
u_p v_q w \sim v_q u_p w.
\qquad (4.5.26)
$$

To illustrate this in a precise way, let us assume that $k = 1$, that is, $u_n v = 0$ for $n \geq 1$, or

$$
(x_1 - x_2)[Y(u, x_1), Y(v, x_2)] = 0.
\qquad (4.5.27)
$$

In this case, $q(x_1, x_2)$ is given simply by

$$
q(x_1, x_2) = \sum_{i=0}^{m} x_1^{-1-i} x_2^i
\qquad (4.5.28)
$$

and so

$$
q(x_1, x_2)(x_1 - x_2)^k = q(x_1, x_2)(x_1 - x_2) = 1 - \left(\frac{x_2}{x_1}\right)^{m+1},
\qquad (4.5.29)
$$

where m is as in (4.5.19). Thus (4.5.20) becomes precisely the simple formula

$$
u_p v_q w = v_q u_p w - v_{q+m+1} u_{p-m-1} w.
\qquad (4.5.30)
$$

This fact arose for example in a special case in [Ge2], Lemma 3.1.

Now we return to our discussion of submodules and related material. It will sometimes be convenient (although not necessary) to invoke Proposition 4.5.6.

For a subset S of V, we denote by (S) the smallest ideal of V containing S. Then (S) is the intersection of all ideals containing S. In view of Remark 4.5.4, Proposition 4.5.6 together with (3.1.32) immediately yields:

Corollary 4.5.10 *Let S be a subset of V. Then*

$$(S) = \text{span } \{v_n \mathcal{D}^i(u) \mid v \in V,\ n \in \mathbb{Z},\ i \geq 0,\ u \in S\}. \tag{4.5.31}$$

If V is a vertex operator algebra, then

$$(S) = \text{span } \{v_n u \mid v \in V,\ n \in \mathbb{Z},\ u \in S\}. \quad \square \tag{4.5.32}$$

Let W be a V-module. We now discuss relations between ideals of V and submodules of W. For a subset T of W we define

$$\mathcal{I}_V(T) = \{v \in V \mid Y(v, x)w = 0 \quad \text{for } w \in T\}, \tag{4.5.33}$$

the *annihilator of T in V*. Then we have:

Proposition 4.5.11 *For a subset T of W, the annihilator $\mathcal{I}_V(T)$ is an ideal of V. Furthermore,*

$$\mathcal{I}_V(T) = \mathcal{I}_V(\langle T \rangle). \tag{4.5.34}$$

Proof. Clearly, $\mathcal{I}_V(T)$ is a subspace of V. The relation $\mathcal{D}\mathcal{I}_V(T) \subset \mathcal{I}_V(T)$ holds because for $v \in \mathcal{I}_V(T)$, $w \in T$,

$$Y(\mathcal{D}v, x)w = \frac{d}{dx}Y(v, x)w = 0.$$

Let $u \in V$, $v \in \mathcal{I}_V(T)$. For $w \in T$, let l be a nonnegative integer such that the weak associativity relation (4.3.1) holds. From

$$(x_0 + x_2)^l Y(Y(u, x_0)v, x_2)w = (x_0 + x_2)^l Y(u, x_0 + x_2)Y(v, x_2)w = 0$$

we have

$$Y(Y(u, x_0)v, x_2)w = 0$$

(since $Y(Y(u, x_0)v, x_2)w$ is truncated from below in powers of x_0), so that $Y(u, x)v \in \mathcal{I}_V(T)((x))$. In view of Remark 3.9.8, $\mathcal{I}_V(T)$ is an ideal of V.

With the inclusion $\mathcal{I}_V(\langle T \rangle) \subset \mathcal{I}_V(T)$ being obvious, we need only prove $\mathcal{I}_V(T) \subset \mathcal{I}_V(\langle T \rangle)$. Let $v \in \mathcal{I}_V(T)$. For $u \in V$, $w \in T$, let k a nonnegative integer such that the weak commutativity relation (4.2.1) holds. Then

$$(x_1 - x_2)^k Y(v, x_1)Y(u, x_2)w = (x_1 - x_2)^k Y(u, x_2)Y(v, x_1)w = 0, \tag{4.5.35}$$

which implies that $Y(v, x_1)Y(u, x_2)w = 0$. It follows from Proposition 4.5.6 that $v \in \mathcal{I}_V(\langle T \rangle)$, concluding the proof. $\square$

Set

$$I_W = \mathcal{I}_V(W) = \{v \in V \mid Y_W(v, x) = 0\} \qquad (4.5.36)$$

(the kernel of the linear map Y_W), the annihilator of the module W. In view of Proposition 4.5.11, I_W is an ideal of V.

Definition 4.5.12 The module W is *faithful* if Y_W is injective, i.e., if $I_W = 0$.

Remark 4.5.13 The linear map Y_W gives rise to a linear isomorphism f from V/I_W onto the space

$$Y_W(V) = \{Y_W(v, x) \mid v \in V\} \subset (\text{End } W)[[x, x^{-1}]]. \qquad (4.5.37)$$

We know from Remark 3.9.9 that V/I_W has a natural (quotient) vertex algebra structure. By transport of structure, there is a vertex algebra structure on $Y_W(V)$ such that f is a vertex algebra isomorphism and Y_W is a vertex algebra homomorphism from V to $Y_W(V)$. A further study will be carried out in Chapter 5; cf. Remark 3.1.7.

Let W be a V-module as before and let S be a subset of V. Define

$$\mathcal{M}_W(S) = \{w \in W \mid Y(v, x)w = 0 \quad \text{for } v \in S\}, \qquad (4.5.38)$$

the *annihilator of S in W*. Then we have (cf. Proposition 4.5.11):

Proposition 4.5.14 *For a subset S of V, the annihilator $\mathcal{M}_W(S)$ is a submodule of W. Furthermore,*

$$\mathcal{M}_W(S) = \mathcal{M}_W((S)), \qquad (4.5.39)$$

(recall that (S) is the ideal of V generated by S).

Proof. Clearly, $\mathcal{M}_W(S)$ is a subspace of W. Let $v \in V$, $w \in \mathcal{M}_W(S)$. For $u \in S$, let k be a nonnegative integer such that the weak commutativity relation (4.2.1) holds. Then

$$(x_1 - x_2)^k Y(u, x_1)Y(v, x_2)w = (x_1 - x_2)^k Y(v, x_2)Y(u, x_1)w = 0,$$

so that $Y(u, x_1)Y(v, x_2)w = 0$. Thus $Y(v, x)w \in \mathcal{M}_W(S)((x))$, and so $\mathcal{M}_W(S)$ is a submodule.

For the second part we need only prove $\mathcal{M}_W(S) \subset \mathcal{M}_W((S))$. Let $w \in \mathcal{M}_W(S)$. By Proposition 4.5.11, the annihilator $\mathcal{I}_V(\{w\})$ of $\{w\}$ in V is an ideal, and this ideal includes S and hence includes (S). Thus $w \in \mathcal{M}_W((S))$ and so $\mathcal{M}_W(S) \subset \mathcal{M}_W((S))$.
    $\square$

The following result (cf. [DL3], Proposition 11.9) is an immediate consequence of Proposition 4.5.11:

Corollary 4.5.15 *Suppose that V is simple and let W be a V-module. Then*

$$Y(v, x)w \neq 0 \quad for\ 0 \neq v \in V,\ 0 \neq w \in W. \quad \square \qquad (4.5.40)$$

Now, let $T = U$ be a vector subspace of our V-module W rather than a subset. We define

$$\mathcal{S}_V(U) = \{v \in V \mid v_n U \subset U \quad \text{for all } n \in \mathbb{Z}\}, \qquad (4.5.41)$$

to be the *stablizer of U in V*. We have $\mathcal{I}_V(U) \subset \mathcal{S}_V(U)$. It is clear that $\mathcal{S}_V(U)$ is a subspace of V and that $\mathbf{1} \in \mathcal{S}_V(U)$. Using the iterate formula (3.8.16) we see that

$$v_m v' \in \mathcal{S}_V(U) \quad \text{for } v, v' \in \mathcal{S}_V(U),\ m \in \mathbb{Z}. \qquad (4.5.42)$$

Thus $\mathcal{S}_V(U)$ is a vertex subalgebra of V, and we have proved:

Proposition 4.5.16 *Let U be a subspace of a V-module W. Then the stablizer $\mathcal{S}_V(U)$ of U in V is a vertex subalgebra of V.* $\quad \square$

In terms of the notion of stablizer, we observe that a subspace U of W is a submodule if and only if $\mathcal{S}_V(U) = V$. The following result will be useful in Chapter 6:

Proposition 4.5.17 *Let S be a generating subset of V and let W be a V-module. Let A be the associative subalgebra of $\mathrm{End}\ W$ generated by the operators a_n for $a \in S$, $n \in \mathbb{Z}$. Then the V-submodules of W are exactly the A-submodules of W.*

Proof. What we must prove is that if U is a subspace of W stable under the action of A, or equivalently,

$$a_n U \subset U \quad \text{for } a \in S,\ n \in \mathbb{Z}, \qquad (4.5.43)$$

then U is a V-submodule, i.e.,

$$v_n U \subset U \quad \text{for all } v \in V,\ n \in \mathbb{Z}. \qquad (4.5.44)$$

That is, we must prove that $S \subset \mathcal{S}_V(U)$ implies that $V = \mathcal{S}_V(U)$. But this holds because $\mathcal{S}_V(U)$ is a vertex subalgebra of V by Proposition 4.5.16 and S generates V. $\quad \square$

In classical associative algebra theory, a nilpotent element acts nilpotently on any module. This is also true for weakly nilpotent elements in vertex algebra theory. First, the proof of Proposition 3.10.2 immediately gives:

Proposition 4.5.18 *Let (W, Y_W) be a V-module. Then*

$$\substack{\circ\\\circ} Y_W(v, x)^r \substack{\circ\\\circ} = Y_W((v_{-1})^r \mathbf{1}, x) \qquad (4.5.45)$$

for $v \in V$, $r > 0$. $\quad \square$

Let v be a weakly nilpotent element of V. By Proposition 3.10.2, $(v_{-1})^r \mathbf{1} = 0$ for some $r > 0$. In view of Proposition 4.5.18, we then have ${}^\circ_\circ Y_W(v, x)^r {}^\circ_\circ = 0$. On the other hand, assume that W is faithful and that ${}^\circ_\circ Y_W(v, x)^r {}^\circ_\circ = 0$. Then it follows from (4.5.45) that $(v_{-1})^r \mathbf{1} = 0$. This in turn implies that v is weakly nilpotent. Thus we conclude:

Proposition 4.5.19 *Let (W, Y_W) be a V-module. Then for any weakly nilpotent element v of V, with $r > 0$ chosen so that $(v_{-1})^r \mathbf{1} = 0$ we have ${}^\circ_\circ Y_W(v, x)^r {}^\circ_\circ = 0$. On the other hand, if W is a faithful V-module and $v \in V$ satisfies the condition that ${}^\circ_\circ Y_W(v, x)^r {}^\circ_\circ = 0$ for some $r > 0$, then $(v_{-1})^r \mathbf{1} = 0$, and in particular, v is weakly nilpotent.* $\square$

Remark 4.5.20 Corollary 3.10.4, for the commuting case $v_n v = 0$ for $n \geq 0$, applies as well to modules.

4.6 Tensor product modules for tensor product vertex algebras

Consider a tensor product vertex algebra

$$V = V_1 \otimes \cdots \otimes V_r \tag{4.6.1}$$

as in (3.12.7)–(3.12.9) and Proposition 3.12.5. For $i = 1, \ldots, r$, let W_i be a V_i-module. Then the space

$$W = W_1 \otimes \cdots \otimes W_r \tag{4.6.2}$$

becomes a V-module via the natural definition

$$Y_W(v^{(1)} \otimes \cdots \otimes v^{(r)}, x) = Y_{W_1}(v^{(1)}, x) \otimes \cdots \otimes Y_{W_r}(v^{(r)}, x) \tag{4.6.3}$$

for $v^{(i)} \in V_i$:

Proposition 4.6.1 *The structure (W, Y_W) is a V-module.*

Proof. The proof of Proposition 3.12.5, based on weak commutativity and weak associativity, carries over immediately to the present situation. $\square$

Suppose now that each V_i is a vertex operator algebra for $i = 1, \ldots, r$, so that V is again a vertex operator algebra, by (3.12.21)–(3.12.23) and Proposition 3.12.8. Assume that W_i is a module for V_i (viewed as a vertex operator algebra) for $i = 1, \ldots, r$, and impose the tensor product grading on $W = W_1 \otimes \cdots \otimes W_r$ (as in (3.12.21)). This grading is compatible with the natural tensor product action of $L(0)$. In order for the vertex algebra module $W = W_1 \otimes \cdots \otimes W_r$ to be a vertex operator algebra module, we need to ensure the finite dimensionality of the homogeneous subspaces of W (recall (4.1.16)); the truncation of the grading of W (recall (4.1.17)) is automatic. This finite dimensionality certainly holds in case each W_i is an irreducible V_i-module, by (4.1.22). Thus we have:

Proposition 4.6.2 *Let $V_1, \ldots, V_r$ be vertex operator algebras, let W_i be a V_i-module, $i = 1, \ldots, r$, and impose the tensor product grading on $W = W_1 \otimes \cdots \otimes W_r$. Assume that each homogeneous subspace of W is finite dimensional; this holds in particular if each W_i is an irreducible V_i-module. Then W is a module for the vertex operator algebra $V = V_1 \otimes \cdots \otimes V_r$.* $\square$

Remark 4.6.3 In the context of either vertex algebras or vertex operator algebras, given modules $W_1^{(1)}, \ldots, W_r^{(1)}$ and $W_1^{(2)}, \ldots, W_r^{(2)}$ for $V_1, \ldots, V_r$, respectively, and V_i-module maps $W_i^{(1)} \to W_i^{(2)}$ for $i = 1, \ldots, r$, the tensor product of these maps is a V-module map from $W_1^{(1)} \otimes \cdots \otimes W_r^{(1)}$ to $W_1^{(2)} \otimes \cdots \otimes W_r^{(2)}$ (assuming the finite dimensionality of the homogeneous subspaces in the vertex operator algebra case).

Remark 4.6.4 We emphasize that we are discussing tensor product modules for *tensor product* vertex algebras or vertex operator algebras. There is a general theory of tensor product modules for a single vertex operator algebra ([HL4]–[HL7], [Hua8]), but this is a completely different story (cf. Remark 3.3.16).

Remark 4.6.5 The reader can consult Section 4.7 of [FHL] for the following results on irreducibility. Under suitable conditions, the V-module $W_1 \otimes \cdots \otimes W_r$ is irreducible if and only if each W_i is an irreducible V_i-module. In particular, $V_1 \otimes \cdots \otimes V_r$ is a simple vertex operator algebra if and only if each V_i is a simple vertex operator algebra. More substantially, an irreducible module for the tensor product $V_1 \otimes \cdots \otimes V_r$ of vertex operator algebras V_i is necessarily a tensor product $W_1 \otimes \cdots \otimes W_r$ of modules W_i (necessarily irreducible) for the V_i.

4.7 Vacuum-like vectors

In this section, we introduce the notion of "vacuum-like vector" in a V-module and discuss certain useful properties of such vectors. This section is based on [Li1].

From the creation property (3.1.5), we have $v_n \mathbf{1} = 0$ for $v \in V$, $n \geq 0$. Motivated by this we define the following notion:

Definition 4.7.1 Let W be a V-module. A vector $w \in W$ is said to be *vacuum-like* if

$$v_n w = 0 \quad \text{for } v \in V, \ n \geq 0, \tag{4.7.1}$$

that is, $Y(v, x)w$ involves only nonnegative (integral) powers of x for all $v \in V$.

With this definition, the vacuum vector $\mathbf{1}$ (in the adjoint module V) is vacuum-like. Furthermore, in view of Remark 3.2.4, an element w of the adjoint module is vacuum-like if and only if w lies in the center $C(V)$ of V (recall (3.11.2)), that is, the commutator $[Y(v, x_1), Y(w, x_2)]$ vanishes for $v \in V$.

Remark 4.7.2 For a (general) V-module W, an equivalent condition for $w \in W$ to be vacuum-like can be naturally formulated in terms of associators: An element w of W is vacuum-like if and only if the associator

$$Y(Y(u, x_0)v, x_2)w - Y(u, x_0 + x_2)Y(v, x_2)w \qquad (4.7.2)$$

vanishes for $u, v \in V$. Indeed, the "only if" part follows from the associator formula
(3.1.13) and the "if" part holds because the associator formula (3.1.14) applied to w
with $v = 1$ gives

$$(Y(u, x_2 + x_0) - Y(u, x_0 + x_2))w = 0,$$

which implies $u_n w = 0$ for $u \in V$, $n \geq 0$. In the case $W = V$, this implies that the
commutator $[Y(u, x_1), Y(w, x_2)]$ vanishes for all $u \in V$ if and only if the associator
(4.7.2) vanishes for all $u, v \in V$. This reflects the equivalence between the commutator
formula and the associator formula discussed in Remark 3.7.3.

To see if a vector is vacuum-like, one in fact only needs to check (4.7.1) for v in a
generating set of V:

Proposition 4.7.3 *Let W be a V-module and let $w \in W$ be such that*

$$v_n w = 0 \quad \text{for } v \in S, \ n \geq 0, \qquad (4.7.3)$$

where S is a generating set of V. Then w is a vacuum-like vector.

Proof. Set

$$K = \{v \in V \mid Y(v, x)w \in W[[x]]\}. \qquad (4.7.4)$$

Then it suffices to prove $K = V$. Clearly, K is a subspace containing $S \cup \mathbb{C}1$. Let
$v, v' \in K$, $m \in \mathbb{Z}$. From the definition,

$$Y(v, x_1)w \in W[[x_1]], \quad Y(v', x)w \in W[[x]].$$

Then using the iterate formula (3.8.16) we get

$$\begin{aligned}
&Y(v_m v', x)w \\
&= \mathrm{Res}_{x_1}\left((x_1 - x)^m Y(v, x_1)Y(v', x)w - (-x + x_1)^m Y(v', x)Y(v, x_1)w\right) \\
&= \mathrm{Res}_{x_1}(x_1 - x)^m Y(v, x_1)Y(v', x)w \\
&\in W[[x]], \qquad (4.7.5)
\end{aligned}$$

so that $v_m v' \in K$. Thus K is a vertex subalgebra of V, and so $K = V$. $\quad\square$

The following result gives a sufficient condition for w to be vacuum-like.

Proposition 4.7.4 *Let (W, Y_W, d) be a V-module (in the sense of Remark 4.1.4) and
let $w \in W$ be such that $dw = 0$. Then*

$$Y_W(v, x)w = e^{xd}v_{-1}w \quad \text{for } v \in V. \qquad (4.7.6)$$

In particular, w is a vacuum-like vector.

Proof. For $v \in V$, if $Y_W(v, x)w = 0$, (4.7.6) is clear. Assume $Y_W(v, x)w \neq 0$. Then there is an integer k such that

$$v_k w \neq 0 \quad \text{and} \quad v_n w = 0 \quad \text{for } n > k. \tag{4.7.7}$$

From the d-bracket formula (4.1.7), we have

$$[d, v_n] = -nv_{n-1} \quad \text{for } n \in \mathbb{Z}. \tag{4.7.8}$$

Then $-(k+1)v_k w = [d, v_{k+1}]w = 0$. If $k \neq -1$, we must have $v_k w = 0$, which contradicts (4.7.7). Thus $k = -1$. Therefore $v_n w = 0$ for $n \geq 0$, that is, w is vacuum-like. Furthermore, using $dw = 0$ and the d-bracket formula (4.1.7), we get

$$e^{xd}Y_W(v, x_1)w = e^{xd}Y_W(v, x_1)e^{-xd}w = Y_W(v, x_1 + x)w. \tag{4.7.9}$$

Because $Y_W(v, x_1)w$ involves only nonnegative integral powers of x_1, we may set x_1 to zero to obtain (4.7.6). $\square$

Remark 4.7.5 Of course, when $W = V$, $d = \mathcal{D}$ and $w = \mathbf{1}$, formula (4.7.6) becomes the usual formula $Y(v, x)\mathbf{1} = e^{x\mathcal{D}}v$ (recall (3.1.29)).

Now let V be a vertex operator algebra. If w is vacuum-like, $L(-1)w = \omega_0 w = 0$. Conversely, if $L(-1)w = 0$, it follows from Proposition 4.7.4 (and Proposition 4.1.5) that w is vacuum-like. Thus we have:

Corollary 4.7.6 *Let V be a vertex operator algebra, W a module for V viewed as a vertex algebra and $w \in W$. Then w is vacuum-like if and only if $L(-1)w = 0$.* $\square$

The following is the main result on vacuum-like vectors:

Proposition 4.7.7 *Let (W, Y_W) be a V-module and let $w \in W$ be a vacuum-like vector. Then the linear map*

$$f : V \to W$$
$$v\ (= v_{-1}\mathbf{1}) \mapsto v_{-1}w\ (= (\text{constant term of } Y(v, x))w) \tag{4.7.10}$$

is a V-homomorphism, uniquely determined by the condition $\mathbf{1} \mapsto w$.

Proof. Let $u, v \in V$. From Remark 4.7.2, we have

$$Y(Y(u, x_0)v, x_2)w = Y(u, x_0 + x_2)Y(v, x_2)w. \tag{4.7.11}$$

Then

$$\begin{aligned}
f(Y(u, x_0)v) &= \operatorname{Res}_{x_2} x_2^{-1} Y(Y(u, x_0)v, x_2)w \\
&= \operatorname{Res}_{x_2} x_2^{-1} Y(u, x_0 + x_2)Y(v, x_2)w \\
&= \lim_{x_2 \to 0} Y(u, x_0 + x_2)Y(v, x_2)w \\
&= Y(u, x_0)v_{-1}w \\
&= Y(u, x_0)f(v).
\end{aligned} \tag{4.7.12}$$

This proves that f is a V-module homomorphism. $\square$

Combining Proposition 4.7.3 with Proposition 4.7.7 we immediately have:

Corollary 4.7.8 *Let S be a subset of V such that S generates V as a vertex algebra, let (W, Y_W) be a V-module and let $w \in W$ be such that*

$$v_n w = 0 \quad \text{for } v \in S, \ n \geq 0. \tag{4.7.13}$$

Then the linear map

$$f : V \to W$$
$$v \ (= v_{-1}\mathbf{1}) \mapsto v_{-1} w \tag{4.7.14}$$

is a V-homomorphism, uniquely determined by the condition $\mathbf{1} \mapsto w$. $\square$

The following result gives conditions ensuring that f is actually an isomorphism:

Proposition 4.7.9 *Let (W, Y_W, d) be a faithful V-module and let $w \in W$ be such that w generates W as a V-module and $dw = 0$. Then the linear map $f : V \to W$, $v \mapsto v_{-1}w$ is a V-isomorphism.*

Proof. In view of Propositions 4.7.4 and 4.7.7, f is a V-homomorphism. Since w generates W, f is surjective. Let $v \in V$ be such that $f(v) = 0$. It follows from Proposition 4.7.4 that $Y(v, x)w = 0$. That is, $w \in \mathcal{M}(\{v\})$ (recall (4.5.38)). Since w generates W as a V-module, using Proposition 4.5.14 we have

$$W = \langle \{w\} \rangle = \mathcal{M}(\{v\}).$$

That is, $Y(v, x) = 0$ on W, and since W is assumed faithful, we must have $v = 0$. This proves that f is injective, completing the proof. $\square$

Remark 4.7.10 For an associative algebra (or ring) A, A is a free A-module. In vertex operator algebra theory, we do not have a notion of free module in the usual sense. In fact, free V-modules in the usual sense (without restricitions) do not exist, just as free vertex algebras in the usual sense do not exist (see [B1], [Roi1]). Proposition 4.7.7 says that V is "relatively free" with generator $\mathbf{1}$ in the sense that for any V-module W and any vacuum-like vector w in W, there exists a unique V-homomorphism from V to W sending $\mathbf{1}$ to w.

4.8 Adjoining a module to a vertex algebra

In this section, based on [Li1], we present an analogue for vertex algebras of the notion of the semidirect product Lie algebra formed by adjoining a Lie-algebra module to a given Lie algebra (cf. [J]).

Let V be a vertex algebra and let (W, Y, d) be a V-module as defined in Remark 4.1.4. That is, (W, Y) is a V-module and d is an endomorphism of W such that

$$[d, Y(v, x)] = Y(\mathcal{D}v, x) = \frac{d}{dx} Y(v, x) \tag{4.8.1}$$

for $v \in V$. We shall define a natural vertex algebra structure on the vector space

$$U = V \oplus W. \tag{4.8.2}$$

To define a linear map Y_U from U to (End $U)[[x, x^{-1}]]$ we must define $Y_U(v, x)$ for $v \in V$ and $Y_U(w, x)$ for $w \in W$. For $v \in V$, we define $Y_U(v, x)$ by

$$Y_U(v, x)(v' + w') = Y(v, x)v' + Y(v, x)w' \quad \text{for } v' \in V, \ w' \in W; \tag{4.8.3}$$

that is, we use the algebra structure of V and the V-module structure of W. For $w \in W$, we define $Y_U(w, x)$ by

$$Y_U(w, x)(v' + w') = e^{xd} Y(v', -x)w \quad \text{for } v' \in V, \ w' \in W; \tag{4.8.4}$$

that is, we use "skew symmetry" for the action of W on V and zero for the action of W on itself. In particular,

$$Y_U(w, x)\mathbf{1} = e^{xd} w. \tag{4.8.5}$$

By linearity we have our linear map $Y_U : U \to$ (End $U)[[x, x^{-1}]]$.

The truncation condition clearly holds, $Y_U(\mathbf{1}, x) = 1$ (the identity operator on U), where $\mathbf{1}$ is the vacuum vector of V, and the creation property holds, in view of (4.8.5).

Define $\mathcal{D}_U \in$ End U by

$$\mathcal{D}_U(v + w) = (v + w)_{-2}\mathbf{1} \quad \text{for } v \in V, \ w \in W, \tag{4.8.6}$$

as in (3.1.24). From (4.8.5) we have

$$\mathcal{D}_U(v + w) = v_{-2}\mathbf{1} + w_{-2}\mathbf{1} = \mathcal{D}(v) + dw, \tag{4.8.7}$$

where $\mathcal{D}$ is the $\mathcal{D}$-operator for V.

Skew symmetry (3.1.30) holds, since

$$\begin{aligned}
Y_U(v + w, x)(v' + w') &= Y(v, x)v' + Y(v, x)w' + e^{xd} Y(v', -x)w \\
&= e^{x\mathcal{D}} Y(v', -x)v + e^{xd} Y_U(w', -x)v + e^{xd} Y(v', -x)w \\
&= e^{x\mathcal{D}_U} Y_U(v' + w', -x)(v + w).
\end{aligned} \tag{4.8.8}$$

We also have (cf. (3.1.33) and (3.1.25))

$$[\mathcal{D}_U, Y_U(v + w, x)] = \frac{d}{dx} Y_U(v + w, x) = Y_U(\mathcal{D}_U(v + w), x) \tag{4.8.9}$$

because

$$\mathcal{D}_U Y_U(v+w, x)(v'+w') - Y_U(v+w, x)\mathcal{D}_U(v'+w')$$
$$= \mathcal{D}Y(v, x)v' + dY(v, x)w' + de^{xd}Y(v', -x)w$$
$$\quad - Y(v, x)\mathcal{D}(v') - Y(v, x)d(w') - e^{xd}Y(\mathcal{D}(v'), -x)w$$
$$= [\mathcal{D}, Y(v, x)]v' + [d, Y(v, x)]w' + \frac{d}{dx}e^{xd}Y(v', -x)w$$
$$= \frac{d}{dx}Y_U(v+w, x)(v'+w')$$

and

$$\mathcal{D}_U Y_U(v+w, x)(v'+w') - Y_U(v+w, x)\mathcal{D}_U(v'+w')$$
$$= [\mathcal{D}, Y(v, x)]v' + [d, Y(v, x)]w' + e^{xd}Y(v', -x)d(w)$$
$$= Y(\mathcal{D}(v), x)v' + Y(\mathcal{D}(v), x)w' + Y_U(d(w), x)v'$$
$$= Y_U(\mathcal{D}_U(v+w), x)(v'+w');$$

we are also using the $\mathcal{D}$- and d-bracket and derivative formulas.

Finally, we show that Y_U satisfies the Jacobi identity. By linearity, it suffices to verify the Jacobi identity for triples (a, b, c) with $a, b, c \in V \cup W$. If two or three of a, b, c lie in W, from the definition all three terms of the Jacobi identity for (a, b, c) are zero, and so the Jacobi identity for (a, b, c) holds. Since U is a V-module, the Jacobi identity for (a, b, c) holds if $a, b \in V$. In view of the S_3-symmetry of the Jacobi identity (Proposition 3.7.1), since the assumptions—skew symmetry and the $\mathcal{D}$-derivative and bracket properties—have been justified, the Jacobi identity for (a, b, c) holds if any two of a, b, c lie in V. Thus the Jacobi identity always holds, and $(U, Y_U, \mathbf{1})$ carries the structure of a vertex algebra. Furthermore, it is clear that $(V, Y, \mathbf{1})$ is a vertex subalgebra and that W is an ideal. Summarizing, we have:

Proposition 4.8.1 *Let $(V, Y, \mathbf{1})$ be a vertex algebra and let (W, Y, d) be a V-module. Then the triple $(U, Y_U, \mathbf{1})$ defined above carries the structure of a vertex algebra, with V a subalgebra and W an ideal.* $\square$

5

Representations of Vertex Algebras and the Construction of Vertex Algebras and Modules

In this chapter, the focal chapter of the work, we shall define and study the notion of *representation* of a vertex algebra. Just as in classical algebraic theories, in particular, the theories of Lie algebras and of associative algebras, this notion of representation is distinct from the notion of *module*, but the two notions will turn out to be essentially equivalent.

Of course, the conceptual distinction between the notion of module and the notion of representation is that a module is a space on which an algebraic structure acts, according to axioms similar to the axioms defining the notion of algebra, while a representation is a *homomorphism* of the algebra into a suitable *endomorphism algebra* of the space.

Specifically, in classical associative and Lie algebra theory, for a vector space W, the vector space End W of endomorphisms of W of course has a natural associative and Lie algebra structure, with W as a canonical module. This leads to the classical notion of representation of an associative or Lie algebra, which is equivalent to the notion of module. Aside from their role in the representation theory of associative and Lie algebras, these classical linear algebras are also the primary sources of examples of simple associative and Lie algebras.

Recall that vertex (operator) algebras are, on the one hand, a kind of generalized algebra equipped with an infinite family of products and, on the other hand, "algebras of vertex operators" (a point of view emphasized in [FLM6]), just as classical associative and Lie algebras can be viewed as algebras of linear operators. The notion of module for a vertex (operator) algebra can of course be thought of from either of these points of view, and in this work we are generally emphasizing the latter viewpoint.

However, in vertex algebra theory, given a vector space W, there is no reasonable notion of "endomorphism vertex algebra" of W such that specifying a V-module structure on W amounts to giving a homomorphism from V to such an "endomorphism vertex algebra." This is one of the main reasons why, in vertex algebra theory, unlike the case in classical algebraic theories, it is a nontrivial matter to construct examples, even very basic examples, of vertex algebras and modules.

It turns out that the representation theory presented in this chapter provides the correct natural replacement(s) for the (nonexistent!) notion of endomorphism vertex algebra based on a given vector space, and at the same time provides very powerful and useful methods for constructing examples of vertex (operator) algebras and modules. In Chapter 6 we shall in fact apply the methods developed in this chapter to construct important families of examples of vertex algebras and modules.

The main theme of this chapter, then, is to develop the appropriate vertex algebra analogues of the notion of endomorphism algebra of a vector space and to establish the natural results for these structures. In contrast with the case in classical algebraic theories, the resulting notion of *representation* of a vertex algebra involves subtle issues not arising in the notion of *module*, which we have already studied extensively. We are able to develop a theory that in particular shows how to construct vertex algebras and modules in an efficient and conceptually satisfying way, starting from sets of what are called "mutually local vertex operators" on a given vector space.

This chapter is based on [Li3], and certain results presented here are new.

Now we describe more specifically what we shall be doing in this chapter. Let W be any given vector space. First, we define (abstract) *weak vertex operators* on W to be formal series that are elements of $\operatorname{Hom}(W, W((x)))$; that is, a weak vertex operator on W is a formal series with coefficients in End W which, when applied to any element of W, yields a *truncated* formal series (with coefficients in W). Here the word "weak" reflects the fact that if W is a module for a vertex algebra V, then the (genuine) vertex operators $Y_W(v, x)$ ($v \in V$) satisfy, in addition, the nontrivial "weak commutativity relation" (recall Proposition 4.2.1), which we exploit below. We shall denote by $\mathcal{E}(W)$ the space of weak vertex operators on W, and we think of the symbol "$\mathcal{E}$" as referring to "endomorphisms" in some generalized sense.

With the iterate formula (3.1.11) as motivation, we shall define a linear map $Y_{\mathcal{E}}$ from $\mathcal{E}(W)$ to (End $\mathcal{E}(W))[[x_0, x_0^{-1}]]$, or of course equivalently, a family, parametrized by $\mathbb{Z}$, of linear maps from $\mathcal{E}(W)$ to End $\mathcal{E}(W)$. (Since each element of $\mathcal{E}(W)$ involves the formal variable x, a second formal variable, say x_0, must be used for the map $Y_{\mathcal{E}}$.) We shall prove certain properties, but the space $\mathcal{E}(W)$ equipped with the map $Y_{\mathcal{E}}$ and with the distinguished element 1_W, the identity operator on W, is *not* a vertex algebra if dim $W \geq 2$. Motivated by this fact, we introduce a notion of "weak vertex algebra" and prove that $\mathcal{E}(W)$ equipped with the indicated structures is a weak vertex algebra (Proposition 5.3.9). (Algebraic structures similar to weak vertex algebras were also independently studied in [LZ3], [LZ4] and [K7]. The reason why this notion [but not the name "weak vertex algebra"] was introduced in [Li3] was that $\mathcal{E}(W)$ was proved to be such a structure.) Weak vertex algebras do not have any "deep" properties (they satisfy no substantial axioms) but the notion of weak vertex algebra is convenient because $\mathcal{E}(W)$ is such a structure; we call $\mathcal{E}(W)$ the *canonical weak vertex algebra* (associated with W).

We then define a *representation* of a vertex algebra V on W to be a weak-vertex-algebra homomorphism from V to the canonical weak vertex algebra $\mathcal{E}(W)$ and we prove the equivalence between the notion of module and the notion of representation (Theorem 5.3.15). *The harder part of this result is based on Corollary 4.4.7, which uses the substantial fact that a proposed module action is indeed a module action if weak associativity holds; recall Theorems 3.6.3 and 4.4.5.* In view of the analogy between vertex algebras and commutative associative algebras, the notion of the canonical weak vertex algebra $\mathcal{E}(W)$ should be compared with the notion of the (associative) endomorphism algebra End W, while the notion of vertex algebra itself should be compared with the notion of commutative associative algebra.

Having noticed that the vertex operators associated to elements of a vertex algebra acting on a module satisfy the weak commutativity relation, we shall study "local subsets" S of $\mathcal{E}(W)$ in the sense that for any $a(x), b(x) \in S$ there exists a nonnegative integer k such that

$$(x_1 - x_2)^k a(x_1) b(x_2) = (x_1 - x_2)^k b(x_2) a(x_1).$$

We shall prove that a subalgebra V of the weak vertex algebra $\mathcal{E}(W)$ is a vertex algebra if and only if V is local (Theorems 5.5.11 and 5.5.14). This result is an analogue of the (tautological) classical fact that a subalgebra of End W is a commutative associative algebra if and only if it consists of pairwise mutually commuting linear operators. We call a *self-local* weak vertex operator on W an (abstract) *vertex operator* on W. Note that a weak vertex operator need not automatically be a vertex operator, and keeping in mind the classical analogy, this is like saying that an operator need not "commute with itself," a subtle matter indeed.

The main result is Theorem 5.5.18, which states that any set of mutually local vertex operators on W generates a vertex algebra, with W as a natural module. This result can be viewed as a (nontrivial) analogue of the (trivial) fact that any set of mutually commuting linear operators generates a commutative associative subalgebra of the endomorphism algebra. (And just as there is no canonical maximal commutative associative subalgebra of End W, there is no canonical maximal vertex subalgebra of $\mathcal{E}(W)$.) Such results and their ramifications and refinements, involving the Virasoro algebra and gradings, developed in this chapter make it easy to solve the problem of constructing vertex operator algebras and modules, as will be demonstrated in the next chapter, and as was carried out in [Li3].

In Section 5.7 we present a very useful general construction theorem, Theorem 5.7.1, for vertex algebras obtained by E. Frenkel, V. Kac, A. Radul and W. Wang in [FKRW], and independently by A. Meurman and M. Primc in [MP2] (see also [MP4] and [P5]) and we prove it and a refinement of this result, Theorem 5.7.4, involving the Virasoro algebra and a grading, by using Theorem 5.5.18 and *its* refinement, Theorem 5.6.14. Then, given a vertex algebra V, by using Theorem 5.5.18 we formulate and prove another useful result—a general existence and uniqueness theorem (Theorem 5.7.6) for a V-module structure on a given vector space equipped with "partial" V-module structure. These results are closely related to the method by which families of examples of vertex algebras were constructed in [Li3]. Using the methods of this chapter we also prove a variant of a theorem of X. Xu obtained in [Xu2] (see also [Xu9], [Xu12]) which is in a somewhat similar spirit.

5.1 Weak vertex operators

In this section we define and consider notions of "weak vertex operator" and of "homogeneous weak vertex operator" on a given vector space and discuss the formal differential operator d/dx on the space of weak vertex operators. Weak vertex operators will soon be regarded as elements of vector spaces on which we define suitably generalized vertex algebra structures.

Let W be a vector space (over $\mathbb{C}$, as usual), fixed throughout this chapter.

According to the following definition, the "weak vertex operators on W" are generalizations of the vertex operators $Y_W(v, x)$ for $v \in V$, where V is some vertex algebra acting on W (recall Remark 4.1.2).

Definition 5.1.1 A *weak vertex operator* on W is a formal series

$$a(x) = \sum_{n \in \mathbb{Z}} a_n x^{-n-1} \in (\text{End } W)[[x, x^{-1}]] \tag{5.1.1}$$

(with each $a_n \in \text{End } W$) such that for every $w \in W$,

$$a_n w = 0 \quad \text{for } n \text{ sufficiently large,} \tag{5.1.2}$$

that is, $a(x)w \in W((x))$.

In other words, a formal series $a(x)$ is a weak vertex operator if and only if it lies in $\text{Hom}(W, W((x)))$.

Remark 5.1.2 Here we preview for the reader some important and subtle issues, including the goal of this chapter. While certain formal series $a(x)$ will eventually be identifiable with vertex operators of the form $Y_W(v, x)$ with v an element of some vertex algebra acting on W, it is also the case that certain formal series $a(x)$ will be viewed as *vectors in a vertex algebra*, and the elements of such a vertex algebra will consist of formal series of the form (5.1.1), (5.1.2). For such a vector $a(x)$, we shall be defining and considering the vertex operator $Y(a(x), x_0)$; note that we have used here a second formal variable x_0, different from x. Given two vectors $a(x)$ and $b(x)$ lying in such a "vertex algebra of endomorphisms of the vector space W," we shall of course consider the action $Y(a(x), x_0)b(x)$ on the vector $b(x)$ of the vertex operator associated with the "vector" $a(x)$. Although the notation $Y(a(x), x_0)b(x)$ might at first appear wrong because the formal variable x is repeated, this notation is actually appropriate; the formal series a and b (in the same variable x) are viewed as vectors in a space of formal series in the variable x. This element can also be written as $(Y(a, x_0)b)(x)$; it is a formal Laurent series in both x and x_0. Of course, sometimes we will need to replace the variable x by another variable, as we often have been doing throughout the theory. In this chapter, we shall systematically investigate the extent to which a collection of weak vertex operators on W (Definition 5.1.1) can be interpreted as elements of a "vertex algebra of endomorphisms of W," and in particular, we shall define and study the action $Y(a(x), x_0)b(x)$ (or $(Y(a, x_0)b)(x)$) of a weak vertex operator $a(x)$ on another weak vertex operator $b(x)$. Among our main goals in this chapter will be the construction of suitable "vertex algebras of endomorphisms of the vector space W," the interpretation of the space W as a module for these vertex algebras and the interpretation of any module for any vertex algebra as a vertex algebra homomorphism from the given vertex algebra to a "vertex algebra of endomorphisms" of the module. In this way, we shall construct a precise correspondence between the notion of *module* for a vertex algebra and a notion of *representation* of the vertex algebra.

Throughout this chapter, (W, d) is a pair consisting of our vector space W and a given linear operator d on W. By analogy with Remark 4.1.4, we formalize a canonical relation between such an operator d and a type of vertex operator (in the present case, a weak vertex operator):

Definition 5.1.3 A *weak vertex operator on* the pair (W, d) is a weak vertex operator $a(x)$ on W such that

$$[d, a(x)] = a'(x) \left(= \frac{d}{dx} a(x) \right). \tag{5.1.3}$$

From the definition, the weak vertex operators on W constitute the subspace $\text{Hom}(W, W((x)))$ of $(\text{End } W)[[x, x^{-1}]]$, as we mentioned above. We also alternatively write $\mathcal{E}(W)$ for this space:

$$\mathcal{E}(W) = \text{Hom}(W, W((x))). \tag{5.1.4}$$

We think of the symbol "$\mathcal{E}$" as referring to "endomorphisms." It is also clear that the weak vertex operators on (W, d) form a subspace of $\mathcal{E}(W)$, which we denote by $\mathcal{E}(W, d)$:

$$\mathcal{E}(W, d) = \{a(x) \in \mathcal{E}(W) \mid [d, a(x)] = a'(x)\}. \tag{5.1.5}$$

Of course, End W is a subspace of $\mathcal{E}(W)$.

Set

$$\mathcal{D} = \frac{d}{dx} : (\text{End } W)[[x, x^{-1}]] \to (\text{End } W)[[x, x^{-1}]], \tag{5.1.6}$$

the formal differentiation operator, viewed as an endomorphism of $(\text{End } W)[[x, x^{-1}]]$. It is clear that the operator $\mathcal{D}$ preserves both $\mathcal{E}(W)$ and $\mathcal{E}(W, d)$. In terms of the notation (5.1.6), formula (5.1.3) becomes

$$[d, a(x)] = \mathcal{D}a(x). \tag{5.1.7}$$

Recall from Proposition 3.1.18 that we had already been using the symbol $\mathcal{D}$ for a certain endomorphism of a vertex algebra related to the differentiation operator via (3.1.25). Soon we shall see that the operator (5.1.6) plays an analogous role: The operator (5.1.6), viewed as an endomorphism of the space $\mathcal{E}(W)$, will be related to the differentiation operator with respect to a different variable according to a formula similar to (3.1.25).

Remark 5.1.4 If W is finite dimensional, then

$$\mathcal{E}(W) = \text{Hom}(W, W((x))) = (\text{End } W)((x)), \qquad (5.1.8)$$

that is, every formal series in $\mathcal{E}(W)$ is truncated from below. In particular, if W is one-dimensional, we have $\mathcal{E}(W) = \mathbb{C}((x))$.

When we study representations of vertex operator algebras rather than of vertex algebras, we will need to consider the Virasoro algebra. The following notions will be useful:

Definition 5.1.5 If W is a module for the Virasoro algebra, a weak vertex operator $a(x)$ on W is said to be *homogeneous of weight $h \in \mathbb{C}$* if

$$[L(0), a(x)] = ha(x) + x\frac{d}{dx}a(x) \qquad (5.1.9)$$

(cf. (3.1.53)). If in addition $a(x)$ is a weak vertex operator on the pair $(W, L(-1))$ (recall Definition 5.1.3), then we also have

$$[L(0), a(x)] = ha(x) + x[L(-1), a(x)]. \qquad (5.1.10)$$

Denote by $\mathcal{E}(W, L(-1))_{(h)}$ the space of homogeneous weak vertex operators on $(W, L(-1))$ of weight h and let $\mathcal{E}^o(W, L(-1))$ be the linear span of all the homogeneous weak vertex operators on $(W, L(-1))$, so that

$$\mathcal{E}^o(W, L(-1)) = \coprod_{h \in \mathbb{C}} \mathcal{E}(W, L(-1))_{(h)}. \qquad (5.1.11)$$

(The sum is direct because the subspaces are the eigenspaces for the operator $[L(0), \cdot] - x\frac{d}{dx}$.) For constructing ($\mathbb{Z}$-)graded vertex operator algebras we shall also need the $\mathbb{Z}$-graded subspace

$$\mathcal{E}^o_{\mathbb{Z}}(W, L(-1)) = \coprod_{n \in \mathbb{Z}} \mathcal{E}(W, L(-1))_{(n)} \qquad (5.1.12)$$

of $\mathcal{E}^o(W, L(-1))$. Note that we do not need to assume that the space W is itself graded; later we shall relate the structure defined here to graded modules for the Virasoro algebra.

Remark 5.1.6 The identity operator 1_W on our vector space W is certainly a weak vertex operator on W. If W is a module for the Virasoro algebra, it is clear that 1_W is a homogeneous weak vertex operator on $(W, L(-1))$ of weight zero. Set

$$L_W(x) = \sum_{n \in \mathbb{Z}} L(n)x^{-n-2}. \qquad (5.1.13)$$

If the module W is *restricted* in the sense that for every $w \in W$, $L(n)w = 0$ for n sufficiently large, then $L_W(x)$ is a homogeneous weak vertex operator on $(W, L(-1))$ of weight two because

$$[L(-1), L_W(x)] = \sum_{n\in\mathbb{Z}}(-1-n)L(n-1)x^{-n-2} = L'_W(x), \qquad (5.1.14)$$

$$[L(0), L_W(x)] = \sum_{n\in\mathbb{Z}} -nL(n)x^{-n-2} = 2L_W(x) + xL'_W(x). \qquad (5.1.15)$$

Proposition 5.1.7 *Let W be a module for the Virasoro algebra and let $a(x)$ be a homogeneous weak vertex operator of weight $h \in \mathbb{C}$ on $(W, L(-1))$. Then $a'(x)$ is a homogeneous weak vertex operator of weight $h + 1$. That is, the formal differentiation operator $\mathcal{D}$ maps $\mathcal{E}(W, L(-1))_{(h)}$ into $\mathcal{E}(W, L(-1))_{(h+1)}$.*

Proof. We simply compute:

$$\begin{aligned}
[L(0), a'(x)] &= [L(0), a(x)]' = ha'(x) + a'(x) + xa''(x) \\
&= (h+1)a'(x) + xa''(x). \quad \square
\end{aligned}$$

5.2 The action of weak vertex operators on the space of weak vertex operators

Recall from the last section that one should think of weak vertex operators on W as generalizations of the vertex operators $Y_W(v, x)$ with v some element of a vertex algebra V acting on W (viewed as a V-module). We now want to formulate and examine the structures on the space $\mathcal{E}(W)$ that are suggested by such a module action.

Suppose, then, that W is a faithful V-module for a vertex algebra V. Let us write ι_W for the map

$$\begin{aligned}
\iota_W : V &\to \mathcal{E}(W) \\
v &\mapsto Y_W(v, x).
\end{aligned} \qquad (5.2.1)$$

Since W is faithful, the vertex algebra structure on V can be transported to the image $\iota_W(V)$, a space of weak vertex operators. What does this transported structure look like? Once we have formulated it, we will automatically have, by construction, a vertex algebra isomorphism from V to $\iota_W(V)$, viewed as a vertex algebra.

First, the vacuum vector of $\iota_W(V)$ will be

$$\iota_W(\mathbf{1}) = 1_W, \qquad (5.2.2)$$

the identity operator on the space W.

Next, and more substantially, we need to define the vertex operators on the space $\iota_W(V)$. We shall write these operators as $Y_{\mathcal{E}}(a(x), x_0)$, where $a(x)$ is a weak vertex operator in the space $\iota_W(V)$. The action of this operator on a weak vertex operator $b(x) \in \iota_W(V)$ will be written $Y_{\mathcal{E}}(a(x), x_0)b(x)$ (or $(Y_{\mathcal{E}}(a, x_0)b)(x)$); recall our discussion of such notation in the last section.

Now, since we are forcing the linear map ι_W to be a vertex algebra homomorphism, we must define $Y_{\mathcal{E}}(a(x), x_0)b(x)$ in such a way that

$$Y_{\mathcal{E}}(\iota_W(u), x_0)\iota_W(v) = \iota_W(Y(u, x_0)v) \tag{5.2.3}$$

(as in (3.9.1)), that is,

$$Y_{\mathcal{E}}(Y_W(u, x), x_0)Y_W(v, x) = Y_W(Y(u, x_0)v, x), \tag{5.2.4}$$

for $u, v \in V$. (Recall that the two occurrences of the same formal variable x on the left-hand side of (5.2.4) are in fact appropriate, since we are now regarding $Y_W(u, x)$ and $Y_W(v, x)$ as vectors in our space rather than as operators.) However, the right-hand side of (5.2.4) is given precisely by the iterate formula (3.1.11) for a module action (recall Section 4.1), namely,

$$Y_W(Y(u, x_0)v, x)$$
$$= \mathrm{Res}_{x_1}\left(x_0^{-1}\delta\left(\frac{x_1 - x}{x_0}\right)Y_W(u, x_1)Y_W(v, x) - x_0^{-1}\delta\left(\frac{x - x_1}{-x_0}\right)Y_W(v, x)Y_W(u, x_1)\right).$$
$$\tag{5.2.5}$$

(The role of the formal variable x_2 in (3.1.11) is now played by x.) Combining (5.2.4) and (5.2.5), we now know how we must define $Y_{\mathcal{E}}(a(x), x_0)b(x)$ for $a(x)$ and $b(x)$ of the form $Y_W(u, x)$ and $Y_W(v, x)$. Let us now use this same definition, more generally, for *arbitrary* weak vertex operators $a(x)$ and $b(x)$ on W. For $a(x), b(x) \in \mathcal{E}(W)$, we define

$$Y_{\mathcal{E}}(a(x), x_0)b(x)$$
$$= \mathrm{Res}_{x_1}\left(x_0^{-1}\delta\left(\frac{x_1 - x}{x_0}\right)a(x_1)b(x) - x_0^{-1}\delta\left(\frac{x - x_1}{-x_0}\right)b(x)a(x_1)\right). \tag{5.2.6}$$

Remark 5.2.1 It is clear that $Y_{\mathcal{E}}(a(x), x_0)b(x)$ is well defined; here we are using the hypothesis that $a(x)$ and $b(x)$ are actually weak vertex operators rather than arbitrary formal Laurent series. When the definition (5.2.6) is applied to $a(x)$ and $b(x)$ of the form $Y_W(u, x)$ and $Y_W(v, x)$ (which provided our motivation), we can conclude, by transport-of-structure, that we have equipped the space of operators $\iota_W(V)$ with vertex algebra structure, in such a way that the map (5.2.1) is an isomorphism of vertex algebras from V to the image $\iota_W(V)$. What will be much more interesting is to investigate under what conditions the general definition (5.2.6) gives rise to vertex algebra structure, and to exploit this structure when it exists, that is, to establish the axioms for a vertex algebra without the benefit of having available a vertex algebra V and a module action on W. It will *not* be true that the space of *all* weak vertex operators $\mathcal{E}(W)$ carries vertex algebra structure. Only suitable "small" subspaces of the space of weak vertex operators on W will admit vertex algebra structure.

In addition to knowing that $Y_{\mathcal{E}}(a(x), x_0)b(x)$ is well defined, we would also like to know that the expansion coefficients of $Y_{\mathcal{E}}(a(x), x_0)b(x)$ (in powers of x_0) are again weak vertex operators in x. We are about to prove this, giving us a linear map

$$Y_{\mathcal{E}}(\cdot, x_0) : \mathcal{E}(W) \to (\mathrm{End}\ \mathcal{E}(W))[[x_0, x_0^{-1}]]$$
$$a(x) \mapsto Y(a(x), x_0). \tag{5.2.7}$$

We shall use the usual notation for the expansion coefficients of $Y_{\mathcal{E}}(a(x), x_0)b(x)$ and of $Y_{\mathcal{E}}(a(x), x_0)$: We write

$$Y_{\mathcal{E}}(a(x), x_0) = \sum_{n \in \mathbb{Z}} a(x)_n x_0^{-n-1} \quad \text{for } a(x) \in \mathcal{E}(W), \tag{5.2.8}$$

with each $a(x)_n$ a linear map from $\mathcal{E}(W)$ to $(\text{End } W)[[x, x^{-1}]]$ (we are about to show that $a(x)_n$ takes $\mathcal{E}(W)$ into itself), and we correspondingly write

$$Y_{\mathcal{E}}(a(x), x_0)b(x) = \sum_{n \in \mathbb{Z}} a(x)_n b(x) x_0^{-n-1} \quad \text{for } a(x), b(x) \in \mathcal{E}(W). \tag{5.2.9}$$

We shall also continue to use the variant notation

$$\begin{aligned}
(Y_{\mathcal{E}}(a, x_0)b)(x) &= Y_{\mathcal{E}}(a(x), x_0)b(x) \\
&= \sum_{n \in \mathbb{Z}} a(x)_n b(x) x_0^{-n-1}.
\end{aligned} \tag{5.2.10}$$

Now we can formulate and prove the desired result; note that formula (5.2.11) below has exactly the same form as (3.8.16).

Proposition 5.2.2 *Let $a(x)$ and $b(x)$ be weak vertex operators on W. Then $a(x)_n b(x)$ is again a weak vertex operator for all $n \in \mathbb{Z}$ and is given by the formula*

$$a(x)_n b(x) = \text{Res}_{x_1}\left((x_1 - x)^n a(x_1)b(x) - (-x + x_1)^n b(x)a(x_1) \right). \tag{5.2.11}$$

Proof. Formula (5.2.11) follows immediately from (5.2.9) and (5.2.6), exactly as in (3.8.16). To see that $a(x)_n b(x)$ is again a weak vertex operator, we apply both sides of (5.2.11) to $w \in W$:

$$\begin{aligned}
&a(x)_n b(x)w \\
&= \text{Res}_{x_1}\left((x_1 - x)^n a(x_1)b(x)w - (-x + x_1)^n b(x)a(x_1)w \right) \\
&= \left(\sum_{i \geq 0} \binom{n}{i}(-x)^i a_{n-i} \right) b(x)w - b(x)\left(\sum_{i \geq 0} \binom{n}{i}(-x)^{n-i} a_i w \right). \tag{5.2.12}
\end{aligned}$$

Since $b(x)w \in W((x))$ and since $\sum_{i \geq 0} \binom{n}{i}(-x)^{n-i} a_i w$ is a finite sum, we see that $a(x)_n b(x)w \in W((x))$. (This argument holds equally well if n is negative or nonnegative; in the latter case the result is immediate.) This proves that $a(x)_n b(x) \in \mathcal{E}(W)$.
$\square$

Remark 5.2.3 We remind the reader that the symbols $a(x)_n$ and a_n refer to very different things: $a(x)_n$ is a map on the space $\mathcal{E}(W)$ of weak vertex operators, in the variable x, on W, while a_n is an operator on W itself (recall Definition 5.1.1). It is for this reason that we are avoiding using abbreviated notation such as $(a_n b)(x)$ for $a(x)_n b(x)$.

Remark 5.2.4 It is also interesting that we are not allowed to expand $Y(a(x), x_0)$ ($= Y(\sum a_n x^{-n-1}, x_0)$) in powers of x, as $Y(a(x), x_0) = \sum Y(a_n, x_0) x^{-n-1}$. While the object a_n can indeed be viewed as a constant series in x and hence as a weak vertex operator on W, this tempting expansion formula is not implied by the theory and is incompatible with (5.2.6) or (5.2.11). (Of course, we can certainly expand $Y(a(x), x_0)$ in powers of x_0, as we have been doing.)

Remark 5.2.5 Suppose as above that W is a faithful module for the vertex algebra V. (The special case $W = V$ is important and interesting.) For $a \in V$, let us write

$$a(x) = Y_W(a, x) \tag{5.2.13}$$

(cf. (5.2.6)). For $a, b \in V$ and $n \in \mathbb{Z}$ we have both the component $a_n b \in V$ of $Y(a, x_0)b$ and the component $a(x)_n b(x) \in \mathcal{E}(W)$ of $Y_{\mathcal{E}}(a(x), x_0) b(x)$ (each of these being the coefficient of x_0^{-n-1}), and these are related by:

$$Y_W(a_n b, x) = a(x)_n b(x). \tag{5.2.14}$$

This is nothing but (5.2.4) in component form, and together with the fact that

$$\mathbf{1}(x) = Y_W(\mathbf{1}, x) = 1_W, \tag{5.2.15}$$

it asserts that the map $\iota_W = Y(\cdot, x)$ (recall (5.2.1)) is a homomorphism of vertex algebras from V to $\iota_W(V)$; this of course was our starting point. Formula (5.2.14) is easily iterated: For $a^{(1)}, \ldots, a^{(r)} \in V$ and $n_1, \ldots, n_r \in \mathbb{Z}$,

$$Y_W(a^{(1)}_{n_1} \cdots a^{(r)}_{n_r} b, x) = a^{(1)}(x)_{n_1} \cdots a^{(r)}(x)_{n_r} b(x), \tag{5.2.16}$$

and taking $b = \mathbf{1}$, we have

$$Y_W(a^{(1)}_{n_1} \cdots a^{(r)}_{n_r} \mathbf{1}, x) = a^{(1)}(x)_{n_1} \cdots a^{(r)}(x)_{n_r} 1_W. \tag{5.2.17}$$

If the elements $a^{(i)}$ are assumed to lie in a given generating subset of V, then from Proposition 3.9.3 we see that (5.2.17) expresses the action of an arbitrary element of V in terms of the iterated action $a(x)_{(n)}$, given by (5.2.11), of a weak vertex operator $a(x)$ on the space of weak vertex operators, where the element a ranges through the generating set. We have been assuming here that the V-module W is faithful, but as we shall see, these considerations all hold for an arbitrary V-module W.

Remark 5.2.6 Recall from Section 3.8 that we have interpreted and re-expressed "iterate vertex operators" such as $Y(u_n v, x)$ in terms of normal-ordered products and so on. Using the iterate formula (3.1.11) (or equivalently, formula (3.8.16) for $Y(u_n v, x)$), we obtained formulas such as (3.8.8) and (3.8.9) for $Y(u_n v, x)$ for $n < 0$, and (3.8.14) for $Y(u_n v, x)$ for $n \geq 0$. These formulas involved normal-ordered products or commutators of $Y(u, x)$ and $Y(v, x)$, as appropriate. But since the present definition (5.2.6), or equivalently, (5.2.11), of the action of a weak vertex operator $a(x)$ on a weak vertex operator $b(x)$ precisely agrees with the iterate formula (3.1.11) (or (3.8.16)), all of the considerations of Section 3.8 automatically carry over to the present context. We correspondingly define normal ordering for weak vertex operators, and then quote some material from Section 3.8.

For a weak vertex operator $a(x) = \sum_{n \in \mathbb{Z}} a_n x^{-n-1} \in \mathcal{E}(W)$, we write (cf. (3.8.1))

$$a(x) = a(x)^+ + a(x)^- \tag{5.2.18}$$

where $a(x)^+ = \sum_{n<0} a_n x^{-n-1}$ and $a(x)^- = \sum_{n \geq 0} a_n x^{-n-1}$ are the regular and singular parts of $a(x)$, respectively. Clearly,

$$\left(\frac{d}{dx} a(x)\right)^{\pm} = \frac{d}{dx} a(x)^{\pm}. \tag{5.2.19}$$

We define the normal-ordered product ${}^{\circ}_{\circ} a(x_1) b(x_2) {}^{\circ}_{\circ}$ for weak vertex operators $a(x)$ and $b(x)$ by

$$ {}^{\circ}_{\circ} a(x_1) b(x_2) {}^{\circ}_{\circ} = a(x_1)^+ b(x_2) + b(x_2) a(x_1)^- \tag{5.2.20}$$

(cf. (3.8.5)). As in Section 3.8 (for vertex operators $Y(u, x)$ and $Y(v, x)$), the operation ${}^{\circ}_{\circ} \cdot {}^{\circ}_{\circ}$ puts the regular part of the first factor on the left and its singular part on the right. As a result, we have

$$ {}^{\circ}_{\circ} a(x_1) b(x_2) {}^{\circ}_{\circ} \in \mathrm{Hom}(W, W((x_1, x_2))), \tag{5.2.21}$$

so that we are allowed to set $x_1 = x_2$ in the normal-ordered product ${}^{\circ}_{\circ} a(x_1) b(x_2) {}^{\circ}_{\circ}$ (in contrast with the situation for the ordinary product $a(x_1) b(x_2)$). Doing this, we get an element of $\mathrm{Hom}(W, W((x)))$, i.e., a weak vertex operator, namely,

$$ {}^{\circ}_{\circ} a(x) b(x) {}^{\circ}_{\circ} = a(x)^+ b(x) + b(x) a(x)^- \tag{5.2.22}$$

(cf. (3.8.6)).

From (5.2.11), for $n \geq 0$ we have

$$a(x)_n b(x) = \mathrm{Res}_{x_1} (x_1 - x)^n [a(x_1), b(x)] \tag{5.2.23}$$

(cf. (3.8.14)). In particular,

$$a(x)_0 b(x) = [a_0, b(x)]. \tag{5.2.24}$$

Also, for $m \geq 1$ we have

$$a(x)_{-m} b(x) = \frac{1}{(m-1)!} {}^{\circ}_{\circ} \left(\left(\frac{d}{dx}\right)^{m-1} a(x)\right) b(x) {}^{\circ}_{\circ} \tag{5.2.25}$$

$$= \frac{1}{(m-1)!} {}^{\circ}_{\circ} (\mathcal{D}^{m-1} a(x)) b(x) {}^{\circ}_{\circ} \tag{5.2.26}$$

(cf. (3.8.8), (3.8.9)). In particular,

$$a(x)_{-1} b(x) = {}^{\circ}_{\circ} a(x) b(x) {}^{\circ}_{\circ}. \tag{5.2.27}$$

Example 5.2.7 Let $a(x) = A$, $b(x) = B \in \text{End } W$ ($\subset \mathcal{E}(W)$). Then using (5.2.23) and (5.2.25) we get

$$a(x)_n b(x) = 0 \quad \text{for } n \neq -1, \tag{5.2.28}$$

$$a(x)_{-1} b(x) = AB. \tag{5.2.29}$$

That is,

$$Y_{\mathcal{E}}(a(x), x_0)b(x) = AB. \tag{5.2.30}$$

Remark 5.2.8 Example 5.2.7 shows that the normal-ordered product (5.2.22) (or (5.2.27)) is not in general commutative. It is also easy to construct examples to show that the normal-ordered product of weak vertex operators is not in general associative. (For instance, take $a(x) = xA$, $b(x) = x^{-1}B$, $c(x) = C$, where $A, B, C \in \text{End } W$.)

Remark 5.2.9 Note that the definition (5.2.6) of $Y_{\mathcal{E}}$ is in some sense analogous to the classical definition of the commutator product of linear operators ($[A, B]w = ABw - BAw$ for $A, B \in \text{End } W$, $w \in W$). On the other hand, from the comments just before Corollary 4.4.7, (5.2.6) implies the weak associativity relation. In view of this, the definition of $Y_{\mathcal{E}}$ is *in addition* analogous, in a significant way, to the ordinary classical *associative* product of linear operators ($(AB)w = A(Bw)$). Again we see subtle analogies between vertex algebras and *both* Lie algebras *and* associative algebras.

5.3 The canonical weak vertex algebra $\mathcal{E}(W)$ and the equivalence between modules and representations

We have constructed a canonical linear map $Y_{\mathcal{E}}(\cdot, x_0)$ from $\mathcal{E}(W)$ to the space $(\text{End } \mathcal{E}(W))[[x_0, x_0^{-1}]]$. In view of this construction and its motivation, it is natural to ask to what extent the triple $(\mathcal{E}(W), Y_{\mathcal{E}}, 1_W)$ carries a vertex algebra structure, where, as we recall from (5.2.2), 1_W is the identity operator on W. It will turn out that the vacuum property, the creation property and the $\mathcal{D}$-bracket-derivative formulas hold for $(\mathcal{E}(W), Y_{\mathcal{E}}, 1_W)$. But the truncation property $a(x)_n b(x) = 0$ for sufficiently large n will *not* hold in general for weak vertex operators $a(x)$, $b(x)$, and correspondingly, the Jacobi identity cannot even be formulated for general elements $a(x)$, $b(x)$ of $\mathcal{E}(W)$. In order to have these stronger properties we will need to restrict ourselves to "mutually local" operators, as we do later. For now, we collect the properties that we *can* in fact establish for the space of weak vertex operators, and we call the resulting structure a "weak vertex algebra" (using, here, the formal variable x rather than x_0):

Definition 5.3.1 A *weak vertex algebra* consists of a vector space V equipped with a linear map

$$\begin{aligned} Y(\cdot, x) : V &\to (\text{End } V)[[x, x^{-1}]] \\ v &\mapsto Y(v, x) = \sum_{n \in \mathbb{Z}} v_n x^{-n-1} \end{aligned} \tag{5.3.1}$$

and a distinguished element $\mathbf{1}$ of V, such that the vacuum property and the creation property hold, that is,

$$Y(\mathbf{1}, x) = 1; \tag{5.3.2}$$

$$Y(v, x)\mathbf{1} \in V[[x]] \quad \text{and} \quad \lim_{x \to 0} Y(v, x)\mathbf{1} = v \tag{5.3.3}$$

for $v \in V$, and such that all the $\mathcal{D}$-bracket and derivative properties hold, that is,

$$[\mathcal{D}, Y(v, x)] = Y(\mathcal{D}v, x) = \frac{d}{dx} Y(v, x), \tag{5.3.4}$$

where $\mathcal{D} \in \operatorname{End} V$ is defined by

$$\mathcal{D}(v) = v_{-2}\mathbf{1} \quad \text{for } v \in V. \tag{5.3.5}$$

We sometimes refer to such a weak vertex algebra by the triple $(V, Y, \mathbf{1})$.

Remark 5.3.2 We know that any vertex algebra is a weak vertex algebra, by (3.1.25) and (3.1.33).

Remark 5.3.3 Exactly as in the proof of (3.1.29), we have the following enhanced form of the creation property:

$$Y(v, x)\mathbf{1} = e^{x\mathcal{D}}v \quad \text{for } v \in V. \tag{5.3.6}$$

Remark 5.3.4 The notion of weak vertex algebra is somewhat analogous to the notion of nonassociative algebra, and is not inherently very interesting, except that $(\mathcal{E}(W), Y_{\mathcal{E}}, 1_W)$ will in fact be such a structure. See also Remark 5.3.11, and recall Remarks 3.1.6 and 3.4.5, where commutative associative algebras were observed to be vertex algebras.

The notion of homomorphism of weak vertex algebras is defined in the obvious way. A *homomorphism* from U to V (both weak vertex algebras) is a linear map ψ such that $\psi(u_n v) = \psi(u)_n \psi(v)$ for $u, v \in U$, $n \in \mathbb{Z}$ and such that $\psi(\mathbf{1}) = \mathbf{1}$.

The notion of weak vertex subalgebra is also defined in the obvious way:

Definition 5.3.5 A *weak vertex subalgebra* of a weak vertex algebra $(V, Y, \mathbf{1})$ is a subspace A of V containing $\mathbf{1}$ such that $a_n b \in A$ for $a, b \in A$, $n \in \mathbb{Z}$; equivalently, $(A, Y, \mathbf{1})$ is itself a weak vertex algebra.

Remark 5.3.6 It is clear that the image of a weak vertex algebra homomorphism is a weak vertex subalgebra.

Definition 5.3.7 The *weak vertex subalgebra generated by a subset S of V* is the smallest weak vertex subalgebra of V containing S, or equivalently, the intersection of all weak vertex subalgebras containing S. It is denoted by $\langle S \rangle$.

Remark 5.3.8 Below we shall prove that $(\mathcal{E}(W), Y_\mathcal{E}, 1_W)$ is a weak vertex algebra. Then we shall show that for a vertex algebra V, having a V-module structure on W is equivalent to having a homomorphism of weak vertex algebras from V to $\mathcal{E}(W)$. In the next section we shall extend this discussion to V-modules (W, Y_W, d) (i.e., with the d-bracket property included; recall Remark 4.1.4). Having such a structure will turn out to be equivalent to having a weak vertex algebra homomorphism from V to $\mathcal{E}(W, d)$ (recall (5.1.5)), which we shall show to be a weak vertex subalgebra of $\mathcal{E}(W)$. In this way we shall exhibit the equivalence between the notion of *module* (in different variants) for a vertex algebra and the notion of *representation* of the vertex algebra. Such equivalence is in the spirit of classical algebraic theories, but in the present theory this equivalence is not trivial.

Now we formulate and prove:

Proposition 5.3.9 *The triple* $(\mathcal{E}(W), Y_\mathcal{E}, 1_W)$ *carries the structure of a weak vertex algebra. Moreover, the operator* $\mathcal{D}$ *defined in (5.3.5) is the differentiation operator (5.1.6).*

Proof. First we show that

$$Y_\mathcal{E}(1_W, x_0)a(x) = a(x) \tag{5.3.7}$$

for $a(x) \in \mathcal{E}(W)$: Using (5.2.6) together with (2.3.17) and (2.3.18), we get

$$
\begin{aligned}
&Y_\mathcal{E}(1_W, x_0)a(x) \\
&= \mathrm{Res}_{x_1}\left(x_0^{-1}\delta\left(\frac{x_1 - x}{x_0}\right) a(x) - x_0^{-1}\delta\left(\frac{x - x_1}{-x_0}\right) a(x)\right) \\
&= \mathrm{Res}_{x_1} x^{-1}\delta\left(\frac{x_1 - x_0}{x}\right) a(x) \\
&= \mathrm{Res}_{x_1} x_1^{-1}\delta\left(\frac{x + x_0}{x_1}\right) a(x) \\
&= a(x).
\end{aligned}
$$

Next we show that

$$Y_\mathcal{E}(a(x), x_0)1_W = e^{x_0 \mathcal{D}}a(x) \quad (= a(x + x_0)), \tag{5.3.8}$$

where $\mathcal{D}$ is the differentiation operator d/dx (recall (5.1.6)):

$$
\begin{aligned}
&Y_\mathcal{E}(a(x), x_0)1_W \\
&= \mathrm{Res}_{x_1}\left(x_0^{-1}\delta\left(\frac{x_1 - x}{x_0}\right) a(x_1) - x_0^{-1}\delta\left(\frac{x - x_1}{-x_0}\right) a(x_1)\right)
\end{aligned}
$$

$$= \mathrm{Res}_{x_1} x^{-1}\delta\left(\frac{x_1 - x_0}{x}\right) a(x_1)$$

$$= \mathrm{Res}_{x_1} x_1^{-1}\delta\left(\frac{x + x_0}{x_1}\right) a(x_1)$$

$$= \mathrm{Res}_{x_1} x_1^{-1}\delta\left(\frac{x + x_0}{x_1}\right) a(x + x_0)$$

$$= a(x + x_0)$$

$$= e^{x_0 \frac{d}{dx}} a(x)$$

$$= e^{x_0 \mathcal{D}} a(x). \tag{5.3.9}$$

It follows that $Y_{\mathcal{E}}(a(x), x_0)1_W \in \mathcal{E}(W)[[x_0]]$ and that

$$\lim_{x_0 \to 0} Y_{\mathcal{E}}(a(x), x_0)1_W = a(x) \tag{5.3.10}$$

and

$$a(x)_{-2}1_W = \mathcal{D}(a(x)) \ (= a'(x)); \tag{5.3.11}$$

in particular, $\mathcal{D}\,(= d/dx)$ coincides with $\mathcal{D}$ in the definition of the notion of weak vertex operator (recall (5.3.5)).

Finally, we need to prove (5.3.4). Let $a(x), b(x) \in \mathcal{E}(W)$. Then

$$\frac{\partial}{\partial x_0} Y_{\mathcal{E}}(a(x), x_0)b(x) = Y_{\mathcal{E}}(\mathcal{D}(a(x)), x_0)b(x) = [\mathcal{D}, Y_{\mathcal{E}}(a(x), x_0)]b(x). \tag{5.3.12}$$

Indeed, using (2.3.20) and (2.2.6), we obtain

$$\frac{\partial}{\partial x_0} Y_{\mathcal{E}}(a(x), x_0)b(x)$$

$$= \mathrm{Res}_{x_1}\left(\left(\frac{\partial}{\partial x_0}\left(x_0^{-1}\delta\left(\frac{x_1 - x}{x_0}\right)\right)\right) a(x_1)b(x) - \frac{\partial}{\partial x_0}\left(x_0^{-1}\delta\left(\frac{x - x_1}{-x_0}\right)\right) b(x)a(x_1)\right)$$

$$= \mathrm{Res}_{x_1}\left(\left(-\frac{\partial}{\partial x_1}x_0^{-1}\delta\left(\frac{x_1 - x}{x_0}\right)\right) a(x_1)b(x) + \left(\frac{\partial}{\partial x_1}x_0^{-1}\delta\left(\frac{x - x_1}{-x_0}\right)\right) b(x)a(x_1)\right)$$

$$= \mathrm{Res}_{x_1}\left(x_0^{-1}\delta\left(\frac{x_1 - x}{x_0}\right) a'(x_1)b(x) - x_0^{-1}\delta\left(\frac{x - x_1}{-x_0}\right) b(x)a'(x_1)\right)$$

$$= Y_{\mathcal{E}}(\mathcal{D}(a(x)), x_0)b(x), \tag{5.3.13}$$

and

$$[\mathcal{D}, Y_{\mathcal{E}}(a(x), x_0)]b(x)$$
$$= \mathcal{D}(Y_{\mathcal{E}}(a(x), x_0)b(x)) - Y_{\mathcal{E}}(a(x), x_0)\mathcal{D}(b(x))$$
$$= \frac{\partial}{\partial x}(Y_{\mathcal{E}}(a(x), x_0)b(x)) - Y_{\mathcal{E}}(a(x), x_0)b'(x)$$
$$= \mathrm{Res}_{x_1}\left(\left(\frac{\partial}{\partial x}x_0^{-1}\delta\left(\frac{x_1 - x}{x_0}\right)\right)a(x_1)b(x) - \left(\frac{\partial}{\partial x}x_0^{-1}\delta\left(\frac{x - x_1}{-x_0}\right)\right)b(x)a(x_1)\right)$$
$$= \mathrm{Res}_{x_1}\left(\frac{\partial}{\partial x_0}\left(x_0^{-1}\delta\left(\frac{x_1 - x}{x_0}\right)\right)a(x_1)b(x) - \frac{\partial}{\partial x_0}\left(x_0^{-1}\delta\left(\frac{x - x_1}{-x_0}\right)\right)b(x)a(x_1)\right)$$
$$= \frac{\partial}{\partial x_0}Y_{\mathcal{E}}(a(x), x_0)b(x). \tag{5.3.14}$$

This completes the proof of weak vertex algebra structure on $\mathcal{E}(W)$. $\square$

Definition 5.3.10 We shall refer to $(\mathcal{E}(W), Y_{\mathcal{E}}, 1_W)$ as the *canonical weak vertex algebra* associated to our vector space W.

Remark 5.3.11 In view of Example 5.2.7, the associative algebra End W is a weak vertex subalgebra of the canonical weak vertex algebra $\mathcal{E}(W)$; in particular, End W is a weak vertex algebra in a natural way, with $\mathcal{D} = 0$. Indeed, *any nonassociative algebra with a left- and right-identity element* is also a weak vertex algebra (with $\mathcal{D} = 0$). In particular, any associative algebra (with 1), when viewed as a subalgebra of End W (for some vector space W), is a weak vertex subalgebra of $\mathcal{E}(W)$.

Now we relate this structure to module structure on W. Suppose that V is a vertex algebra and that W is a V-module. We continue to use the notation (5.2.1) (the case of a faithful module) for the map

$$\iota_W : V \to \mathcal{E}(W)$$
$$v \mapsto Y_W(v, x). \tag{5.3.15}$$

Then as in (5.2.2),

$$\iota_W(\mathbf{1}) = 1_W, \tag{5.3.16}$$

and by the same considerations as in (5.2.3)–(5.2.6), we see that ι_W is a homomorphism of weak vertex algebras from V to the canonical weak vertex algebra $\mathcal{E}(W)$.

Remark 5.3.12 Before we discuss the converse, we note that all of the considerations and formulas in Remark 5.2.5 immediately carry over from the case in which W is a faithful V-module to the general case.

Conversely, let ϕ be a weak vertex algebra homomorphism from V to $\mathcal{E}(W)$. The structure $\mathcal{E}(W)$ being associated with the formal variable x, we write ϕ_x for ϕ to indicate this dependence, and as usual we shall have occasion to consider ϕ_{x_1} for formal variables x_1 different from x. We define

$$Y_W(v, x) = \phi_x(v) \quad \text{for } v \in V. \tag{5.3.17}$$

Then for $u \in V$ and $w \in W$, we have $Y_W(u, x)w \in W((x))$ since $\phi_x(u)$ is a weak vertex operator. By definition,

$$Y_W(\mathbf{1}, x) = \phi_x(\mathbf{1}) = 1_W. \tag{5.3.18}$$

Furthermore, for $u, v \in V$, we have

$$\begin{aligned}
Y_W&(Y(u, x_0)v, x_2)\\
&= \phi_{x_2}(Y(u, x_0)v)\\
&= Y_{\mathcal{E}}(\phi_{x_2}(u), x_0)\phi_{x_2}(v)\\
&= \mathrm{Res}_{x_1}\left(x_0^{-1}\delta\left(\frac{x_1 - x_2}{x_0}\right)\phi_{x_1}(u)\phi_{x_2}(v) - x_0^{-1}\delta\left(\frac{x_2 - x_1}{-x_0}\right)\phi_{x_2}(v)\phi_{x_1}(u)\right)\\
&= \mathrm{Res}_{x_1}\left(x_0^{-1}\delta\left(\frac{x_1 - x_2}{x_0}\right)Y_W(u, x_1)Y_W(v, x_2)\right.\\
&\qquad\qquad \left. -x_0^{-1}\delta\left(\frac{x_2 - x_1}{-x_0}\right)Y_W(v, x_2)Y_W(u, x_1)\right).
\end{aligned} \tag{5.3.19}$$

Thus by Corollary 4.4.7, which characterizes V-modules in terms of the iterate formula, (W, Y_W) is a V-module. (Recall that Corollary 4.4.7 is a substantial result, using, as it does, the fact that weak associativity for a proposed module action implies that we indeed have a module action; recall Theorems 3.6.3 and 4.4.5.)

Remark 5.3.13 Consider the important special situation in which the vertex algebra V is actually a substructure of $\mathcal{E}(W)$, i.e., a weak vertex subalgebra of $\mathcal{E}(W)$ which is a vertex algebra. Then the module structure of W has a particularly simple form. Applying (5.3.17) to $v = a(x) \in V$ ($\subset \mathcal{E}(W)$) and replacing x by x_0 in (5.3.17), we see that

$$Y_W(a(x), x_0) = a(x_0), \tag{5.3.20}$$

since ϕ is the embedding map of V into $\mathcal{E}(W)$, and we are (necessarily) using the second variable, x_0, in the expression $Y_W(a(x), x_0)$.

We make the following natural definition:

Definition 5.3.14 A *representation* of a vertex algebra V on a vector space W is a weak vertex algebra homomorphism from V to the canonical weak vertex algebra $\mathcal{E}(W)$.

We have proved:

Theorem 5.3.15 *Let V be a vertex algebra and let W be a vector space. Then giving a V-module structure on W is equivalent to giving a representation of V on W, with the correspondences between the structures as specified above.* $\quad\square$

Remark 5.3.16 Under these conditions, the vertex algebra V is "represented by" the weak vertex subalgebra $\iota_W(V) = \phi_x(V)$ of $\mathcal{E}(W)$ (recall Remark 5.3.6), and this weak vertex subalgebra is actually a geniune vertex subalgebra. Indeed, the missing properties—the truncation condition and the Jacobi identity—clearly hold on this image; for this, simply apply ι_W to the truncation condition and to the Jacobi identity for V in order to deduce them for $\iota_W(V)$.

Example 5.3.17 Consider the simplest nontrivial case, where W is one-dimensional (recall Remark 5.1.4); we have $\mathcal{E}(W) = \mathbb{C}((x))$. Since $\mathbb{C}((x))$ is a commutative associative algebra, by (5.2.23) and (5.2.25) we have

$$f(x)_n g(x) = 0 \quad \text{for } n \geq 0, \tag{5.3.21}$$

$$f(x)_n g(x) = \frac{1}{(-n-1)!}(\mathcal{D}^{(-n-1)} f(x))g(x) \quad \text{for } n < 0, \tag{5.3.22}$$

for $f(x), g(x) \in \mathbb{C}((x))$. Thus

$$Y_{\mathcal{E}}(f(x), x_0)g(x) = \left(e^{x_0 \frac{d}{dx}} f(x)\right) g(x). \tag{5.3.23}$$

From Example 3.4.6, we see that the weak vertex algebra $(\mathbb{C}((x)), Y_{\mathcal{E}}, 1_W)$ is actually a vertex algebra—the vertex algebra associated to the commutative associative algebra $\mathbb{C}((x))$ with the derivation $\mathcal{D} = d/dx$.

Remark 5.3.18 On the other hand, suppose that $\dim W \geq 2$. Then the weak vertex algebra $\mathcal{E}(W)$ is *not* a vertex algebra. Indeed, if it were, then its weak vertex subalgebra End W (recall Example 5.2.7 and Remark 5.3.11) would be a vertex algebra, but this cannot be so because weak commutativity for this structure fails to hold, since the left-multiplication operations on the algebra End W fail to commute in general. Another way to see that the weak vertex subalgebra End W of $\mathcal{E}(W)$ is not a vertex algebra is to use Theorem 5.3.15. If End W were a vertex algebra, then the embedding End $W \to \mathcal{E}(W)$, which is a homomorphism of weak vertex algebras, would give a representation of the vertex algebra End W on W (recall Definition 5.3.14). Then by Theorem 5.3.15, W would be an End W-module, with $A \in$ End W acting as $Y_W(A, x) = A$ on W (according to Remark 5.3.13). However, for any module for a vertex algebra, we must have weak commutativity for the module action, by Proposition 4.2.1. But this would imply that any pair of endomorphisms of W would commute, contradicting our assumption that $\dim W \geq 2$. Notice that in the first of these two different arguments showing that $\mathcal{E}(W)$ is not a vertex algebra, we showed that weak commutativity for the *algebra* fails, while in the second of the arguments, we showed that weak commutativity for the *module action* fails. In the first argument, the point is that the *left multiplication operations on* End W fail to commute, while in the second argument, the point is that the *elements of* End W fail to commute. These issues will be central to the theory, as we shall see in Section 5.5.

5.4 Subalgebras of $\mathcal{E}(W)$

Recall that in Section 5.1 we considered the space $\mathcal{E}(W, d)$ of weak vertex operators on the pair (W, d), where d is a given linear operator on our space W (see Definition 5.1.3 and (5.1.5)). We also considered the space $\mathcal{E}(W, L(-1))$ and certain of its subspaces in case W is a module for the Virasoro algebra (recall Definition 5.1.5). In this section we extend the considerations of the last section to these contexts.

Proposition 5.4.1 *Let d be a linear operator on W. Then the subspace $\mathcal{E}(W, d)$ of $\mathcal{E}(W)$ is a weak vertex subalgebra.*

Proof. Since $1_W \in \mathcal{E}(W, d)$, what we must show is that

$$[d, Y_{\mathcal{E}}(a(x), x_0)b(x)] = \frac{\partial}{\partial x}(Y_{\mathcal{E}}(a(x), x_0)b(x)) \tag{5.4.1}$$

for $a(x), b(x) \in \mathcal{E}(W, d)$. But using (2.3.20) and (2.2.6), we obtain

$$\frac{\partial}{\partial x}(Y_{\mathcal{E}}(a(x), x_0)b(x))$$

$$= \frac{\partial}{\partial x}\mathrm{Res}_{x_1}\left(x_0^{-1}\delta\left(\frac{x_1 - x}{x_0}\right)a(x_1)b(x) - x_0^{-1}\delta\left(\frac{x - x_1}{-x_0}\right)b(x)a(x_1)\right)$$

$$= \mathrm{Res}_{x_1}\left(\left(\frac{\partial}{\partial x}x_0^{-1}\delta\left(\frac{x_1 - x}{x_0}\right)\right)a(x_1)b(x) - \left(\frac{\partial}{\partial x}x_0^{-1}\delta\left(\frac{x - x_1}{-x_0}\right)\right)b(x)a(x_1)\right)$$

$$+ \mathrm{Res}_{x_1}\left(x_0^{-1}\delta\left(\frac{x_1 - x}{x_0}\right)a(x_1)b'(x) - x_0^{-1}\delta\left(\frac{x - x_1}{-x_0}\right)b'(x)a(x_1)\right)$$

$$= -\mathrm{Res}_{x_1}\left(\left(\frac{\partial}{\partial x_1}x_0^{-1}\delta\left(\frac{x_1 - x}{x_0}\right)\right)a(x_1)b(x) - \left(\frac{\partial}{\partial x_1}x_0^{-1}\delta\left(\frac{x - x_1}{-x_0}\right)\right)b(x)a(x_1)\right)$$

$$+ \mathrm{Res}_{x_1}\left(x_0^{-1}\delta\left(\frac{x_1 - x}{x_0}\right)a(x_1)b'(x) - x_0^{-1}\delta\left(\frac{x - x_1}{-x_0}\right)b'(x)a(x_1)\right)$$

$$= \mathrm{Res}_{x_1}\left(x_0^{-1}\delta\left(\frac{x_1 - x}{x_0}\right)a'(x_1)b(x) - x_0^{-1}\delta\left(\frac{x - x_1}{-x_0}\right)b(x)a'(x_1)\right)$$

$$+ \mathrm{Res}_{x_1}\left(x_0^{-1}\delta\left(\frac{x_1 - x}{x_0}\right)a(x_1)b'(x) - x_0^{-1}\delta\left(\frac{x - x_1}{-x_0}\right)b'(x)a(x_1)\right)$$

$$= [d, Y_{\mathcal{E}}(a(x), x_0)b(x)]. \qquad \square$$

We use this structure to extend the correspondence between modules and representations. The following result is immediate (recall Remark 4.1.4).

Theorem 5.4.2 *Let V be a vertex algebra and let W be a vector space equipped with a linear operator d. Then giving a V-module structure (W, Y_W, d) on W compatible with d is equivalent to giving a weak vertex algebra homomorphism from V to the weak vertex algebra $\mathcal{E}(W, d)$.* $\quad\square$

We also have:

Proposition 5.4.3 *Suppose W is a module for the Virasoro algebra. The space $\mathcal{E}^o(W, L(-1))$ is then a weak vertex subalgebra of $\mathcal{E}(W, L(-1))$. Furthermore, if $a(x)$ and $b(x)$ are homogeneous weak vertex operators on $(W, L(-1))$ of weights α and β in $\mathbb{C}$, respectively, then for $n \in \mathbb{Z}$, $a(x)_n b(x)$ is a homogeneous weak vertex operator of weight $\alpha + \beta - n - 1$ on $(W, L(-1))$.*

Proof. We know from Remark 5.1.6 that $1_W \in \mathcal{E}^o(W, L(-1))$, so it is sufficient to prove the last assertion of the proposition, or equivalently, in terms of generating functions, that

$$[L(0), Y_{\mathcal{E}}(a(x), x_0)b(x)] \quad \left(= \sum_{n\in\mathbb{Z}} [L(0), a(x)_n b(x)] x_0^{-n-1} \right)$$

$$= (\alpha + \beta) Y_{\mathcal{E}}(a(x), x_0)b(x) + x_0 \frac{\partial}{\partial x_0}(Y_{\mathcal{E}}(a(x), x_0)b(x)) + x \frac{\partial}{\partial x}(Y_{\mathcal{E}}(a(x), x_0)b(x))$$

$$\left(= \sum_{n\in\mathbb{Z}} \left((\alpha + \beta - n - 1)a(x)_n b(x) + x \frac{d}{dx}(a(x)_n b(x)) \right) x_0^{-n-1} \right) \tag{5.4.2}$$

(recall (5.1.9)). But again using (2.3.20) and (2.2.6), we have

$$[L(0), Y_{\mathcal{E}}(a(x), x_0)b(x)]$$

$$= \operatorname{Res}_{x_1} \left(x_0^{-1}\delta\left(\frac{x_1 - x}{x_0}\right)[L(0), a(x_1)b(x)] - x_0^{-1}\delta\left(\frac{x - x_1}{-x_0}\right)[L(0), b(x)a(x_1)] \right)$$

$$= \operatorname{Res}_{x_1} x_0^{-1}\delta\left(\frac{x_1 - x}{x_0}\right)(a(x_1)[L(0), b(x)] + [L(0), a(x_1)]b(x))$$

$$\quad - \operatorname{Res}_{x_1} x_0^{-1}\delta\left(\frac{x - x_1}{-x_0}\right)(b(x)[L(0), a(x_1)] + [L(0), b(x)]a(x_1))$$

$$= \operatorname{Res}_{x_1} x_0^{-1}\delta\left(\frac{x_1 - x}{x_0}\right)(\beta a(x_1)b(x) + xa(x_1)b'(x) + \alpha a(x_1)b(x) + x_1 a'(x_1)b(x))$$

$$\quad - \operatorname{Res}_{x_1} x_0^{-1}\delta\left(\frac{x - x_1}{-x_0}\right)(\alpha b(x)a(x_1) + x_1 b(x)a'(x_1) + \beta b(x)a(x_1) + xb'(x)a(x_1))$$

$$= (\alpha + \beta) Y_{\mathcal{E}}(a(x), x_0)b(x)$$

$$\quad + \operatorname{Res}_{x_1} \left(x_0^{-1}\delta\left(\frac{x_1 - x}{x_0}\right)xa(x_1)b'(x) - x_0^{-1}\delta\left(\frac{x - x_1}{-x_0}\right)xb'(x)a(x_1) \right)$$

$$\quad - \operatorname{Res}_{x_1} \left(\frac{\partial}{\partial x_1}\left(x_1 x_0^{-1}\delta\left(\frac{x_1 - x}{x_0}\right) \right) \right) a(x_1)b(x)$$

$$\quad - \frac{\partial}{\partial x_1}\left(x_1 x_0^{-1}\delta\left(\frac{x - x_1}{-x_0}\right) \right) b(x)a(x_1) \Big)$$

$$= (\alpha + \beta) Y_{\mathcal{E}}(a(x), x_0)b(x)$$

$$\quad + \operatorname{Res}_{x_1} \left(x_0^{-1}\delta\left(\frac{x_1 - x}{x_0}\right)xa(x_1)b'(x) - x_0^{-1}\delta\left(\frac{x - x_1}{-x_0}\right)xb'(x)a(x_1) \right)$$

$$\quad - \operatorname{Res}_{x_1}(x_0 + x)\left(\frac{\partial}{\partial x_1}x_0^{-1}\delta\left(\frac{x_1 - x}{x_0}\right) \right) a(x_1)b(x)$$

$$+\mathrm{Res}_{x_1}(x_0+x)\left(\frac{\partial}{\partial x_1}x_0^{-1}\delta\left(\frac{x-x_1}{-x_0}\right)\right)b(x)a(x_1)$$

$$= (\alpha+\beta)Y_{\mathcal{E}}(a(x),x_0)b(x)$$

$$+\mathrm{Res}_{x_1}\left(x_0^{-1}\delta\left(\frac{x_1-x}{x_0}\right)xa(x_1)b'(x)-x_0^{-1}\delta\left(\frac{x-x_1}{-x_0}\right)xb'(x)a(x_1)\right)$$

$$+\mathrm{Res}_{x_1}x_0\left(\left(\frac{\partial}{\partial x_0}x_0^{-1}\delta\left(\frac{x_1-x}{x_0}\right)\right)a(x_1)b(x)\right.$$

$$\left.-\left(\frac{\partial}{\partial x_0}x_0^{-1}\delta\left(\frac{x-x_1}{-x_0}\right)\right)b(x)a(x_1)\right)$$

$$+\mathrm{Res}_{x_1}x\left(\left(\frac{\partial}{\partial x}x_0^{-1}\delta\left(\frac{x_1-x}{x_0}\right)\right)a(x_1)b(x)-\left(\frac{\partial}{\partial x}x_0^{-1}\delta\left(\frac{x-x_1}{-x_0}\right)\right)b(x)a(x_1)\right)$$

$$= (\alpha+\beta)Y_{\mathcal{E}}(a(x),x_0)b(x)+x\frac{\partial}{\partial x}\left(Y_{\mathcal{E}}(a(x),x_0)b(x)\right)+x_0\frac{\partial}{\partial x_0}\left(Y_{\mathcal{E}}(a(x),x_0)b(x)\right).$$

This concludes the proof. $\square$

From Proposition 5.4.3 we immediately have, recalling the notation (5.1.12):

Corollary 5.4.4 *Suppose that W is a module for the Virasoro algebra. Then $\mathcal{E}^o_{\mathbb{Z}}(W,L(-1))$ is a ($\mathbb{Z}$-graded) weak vertex subalgebra of $\mathcal{E}^o(W,L(-1))$.* $\square$

5.5 Local subalgebras and vertex subalgebras of $\mathcal{E}(W)$

We have seen in Remark 5.3.18 that the canonical weak vertex algebra $(\mathcal{E}(W),Y_{\mathcal{E}},1_W)$ of weak vertex operators on W is usually not a vertex algebra, for either of two different reasons: Weak commutativity for $\mathcal{E}(W)$ fails to hold, and in addition, weak commutativity for the action of $\mathcal{E}(W)$ on W fails to hold. But vertex algebras are our main interest, and in this section we shall determine which weak vertex subalgebras of $\mathcal{E}(W)$ are in fact vertex algebras. It turns out that the key is to focus on the weak commutativity condition for pairs of elements of $\mathcal{E}(W)$ (weak vertex operators on W) acting on W—that is, to focus on the condition, on a given pair of elements $a(x)$, $b(x)$ (possibly equal) of $\mathcal{E}(W)$, that there should exist a nonnegative integer k such that

$$(x_1-x_2)^k a(x_1)b(x_2) = (x_1-x_2)^k b(x_2)a(x_1). \tag{5.5.1}$$

In Definition 5.5.1, this condition is called the "mutual locality" of the elements $a(x)$ and $b(x)$ of $\mathcal{E}(W)$. In fact, it turns out that the vertex subalgebras of the canonical weak vertex algebra $\mathcal{E}(W)$ are precisely those weak vertex operator subalgebras all of whose elements are mutually local with respect to each other.

This picture is of course already previewed by the now familiar situation that arises when the space W is a module for some vertex algebra V: Using the usual notation $\iota_W : V \to \mathcal{E}(W)$ for the corresponding representation of V on W, we know that the image $\iota_W(V)$ in the weak vertex algebra $\mathcal{E}(W)$ is a vertex algebra, and also that the elements of this image are mutually local with respect to one another, by weak commutativity for the module action (Proposition 4.2.1).

Moreover, and even more significantly, it turns out that these considerations give us a practical and powerful method for constructing concrete and nontrivial examples of vertex algebras and of vertex operator algebras: We start with any collection of mutually local weak vertex operators on W, and we form the weak vertex subalgebra of operators on W generated by these operators. The result is in fact a vertex algebra of operators on W, and W turns out to be a module for this vertex algebra.

Here is another, very simple, reason to consider the mutual locality condition (5.5.1) on a pair of elements $a(x)$, $b(x)$ of $\mathcal{E}(W)$: As was mentioned at the beginning of Section 5.3, the truncation condition $a(x)_n b(x) = 0$ for sufficiently large n does not hold in general, but if $a(x)$ and $b(x)$ are mutually local, then from the formula (5.2.11) for $a(x)_n b(x)$, it is clear that the truncation condition on the pair $a(x)$, $b(x)$ in fact holds. Of course, we will need this truncation condition if $a(x)$ and $b(x)$ are to be elements of a vertex subalgebra of $\mathcal{E}(W)$.

In this section, we define the notion of mutual locality for weak vertex operators on W, and define a "vertex operator" on W to be a self-local weak vertex operator. We then define and study the notions of local subset (a set of pairwise mutually local vertex operators on W), local subspace and local (weak vertex) subalgebra of $\mathcal{E}(W)$. We prove that the local subalgebras of $\mathcal{E}(W)$ are exactly those weak vertex subalgebras that are in fact vertex algebras and that W is a module for such a vertex algebra in a natural way.

As the main result of this section and also of this chapter, we then prove that any local subset of $\mathcal{E}(W)$ canonically generates in $\mathcal{E}(W)$ a vertex algebra, with W as a natural module. This result is very useful in constructing examples of vertex algebras.

Continuing to use W as our underlying vector space, we define the following notion of locality:

Definition 5.5.1 Weak vertex operators $a(x)$ and $b(x)$ on W are said to be *mutually local* if there exists a nonnegative integer k such that

$$(x_1 - x_2)^k a(x_1)b(x_2) = (x_1 - x_2)^k b(x_2)a(x_1), \tag{5.5.2}$$

or equivalently,

$$(x_1 - x_2)^k [a(x_1), b(x_2)] = 0. \tag{5.5.3}$$

Remark 5.5.2 The use of the term "locality" is historical. In conformal field theory (see [Go1]), "locality" for two "fields" $\phi(z)$ and $\psi(z)$ amounts to the condition that $\phi(z_1)\psi(z_2)$ is equal to $\psi(z_2)\phi(z_1)$ up to a certain analytic continuation; in mathematical language, this is the analytic commutativity property given in Corollary 3.2.10 above. After the algebraic condition, weak commutativity, was discovered and proved to be applicable as the main axiom for the notion of vertex operator algebra in [DL3], the present notion of mutual locality was defined and used in [Li3].

Remark 5.5.3 As usual, the assertion of (5.5.2) for some fixed $k \geq 0$ implies the assertion for all larger k. The assertion of (5.5.2) for $k = 0$ amounts to the condition that every component of $a(x)$ commutes with every component of $b(x)$.

Definition 5.5.4 A weak vertex operator $a(x)$ on W or (W, d) (recall Definition 5.1.3) is called a *vertex operator* on W or (W, d) if it is local with itself, that is, if there exists a nonnegative integer k such that

$$(x_1 - x_2)^k a(x_1)a(x_2) = (x_1 - x_2)^k a(x_2)a(x_1). \tag{5.5.4}$$

Remark 5.5.5 Let V be a vertex algebra and assume that W is a V-module, with module action Y_W. In view of Proposition 4.2.1, for every $v \in V$, $Y_W(v, x)$ is indeed a vertex operator on W in the present sense. On the other hand, it will be proved in Corollary 5.5.19 that every vertex operator on W in this sense can be realized as the vertex operator $Y_W(v, x)$ associated to some element v of some vertex algebra acting on W. This justifies the current definition of "vertex operator."

Example 5.5.6 Suppose that W is a restricted module for the Virasoro algebra (as defined in Remark 5.1.6) of central charge ℓ (recall Remark 3.1.23). It was shown that $L_W(x) = \sum_{n \in \mathbb{Z}} L(n)x^{-n-2}$ is a weak vertex operator on $(W, L(-1))$ (homogeneous of weight two). Now we show that $L_W(x)$ is in fact a vertex operator on $(W, L(-1))$. In terms of generating functions, the Virasoro algebra relations (3.1.46) can be equivalently written as

$$[L_W(x_1), L_W(x_2)]$$

$$= \sum_{m,n \in \mathbb{Z}} (m - n)L(m + n)x_1^{-m-2}x_2^{-n-2} + \sum_{m \in \mathbb{Z}} \frac{\ell}{12}(m^3 - m)x_1^{-m-2}x_2^{m-2}$$

$$= \sum_{m,n \in \mathbb{Z}} (-m - n - 2)L(m + n)x_2^{-m-n-3}(x_1^{-m-2}x_2^{m+1})$$

$$+ \sum_{m,n \in \mathbb{Z}} 2(m + 1)L(m + n)x_2^{-m-n-2}(x_1^{-m-2}x_2^{m})$$

$$+ \sum_{m \in \mathbb{Z}} \frac{\ell}{12}(m^3 - m)x_1^{-m-2}x_2^{m-2}$$

$$= L'_W(x_2)x_1^{-1}\delta\left(\frac{x_2}{x_1}\right) - 2L_W(x_2)\frac{\partial}{\partial x_1}x_1^{-1}\delta\left(\frac{x_2}{x_1}\right) - \frac{\ell}{12}\left(\frac{\partial}{\partial x_1}\right)^3 x_1^{-1}\delta\left(\frac{x_2}{x_1}\right)$$

$$= L'_W(x_2)x_2^{-1}\delta\left(\frac{x_1}{x_2}\right) - 2L_W(x_2)\frac{\partial}{\partial x_1}x_2^{-1}\delta\left(\frac{x_1}{x_2}\right) - \frac{\ell}{12}\left(\frac{\partial}{\partial x_1}\right)^3 x_2^{-1}\delta\left(\frac{x_1}{x_2}\right). \tag{5.5.5}$$

In view of (2.3.13), we thus have

$$(x_1 - x_2)^k[L_W(x_1), L_W(x_2)] = 0 \quad \text{for } k \geq 4, \tag{5.5.6}$$

and so $L_W(x)$ is a vertex operator. Of course, the generating function relation (5.5.5) is the same relation that we have in the case of a vertex operator algebra, when we write the Virasoro algebra relations (with central charge $\ell = c_V$) using the commutator formula (3.1.8), the relations (3.1.70)–(3.1.74) and the $L(-1)$-derivative property (3.1.45), as indicated at the end of Section 3.1; recall that the relations (3.1.70)–(3.1.74) describe the singular part of $Y(\omega, x)\omega$.

Remark 5.5.7 It is easy to see that if dim $W \geq 2$, then the sum of two vertex operators on W is not necessarily a vertex operator, so that the vertex operators on W do not form a subspace of $\mathcal{E}(W)$. However, it is also easy to see that if $a(x)$ and $b(x)$ are *mutually local* vertex operators on W, then any linear combination of $a(x)$ and $b(x)$ is again a vertex operator.

Definition 5.5.8 A subset or subspace A of $\mathcal{E}(W)$ is said to be *local* if for any $a(x)$, $b(x) \in A$, $a(x)$ and $b(x)$ are mutually local; in this case, A consists of vertex operators. A *local (weak vertex) subalgebra* of $\mathcal{E}(W)$ is a weak vertex subalgebra which is a local subspace. For the case in which W is a module for the Virasoro algebra, the notions of *graded local subspace* and *graded local subalgebra* of the (graded) weak vertex algebra $\mathcal{E}^o(W, L(-1))$ are defined in the obvious ways (recall Definition 5.1.5 and Proposition 5.4.3). (Such graded structures will be considered in the next section.)

Remark 5.5.9 It is clear that the linear span of a local subset of $\mathcal{E}(W)$ is a local subspace.

Remark 5.5.10 Let V be a vertex algebra and suppose that (W, Y_W) is a V-module. Since the linear map ι_W taking v to $Y_W(v, x)$ (recall (5.3.15)) is a weak vertex algebra homomorphism (Theorem 5.3.15), $\iota_W(V)$ is a weak vertex subalgebra of $\mathcal{E}(W)$, and it follows from Proposition 4.2.1 that $\iota_W(V)$ is a local subalgebra. In addition, as we have seen in Remark 5.3.16, $\iota_W(V)$ is actually a vertex subalgebra of $\mathcal{E}(W)$.

We shall prove that the vertex subalgebras of the weak vertex algebra $\mathcal{E}(W)$ are exactly the local subalgebras of $\mathcal{E}(W)$. First we formulate one direction of this assertion (the first assertion of the following theorem was also independently obtained by Rosellen in [Ros]).

Theorem 5.5.11 *Let V be a vertex subalgebra of $\mathcal{E}(W)$. Then V is a local subalgebra. Moreover, W is a module for V viewed as a vertex algebra, with*

$$Y_W(a(x), x_0) = a(x_0) \tag{5.5.7}$$

for $a(x) \in V$. In particular, the module W is faithful.

Proof. We have essentially already proved this, in the discussion before and contained in Remark 5.3.13. In the present case, the proof takes the following form:

Since $V \subset \mathcal{E}(W)$, for $a(x), b(x) \in V$, $w \in W$, by (5.2.6) we have

$$(Y_{\mathcal{E}}(a(x), x_0)b(x))w$$
$$= \mathrm{Res}_{x_1}\left(x_0^{-1}\delta\left(\frac{x_1 - x}{x_0}\right)a(x_1)b(x)w - x_0^{-1}\delta\left(\frac{x - x_1}{-x_0}\right)b(x)a(x_1)w\right).$$

$$(5.5.8)$$

By Corollary 4.4.7, W is a V-module with $Y_W(a(x), x_0) = a(x_0)$. In particular, weak commutativity holds for Y_W (Proposition 4.2.1). This amounts to saying that any $a(x), b(x) \in V$ are mutually local. Thus V is a local subalgebra. $\square$

For proving the converse, that any local subalgebra of $\mathcal{E}(W)$ is a vertex algebra, we first observe that we have the lower truncation condition:

Proposition 5.5.12 *Let $a(x), b(x) \in \mathcal{E}(W)$ be mutually local, with $k \geq 0$ such that*

$$(x_1 - x_2)^k a(x_1)b(x_2) = (x_1 - x_2)^k b(x_2)a(x_1) \tag{5.5.9}$$

on W. Then

$$a(x)_n b(x) = 0 \quad \text{for } n \geq k. \tag{5.5.10}$$

In particular, if V is a local subalgebra of $\mathcal{E}(W)$, then for $a(x), b(x) \in V$, $a(x)_n b(x) = 0$ for n sufficiently large.

Proof. This immediately follows from (5.2.11) and (5.5.9). $\square$

The next result gives weak commutativity for a local subalgebra.

Proposition 5.5.13 *Let $a(x), b(x) \in \mathcal{E}(W)$ be mutually local, with $k \geq 0$ such that*

$$(x_1 - x_2)^k a(x_1)b(x_2) = (x_1 - x_2)^k b(x_2)a(x_1) \tag{5.5.11}$$

on W. Then

$$(x_1 - x_2)^k Y_{\mathcal{E}}(a(x), x_1)Y_{\mathcal{E}}(b(x), x_2) = (x_1 - x_2)^k Y_{\mathcal{E}}(b(x), x_2)Y_{\mathcal{E}}(a(x), x_1)$$

$$(5.5.12)$$

on $\mathcal{E}(W)$. In particular, if V is a local subalgebra of $\mathcal{E}(W)$, then weak commutativity holds for $Y_{\mathcal{E}}$ on V.

Proof. Let $c(x) \in \mathcal{E}(W)$. From the definition (5.2.6), we have

$$Y_{\mathcal{E}}(a(x), x_1)Y_{\mathcal{E}}(b(x), x_2)c(x)$$
$$= \mathrm{Res}_{x_3}x_1^{-1}\delta\left(\frac{x_3 - x}{x_1}\right)a(x_3)(Y_{\mathcal{E}}(b(x), x_2)c(x))$$
$$- x_1^{-1}\delta\left(\frac{x - x_3}{-x_1}\right)(Y_{\mathcal{E}}(b(x), x_2)c(x))a(x_3)$$

$$= \mathrm{Res}_{x_3}\mathrm{Res}_{x_4}x_1^{-1}\delta\left(\frac{x_3-x}{x_1}\right)x_2^{-1}\delta\left(\frac{x_4-x}{x_2}\right)a(x_3)b(x_4)c(x)$$

$$-\mathrm{Res}_{x_3}\mathrm{Res}_{x_4}x_1^{-1}\delta\left(\frac{x_3-x}{x_1}\right)x_2^{-1}\delta\left(\frac{x-x_4}{-x_2}\right)a(x_3)c(x)b(x_4)$$

$$-\mathrm{Res}_{x_3}\mathrm{Res}_{x_4}x_1^{-1}\delta\left(\frac{x-x_3}{-x_1}\right)x_2^{-1}\delta\left(\frac{x_4-x}{x_2}\right)b(x_4)c(x)a(x_3)$$

$$+\mathrm{Res}_{x_3}\mathrm{Res}_{x_4}x_1^{-1}\delta\left(\frac{x-x_3}{-x_1}\right)x_2^{-1}\delta\left(\frac{x-x_4}{-x_2}\right)c(x)b(x_4)a(x_3). \qquad (5.5.13)$$

Similarly, we have (reversing the roles of a and b, of x_1 and x_2, and of x_3 and x_4)

$$Y_{\mathcal{E}}(b(x),x_2)Y_{\mathcal{E}}(a(x),x_1)c(x)$$

$$= \mathrm{Res}_{x_3}\mathrm{Res}_{x_4}x_1^{-1}\delta\left(\frac{x_3-x}{x_1}\right)x_2^{-1}\delta\left(\frac{x_4-x}{x_2}\right)b(x_4)a(x_3)c(x)$$

$$-\mathrm{Res}_{x_3}\mathrm{Res}_{x_4}x_1^{-1}\delta\left(\frac{x-x_3}{-x_1}\right)x_2^{-1}\delta\left(\frac{x_4-x}{x_2}\right)b(x_4)c(x)a(x_3)$$

$$-\mathrm{Res}_{x_3}\mathrm{Res}_{x_4}x_1^{-1}\delta\left(\frac{x_3-x}{x_1}\right)x_2^{-1}\delta\left(\frac{x-x_4}{-x_2}\right)a(x_3)c(x)b(x_4)$$

$$+\mathrm{Res}_{x_3}\mathrm{Res}_{x_4}x_1^{-1}\delta\left(\frac{x-x_3}{-x_1}\right)x_2^{-1}\delta\left(\frac{x-x_4}{-x_2}\right)c(x)a(x_3)b(x_4). \qquad (5.5.14)$$

Using (2.3.56) we get

$$(x_1-x_2)^k x_1^{-1}\delta\left(\frac{x_3-x}{x_1}\right)x_2^{-1}\delta\left(\frac{x_4-x}{x_2}\right)$$

$$= (x_3-x_4)^k x_1^{-1}\delta\left(\frac{x_3-x}{x_1}\right)x_2^{-1}\delta\left(\frac{x_4-x}{x_2}\right)$$

(replacing x_1 and x_2 by x_3-x and x_4-x, respectively), and we have similar identities involving the other products of delta functions. Then (5.5.12) follows from (5.5.13), (5.5.14) and (5.5.11). $\square$

We now have:

Theorem 5.5.14 *Any local subalgebra V of $\mathcal{E}(W)$ is a vertex algebra and W is a faithful module, where $Y_W(a(x),x_0) = a(x_0)$ for $a(x) \in V$. In particular, the local subalgebras of $\mathcal{E}(W)$ are precisely the vertex subalgebras.*

Proof. It follows from Theorem 3.5.1 and Propositions 5.3.9, 5.5.12 and 5.5.13 that V is a vertex algebra. From Theorem 5.5.11 we already know that W is a module as indicated. $\square$

Now that we know that vertex subalgebras and local subalgebras of the weak vertex algebra $\mathcal{E}(W)$ are the same, for the rest of this section we focus our attention on how to obtain local subalgebras of $\mathcal{E}(W)$. Our main goal is to prove that any local subset of $\mathcal{E}(W)$ generates a local subalgebra, which is in particular a vertex subalgebra.

First, we have the following key result ([Li3], where the proof was credited to C. Dong):

Proposition 5.5.15 *Let $a(x)$, $b(x)$ and $c(x)$ be pairwise mutually local weak vertex operators on W. Then $a(x)_n b(x)$ and $c(x)$ are mutually local for all $n \in \mathbb{Z}$.*

Proof. Given $n \in \mathbb{Z}$, let r be a nonnegative integer $\geq -n$ such that the following locality identities hold:

$$(x_1 - x_2)^r a(x_1)b(x_2) = (x_1 - x_2)^r b(x_2)a(x_1),$$
$$(x_1 - x_2)^r a(x_1)c(x_2) = (x_1 - x_2)^r c(x_2)a(x_1),$$
$$(x_1 - x_2)^r b(x_1)c(x_2) = (x_1 - x_2)^r c(x_2)b(x_1).$$

From (5.2.11),

$$a(x)_n b(x) = \text{Res}_{x_1} \left((x_1 - x)^n a(x_1)b(x) - (-x + x_1)^n b(x)a(x_1) \right). \quad (5.5.15)$$

Noticing that $(x - x_2)^{4r} = ((x - x_1) + (x_1 - x_2))^{3r}(x - x_2)^r$ and then using the locality identities above, we get

$$(x - x_2)^{4r} \left((x_1 - x)^n a(x_1)b(x)c(x_2) - (-x + x_1)^n b(x)a(x_1)c(x_2) \right)$$

$$= \sum_{s=0}^{3r} \binom{3r}{s}(x - x_1)^{3r-s}(x_1 - x_2)^s(x - x_2)^r \cdot$$

$$\left((x_1 - x)^n a(x_1)b(x)c(x_2) - (-x + x_1)^n b(x)a(x_1)c(x_2) \right)$$

$$= \sum_{s=r+1}^{3r} \binom{3r}{s}(x - x_1)^{3r-s}(x_1 - x_2)^s(x - x_2)^r \cdot$$

$$\left((x_1 - x)^n a(x_1)b(x)c(x_2) - (-x + x_1)^n b(x)a(x_1)c(x_2) \right)$$
$$\text{(since } (x_1 - x)^{r+(r+n)}[a(x_1), b(x)] = 0)$$

$$= \sum_{s=r+1}^{3r} \binom{3r}{s}(x - x_1)^{3r-s}(x_1 - x_2)^s(x - x_2)^r \cdot$$

$$\left((x_1 - x)^n c(x_2)a(x_1)b(x) - (-x + x_1)^n c(x_2)b(x)a(x_1) \right)$$

$$= \sum_{s=0}^{3r} \binom{3r}{s}(x - x_1)^{3r-s}(x_1 - x_2)^s(x - x_2)^r \cdot$$

$$\left((x_1 - x)^n c(x_2)a(x_1)b(x) - (-x + x_1)^n c(x_2)b(x)a(x_1) \right)$$

$$= (x - x_2)^{4r} \left((x_1 - x)^n c(x_2)a(x_1)b(x) - (-x + x_1)^n c(x_2)b(x)a(x_1) \right).$$

$$(5.5.16)$$

Taking Res_{x_1} we get

$$(x - x_2)^{4r}(a(x)_n b(x))c(x_2) = (x - x_2)^{4r} c(x_2)(a(x)_n b(x)),$$

showing that $a(x)_n b(x)$ and $c(x)$ are mutually local. $\square$

Remark 5.5.16 In the classical case we of course have that for $A, B, C \in \text{End } W$, if both A and B commute with C, then AB commutes with C. But note that the proof of Proposition 5.5.15 uses the mutual locality of $a(x)$ and $b(x)$ as well as the mutual locality of $c(x)$ and both $a(x)$ and $b(x)$.

As a consequence of Proposition 5.5.15 we have:

Theorem 5.5.17 *Any maximal local subspace of $\mathcal{E}(W)$ is a vertex subalgebra with W as a faithful module.*

Proof. Let A be a maximal local subspace of $\mathcal{E}(W)$. Since 1_W is local with any weak vertex operator on W, $A + \mathbb{C}1_W$ is a local subspace of $\mathcal{E}(W)$. The space A being maximal, we must have $A = A + \mathbb{C}1_W$, and so $1_W \in A$.

Let $a(x), b(x) \in A$, $n \in \mathbb{Z}$. It follows from Proposition 5.5.15 that $a(x)_n b(x)$ is local with any vertex operator in A. In particular, $a(x)_n b(x)$ is local with $a(x)$ and $b(x)$. Using Proposition 5.5.15 again, we see that $a(x)_n b(x)$ is local with itself. Thus $A + \mathbb{C}a(x)_n b(x)$ is a local subspace of $\mathcal{E}(W)$, and again from the maximality of A we have $A = A + \mathbb{C}a(x)_n b(x)$, so that $a(x)_n b(x) \in A$. This proves that A is a weak vertex subalgebra of $\mathcal{E}(W)$ (recall Definition 5.3.5). Thus A is a local subalgebra, and hence, by Theorem 5.5.14, a vertex algebra, and W is a module. $\quad\square$

In the classical case, any set of pairwise commuting linear operators on W of course generates a commutative associative subalgebra of $\text{End } W$. We now have (recalling Definition 5.3.7) the following analogue, which will give us a valuable practical tool for constructing concrete vertex algebras.

Theorem 5.5.18 *Let S be a set of pairwise mutually local vertex operators on W, i.e., a local subset of $\mathcal{E}(W)$. Then S can be embedded in a vertex subalgebra of $\mathcal{E}(W)$, and in fact, the weak vertex subalgebra $\langle S \rangle$ generated by S is a vertex algebra, with W as a natural faithful module. Furthermore, $\langle S \rangle$ is the linear span of the elements of the form*

$$a^{(1)}(x)_{n_1} \cdots a^{(r)}(x)_{n_r} 1_W \tag{5.5.17}$$

for $a^{(i)}(x) \in S$, $n_1, \ldots, n_r \in \mathbb{Z}$, with $r \geq 0$. In particular, this linear span carries the structure of a vertex algebra.

Proof. It follows from Zorn's lemma that there exists a maximal local subspace V containing S. By Theorem 5.5.17, V is a vertex subalgebra, with W as a natural module. Since $\langle S \rangle$ is a (weak) subalgebra of V, $\langle S \rangle$ is necessarily a vertex subalgebra and W is an $\langle S \rangle$-module. The rest follows immediately from Proposition 3.9.3. $\quad\square$

Let $a(x)$ be a vertex operator on W. Then $S = \{a(x)\}$ is a local subset of $\mathcal{E}(W)$, and in view of Theorem 5.5.18 $\langle S \rangle$ is a vertex subalgebra with W as a module, with $Y_W(a(x), x_0) = a(x_0)$. Thus we have proved:

Corollary 5.5.19 *Any vertex operator $a(x)$ on W is a vertex operator associated to an element of some vertex algebra acting on W.* $\quad\square$

5.6 Vertex subalgebras of $\mathcal{E}(W)$ associated with the Virasoro algebra

Since many of the most important vertex algebras are vertex operator algebras, equipped with a grading, and with graded modules and structure associated with the Virasoro algebra as usual, we want to extend and refine the results of the last section to include these structures.

We shall be focusing on vertex subalgebras of $\mathcal{E}(W, d)$, where d is a fixed linear operator on W (recall (5.1.5)). In particular, we shall discuss vertex operator subalgebras of $\mathcal{E}(W, L(-1))$ for W a restricted module for the Virasoro algebra. One of our aims is to prove that any graded local subalgebra of $\mathcal{E}^{o}(W, L(-1))$ (recall Definitions 5.1.5 and 5.5.8) containing $L_W(x)$ is a vertex operator algebra with $L_W(x)$ as the conformal vector (called ω in Definition 3.1.22) but without the two grading restrictions (3.1.38) and (3.1.39).

First, we have the following refinement of Theorem 5.5.18.

Theorem 5.6.1 *Let S be any local subset of $\mathcal{E}(W, d)$. Then the weak vertex subalgebra $\langle S \rangle$ of $\mathcal{E}(W)$ generated by S is a vertex subalgebra of $\mathcal{E}(W, d)$ and (W, d) is a faithful module for $\langle S \rangle$, with $Y_W(a(x), x_0) = a(x_0)$ for $a(x) \in \langle S \rangle$. In particular, the d-bracket property holds for $a(x) \in \langle S \rangle$:*

$$Y_W(\mathcal{D}a(x), x_0) = [d, Y_W(a(x), x_0)] \tag{5.6.1}$$

(recall Proposition 5.3.9).

Proof. By Theorem 5.5.18, $\langle S \rangle$ is a vertex subalgebra $\mathcal{E}(W)$ and W is a module for $\langle S \rangle$. Since $S \subset \mathcal{E}(W, d)$ and $\mathcal{E}(W, d)$ is a weak vertex subalgebra of $\mathcal{E}(W)$ by Proposition 5.4.1, $\langle S \rangle$ is a subalgebra of $\mathcal{E}(W, d)$. The rest follows immediately from Theorem 5.4.2. $\square$

As a further refinement, we have the following immediate consequence of Theorem 5.6.1, Proposition 5.4.3 and Corollary 5.4.4.

Theorem 5.6.2 *Suppose that W is a module for the Virasoro algebra and that S is a set of mutually local homogeneous vertex operators on $(W, L(-1))$, so that the linear span of S is a graded local subspace of $\mathcal{E}^{o}(W, L(-1))$. Then the weak vertex subalgebra $\langle S \rangle$ of $\mathcal{E}(W)$ generated by S is a graded vertex subalgebra of $\mathcal{E}^{o}(W, L(-1))$ and $(W, L(-1))$ is a faithful module for $\langle S \rangle$, with $Y_W(a(x), x_0) = a(x_0)$ for $a(x) \in \langle S \rangle$; in particular, the $L(-1)$-bracket property holds for $a(x) \in \langle S \rangle$:*

$$Y_W(\mathcal{D}a(x), x_0) = [L(-1), Y_W(a(x), x_0)]. \tag{5.6.2}$$

If in addition the weights of the elements of S are integers, then $\langle S \rangle$ is a graded vertex subalgebra of $\mathcal{E}^{o}_{\mathbb{Z}}(W, L(-1))$ (recall the notation (5.1.12)); that is, $\langle S \rangle$ is $\mathbb{Z}$-graded.

$\square$

Remark 5.6.3 Suppose that W is in fact a restricted module for the Virasoro algebra. Taking S to consist of the single element $L_W(x)$, which we know to be a homogeneous vertex operator on $(W, L(-1))$ of weight 2 by Remark 5.1.6 and Example 5.5.6, we see that $\langle L_W(x) \rangle$ is a ($\mathbb{Z}$-)graded vertex subalgebra of $\mathcal{E}_{\mathbb{Z}}^o(W, L(-1))$.

Now we focus on vertex operator algebra structure. First we discuss what happens if we start with such an algebra.

Remark 5.6.4 Suppose that $(V, Y, \mathbf{1}, \omega)$ is a vertex operator algebra of central charge ℓ and that (W, Y_W) is a module for V viewed as a vertex algebra (so that in particular, W is not assumed to be graded). In view of Proposition 4.1.5, $\iota_W(\omega) = Y_W(\omega, x)$ gives rise to a representation of the Virasoro algebra on W of central charge ℓ. It follows from the truncation condition (4.1.2) (in the definition of the notion of module) that W is a restricted module for the Virasoro algebra as defined in Remark 5.1.6. Furthermore, from Remark 5.5.10, $\iota_W(V)$ is a (local) vertex subalgebra of $\mathcal{E}(W)$, and by Proposition 4.1.5, $\iota_W(V)$ is a vertex subalgebra of $\mathcal{E}(W, L(-1))$ (recall (5.1.5)). Notice that the formula (4.1.18) for $[L(0), Y_W(v, x)]$ holds (even though W is just a module for V viewed as a vertex algebra). Then it follows from (4.1.18) and (4.1.12) that for $v \in V_{(n)}$,

$$Y_W(v, x) \in \mathcal{E}(W, L(-1))_{(n)} \tag{5.6.3}$$

(recall the definition (5.1.9)). Therefore, $\iota_W(V)$ is a graded (local) vertex subalgebra of $\mathcal{E}^o(W, L(-1))$, and in fact, of the integrally graded subalgebra $\mathcal{E}_{\mathbb{Z}}^o(W, L(-1))$ (recall (5.1.12)). Note that even if V is a vertex operator algebra without the two grading restrictions (3.1.38) and (3.1.39), all the assertions here still hold. Consider the special case in which ι_W is the embedding map, so that the vertex operator algebra V (without grading restrictions) is a weak vertex subalgebra of $\mathcal{E}(W)$, as in Remark 5.3.13 or Theorem 5.5.11. Then W is a restricted module for the Virasoro algebra with central charge equal to the central charge of V. Moreover, from (5.6.3) we see that the given grading on V coincides with the grading on V viewed as a subspace of $\mathcal{E}_{\mathbb{Z}}^o(W, L(-1))$: For $n \in \mathbb{Z}$,

$$V_{(n)} = V \cap \mathcal{E}(W, L(-1))_{(n)}. \tag{5.6.4}$$

Our goal is now to obtain vertex operator algebra structure from a given representation of the Virasoro algebra on W, and in view of Remark 5.6.4, we assume that W is a restricted module of central charge ℓ. We shall need a conformal vector, in the sense of Definition 3.1.22. If V is a graded vertex subalgebra of $\mathcal{E}_{\mathbb{Z}}^o(W, L(-1))$ containing $L_W(x)$ (using the notation of Example 5.5.6), an obvious candidate for a conformal vector for V is $L_W(x)$. (We have seen in Remark 5.6.3 that there indeed exists a vertex subalgebra of $\mathcal{E}_{\mathbb{Z}}^o(W, L(-1))$ containing $L_W(x)$.) We shall prove that $L_W(x)$ is indeed a conformal vector of V, which will be a vertex operator algebra except that the two grading restrictions (3.1.38) and (3.1.39) might not hold.

For this we must first prove that the components of $Y_{\mathcal{E}}(L_W(x), x_0)$ give rise to a representation of the Virasoro algebra of central charge ℓ. Since W is a faithful V-module, it will suffice to prove the abstract result (see Proposition 5.6.8) that if W

is a faithful module for a vertex algebra V and if ω is a vector in V such that the components of the vertex operator $Y_W(\omega, x)$ (on W) give rise to a representation of the Virasoro algebra of central charge ℓ, then the components of the vertex operator $Y(\omega, x)$ on V also give rise to a representation of the Virasoro algebra of central charge ℓ.

To prove this, we shall need the following simple fact ([Li3]; cf. [FLM6], Proposition 8.1.3):

Proposition 5.6.5 *Let U be a vector space and let $h(x_1, x_2) \in U[[x_1, x_2, x_1^{-1}, x_2^{-1}]]$. An expression of $h(x_1, x_2)$ as a finite sum $\sum_{i=0}^{n} g_i(x_2) \left(\frac{\partial}{\partial x_1}\right)^i x_2^{-1} \delta\left(\frac{x_1}{x_2}\right)$ with $g_i(x_2) \in U[[x_2, x_2^{-1}]]$ is unique if it exists.*

Proof. It is sufficient to prove that if

$$g_0(x_2)x_2^{-1}\delta\left(\frac{x_1}{x_2}\right) + g_1(x_2)\frac{\partial}{\partial x_1}x_2^{-1}\delta\left(\frac{x_1}{x_2}\right) + \cdots$$
$$+ g_n(x_2)\left(\frac{\partial}{\partial x_1}\right)^n x_2^{-1}\delta\left(\frac{x_1}{x_2}\right) = 0, \tag{5.6.5}$$

then $g_i(x_2) = 0$ for all i. Suppose instead that $g_n \neq 0$. Multiplying (5.6.5) by $(x_1 - x_2)^n$ and then using (2.3.12) and (2.3.13) we obtain

$$0 = (x_1 - x_2)^n g_n(x_2)\left(\frac{\partial}{\partial x_1}\right)^n x_2^{-1}\delta\left(\frac{x_1}{x_2}\right) = (-1)^n n! x_2^{-1}\delta\left(\frac{x_1}{x_2}\right) g_n(x_2), \tag{5.6.6}$$

which immediately implies that $g_n = 0$ (take Res_{x_1}), a contradiction. $\square$

Remark 5.6.6 In a similar fashion, an expression of $h(x_1, x_2)$ as a finite sum $\sum_{i=0}^{n} g_i(x_2) \left(\frac{\partial}{\partial x_2}\right)^i x_2^{-1}\delta\left(\frac{x_1}{x_2}\right)$ is unique if it exists (recall Proposition 2.3.6).

We shall also use the following result (cf. [Li3]):

Proposition 5.6.7 *Let V be a vertex algebra, let $u, v, w^{(0)}, \ldots, w^{(k)} \in V$ and let (W, Y_W) be a faithful V-module. Then*

$$[Y(u, x_1), Y(v, x_2)] = \sum_{i=0}^{k} \frac{(-1)^i}{i!} Y(w^{(i)}, x_2)\left(\frac{\partial}{\partial x_1}\right)^i x_2^{-1}\delta\left(\frac{x_1}{x_2}\right) \tag{5.6.7}$$

on V if and only if the analogous formula holds on W:

$$[Y_W(u, x_1), Y_W(v, x_2)] = \sum_{i=0}^{k} \frac{(-1)^i}{i!} Y_W(w^{(i)}, x_2)\left(\frac{\partial}{\partial x_1}\right)^i x_2^{-1}\delta\left(\frac{x_1}{x_2}\right). \tag{5.6.8}$$

In this case, the elements $w^{(i)}$ are given by

$$w^{(i)} = u_i v \quad for \quad i = 0, \ldots, k, \tag{5.6.9}$$

and we also have

$$u_i v = 0 \quad \text{for} \quad i > k. \tag{5.6.10}$$

Moreover, under the same hypotheses but with the V-module W not necessarily faithful, (5.6.7) implies (5.6.8), (5.6.9) and (5.6.10).

Proof. The commutator formula (3.1.8), for a vertex algebra and its modules, asserts that

$$[Y(u, x_1), Y(v, x_2)] = \sum_{i=0}^{n} \frac{(-1)^i}{i!} Y(u_i v, x_2) \left(\frac{\partial}{\partial x_1} \right)^i x_2^{-1} \delta \left(\frac{x_1}{x_2} \right), \tag{5.6.11}$$

where n is arbitrary such that $u_i v = 0$ for $i > n$, and in particular, this holds on V and on W. Thus (5.6.7) and (5.6.8) both hold if (5.6.9) and (5.6.10) hold, where we take $n = k$. Conversely, if (5.6.7) holds or if W is faithful and (5.6.8) holds, then since V is a faithful V-module, we see from Proposition 5.6.5 that (5.6.9) and (5.6.10) hold. Thus if W is faithful, (5.6.7) and (5.6.8) are equivalent. $\square$

This result will now enable us to transport the Virasoro algebra bracket relations from a faithful V-module to V itself.

Proposition 5.6.8 *Let V be a vertex algebra equipped with a vector $\omega \in V$ and let (W, Y_W) be a faithful V-module. Then the components of $Y(\omega, x)$ give rise to a representation of the Virasoro algebra of central charge ℓ on V if and only if the components of $Y_W(\omega, x)$ give rise to a representation of the Virasoro algebra of central charge ℓ on W (in the usual sense of (3.1.40), (3.1.41) and (4.1.10), and of formula (5.5.5) in Example 5.5.6).*

Proof. In view of the $\mathcal{D}$-derivative property for both vertex algebras and modules (recall Propositions 3.1.18 and 4.1.3), the Virasoro algebra relations in the generating function form (5.5.5) amount to the relation

$$[Y(\omega, x_1), Y(\omega, x_2)] = Y(\mathcal{D}\omega, x_2) x_2^{-1} \delta \left(\frac{x_1}{x_2} \right) - 2Y(\omega, x_2) \frac{\partial}{\partial x_1} x_2^{-1} \delta \left(\frac{x_1}{x_2} \right)$$
$$- \frac{\ell}{12} \left(\frac{\partial}{\partial x_1} \right)^3 x_2^{-1} \delta \left(\frac{x_1}{x_2} \right), \tag{5.6.12}$$

both for the algebra V and for the module W (in which case Y is replaced by Y_W). The result now follows immediately from Proposition 5.6.7, where we take $k = 3$ and $w^{(0)} = \mathcal{D}\omega$, $w^{(1)} = 2\omega$, $w^{(2)} = 0$ and $w^{(3)} = (\ell/2)\mathbf{1}$. $\square$

Remark 5.6.9 Of course, the relations (5.6.9) and (5.6.10) in this case are just the familiar relations (3.1.70)–(3.1.74); recall Example 5.5.6.

Remark 5.6.10 Results analogous to Proposition 5.6.8 also hold for other families of Lie algebras, including affine Lie algebras, as we discuss in the next chapter.

We now have the following consequence, in our present situation:

Theorem 5.6.11 *Let W be a restricted module for the Virasoro algebra of central charge ℓ and let V be any graded vertex subalgebra of $\mathcal{E}^o(W, L(-1))$ containing*

$$
L_W(x) = \sum_{n \in \mathbb{Z}} L_W(n) x^{-n-2} \left(= \sum_{n \in \mathbb{Z}} L(n) x^{-n-2} \right). \tag{5.6.13}
$$

Then $(V, Y_{\mathcal{E}}, 1_W, L_W(x))$ satisfies all the axioms in the definition of the notion of vertex operator algebra of central charge ℓ, except that V is $\mathbb{C}$-graded rather than $\mathbb{Z}$-graded, and in addition, the two grading restrictions (3.1.38) and (3.1.39) might not hold. The grading on V as a subspace of $\mathcal{E}^o(W, L(-1))$ coincides with its vertex operator algebra grading (by $L(0)$-eigenvalues). If V is a graded vertex subalgebra of $\mathcal{E}^o_{\mathbb{Z}}(W, L(-1))$, then V is indeed $\mathbb{Z}$-graded.

Proof. We know from Theorem 5.5.11 that W is a faithful V-module, with $Y_W(L_W(x), x_0) = L_W(x_0)$, where the components of $L_W(x_0)$ are assumed to give rise to a representation on W of the Virasoro algebra of central charge ℓ. Then by Proposition 5.6.8, the components of $Y_{\mathcal{E}}(L_W(x), x_0)$ give rise to a representation on V of the Virasoro algebra of central charge ℓ.

Set

$$
L_V(x_0) = \sum_{n \in \mathbb{Z}} L_V(n) x_0^{-n-2} = Y_{\mathcal{E}}(L_W(x), x_0). \tag{5.6.14}
$$

Then for any $a(x) \in V$, from the definitions (5.2.11), (5.1.5) and (5.1.6), we have

$$
L_V(-1)a(x) = L_W(x)_0 a(x) = [L_W(-1), a(x)] = a'(x) = \mathcal{D}a(x), \tag{5.6.15}
$$
$$
L_V(0)a(x) = L_W(x)_1 a(x) = [L_W(0), a(x)] - x[L_W(-1), a(x)], \tag{5.6.16}
$$

and this in turn equals $ha(x)$ if $a(x) \in V_{(h)}$. Thus the grading on $\mathcal{E}^o(W, L(-1))$ given in Definition 5.1.5 provides a grading on V by $L_V(0)$-eigenvalues. Also, since by (5.6.15) $L_V(-1) = \mathcal{D}$ on V, we know from Proposition 5.3.9 that the $L(-1)$-derivative property (3.1.45) holds on V. This proves that V has all the properties of a vertex operator algebra of central charge ℓ except for the two grading restrictions and the $\mathbb{C}$-grading. The last assertion of the proposition is clear. $\square$

Remark 5.6.12 It is interesting that so far we have not needed to assume that the module W is graded. All of our gradings have been gradings on spaces and algebras of *operators* on W. In the next result, we relate our operator grading on V to a given grading on W.

Theorem 5.6.13 *In the setting of Theorem 5.6.11, suppose also that W is $\mathbb{C}$-graded by $L_W(0)$-eigenvalues. Then W is a (faithful) module for the vertex operator algebra V in the sense of Definition 4.1.6, except for the grading restrictions (4.1.16) and (4.1.17) on W (and except for the grading restrictions on V).*

Proof. In view of Theorem 5.5.11, W is a module for V viewed as a vertex algebra, and $Y_W(L_W(x), x_0) = L_W(x_0)$. Thus $L_W(0)$ agrees with the coefficient of x_0^{-2} of $Y_W(L_W(x), x_0)$, and so W is graded by eigenvalues of this operator. Thus W satisfies the desired conditions. $\square$

We are now prepared to present the following main result, which is a natural refinement of Theorem 5.5.18:

Theorem 5.6.14 *Let W be a restricted module for the Virasoro algebra of central charge ℓ and let S be a set of pairwise mutually local homogeneous vertex operators on $(W, L(-1))$, containing $L_W(x) = \sum_{n \in \mathbb{Z}} L_W(n) x^{-n-2} \left(= \sum_{n \in \mathbb{Z}} L(n) x^{-n-2} \right)$. Let V be the subspace of $\mathcal{E}(W)$ linearly spanned by the elements*

$$a^{(1)}(x)_{n_1} \cdots a^{(r)}(x)_{n_r} 1_W \tag{5.6.17}$$

for $r \geq 0$, $a^{(1)}(x), \ldots, a^{(r)}(x) \in S$ and $n_1, \ldots, n_r \in \mathbb{Z}$, or equivalently, the weak vertex subalgebra of $\mathcal{E}(W)$ generated by S. Then $(V, Y_{\mathcal{E}}, 1_W, L_W(x))$ satisfies all the axioms in the definition of the notion of vertex operator algebra of central charge ℓ, except that V is $\mathbb{C}$-graded rather than $\mathbb{Z}$-graded, and in addition, the two grading restrictions (3.1.38) and (3.1.39) might not hold. The grading on V by $L(0)$-eigenvalues coincides with its grading as a subspace of $\mathcal{E}^o(W, L(-1))$. If S consists of homogeneous vertex operators of integral weights, then V is indeed $\mathbb{Z}$-graded. Moreover, if W is $\mathbb{C}$-graded by $L_W(0)$-eigenvalues, then W is a faithful module for the vertex operator algebra V, except perhaps for the two grading restrictions on W (and except perhaps for the grading restrictions on V).

Proof. From Theorems 5.5.18 and 5.6.2, V is a graded vertex subalgebra of $\mathcal{E}^o(W, L(-1))$ and is exactly the vertex subalgebra generated by S. Moreover, if the weights of the elements of S are integers, then V is $\mathbb{Z}$-graded. The result now follows immediately from Theorems 5.6.11 and 5.6.13. $\square$

As an important—and immediate—application, we have, recalling Remark 5.6.3:

Corollary 5.6.15 *Let W be a restricted module for the Virasoro algebra of central charge ℓ and let U be the subspace of $\mathcal{E}(W)$ linearly spanned by the elements*

$$L_W(x)_{n_1} \cdots L_W(x)_{n_r} 1_W \tag{5.6.18}$$

for $r \geq 0$ and $n_1, \ldots, n_r \in \mathbb{Z}$. Then $(U, Y_{\mathcal{E}}, 1_W, L_W(x))$ satisfies all the axioms in the definition of the notion of vertex operator algebra of central charge ℓ except perhaps for the two grading restrictions (3.1.38) and (3.1.39). The grading on U by $L(0)$-eigenvalues coincides with its grading as a subspace of $\mathcal{E}^o(W, L(-1))$. If W is $\mathbb{C}$-graded by $L_W(0)$-eigenvalues, then W is a faithful U-module except perhaps for the two grading restrictions on W. $\square$

Remark 5.6.16 In fact, it will take only a small amount of additional effort to show that U is actually a vertex operator algebra of central charge ℓ (that is, with the two grading restrictions); see Section 6.1.

5.7 General construction theorems for vertex algebras and modules

In [Li3], Theorem 5.5.18 was used to construct general classes of vertex operator algebras and modules associated to the Virasoro algebra and to affine Lie algebras. (General classes of vertex operator algebras and modules associated to the Virasoro algebra and affine Lie algebras were also constructed in [FF7], [FZ], [Hua4] and [DL3], by different methods.) The methods used in [Li3] for both the Virasoro algebra and affine Lie algebras were the same, and can be summarized as follows. For any (restricted) module W for the Lie algebra, the set S of natural generating functions for the Lie algebra elements acting on W is a local subset of $\mathcal{E}(W)$ (cf. Example 5.5.6), and so by Theorem 5.5.18, $\langle S \rangle$ is a vertex algebra with W as a module. Furthermore, the vertex algebra $\langle S \rangle$ is itself also a natural module for the Lie algebra (cf. Proposition 5.6.8 and Remark 5.6.10). If our space W is of a certain special type, it follows that W and $\langle S \rangle$ are naturally isomorphic modules for the Lie algebra, so that our vector space W has a natural vertex algebra structure transported from $\langle S \rangle$. For constructing modules for such a vertex algebra, Theorem 5.5.18 was used once again.

In the same spirit there is a very useful general construction theorem for vertex algebras obtained in [FKRW] and [MP2] that asserts that given a vector space V equipped with a vector playing the role of vacuum vector and with a set of mutually local vertex operators with certain properties, one obtains a canonical vertex algebra structure on V suitably generated by the given operators. This theorem extends and generalizes the construction results of [Li3] for vertex operator algebras based on the Virasoro algebra and on affine Lie algebras.

In this section we shall present the theorem of [FKRW] and [MP2] (Theorem 5.7.1) and a refinement incorporating the Virasoro algebra and a grading (Theorem 5.7.4). We prove these results using the method summarized above by which vertex algebras were constructed in [Li3], specifically, by using Theorem 5.5.18 and Proposition 4.7.9 on vacuum-like vectors (see [Li1]). Then, using Theorem 5.5.18 in its full generality (for modules), we prove a general existence and uniqueness result, Theorem 5.7.6, for a module structure on a given vector space for a given vertex algebra. At the end of this section we prove a variant of a theorem of Xu ([Xu2], [Xu9], [Xu12]), which is also in a similar spirit; see Theorems 5.7.11 and 5.7.13.

The following is the general construction theorem for vertex algebras obtained (except for the last assertion, which is based on Theorem 5.5.18) in [FKRW] and [MP2] (cf. [F5], [FB], [K7], [P5]):

Theorem 5.7.1 *Let V be a vector space equipped with a distinguished vector $\mathbf{1}$ and with a linear operator d such that $d\mathbf{1} = 0$, and let T be a subset of V equipped with a map*

$$Y_0(\cdot, x) : T \to \mathrm{Hom}(V, V((x)))$$
$$a \mapsto Y_0(a, x) = \sum_{n \in \mathbb{Z}} a_n x^{-n-1}. \tag{5.7.1}$$

Assume that the following conditions hold: For $a \in T$,

$$Y_0(a, x)\mathbf{1} \in V[[x]] \quad \text{and} \quad \lim_{x \to 0} Y_0(a, x)\mathbf{1} \ (= a_{-1}\mathbf{1}) = a, \tag{5.7.2}$$

$$[d, Y_0(a, x)] = \frac{d}{dx} Y_0(a, x); \tag{5.7.3}$$

for $a, b \in T$, there exists a nonnegative integer k such that

$$(x_1 - x_2)^k [Y_0(a, x_1), Y_0(b, x_2)] = 0; \tag{5.7.4}$$

and V is linearly spanned by the vectors

$$a_{n_1}^{(1)} \cdots a_{n_r}^{(r)} \mathbf{1} \tag{5.7.5}$$

for $r \geq 0$, $a^{(i)} \in T$, $n_i \in \mathbb{Z}$. Then Y_0 can be uniquely extended to a linear map Y from V to $\mathrm{Hom}(V, V((x)))$ such that $(V, Y, \mathbf{1})$ carries the structure of a vertex algebra. The vertex operator map Y for this vertex algebra structure is given by

$$Y(a_{n_1}^{(1)} \cdots a_{n_r}^{(r)} \mathbf{1}, x) = a^{(1)}(x)_{n_1} \cdots a^{(r)}(x)_{n_r} 1_V \tag{5.7.6}$$

(recall formula (5.2.11) for the n-th action of a weak vertex operator $a(x)$ on a weak vertex operator $b(x)$), where for $a \in T$ we write

$$a(x) = Y_0(a, x). \tag{5.7.7}$$

The operator d on V agrees with the operator $\mathcal{D}$ on V defined in Proposition 3.1.18:

$$d = \mathcal{D} \quad \text{on } V. \tag{5.7.8}$$

Furthermore, writing

$$T(x) = \{a(x) \mid a \in T\} \subset \mathrm{Hom}(V, V((x))), \tag{5.7.9}$$

we have that the linear map

$$\psi : \langle T(x) \rangle \to V$$
$$\alpha(x) \mapsto \mathrm{Res}_x x^{-1} \alpha(x)\mathbf{1} \quad (= \text{constant term of } \alpha(x)\mathbf{1}) \tag{5.7.10}$$

(where $\alpha(x)$ is an arbitrary element of $\langle T(x) \rangle$) is an isomorphism of vertex algebras, $\langle T(x) \rangle$ being the vertex algebra constructed in Theorem 5.5.18.

Proof. Uniqueness and formula (5.7.6) follow immediately from Remark 5.2.5 and the spanning hypothesis. By (5.7.1), (5.7.3) and (5.7.4), $T(x)$ is a local subset of $\mathcal{E}(V, d)$, and so by Theorem 5.6.1 $\langle T(x) \rangle$ is a vertex subalgebra of $\mathcal{E}(V, d)$ with (V, d) as a module, with the module action Y_V given by $Y_V(\alpha(x), x_0) = \alpha(x_0)$ for $\alpha(x) \in \langle T(x) \rangle$. Note that the map ψ in (5.7.10) can be expressed as

$$\psi(\alpha(x)) = \mathrm{Res}_{x_0} x_0^{-1} \alpha(x_0)\mathbf{1} = \mathrm{Res}_{x_0} x_0^{-1} Y_V(\alpha(x), x_0)\mathbf{1}. \tag{5.7.11}$$

Thus it follows from Proposition 4.7.9 that ψ is a $\langle T(x) \rangle$-module isomorphism since (V, d) is a faithful $\langle T(x) \rangle$-module, $d\mathbf{1} = 0$, and V is spanned by the indicated elements. Hence V has a vertex algebra structure transported from $\langle T(x) \rangle$, where we force ψ to be a vertex algebra isomorphism. It remains to determine the vacuum vector and the vertex operator map for V, and to prove (5.7.8). First, $\mathbf{1}$ is the vacuum vector of V because

$$\psi(1_V) = \text{Res}_{x_0} x_0^{-1} Y_V(1_V, x_0)\mathbf{1} = \text{Res}_{x_0} x_0^{-1} 1_V(x_0)\mathbf{1} = 1_V \mathbf{1} = \mathbf{1}.$$

$$(5.7.12)$$

Denote by $Y(\cdot, x)$ the vertex operator map for the vertex algebra V and let $a \in T$. Since

$$\psi(a(x)) = \lim_{x \to 0} a(x)\mathbf{1} = a, \qquad (5.7.13)$$

by (5.7.2), and since ψ is a $\langle T(x) \rangle$-module homomorphism, we have

$$Y(a, x_0) = \psi Y_{\mathcal{E}}(\psi^{-1}a, x_0)\psi^{-1} = Y_V(\psi^{-1}a, x_0) = Y_V(a(x), x_0) = a(x_0).$$

$$(5.7.14)$$

That is, the map Y from V to $\text{Hom}(V, V((x)))$ extends the given map Y_0 on T. Finally, to show that $d = \mathcal{D}$ we note that V is linearly spanned by the coefficients of all the monomials in the expressions

$$Y(a^{(1)}, x_1) Y(a^{(2)}, x_2) \cdots Y(a^{(r)}, x_r)\mathbf{1} \qquad (5.7.15)$$

for $a^{(i)} \in T$, and in view of (3.1.32) and (5.7.3) and the formulas $\mathcal{D}\mathbf{1} = 0 = d\mathbf{1}$, the operators d and $\mathcal{D}$ act the same way on (5.7.15). $\square$

Remark 5.7.2 Theorem 5.7.1 (or rather, Theorem 5.7.1 minus the last assertion, which is based on Theorem 5.5.18) is actually a bit more general than the corresponding theorems in [FKRW] and [MP2], [MP4], while the corresponding theorem in [P5] has the full generality of Theorem 5.7.1, with V a superspace. In the theorem of [FKRW], V is assumed to be $\mathbb{N}$-graded and linearly spanned by the elements (5.7.5) with $n_i < 0$; for the corresponding theorem in [K7], V is a superspace and the same spanning property is assumed. The proof in [FKRW] and [K7] (the same proof) used [Li3], in particular, Proposition 5.5.15 (which is Proposition 3.2.7 in [Li3]). In the corresponding theorem in [MP2], [MP4], which was obtained independently of [Li3], V is assumed to be $\mathbb{Z}$-graded with the grading truncated from below, allowing the use of formal rational functions, in the spirit of Proposition 3.2.7. Proposition 5.5.15 for $W = V$ was independently obtained in [MP4] and this result was used in the proof of the theorem of [MP2].

Remark 5.7.3 To illustrate how natural the use of Theorem 5.5.18—the exploitation of a vertex algebra of "endomorphisms"— is in the proof of Theorem 5.7.1 (even with its last assertion, explicitly invoking Theorem 5.5.18, omitted), consider the following (relatively easy) analogue of Theorem 5.7.1 in the classical context of commutative associative algebras in place of vertex algebras. *Let A be a vector space equipped with a distinguished vector $\mathbf{1}$ and a subset T together with a map*

$$L : T \to \text{End } A$$

$$a \mapsto L_a. \tag{5.7.16}$$

Assume that $L_a 1 = a$ and $[L_a, L_b] = 0$ for $a, b \in T$, and that A is spanned by the vectors $L_{a_1} \cdots L_{a_r} 1$ with $a_i \in T$. Then L can be uniquely extended to a linear map $L : A \to \text{End } A$ such that A is a commutative associative algebra with identity element 1 and with L the (left-) multiplication operation. The formula

$$L_{L_{a_1} \cdots L_{a_r} 1} = L_{a_1} \cdots L_{a_r} \tag{5.7.17}$$

determines the left-multiplication operation on A. We prove this result by analogy with the proof of Theorem 5.5.18: First, note that the uniqueness and formula (5.7.17) are clear. Consider the linear map

$$\psi : \langle L_T \rangle \to A$$

$$\alpha \mapsto \alpha \mathbf{1}, \tag{5.7.18}$$

where $\langle L_T \rangle$ is the commutative (associative) subalgebra of End A generated by the (commuting) endomorphisms L_a for $a \in T$. The space A is a natural faithful $\langle L_T \rangle$-module, ψ is clearly a map of $\langle L_T \rangle$-modules, and since $\mathbf{1}$ generates A as an $\langle L_T \rangle$-module in view of the spanning assumption, ψ is surjective. If $\alpha \mathbf{1} = 0$, then

$$\alpha A = \alpha \langle L_T \rangle \mathbf{1} = \langle L_T \rangle \alpha \mathbf{1} = 0, \tag{5.7.19}$$

so that α annihilates A and hence is 0. Thus ψ is an $\langle L_T \rangle$-module isomorphism. (This is the analogue of the use of Proposition 4.7.9.) Hence A has a commutative associative algebra structure transported from $\langle L_T \rangle$, with ψ an isomorphism of commutative associative algebras. The identity element of A is $\psi(1_A) = 1_A \mathbf{1} = \mathbf{1}$. To determine the product on A, note that for $a \in T$, we have $\psi(L_a) = L_a \mathbf{1} = a$ by assumption, and since ψ is an $\langle L_T \rangle$-module isomorphism, for $a \in T$ and $b \in A$ we have

$$ab = \psi((\psi^{-1}a)(\psi^{-1}b)) = (\psi^{-1}a)b = L_a b, \tag{5.7.20}$$

showing that L can indeed be extended from T to A as desired. Notice that this proof continues to highlight the commutative-associative algebra features of vertex algebras as opposed to the Lie algebra features, but notice also that the formulation and proof of Theorems 5.5.18 and 5.7.1 and of the results used in their proofs are far more subtle than for this classical analogue.

If V in Theorem 5.7.1 is a restricted module for the Virasoro algebra of central charge ℓ with suitable assumptions, we naturally expect that V will be a vertex operator algebra of central charge ℓ. We prove the following refinement of Theorem 5.7.1 using the corresponding refinement, Theorem 5.6.14, of Theorem 5.5.18:

Theorem 5.7.4 *In the setting of Theorem 5.7.1, let us assume in addition that V is a restricted module for the Virasoro algebra of central charge ℓ such that $L(-1) = d$ (so* *that in particular, $L(-1)\mathbf{1} = 0$) and $L(-2)\mathbf{1} \in T$, with $Y_0(L(-2)\mathbf{1}, x)$ providing the given representation of the Virasoro algebra:*

$$Y_0(L(-2)\mathbf{1}, x) = L_V(x) = \sum_{n \in \mathbb{Z}} L(n)x^{-n-2}, \tag{5.7.21}$$

in the notation of Remark 5.1.6. We also assume that for every $a \in T$, there exists $m \in \mathbb{Z}$ such that

$$[L(0), a(x)] = ma(x) + x\frac{d}{dx}a(x), \tag{5.7.22}$$

or equivalently,

$$[L(0), a_n] = (m - n - 1)a_n \quad \text{for } n \in \mathbb{Z}. \tag{5.7.23}$$

Then Y_0 can be uniquely extended to a linear map Y from V to $\mathrm{Hom}(V, V((x)))$ such that $(V, Y, \mathbf{1}, L(-2)\mathbf{1})$ carries the structure of a vertex operator algebra of central charge ℓ but without the two grading restrictions (3.1.38) and (3.1.39). Each element $a \in T$ is a homogeneous element of V and the weight of a is m, in the notation of (5.7.22).

Proof. Since $L(-1) = d$, we have from (5.7.3) that

$$[L(-1), a(x)] = \frac{d}{dx}a(x) \quad \text{for } a \in T, \tag{5.7.24}$$

so that $T(x) \subset \mathcal{E}(V, L(-1))$. The equivalence between (5.7.22) and (5.7.23) is clear, and for $a \in T$ as in (5.7.22), $a(x)$ is a homogeneous weak vertex operator on $(V, L(-1))$ of (integral) weight m (recall Definition 5.1.5). By Theorem 5.6.14, $(\langle T(x)\rangle, Y_\mathcal{E}, 1_V, L_V(x))$ is a vertex operator algebra of central charge ℓ except for the two grading restrictions, and the grading on $\langle T(x)\rangle$ by $L(0)$-eigenvalues coincides with its grading as a subspace of $\mathcal{E}^o(W, L(-1))$. In particular, $a(x)$ as in (5.7.22) is homogeneous of weight m in the vertex operator algebra $\langle T(x)\rangle$. In view of Theorem 5.7.1, the linear map ψ defined in (5.7.10) is a vertex algebra isomorphism from $\langle T(x)\rangle$ to V, so that forcing ψ to be an isomorphism of vertex operator algebras (without grading restrictions) gives us such a structure on V. Then V is a vertex operator algebra of central charge ℓ without grading restrictions (and is in particular $\mathbb{Z}$-graded by $L(0)$-eigenvalues), with $L(-2)\mathbf{1}$ the conformal vector, since

$$\psi(L_V(x)) = \mathrm{Res}_{x_0}x_0^{-1}Y_V(L_V(x), x_0)\mathbf{1} = \mathrm{Res}_{x_0}x_0^{-1}L_V(x_0)\mathbf{1} = L(-2)\mathbf{1}. \tag{5.7.25}$$

This formula is of course a special case of (5.7.13), which asserts that $\psi(a(x)) = a$ for $a \in T$. Since $a(x)$ in (5.7.22) is homogeneous of weight m in the vertex operator algebra $\langle T(x)\rangle$, $\psi(a(x)) = a$ is homogeneous of weight m in V. $\square$

Now Theorem 5.7.1 and its refinement, Theorem 5.7.4, deal with constructing only algebras and not modules, and it is very interesting to find analogues of such construction theorems for modules for a given vertex algebra. Let V be a vertex algebra and let W be a vector space equipped with a map Y_W^0 from T to $\mathrm{Hom}(W, W((x)))$, where T is a given generating subset of V. The natural question that arises is: Under what conditions can Y_W^0 be (uniquely) extended to a linear map Y_W from V to $\mathrm{Hom}(W, W((x)))$ such that (W, Y_W) is a V-module?

This question seems too general to answer, even for its classical analogue in associative algebra theory. Let us look at this question from a different angle. By Theorem 5.3.15, giving a V-module (W, Y_W) amounts to giving a weak vertex algebra homomorphism ι_W from V to $\mathcal{E}(W)$, where ι_W and Y_W are related by $\iota_W(v) = Y_W(v, x)$, and such an extension is clearly unique; recall that $\mathcal{E}(W) = \mathrm{Hom}(W, W((x)))$. Thus we have:

Theorem 5.7.5 *Let V be a given vertex algebra and let W be a vector space equipped with a map Y_W^0 from T to $\mathrm{Hom}(W, W((x)))$ $(= \mathcal{E}(W))$, where T is a given generating subset of V. Also write Y_W^0 as $\iota_W^0 : T \to \mathcal{E}(W)$, with $\iota_W^0(a) = Y_W^0(a, x)$ for $a \in T$. Then the map Y_W^0 can be extended to a linear map Y_W from V to $\mathrm{Hom}(W, W((x)))$ such that (W, Y_W) carries the structure of a V-module if and only if the map ι_W^0 on T can be extended to a weak vertex algebra homomorphism ι_W from V into $\mathcal{E}(W)$. The correspondence is given by $Y_W(v, x) = \iota_W(v)$, and such an extension is unique.* $\square$

Whenever this happens, the weak vertex subalgebra $\langle \iota_W^0(T) \rangle$ of $\mathcal{E}(W)$, as the image of the vertex algebra V, is a vertex subalgebra of $\mathcal{E}(W)$. Then in view of Theorem 5.5.11, $\iota_W^0(T)$ must be a local subset of $\mathcal{E}(W)$. We have the following general and useful criterion for extendability:

Theorem 5.7.6 *Let V be a vertex algebra and let W be a vector space equipped with a map*

$$Y_W^0 : T \to \mathrm{Hom}(W, W((x)))   (= \mathcal{E}(W))$$
$$a \mapsto Y_W^0(a, x) = a_W(x), \tag{5.7.26}$$

where T is a given generating subset of V. Also write Y_W^0 as $\iota_W^0 : T \to \mathcal{E}(W)$, with $\iota_W^0(a) = Y_W^0(a, x) = a_W(x)$. Then the linear map Y_W^0 can be extended to a linear map

$$Y_W : V \to \mathrm{Hom}(W, W((x)))$$

such that (W, Y_W) carries the structure of a V-module if and only if the map ι_W^0 can be extended to a linear map $\iota_W : V \to \mathcal{E}(W)$ such that

$$\iota_W(\mathbf{1}) = 1_W \tag{5.7.27}$$

and

$$\iota_W(a_n v) = a_W(x)_n \iota_W(v)   \text{for } a \in T,\ v \in V,\ n \in \mathbb{Z}, \tag{5.7.28}$$

and in addition, for any $a, b \in T$ there exists a nonnegative integer k such that

$$(x_1 - x_2)^k [a_W(x_1), b_W(x_2)] = 0. \tag{5.7.29}$$

In this case, the correspondence is given by $Y_W(v, x) = \iota_W(v)$ for $v \in V$, and such an extension is unique. Moreover, the map

$$Y_W(\cdot, x) = \iota_W : V \to \mathcal{E}(W) \tag{5.7.30}$$

is a homomorphism of weak vertex algebras.

Proof. If Y_W^0 can be extended to the module action Y_W, then ι_W^0 can be correspondingly extended to a weak vertex algebra homomorphism $\iota_W : V \to \mathcal{E}(W)$, and (5.7.29) must hold, as we have mentioned. Such an extension is unique. What we must prove is that a linear map $\iota_W : V \to \mathcal{E}(W)$ with the indicated properties is a weak vertex algebra homomorphism. (This will in particular imply its uniqueness.)

Let ι_W be such a linear map. Since $\iota_W^0(T)$ is assumed to be a local subset of $\mathcal{E}(W)$, by Theorem 5.5.18 the weak vertex subalgebra $\langle \iota_W^0(T) \rangle$ of $\mathcal{E}(W)$ is a vertex algebra. We shall want to use the Jacobi identity in the image $\iota_W(V)$, so we show that

$$\iota_W(V) \subset \langle \iota_W^0(T) \rangle. \tag{5.7.31}$$

This follows immediately from (5.7.27), (5.7.28) and the fact that V is spanned by the elements $a_{n_1}^{(1)} \cdots a_{n_r}^{(r)} \mathbf{1}$ with $a^{(1)}, \ldots, a^{(r)} \in T$ and $n_1, \ldots, n_r \in \mathbb{Z}$, by Proposition 3.9.3. Thus $\iota_W(V)$ lies in a vertex subalgebra of the weak vertex algebra $\mathcal{E}(W)$.

Let

$$U = \{v \in V \mid \iota_W(Y(v, x_0)w) = Y_{\mathcal{E}}(\iota_W(v), x_0)\iota_W(w) \quad \text{for } w \in V\}. \tag{5.7.32}$$

We must show that $U = V$. Since $T \subset U$ and T generates V, it suffices to show that U is a vertex subalgebra of V. But $\mathbf{1} \in U$, so it remains to show that for $u, v \in U$ and $m \in \mathbb{Z}$ we have that $u_m v \in U$, i.e., that

$$\iota_W(Y(u_m v, x_2)w) = Y_{\mathcal{E}}(\iota_W(u_m v), x_2)\iota_W(w) \tag{5.7.33}$$

for $w \in V$. But

$$\begin{aligned}
&\iota_W(Y(u_m v, x_2)w) \\
&= \mathrm{Res}_{x_1}((x_1 - x_2)^m \iota_W(Y(u, x_1)Y(v, x_2)w) \\
&\quad -(-x_2 + x_1)^m \iota_W(Y(v, x_2)Y(u, x_1)w)) \\
&= \mathrm{Res}_{x_1}(x_1 - x_2)^m Y_{\mathcal{E}}(\iota_W(u), x_1)Y_{\mathcal{E}}(\iota_W(v), x_2)\iota_W(w) \\
&\quad -\mathrm{Res}_{x_1}(-x_2 + x_1)^m Y_{\mathcal{E}}(\iota_W(v), x_2)Y_{\mathcal{E}}(\iota_W(u), x_1)\iota_W(w),
\end{aligned} \tag{5.7.34}$$

and since the Jacobi identity holds in the image $\iota_W(V)$, this expression equals

$$Y_{\mathcal{E}}(\iota_W(u)_m \iota_W(v), x_2)\iota_W(w) = Y_{\mathcal{E}}(\iota_W(u_m v), x_2)\iota_W(w) \tag{5.7.35}$$

(since $u \in U$), as desired. $\square$

Remark 5.7.7 As we shall see in Chapter 6, Theorem 5.7.6 is very useful in constructing modules for a given vertex algebra. In many situations, we shall start with a certain associative algebra A such as the universal enveloping algebra of the Virasoro algebra or an affine Lie algebra, with V a generalized Verma module, in a certain sense. Furthermore, analogues of Proposition 5.6.8 also hold (recall Remark 5.6.10), so that $\langle \iota_W^0(T) \rangle$ is also an A-module of a certain type. Then it will follow from the universal property of V that there exists a natural A-module homomorphism from V to $\langle \iota_W^0(T) \rangle$, giving rise to the desired linear map Y_W (and homomorphism of weak vertex algebras) from V to $\mathrm{Hom}(W, W((x)))$ extending the map Y_W^0 on T.

Remark 5.7.8 The application of Theorem 5.5.18 in [Li3] to construct vertex (operator) algebras and modules associated to the Virasoro algebra and to an affine algebra is briefly summarized as follows: Let M be any restricted module (suitably defined) for the Lie algebra and let V be a generalized Verma module of a certain type. Set $W = V \oplus M$, which is again a restricted module for the Lie algebra. The set S of natural generating functions for the Lie algebra acting on W is a local subset of $\mathcal{E}(W)$, so by Theorem 5.5.18 S generates a vertex algebra $\langle S \rangle$ with W as a natural module and with M as a submodule. An analogue of Proposition 5.6.8 asserts that $\langle S \rangle$ is a natural module of the same type as V for the Lie algebra. Furthermore, it follows from Proposition 4.7.7 that there is a natural $\langle S \rangle$-module homomorphism F from $\langle S \rangle$ onto V, and then it follows from the universal property of V that F is an isomorphism, so that the vertex algebra structure on $\langle S \rangle$ can be transported to V. Consequently, V has a natural vertex algebra structure (isomorphic to $\langle S \rangle$) with M as a module. To a certain extent, Theorems 5.7.1 and 5.7.6 summarize and generalize the application of Theorem 5.5.18 in [Li3].

Implicit in the proof of Theorem 5.7.6 is the following simple general principle that will be convenient to have available in Chapter 6 and will again be essentially used in the proof of Theorem 5.7.11:

Proposition 5.7.9 *Let f be a linear map from a vertex algebra V_1 to a vertex algebra V_2 such that $f(1) = 1$ and such that*

$$f(Y(a, x)v) = Y(f(a), x)f(v) \quad \textit{for } a \in T, \ v \in V_1, \tag{5.7.36}$$

where T is a given generating subset of V_1. Then f is a vertex algebra homomorphism. In addition, if V_1 and V_2 are vertex operator algebras and if $f(\omega) = \omega$, then f is a vertex operator algebra homomorphism.

Proof. Set

$$U = \{u \in V_1 \mid f(Y(u, x)v) = Y(f(u), x)f(v) \quad \text{for } v \in V_1\}. \tag{5.7.37}$$

We must show that $U = V_1$. From the assumptions we have $1 \in U$ (since $f(1) = 1$) and $T \subset U$. Since T generates V_1, it suffices to prove that U is a vertex subalgebra, and this follows from the last paragraph of the proof of Theorem 5.7.6. The last assertion is clear. $\square$

Theorems 5.7.1 and 5.7.6 assert that a certain partial vertex algebra structure or a certain partial module structure for a given vertex algebra can be uniquely extended to a vertex algebra structure or module structure. Closely related to these two theorems is a theorem of X. Xu obtained in [Xu2], [Xu9], [Xu12]. Xu's original result ([Xu12], Theorem 3.3.10) deals with both algebras and modules, in the setting of vertex superalgebras and other settings slightly more general than that of vertex algebras. Here we split it into two theorems—one for algebras and the other for modules—and we prove these using the machinery and techniques we have developed; as usual, we only consider the vertex algebra case.

Before presenting (a variant of) Xu's results, we first prove the following:

Proposition 5.7.10 *Let V and W be vector spaces and let Y and Y_W be linear maps from V to $\mathrm{Hom}(V, V((x)))$ and from V to $\mathrm{Hom}(W, W((x))) (= \mathcal{E}(W))$, respectively. Also write Y_W as $\iota_W : V \to \mathcal{E}(W)$, with $\iota_W(v) = Y_W(v, x)$ for $v \in V$. Let $u, v \in V$. Then the Jacobi identity*

$$x_0^{-1}\delta\left(\frac{x_1 - x_2}{x_0}\right) Y_W(u, x_1)Y_W(v, x_2) - x_0^{-1}\delta\left(\frac{x_2 - x_1}{-x_0}\right) Y_W(v, x_2)Y_W(u, x_1)$$

$$= x_2^{-1}\delta\left(\frac{x_1 - x_0}{x_2}\right) Y_W(Y(u, x_0)v, x_2) \tag{5.7.38}$$

holds if and only if

$$\iota_W(Y(u, x_0)v) = Y_{\mathcal{E}}(\iota_W(u), x_0)\iota_W(v) \tag{5.7.39}$$

and there exists a nonnegative integer k such that

$$(x_1 - x_2)^k[Y_W(u, x_1), Y_W(v, x_2)] = 0. \tag{5.7.40}$$

Proof. From the definition (5.2.6) of $Y_{\mathcal{E}}$, (5.7.39) amounts to the iterate formula

$$Y_W(Y(u, x_0)v, x_2)$$
$$= \mathrm{Res}_{x_1}\left(x_0^{-1}\delta\left(\frac{x_1 - x_2}{x_0}\right) Y_W(u, x_1)Y_W(v, x_2)\right.$$
$$\left. - x_0^{-1}\delta\left(\frac{x_2 - x_1}{-x_0}\right) Y_W(v, x_2)Y_W(u, x_1)\right) \tag{5.7.41}$$

(cf. (4.4.3)). We know that (5.7.38) implies (5.7.41) and (5.7.40), and so we must show the converse. As was pointed out in Section 4.4 just before Corollary 4.4.7, (5.7.41), when applied to a vector $w \in W$, implies the weak associativity relation (4.3.1) for some $l \in \mathbb{N}$, and so by Proposition 3.4.3, (5.7.38) follows from (5.7.40) and (5.7.41). (The truncation conditions needed in Proposition 3.4.3 are ensured by our stated hypotheses.) $\square$

The following is a variant of the "algebra" half of Xu's result, Theorem 3.3.10 of [Xu12]:

Theorem 5.7.11 *Let V be a vector space equipped with a distinguished vector $\mathbf{1}$ and with a linear map*

$$Y : V \to \mathrm{Hom}(V, V((x)))$$
$$v \mapsto Y(v, x) = \sum_{n \in \mathbb{Z}} v_n x^{-n-1} \tag{5.7.42}$$

such that the vacuum property (3.1.4) and the creation property (3.1.5) hold, and let T be a subset of V such that V is linearly spanned by the vectors

$$a_{n_1}^{(1)} \cdots a_{n_r}^{(r)} \mathbf{1} \tag{5.7.43}$$

for $r \geq 0$, $a^{(i)} \in T$, $n_i \in \mathbb{Z}$. Assume that for any $a \in T$, $v \in V$ there exists a nonnegative integer k such that

$$(x_1 - x_2)^k [Y(a, x_1), Y(v, x_2)] = 0 \tag{5.7.44}$$

and also that for any $a \in T$, $v, v' \in V$, there exists a nonnegative integer l such that

$$(x_0 + x_2)^l Y(a, x_0 + x_2) Y(v, x_2) v' = (x_0 + x_2)^l Y(Y(a, x_0)v, x_2) v'. \tag{5.7.45}$$

Then V is a vertex algebra.

Proof. What we must prove is the Jacobi identity (and in doing so, we shall not need to use the creation property hypothesis). We shall use a modification of the proof of Theorem 5.7.6 with $W = V$. Let ι_V be the linear map from V to $\mathcal{E}(V)$ defined by $\iota_V(v) = Y(v, x)$. For $a \in T$, $v, v' \in V$, by Proposition 3.4.3 and Remark 3.4.4 the weak commutativity relation (5.7.44) and the weak associativity relation (5.7.45) for the triple (a, v, v') amount to the Jacobi identity

$$x_0^{-1} \delta \left(\frac{x_1 - x_2}{x_0} \right) Y(a, x_1) Y(v, x_2) v' - x_0^{-1} \delta \left(\frac{x_2 - x_1}{-x_0} \right) Y(v, x_2) Y(a, x_1) v'$$
$$= x_2^{-1} \delta \left(\frac{x_1 - x_0}{x_2} \right) Y(Y(a, x_0)v, x_2) v'. \tag{5.7.46}$$

From Proposition 5.7.10 we have

$$\iota_V(Y(a, x_0)v) = Y_{\mathcal{E}}(\iota_V(a), x_0)\iota_V(v) \quad \text{for } a \in T, \ v \in V \tag{5.7.47}$$

(cf. (5.7.28)). Since $\iota_V(T)$ is a local subset of $\mathcal{E}(V)$ by (5.7.44), by Theorem 5.5.18 the weak vertex subalgebra $\langle \iota_V(T) \rangle$ of $\mathcal{E}(V)$ generated by $\iota_V(T)$ is a vertex subalgebra. It also follows from (5.7.47), the spanning assumption (5.7.43) and the assumption $Y(\mathbf{1}, x) = 1_V$ that

$$\iota_V(V) \subset \langle \iota_V(T) \rangle. \tag{5.7.48}$$

By Theorem 5.5.11 $\langle \iota_V(T) \rangle$ is a local subalgebra, so $\iota_V(V)$, as a subset of $\langle \iota_V(T) \rangle$, is a local subset. That is, for any $u, v \in V$, there exists a nonnegative integer k such that

$$(x_1 - x_2)^k [Y(u, x_1), Y(v, x_2)] = 0. \tag{5.7.49}$$

In view of Proposition 5.7.10, we must prove that

$$\iota_V(Y(u, x_0)v) = Y_\mathcal{E}(\iota_V(u), x_0)\iota_V(v) \quad \text{for all } u, v \in V. \tag{5.7.50}$$

Set

$$U = \{u \in V \mid \iota_V(Y(u, x_0)v) = Y_\mathcal{E}(\iota_V(u), x_0)\iota_V(v) \quad \text{for all } v \in V\}. \tag{5.7.51}$$

We need to show that $U = V$. We have $\mathbf{1} \in U$ because $Y(\mathbf{1}, x) = 1_V$, and by (5.7.47), $T \subset U$. Let $a \in T \subset U$, $u \in U$, $v \in V$. Note that $\iota_V(Y(a, x_0)u) = Y_\mathcal{E}(\iota_V(a), x_0)\iota_V(u)$ holds, and as in the proof of Proposition 5.7.10, this amounts to:

$$Y(Y(a, x_0)u, x_2)$$
$$= \mathrm{Res}_{x_1} \left(x_0^{-1}\delta\left(\frac{x_1 - x_2}{x_0}\right) Y(a, x_1)Y(u, x_2) - x_0^{-1}\delta\left(\frac{x_2 - x_1}{-x_0}\right) Y(u, x_2)Y(a, x_1) \right). \tag{5.7.52}$$

Using this formula together with (5.7.46), (5.7.51) and the Jacobi identity for $(\langle \iota_V(T) \rangle, Y_\mathcal{E})$ we get

$$\iota_V(Y(Y(a, x_0)u, x_2)v)$$
$$= \mathrm{Res}_{x_1} \left(x_0^{-1}\delta\left(\frac{x_1 - x_2}{x_0}\right) \iota_V(Y(a, x_1)Y(u, x_2)v) \right.$$
$$\left. - x_0^{-1}\delta\left(\frac{x_2 - x_1}{-x_0}\right) \iota_V(Y(u, x_2)Y(a, x_1)v) \right)$$
$$= \mathrm{Res}_{x_1} x_0^{-1}\delta\left(\frac{x_1 - x_2}{x_0}\right) Y_\mathcal{E}(\iota_V(a), x_1)Y_\mathcal{E}(\iota_V(u), x_2)\iota_V(v)$$
$$- \mathrm{Res}_{x_1} x_0^{-1}\delta\left(\frac{x_2 - x_1}{-x_0}\right) Y_\mathcal{E}(\iota_V(u), x_2)Y_\mathcal{E}(\iota_V(a), x_1)\iota_V(v)$$
$$= Y_\mathcal{E}(Y_\mathcal{E}(\iota_V(a), x_0)\iota_V(u), x_2)\iota_V(v)$$
$$= Y_\mathcal{E}(\iota_V(Y(a, x_0)u), x_2)\iota_V(v). \tag{5.7.53}$$

That is, $a_m u \in U$ for $a \in T$, $m \in \mathbb{Z}$, $u \in U$. It now follows from the spanning assumption (5.7.43) that $U = V$, as desired. $\square$

Remark 5.7.12 Notice that the idea and the proof of Proposition 5.7.9 have been used here, although we cannot simply quote Proposition 5.7.9 because we do not know that V is a vertex algebra.

The following is a variant of the "module" half of Xu's result:

Theorem 5.7.13 *Let V be a vertex algebra and let W be a vector space equipped with a linear map*

$$Y_W : V \to \mathrm{Hom}(W, W((x)))$$
$$v \mapsto Y_W(v, x) = \sum_{n \in \mathbb{Z}} v_n x^{-n-1} \tag{5.7.54}$$

such that $Y_W(\mathbf{1}, x) = 1_W$. Let T be a generating subset of V and assume that for any $a \in T$, $v \in V$ there exists a nonnegative integer k such that

$$(x_1 - x_2)^k [Y_W(a, x_1), Y_W(v, x_2)] = 0 \tag{5.7.55}$$

and also that for any $a \in T$, $v \in V$, $w \in W$, there exists a nonnegative integer l such that

$$(x_0 + x_2)^l Y_W(a, x_0 + x_2) Y_W(v, x_2) w = (x_0 + x_2)^l Y_W(Y(a, x_0)v, x_2) w. \tag{5.7.56}$$

Then (W, Y_W) carries the structure of a V-module.

Proof. As before, let ι_W be the linear map from V to $\mathcal{E}(W)$ defined by $\iota_W(v) = Y_W(v, x)$. In view of Theorem 5.7.6, we must prove that

$$\iota_W(Y(a, x_0)v) = Y_{\mathcal{E}}(\iota_W(a), x_0)\iota_W(v) \tag{5.7.57}$$

for $a \in T$, $v \in V$, since we already have $\iota_W(\mathbf{1}) = 1_W$ and (5.7.55). But by Proposition 3.4.3 and Remark 3.4.4, (5.7.55) and (5.7.56) amount to the Jacobi identity for the triple (a, v, w). In particular, the Jacobi identity holds for the pair (a, v) (since $w \in W$ is arbitrary), and (5.7.57) follows from Proposition 5.7.10. $\square$

6

Construction of Families of Vertex Operator Algebras and Modules

We have developed the fundamental theory of vertex operator algebras and modules in Chapters 1 through 5. The reader has certainly noticed that so far we have not constructed or exhibited any examples of vertex (operator) algebras other than the vertex (operator) algebras based on commutative associative algebras. Unlike in classical algebraic theories such as the theory of Lie or associative algebras, nontrivial examples of vertex (operator) algebras and modules for them cannot be easily presented right after the definitions. But now, with the general representation theory having been developed in Chapter 5, we are fully prepared to present an array of interesting examples of vertex operator algebras and modules, by systematically invoking this general representation theory.

Among the most important vertex (operator) algebras are those associated with the Virasoro Lie algebra, those associated with affine Lie algebras, including Heisenberg Lie algebras and affine Kac–Moody Lie algebras, and those associated with nondegenerate even lattices. Such vertex operator algebras and modules have been constructed and studied in [FZ], [DMZ], [Hua4], [Wa1] and [Li3] for the Virasoro algebra, in [FF7], [FZ], [DL3], [Lia] and [Li3] for affine Lie algebras and in [B1], [FLM6], [D2], [Xu12] and [BDT] for nondegenerate even lattices. The study of vertex operator algebras associated with the Virasoro algebra is the algebraic foundation of the study of the "minimal models" in conformal field theory (cf. [BPZ]) while the study of vertex operator algebras associated with affine Lie algebras is analogously related to the Wess–Zumino–Novikov–Witten model (cf. [Wi1], [KZ], [GW], [Wi2]) in conformal field theory.

Certain of these vertex operator algebras associated with even lattices, with the Virasoro algebra and with affine Lie algebras enter crucially into the construction of Frenkel–Lepowsky–Meurman's moonshine module vertex operator algebra $V^\natural$ and the construction of the Monster as its automorphism group (see [FLM6]), but even with these "ingredient" vertex operator algebras constructed it is still a very elaborate matter to construct $V^\natural$ and its automorphism group (again see [FLM6]), and the representation theory presented in Chapter 5 does not serve to simplify Frenkel–Lepowsky–Meurman's main results in [FLM6]. But we emphasize that the representation theory presented in Chapter 5 does indeed give us an elegant, general and powerful method for the construction of the many vertex operator algebras and modules to which this chapter is devoted.

In this chapter, then, we shall apply the theory developed in the previous chapters, especially Chapter 5, to construct the vertex operator algebras that we have mentioned and also their irreducible modules. With Theorems 5.5.18, 5.6.14, 5.7.1 and 5.7.6 on hand it is now quite easy to construct vertex operator algebras and modules from Lie algebras such as the Virasoro algebra and affine Lie algebras. In Section 6.1 we construct a family of vertex operator algebras $V_{Vir}(\ell, 0)$ associated with the Virasoro algebra and with any complex number ℓ and we classify their irreducible modules. Section 6.2 is parallel to Section 6.1; in Section 6.2 we construct a family of vertex operator algebras $V_{\hat{\mathfrak{g}}}(\ell, 0)$ from a finite-dimensional Lie algebra $\mathfrak{g}$ equipped with a nondegenerate symmetric invariant bilinear form $\langle \cdot, \cdot \rangle$, where ℓ is any complex number except for one specific value, and we also classify their irreducible modules. In Section 6.3 we study the vertex operator algebras $V_{\hat{\mathfrak{g}}}(\ell, 0)$ and their irreducible modules when $\mathfrak{g}$ is taken to be an abelian Lie algebra, which we call $\mathfrak{h}$, and with ℓ a nonzero complex number; in this case, the affine Lie algebra $\hat{\mathfrak{h}}$ is essentially a Heisenberg algebra. We prove that all the vertex operator algebras $V_{\hat{\mathfrak{h}}}(\ell, 0)$ are simple and isomorphic to one another as ℓ varies. This section also serves as a preparation for the next subject—the construction of the vertex (operator) algebras and their irreducible modules associated with nondegenerate even lattices.

Sections 6.4 and 6.5 are devoted to the construction of the vertex (operator) algebra V_{L_0} associated with a nondegenerate even lattice L_0 and the construction and classification of the irreducible V_{L_0}-modules. The construction of the vertex (operator) algebra V_{L_0} is relatively easy after the setting is defined in Section 6.4, while the construction and the classification of the irreducible V_{L_0}-modules occupy most of Section 6.5. For the construction and the classification of the irreducible V_{L_0}-modules we introduce and use an associative algebra we call $A(L_0)$, which replaces, and can be thought of as analogous to, the Virasoro algebra and the affine Lie algebras used in Sections 6.1–6.3. (In the construction in [BDT] of the irreducible V_{L_0}-modules, Berman, Dong and Tan also introduced and used certain associative algebras.) This approach to the construction of V_{L_0}-modules is somewhat different from the other approaches (see [FLM6], Chapters 1 to 8, [Xu12] and [BDT]), and the idea of introducing and exploiting associative algebras like $A(L_0)$ is expected to be useful in studying modules for still more general classes of vertex (operator) algebras. In the classification of the irreducible V_{L_0}-modules, we also use the theory of Z-operators developed in [LW4], [LW5], [LW6], [LP1], [LP2] and [FLM6]; Z-operators are also used similarly in [D2].

In Section 6.6 we study the simple vertex operator algebra $L_{\hat{\mathfrak{g}}}(\ell, 0)$ with $\mathfrak{g}$ taken to be a finite-dimensional simple Lie algebra equipped with the suitably normalized Killing form and with ℓ taken to be a nonnegative integer. We classify the irreducible modules for the vertex operator algebra $L_{\hat{\mathfrak{g}}}(\ell, 0)$ and show that the irreducible $L_{\hat{\mathfrak{g}}}(\ell, 0)$-modules are precisely the standard modules (the integrable highest weight modules) of level ℓ (see [K6]; cf. [Le2]) for the untwisted affine Kac–Moody algebra associated with $\mathfrak{g}$ (see [K6], [MoP]). Such results were originally obtained by I. Frenkel and Zhu in [FZ] and by Dong and Lepowsky in [DL3] (see also [Li3], [MP2], [MP4]). The original methods are closely related to, but different from, those used in this section; our approach here, based on [Li3], uses the theory developed in Chapter 5.

6.1 Vertex operator algebras and modules associated to the Virasoro algebra

In this section we shall construct a natural family of vertex operator algebras associated to the Virasoro algebra along with all their irreducible modules.

As we know (cf. Remark 3.9.6), any vertex operator algebra V has the vertex sub-algebra $\langle \omega \rangle$ generated by the conformal vector ω, and this is in fact the smallest vertex operator subalgebra of V; it is exactly the submodule of V for the Virasoro algebra generated by $\mathbf{1}$:

$$\mathrm{span}\left\{ L(n_1) \cdots L(n_r)\mathbf{1} \mid r \geq 0, \ n_j \in \mathbb{Z} \right\}. \tag{6.1.1}$$

In view of the creation property (3.1.5) we have

$$L(n)\mathbf{1} \ (= \omega_{n+1}\mathbf{1}) = 0 \quad \text{for } n \geq -1 \tag{6.1.2}$$

(recall (3.1.65)).

In this section, we shall construct universal restricted modules for the Virasoro alge-bra, as defined in Remark 5.1.6, satisfying (6.1.2), and we shall prove that they indeed have a natural vertex operator algebra structure. We also prove that the modules for such vertex operator algebras, viewed as vertex algebras, are exactly the restricted modules for the Virasoro algebra, while the modules for such vertex operator algebras (viewed as vertex operator algebras) are exactly those restricted modules for the Virasoro algebra that are $\mathbb{C}$-graded by $L(0)$-eigenvalues, with the usual grading restrictions.

First, recall from Remark 3.1.23 that the Virasoro algebra $\mathcal{L}$ is the Lie algebra with basis $\{L_m \mid m \in \mathbb{Z}\} \cup \{\mathbf{c}\}$ equipped with the bracket relations

$$[L_m, L_n] = (m - n)L_{m+n} + \frac{1}{12}(m^3 - m)\delta_{m+n,0}\mathbf{c} \tag{6.1.3}$$

together with the condition that $\mathbf{c}$ is a central element of $\mathcal{L}$. The Virasoro algebra $\mathcal{L}$ equipped with the $\mathbb{Z}$-grading

$$\mathcal{L} = \coprod_{n \in \mathbb{Z}} \mathcal{L}_{(n)}, \tag{6.1.4}$$

where

$$\mathcal{L}_{(0)} = \mathbb{C}L_0 \oplus \mathbb{C}\mathbf{c} \quad \text{and} \quad \mathcal{L}_{(n)} = \mathbb{C}L_{-n} \text{ for } n \neq 0, \tag{6.1.5}$$

is a $\mathbb{Z}$-graded Lie algebra, and this grading is given by ad L_0-eigenvalues.

Here we are using the standard notion that a Lie algebra $\mathfrak{g}$ equipped with a $\mathbb{Z}$-grading $\mathfrak{g} = \coprod_{n \in \mathbb{Z}} \mathfrak{g}_{(n)}$ is called a *$\mathbb{Z}$-graded Lie algebra* if

$$[\mathfrak{g}_{(m)}, \mathfrak{g}_{(n)}] \subset \mathfrak{g}_{(m+n)} \quad \text{for } m, n \in \mathbb{Z}. \tag{6.1.6}$$

We also recall that a subalgebra $\mathfrak{k}$ of a $\mathbb{Z}$-graded Lie algebra $\mathfrak{g}$ is called a *graded subalgebra* if

$$\mathfrak{k} = \coprod_{n \in \mathbb{Z}} \mathfrak{k}_{(n)}, \quad \text{where } \mathfrak{k}_{(n)} = \mathfrak{k} \cap \mathfrak{g}_{(n)} \text{ for } n \in \mathbb{Z}. \tag{6.1.7}$$

In particular, $\mathfrak{g}_{(0)}, \mathfrak{g}_{(\pm)} = \coprod_{n \geq 1} \mathfrak{g}_{(\pm n)}$ and $\mathfrak{g}_{(0)} \oplus \mathfrak{g}_{(\pm)}$ are graded subalgebras.

Remark 6.1.1 Often the Virasoro algebra $\mathcal{L}$ is graded by *degree* as $\mathcal{L} = \coprod_{n \in \mathbb{Z}} \mathcal{L}(n)$, where $\mathcal{L}(0) = \mathbb{C} L_0 \oplus \mathbb{C} \mathbf{c}$ and $\mathcal{L}_n = \mathbb{C} L_n$ (rather than $\mathbb{C} L_{-n}$) for $n \neq 0$. With the $\mathbb{Z}$-grading we are using, the Virasoro algebra $\mathcal{L}$ is graded by ad L_0-eigenvalues, that is, by *weight*.

For the Virasoro algebra $\mathcal{L}$, equipped with the given grading, we have the graded subalgebras

$$\mathcal{L}_{(\pm)} = \coprod_{n \geq 1} \mathcal{L}_{(\pm n)} = \coprod_{n \geq 1} \mathbb{C} L_{\mp n} \tag{6.1.8}$$

and also $\mathcal{L}_{(0)} \oplus \mathcal{L}_{(-)}$ and $\mathcal{L}_{(0)} \oplus \mathcal{L}_{(+)}$. We also have the graded subalgebras

$$\mathcal{L}_{(\leq 1)} = \coprod_{n \leq 1} \mathcal{L}_{(n)} = \mathcal{L}_{(-)} \oplus \mathbb{C} L_0 \oplus \mathbb{C} \mathbf{c} \oplus \mathbb{C} L_{-1}, \tag{6.1.9}$$

$$\mathcal{L}_{(\geq 2)} = \coprod_{n \geq 2} \mathcal{L}_{(n)} = \coprod_{n \geq 2} \mathbb{C} L_{-n} \subset \mathcal{L}_{(+)}, \tag{6.1.10}$$

and we have the decomposition

$$\mathcal{L} = \mathcal{L}_{(\leq 1)} \oplus \mathcal{L}_{(\geq 2)}. \tag{6.1.11}$$

Let ℓ be any complex number. Consider $\mathbb{C}$ as an $\mathcal{L}_{(\leq 1)}$-module with $\mathbf{c}$ acting as the scalar ℓ and with $\mathcal{L}_{(-)} \oplus \mathbb{C} L_0 \oplus \mathbb{C} L_{-1}$, which is a subalgebra of $\mathcal{L}$, acting trivially (recall (6.1.9)); in particular, L_0 acts trivially on $\mathbb{C}$. Denote this $\mathcal{L}_{(\leq 1)}$-module by $\mathbb{C}_\ell$. Form the induced module

$$V_{Vir}(\ell, 0) = U(\mathcal{L}) \otimes_{U(\mathcal{L}_{(\leq 1)})} \mathbb{C}_\ell, \tag{6.1.12}$$

where $U(\cdot)$ denotes the universal enveloping algebra of a Lie algebra. Clearly, $V_{Vir}(\ell, 0)$ is $\mathbb{Z}$-graded by L_0-eigenvalues, since $\mathcal{L}_{(\leq 1)}$ is a graded subalgebra and $\mathbb{C}_\ell$ is a $\mathbb{Z}$-graded $\mathcal{L}_{(\leq 1)}$-module whose grading is given by L_0-eigenvalues. Here we recall that for a $\mathbb{Z}$-graded Lie algebra $\mathfrak{g} = \coprod_{n \in \mathbb{Z}} \mathfrak{g}_{(n)}$, a *$\mathbb{C}$-graded $\mathfrak{g}$-module* is a $\mathfrak{g}$-module W equipped with a $\mathbb{C}$-grading $W = \coprod_{r \in \mathbb{C}} W_{(r)}$ such that

$$\mathfrak{g}_{(n)} W_{(r)} \subset W_{(n+r)} \quad \text{for } n \in \mathbb{Z}, \; r \in \mathbb{C}; \tag{6.1.13}$$

similarly for a $\mathbb{Z}$-*graded* $\mathfrak{g}$-module. Let $\mathfrak{k}$ be a graded subalgebra of $\mathfrak{g}$ and let U be a $\mathbb{C}$-graded $\mathfrak{k}$-module. Then the induced module $U(\mathfrak{g}) \otimes_{U(\mathfrak{k})} U$ is naturally a $\mathbb{C}$-graded $\mathfrak{g}$-module; similarly for the $\mathbb{Z}$-graded case. In particular, $V_{Vir}(\ell, 0)$ is a $\mathbb{Z}$-graded $\mathcal{L}$-module, and in particular, a $\mathbb{C}$-graded $\mathcal{L}$-module.

Remark 6.1.2 Given a module W for the Virasoro algebra, we shall typically write $L(n)$ for the operator on W corresponding to L_n for $n \in \mathbb{Z}$, as we have been doing in the past.

From the Poincaré–Birkhoff–Witt theorem, as a vector space,

$$V_{Vir}(\ell, 0) = U(\mathcal{L}_{(\geq 2)}) \simeq S(\mathcal{L}_{(\geq 2)}), \tag{6.1.14}$$

where $S(\cdot)$ denotes the symmetric algebra of a vector space, and we may and do naturally consider $\mathbb{C}$ as a subspace of $V_{Vir}(\ell, 0)$. Set

$$\mathbf{1} = 1 \in \mathbb{C} \subset V_{Vir}(\ell, 0). \tag{6.1.15}$$

Then

$$V_{Vir}(\ell, 0) = \coprod_{n \geq 0} V_{Vir}(\ell, 0)_{(n)}, \tag{6.1.16}$$

where $V_{Vir}(\ell, 0)_{(n)}$ has a basis consisting of the vectors

$$L(-m_1) \cdots L(-m_r)\mathbf{1} \tag{6.1.17}$$

for $r \geq 0$, $m_1 \geq m_2 \geq \cdots \geq m_r \geq 2$, with $m_1 + \cdots + m_r = n$. The two grading restriction conditions (3.1.38) and (3.1.39) clearly hold for $V_{Vir}(\ell, 0)$ equipped with its $\mathbb{Z}$-grading. It is also clear that $V_{Vir}(\ell, 0)$ is a restricted module, in the sense of Remark 5.1.6, as we see from the grading.

Remark 6.1.3 As a module for the Virasoro algebra, $V_{Vir}(\ell, 0)$ is generated by $\mathbf{1}$ ($= 1$), with the relations $\mathbf{c} = \ell$ and $L(n)\mathbf{1} = 0$ for $n \geq -1$, and in fact $V_{Vir}(\ell, 0)$ is universal in the sense that for any module W for the Virasoro algebra of central charge ℓ equipped with a vector $e \in W$ such that $L(n)e = 0$ for $n \geq -1$, there exists a unique module homomorphism from $V_{Vir}(\ell, 0)$ to W sending $\mathbf{1}$ to e.

In Corollary 5.6.15, for any restricted module W for the Virasoro algebra of central charge ℓ we obtained a substructure $(U, Y_{\mathcal{E}}, 1_W, L_W(x))$ of $\mathcal{E}(W)$ satisfying all the axioms in the definition of the notion of vertex operator algebra of central charge ℓ except perhaps for the two grading restrictions (3.1.38) and (3.1.39). As was emphasized in Remark 5.6.12, the module W need not even be graded. We mentioned in Remark 5.6.16 that it would take only a small amount of additional effort to show that the two grading restriction conditions (3.1.38) and (3.1.39) are also satisfied, so that U is in fact a vertex operator algebra of central charge ℓ. Now that we have set up the relevant elementary Lie algebra notations and structures, we are ready to justify this statement, obtaining the following refinement of Corollary 5.6.15.

Theorem 6.1.4 *Let W be a restricted module for the Virasoro algebra of central charge ℓ and let U be the subspace of $\mathcal{E}(W)$ linearly spanned by the elements*

$$L_W(x)_{n_1} \cdots L_W(x)_{n_r} 1_W \tag{6.1.18}$$

for $r \geq 0$ and $n_1, \ldots, n_r \in \mathbb{Z}$. Then $(U, Y_{\mathcal{E}}, 1_W, L_W(x))$ carries the structure of a vertex operator algebra of central charge ℓ. The grading on U by $L(0)$-eigenvalues coincides with its grading as a subspace of $\mathcal{E}^o(W, L(-1))$. Furthermore, W is a natural faithful module for U viewed as a vertex algebra. If W is $\mathbb{C}$-graded by $L_W(0)$-eigenvalues, then W is a U-module except perhaps for the two grading restrictions on W.

Proof. From Corollary 5.6.15, U is a module for the Virasoro algebra of central charge ℓ, with L_n acting as $L_W(x)_{n+1}$ for $n \in \mathbb{Z}$ and with the relations $L_W(x)_n 1_W = 0$ for all $n \geq 0$ (from the creation property (3.1.5)). All that we need to prove are the two grading restrictions for U. In view of Remark 6.1.3, there is a unique Virasoro algebra module homomorphism ψ from $V_{Vir}(\ell, 0)$ to U sending 1 to 1_W, and it is clear that ψ is surjective. Since $V_{Vir}(\ell, 0)$ with its $\mathbb{Z}$-grading given by $L(0)$-eigenvalues satisfies the two grading restrictions, so does U, with its $\mathbb{Z}$-grading given by $L(0)$-eigenvalues. $\square$

Note that the vertex operator algebra U obtained in Theorem 6.1.4, viewed as a module for the Virasoro algebra, is generated by $1 = 1_W$, and that it satisfies the relations $L(n)1 = 0$ for $n \geq -1$. In view of Remark 6.1.3, U, as well as the smallest vertex operator subalgebra $\langle \omega \rangle$ of *any* vertex operator algebra V, is a homomorphic image of $V_{Vir}(\ell, 0)$, when viewed as a module for the Virasoro algebra. This naturally leads us to consider homomorphic images of $V_{Vir}(\ell, 0)$, and in particular, $V_{Vir}(\ell, 0)$ itself, as candidates for natural vertex operator algebra structures. In fact, Theorem 5.7.4 indeed guarantees that there exists a natural vertex operator algebra structure on each homomorphic image of $V_{Vir}(\ell, 0)$, including $V_{Vir}(\ell, 0)$ itself, as we show next (cf. [Hua4], [FZ], [Li3]):

Theorem 6.1.5 *Let ℓ be any complex number and let V be a module for the Virasoro algebra of central charge ℓ equipped with a vector 1 such that V is generated by 1 and such that the relations $L(n)1 = 0$ hold for $n \geq -1$. Then there exists a unique vertex operator algebra structure $(V, Y, 1, \omega)$ on V with 1 as the vacuum vector and with $\omega = L(-2)1$ as the conformal vector such that*

$$Y(L(-2)1, x) = L_V(x) \left(= \sum_{n \in \mathbb{Z}} L(n) x^{-n-2} \right). \tag{6.1.19}$$

The vertex operator map Y for this vertex operator algebra structure is given by

$$Y(L(n_1) \cdots L(n_r)1, x) = L_V(x)_{n_1+1} \cdots L_V(x)_{n_r+1} 1_V \tag{6.1.20}$$

for $r \geq 0$ and $n_1, \ldots, n_r \in \mathbb{Z}$. In particular, this holds for $V = V_{Vir}(\ell, 0)$.

Proof. First, as a homomorphic image of $V_{Vir}(\ell, 0)$, V is a restricted module for the Virasoro algebra, and it is $\mathbb{Z}$-graded by $L(0)$-eigenvalues with the two grading restrictions. By assumption we also have

$$V = \text{span}\{L(n_1) \cdots L(n_r)\mathbf{1} \mid r \geq 0, \; n_i \in \mathbb{Z}\}.$$

From Remark 5.1.6 and Example 5.5.6 we have

$$[L(-1), L_V(x)] = \frac{d}{dx} L_V(x), \tag{6.1.21}$$

$$[L(0), L_V(x)] = 2L_V(x) + x\frac{d}{dx} L_V(x), \tag{6.1.22}$$

$$(x_1 - x_2)^4 [L_V(x_1), L_V(x_2)] = 0. \tag{6.1.23}$$

Then the desired result follows immediately from Theorem 5.7.4 with V in Theorem 5.7.4 taken to be the present V and with $d = L(-1)$, $T = \{\omega = L(-2)\mathbf{1}\}$ and $Y_0(L(-2)\mathbf{1}, x) = L_V(x)$. $\square$

Remark 6.1.6 Since the vertex operator algebra V constructed in Theorem 6.1.5, and in particular, $V_{Vir}(\ell, 0)$, is generated by the conformal vector $\omega = L(-2)\mathbf{1}$, V is a *minimal* vertex operator algebra in the sense that it does not have any proper vertex operator subalgebra (with the same conformal vector). On the other hand, in view of Remark 3.9.6, any minimal vertex operator algebra of central charge ℓ is one of these vertex operator algebras, and so the vertex operator algebras constructed in Theorem 6.1.5 are exactly the minimal ones. Furthermore, every minimal vertex operator algebra V of central charge ℓ is a quotient vertex operator algebra of $V_{Vir}(\ell, 0)$. Indeed, in view of Remark 6.1.3, there exists a unique Virasoro algebra module homomorphism ψ from $V_{Vir}(\ell, 0)$ to V such that $\psi(\mathbf{1}) = \mathbf{1}$. Since ψ is surjective and

$$\psi(\omega) = \psi(L(-2)\mathbf{1}) = L(-2)\psi(\mathbf{1}) = L(-2)\mathbf{1} = \omega, \tag{6.1.24}$$

it follows from Proposition 5.7.9 that ψ is a vertex operator algebra homomorphism. Conversely, every quotient vertex operator algebra of $V_{Vir}(\ell, 0)$ is clearly a minimal vertex operator algebra, so that the minimal vertex operator algebras of central charge ℓ are exactly the quotient algebras of $V_{Vir}(\ell, 0)$.

Now for any complex number ℓ, we have constructed a (canonical) universal minimal vertex operator algebra $V_{Vir}(\ell, 0)$. We next construct and study $V_{Vir}(\ell, 0)$-modules.

First, in view of Proposition 4.1.5, any module for $V_{Vir}(\ell, 0)$ viewed as a vertex algebra is necessarily a restricted module for the Virasoro algebra of central charge ℓ, with $L_W(x) = Y_W(\omega, x)$. On the other hand, we would like to know which restricted modules W for the Virasoro algebra of central charge ℓ admit a module structure for $V_{Vir}(\ell, 0)$ viewed as a vertex algebra, with $Y_W(\omega, x) = L_W(x)$. It turns out that they all do. This is exactly a special case of a general problem that we addressed in Section 5.7 and that we solved in Theorem 5.7.6. The reader will see how naturally and easily Theorem 5.7.6 can be applied in proving the following result (and analogous results in later sections):

Theorem 6.1.7 *Let ℓ be any complex number and let W be any restricted module for the Virasoro algebra of central charge ℓ. Then there exists a unique module structure on W for $V_{Vir}(\ell, 0)$ viewed as a vertex algebra such that*

$$Y_W(L(-2)\mathbf{1}, x) = L_W(x) \left(= \sum_{n \in \mathbb{Z}} L(n)x^{-n-2} \right). \tag{6.1.25}$$

The vertex operator map Y_W for this module structure is given by

$$Y_W(L(n_1) \cdots L(n_r)\mathbf{1}, x) = L_W(x)_{n_1+1} \cdots L_W(x)_{n_r+1}1_W \tag{6.1.26}$$

for $r \geq 0$ and $n_1, \ldots, n_r \in \mathbb{Z}$. If W is $\mathbb{C}$-graded by $L(0)$-eigenvalues, then W is a module for $V_{Vir}(\ell, 0)$ viewed as a vertex operator algebra, possibly without the grading restrictions (4.1.16) and (4.1.17).

Proof. Again, from Example 5.5.6 we have

$$(x_1 - x_2)^4[L_W(x_1), L_W(x_2)] = 0. \tag{6.1.27}$$

Set

$$U = \mathrm{span}\{L_W(x)_{n_1} \cdots L_W(x)_{n_r}1_W \mid r \geq 0, \ n_i \in \mathbb{Z}\} \subset \mathcal{E}(W). \tag{6.1.28}$$

From Theorem 6.1.4, U is a vertex operator algebra of central charge ℓ and hence a module for the Virasoro algebra $\mathcal{L}$ of central charge ℓ with L_n $(= \omega_{n+1})$ acting as $L_W(x)_{n+1}$ for $n \in \mathbb{Z}$ and such that $L_W(x)_n 1_W = 0$ for $n \geq 0$. In view of Remark 6.1.3, there exists a unique $\mathcal{L}$-module map ψ from $V_{Vir}(\ell, 0)$ to U such that $\psi(\mathbf{1}) = 1_W$, and $\psi(\omega) = L_W(x)$, as in (6.1.24). Then

$$\psi(\omega_n v) = L_W(x)_n \psi(v) \quad \text{for } n \in \mathbb{Z}, \ v \in V_{Vir}(\ell, 0). \tag{6.1.29}$$

The existence and uniqueness of $V_{Vir}(\ell, 0)$-module structure on W now immediately follows from Theorem 5.7.6 with $T = \{\omega\}$ and $\omega_W(x) = L_W(x)$ (recall that $V_{Vir}(\ell, 0)$ is generated by ω). The rest is also clear. $\square$

Summarizing some basic facts related to Theorem 6.1.7, we immediately have, using Propositions 4.1.5 and 4.5.17:

Theorem 6.1.8 *Every module for $V_{Vir}(\ell, 0)$ viewed as a vertex algebra is naturally a restricted module for the Virasoro algebra of central charge ℓ, with $L_W(x) = Y_W(L(-2)\mathbf{1}, x)$. Conversely, every restricted module W for the Virasoro algebra of central charge ℓ is naturally a module for $V_{Vir}(\ell, 0)$ viewed as a vertex algebra, with*

$$Y_W(L(n_1) \cdots L(n_r)\mathbf{1}, x) = L_W(x)_{n_1+1} \cdots L_W(x)_{n_r+1}1_W \tag{6.1.30}$$

for $r \geq 0$ and $n_1, \ldots, n_r \in \mathbb{Z}$. Under this correspondence, the modules for $V_{Vir}(\ell, 0)$ viewed as a vertex operator algebra ($\mathbb{C}$-graded by $L(0)$-eigenvalues and with the two *grading restrictions (4.1.16) and (4.1.17)) are exactly those restricted modules for the Virasoro algebra of central charge ℓ that are $\mathbb{C}$-graded by $L(0)$-eigenvalues and with the two grading restrictions. Furthermore, for any $V_{Vir}(\ell, 0)$-module W, the $V_{Vir}(\ell, 0)$-submodules of W are exactly the submodules of W for the Virasoro algebra, and these submodules are in particular graded.*   □

Next, we shall modify the construction of the $\mathcal{L}$-module $V_{Vir}(\ell, 0)$ (recall (6.1.12)) to get a certain natural family of restricted $\mathcal{L}$-modules of central charge ℓ that are $\mathbb{C}$-graded by $L(0)$-eigenvalues and satisfy the two grading restrictions. Such Virasoro algebra modules are naturally modules for the vertex operator algebra $V_{Vir}(\ell, 0)$, by Theorem 6.1.8.

We shall construct a family of induced Virasoro algebra modules and also their irreducible quotient modules (cf. [FFu1]–[FFu3], [K6]) and endow them with appropriate vertex operator algebra and module structure. Let ℓ be a complex number as before and let h also be a complex number. Consider $\mathbb{C}$ as an $\mathcal{L}_{(0)}$-module with $\mathbf{c}$ and L_0 acting as the scalars ℓ and h, respectively (recall (6.1.5)). Let $\mathcal{L}_{(-)}$ act trivially on $\mathbb{C}$, making $\mathbb{C}$ an $(\mathcal{L}_{(-)} \oplus \mathcal{L}_{(0)})$-module, which we denote by $\mathbb{C}_{\ell,h}$. (Note that $\mathcal{L}_{(-)}$ is an ideal of the Lie algebra $\mathcal{L}_{(-)} \oplus \mathcal{L}_{(0)}$.) Form the induced module

$$M_{Vir}(\ell, h) = U(\mathcal{L}) \otimes_{U(\mathcal{L}_{(-)} \oplus \mathcal{L}_{(0)})} \mathbb{C}_{\ell,h}. \tag{6.1.31}$$

As with $V_{Vir}(\ell, 0)$, $M_{Vir}(\ell, h)$, a module for the Virasoro algebra of central charge ℓ, is $\mathbb{C}$-graded by $L(0)$-eigenvalues.

Again from the Poincaré–Birkhoff–Witt theorem, as a vector space,

$$M_{Vir}(\ell, h) = U(\mathcal{L}_{(+)}) \otimes \mathbb{C}_{\ell,h} = U(\mathcal{L}_{(+)}) \simeq S(\mathcal{L}_{(+)}). \tag{6.1.32}$$

We naturally consider $\mathbb{C}_{\ell,h}$ as a subspace of $M_{Vir}(\ell, h)$ and we set

$$\mathbf{1}_{(\ell,h)} = 1 \in \mathbb{C}_{\ell,h} \subset M_{Vir}(\ell, h). \tag{6.1.33}$$

Then

$$M_{Vir}(\ell, h) = \coprod_{n \geq 0} M_{Vir}(\ell, h)_{(n+h)}, \tag{6.1.34}$$

where $M_{Vir}(\ell, h)_{(n+h)}$, the $L(0)$-eigenspace of eigenvalue $n + h$, has a basis consisting of the vectors

$$L(-m_1) \cdots L(-m_r)\mathbf{1}_{(\ell,h)} \tag{6.1.35}$$

for $r \geq 0$, $m_1 \geq m_2 \geq \cdots \geq m_r \geq 1$, with $m_1 + \cdots + m_r = n$. Consequently, $M_{Vir}(\ell, h)$ with the given $\mathbb{C}$-grading satisfies the grading restriction conditions (4.1.16) and (4.1.17). This in particular implies that $M_{Vir}(\ell, h)$ is a restricted $\mathcal{L}$-module. Thus from Theorems 6.1.7 and 6.1.8 we immediately have (see also [FZ], [Li3]):

Theorem 6.1.9 *For any complex numbers ℓ and h, $W = M_{Vir}(\ell, h)$ has a unique module structure for the vertex operator algebra $V_{Vir}(\ell, 0)$ such that $Y_W(\omega, x) = L_W(x)$.*
$\square$

Remark 6.1.10 The (Virasoro algebra) module $M_{Vir}(\ell, h)$ is commonly referred to in the literature as the *Verma module* for the Virasoro algebra of central charge ℓ with the lowest weight h. As a module for the Virasoro algebra, $M_{Vir}(\ell, h)$ is generated by $\mathbf{1}_{(\ell,h)}$ with the relations $\mathbf{c} = \ell$, $L(0)\mathbf{1}_{(\ell,h)} = h\mathbf{1}_{(\ell,h)}$ and $L(n)\mathbf{1}_{(\ell,h)} = 0$ for $n \geq 1$. In fact $M_{Vir}(\ell, h)$ is universal in the sense that for any module W for the Virasoro algebra of central charge ℓ equipped with a vector w such that $L(0)w = hw$ and $L(n)w = 0$ for $n \geq 1$, there exists a unique module map from $M_{Vir}(\ell, 0)$ to W sending $\mathbf{1}_{(\ell,h)}$ to w (cf. Remark 6.1.3).

Remark 6.1.11 In particular, the Virasoro algebra module $V_{Vir}(\ell, 0)$ is naturally a quotient module of $M_{Vir}(\ell, 0)$, and from the Poincaré–Birkhoff–Witt theorem, the kernel of the $\mathcal{L}$-module surjection $M_{Vir}(\ell, 0) \rightarrow V_{Vir}(\ell, 0)$, expressed as a surjection $U(\mathcal{L}_{(+)}) \twoheadrightarrow U(\mathcal{L}_{(\geq 2)})$ (see above) is the right ideal $U(\mathcal{L}_{(+)})L_{-1}$ of $U(\mathcal{L}_{(+)})$; a basis for this kernel in $M_{Vir}(\ell, 0)$ consists of the vectors $L(-m_1) \cdots L(-m_r)\mathbf{1}_{(\ell,0)}$ for $r \geq 0$ and $m_1 \geq m_2 \geq \cdots \geq m_r \geq 1$, with m_r required to be 1. By Proposition 4.5.1, this $\mathcal{L}$-module surjection is also a map of modules for the vertex operator algebra $V_{Vir}(\ell, 0)$.

In general, $M_{Vir}(\ell, h)$ as a module for the Virasoro algebra $\mathcal{L}$ may be reducible (and we have just seen that for $h = 0$, it is indeed reducible), in which case it is a reducible $V_{Vir}(\ell, 0)$-module by Theorem 6.1.8 (or Proposition 4.5.17). Since $M_{Vir}(\ell, h)_{(h)}$ $(= \mathbb{C}_{\ell,h})$ generates $M_{Vir}(\ell, h)$, for any proper submodule U, $U_{(h)} = U \cap M_{Vir}(\ell, h)_{(h)} = 0$. Let $T_{Vir}(\ell, h)$ be the sum of all the proper $\mathcal{L}$-submodules of $M_{Vir}(\ell, h)$. Then $T_{Vir}(\ell, h)_{(h)} = 0$, and so $T_{Vir}(\ell, h)$ is also proper and is the largest proper submodule. Set

$$L_{Vir}(\ell, h) = M_{Vir}(\ell, h)/T_{Vir}(\ell, h). \tag{6.1.36}$$

By Theorem 6.1.8, $T_{Vir}(\ell, h)$ is also the (unique) largest proper $V_{Vir}(\ell, 0)$-submodule of $M_{Vir}(\ell, h)$, so that $L_{Vir}(\ell, h)$ is an irreducible $V_{Vir}(\ell, 0)$-module. This proves the first assertion of the following:

Theorem 6.1.12 *For any complex numbers ℓ and h, $L_{Vir}(\ell, h)$ is an irreducible module for the vertex operator algebra $V_{Vir}(\ell, 0)$. Furthermore, the modules $L_{Vir}(\ell, h)$ for $h \in \mathbb{C}$ exhaust the irreducible $V_{Vir}(\ell, 0)$-modules up to equivalence.*

Proof. For the second assertion, let $W = \coprod_{r \in \mathbb{C}} W_{(r)}$ be an irreducible $V_{Vir}(\ell, 0)$-module. From (4.1.22) there exists $r \in \mathbb{C}$ such that $W_{(h)} \neq 0$ and $W_{(h-n)} = 0$ for all positive integers n. Let $0 \neq w \in W_{(h)}$. Then $L(0)w = hw$ and $L(n)w = 0$ for $n \geq 1$. In view of Remark 6.1.10, there is a unique Virasoro algebra module homomorphism ψ from $M_{Vir}(\ell, h)$ to W such that $\psi(\mathbf{1}_{(\ell,h)}) = w$. By Proposition 4.5.1, ψ is a $V_{Vir}(\ell, 0)$-module homomorphism (since ω generates $V_{Vir}(\ell, 0)$). Since W is irreducible and $T_{Vir}(\ell, h)$ is the (unique) largest proper submodule of $M_{Vir}(\ell, h)$, it follows that $\psi(M_{Vir}(\ell, h)) = W$ and that Ker $\psi = T_{Vir}(\ell, h)$. Thus ψ reduces to a $V_{Vir}(\ell, 0)$-module isomorphism from $L_{Vir}(\ell, h)$ onto W, proving the second assertion.

$\square$

Remark 6.1.13 In general, the $\mathcal{L}$-module $V_{Vir}(\ell, 0)$, which as we have just seen is a nontrivial quotient of $M_{Vir}(\ell, 0)$, may be reducible. In this case, from Theorem 6.1.8, $V_{Vir}(\ell, 0)$ is a reducible $V_{Vir}(\ell, 0)$-module, and thus by Remark 4.5.4 $V_{Vir}(\ell, 0)$ is not a simple vertex operator algebra. Since $V_{Vir}(\ell, 0)_{(0)} = \mathbb{C}\mathbf{1}$, by Remark 3.9.11 the sum I_ℓ of all the proper ideals, or equivalently, all the proper submodules for the Virasoro algebra, of $V_{Vir}(\ell, 0)$ is the (unique) largest proper ideal, so that $V_{Vir}(\ell, 0)/I_\ell$ is a simple vertex operator algebra. Since $V_{Vir}(\ell, 0)/I_\ell$ is an irreducible $V_{Vir}(\ell, 0)$-module, from Theorem 6.1.12 $V_{Vir}(\ell, 0)/I_\ell$ must be isomorphic to $L_{Vir}(\ell, 0)$, so that $L_{Vir}(\ell, 0) = V_{Vir}(\ell, 0)/I_\ell$ is a simple vertex operator algebra. (It also follows from Theorem 6.1.5 that $L_{Vir}(\ell, 0)$ has a natural vertex operator algebra structure.) From Remark 6.1.6, $L_{Vir}(\ell, 0)$ is in fact the unique simple minimal vertex operator algebra of central charge ℓ up to isomorphism.

Remark 6.1.14 As was pointed out in [Wa1], it follows from Kac's determinant formula in [K5] and the module structure theorem in [FFu2] that $V_{Vir}(\ell, 0)$ is an irreducible module for the Virasoro algebra if and only if $\ell \neq c_{p,q}$, where

$$c_{p,q} = 1 - \frac{6(p - q)^2}{pq} \tag{6.1.37}$$

for $p, q \in \{2, 3, 4, \dots\}$ with p and q relatively prime. In particular, if $\ell \notin \{c_{p,q}\}$, then $V_{Vir}(\ell, 0) = L_{Vir}(\ell, 0)$. In this case, by Theorem 6.1.12, all the irreducible $L_{Vir}(\ell, 0)$-modules are classified; see [Wa1]. For $\ell = c_{p,q} = 1 - \frac{6(p-q)^2}{pq}$ for some p, q as above, all the irreducible $L_{Vir}(\ell, 0)$-modules have been also classified in [Wa1] (see also [DMZ]). In this case, every irreducible $L_{Vir}(\ell, 0)$-module, also an irreducible $V_{Vir}(\ell, 0)$-module, is isomorphic to $L_{Vir}(\ell, h)$ for some $h \in \mathbb{C}$ by Theorem 6.1.12, but an irreducible $\mathcal{L}$-module of the form $L_{Vir}(\ell, h)$ is *not* necessarily an (irreducible) $L_{Vir}(\ell, 0)$-module; in fact, *only for certain finitely many rational numbers h is $L_{Vir}(\ell, h)$ an (irreducible)* $L_{Vir}(\ell, 0)$-module.

6.2 Vertex operator algebras and modules associated to affine Lie algebras

Here we shall construct a natural family of vertex (operator) algebras, along with all their irreducible modules, associated to the family of affine Lie algebras, among them, the affine Kac–Moody algebras associated to finite-dimensional simple Lie algebras as well as a natural family of Heisenberg Lie algebras.

The method used in this section is the same as in Section 6.1 for the Virasoro algebra; it is to apply Theorems 5.7.1, 5.7.4 and 5.7.6 in natural ways, and the results are parallel.

Just as in Section 6.1, once the necessary Lie algebra notations and structures have been set up the main results follow easily.

We shall start with a general affine Lie algebra $\hat{\mathfrak{g}}$ associated to a finite-dimensional Lie algebra $\mathfrak{g}$ equipped with a nondegenerate symmetric invariant bilinear form $\langle \cdot, \cdot \rangle$. We prove that for any complex number ℓ, a certain "generalized Verma" $\hat{\mathfrak{g}}$-module $V_{\hat{\mathfrak{g}}}(\ell, 0)$ of "level ℓ" has a natural vertex algebra structure, and we prove that any "restricted" $\hat{\mathfrak{g}}$-module of level ℓ has a natural $V_{\hat{\mathfrak{g}}}(\ell, 0)$-module structure.

The vertex algebras $V_{\hat{\mathfrak{g}}}(\ell, 0)$ were constructed and studied by Lian in [Lia] by a different method, generalizing I. Frenkel and Zhu's original construction in [FZ] for affine Lie algebras associated with finite-dimensional simple Lie algebras. For a finite-dimensional simple Lie algebra $\mathfrak{g}$ and for all complex numbers ℓ except for a specific negative value — the negative of the dual Coxeter number of $\mathfrak{g}$, it was known (cf. [FZ]) that the vertex algebras $V_{\hat{\mathfrak{g}}}(\ell, 0)$ are actually vertex operator algebras, where the Virasoro algebra is realized through the well-known Segal–Sugawara construction. For the case in which ℓ equals the negative of the dual Coxeter number of $\mathfrak{g}$, the structure of $V_{\hat{\mathfrak{g}}}(\ell, 0)$ as a $\hat{\mathfrak{g}}$-module has been extensively studied by E. Frenkel in [F1] and by E. Frenkel and Feigin in [FF8], and the structure of $V_{\hat{\mathfrak{g}}}(\ell, 0)$ as a vertex algebra has been studied by E. Frenkel in [F6]. Simple vertex operator algebras associated to $\hat{\mathfrak{g}}$ with $\mathfrak{g}$ a simple Lie algebra and ℓ a positive integer were also constructed and studied in [DL3] by a method different from that used in [FZ] and different from that used in the present section and in Section 6.6. In a more general setting, the existence and uniqueness of a conformal vector with certain properties for $V_{\hat{\mathfrak{g}}}(\ell, 0)$ was studied in [Lia]. In this section we shall consider only the important case in which the Casimir operator Ω associated to $\mathfrak{g}$ acts on $\mathfrak{g}$ as a scalar. We shall also follow [DL3] in using the vertex algebra structure to establish the Virasoro algebra relations, rather than directly calculating the commutators by using the Segal–Sugawara construction.

Let $\mathfrak{g}$ be a (possibly infinite-dimensional) Lie algebra equipped with a symmetric invariant bilinear form $\langle \cdot, \cdot \rangle$. To the pair $(\mathfrak{g}, \langle \cdot, \cdot \rangle)$, we associate the *(untwisted) affine Lie algebra* $\hat{\mathfrak{g}}$, with the underlying vector space

$$\hat{\mathfrak{g}} = \mathfrak{g} \otimes \mathbb{C}[t, t^{-1}] \oplus \mathbb{C}\mathbf{k}, \tag{6.2.1}$$

equipped with the bracket relations

$$[a \otimes t^m, b \otimes t^n] = [a, b] \otimes t^{m+n} + m \langle a, b \rangle \delta_{m+n,0} \mathbf{k}, \tag{6.2.2}$$

for $a, b \in \mathfrak{g}$ and $m, n \in \mathbb{Z}$, together with the condition that $\mathbf{k}$ is a nonzero central element of $\hat{\mathfrak{g}}$.

In many ways, the affine Lie algebra $\hat{\mathfrak{g}}$ is analogous to the Virasoro algebra. (Of course, they are also different in many ways.) First, the affine Lie algebra $\hat{\mathfrak{g}}$, equipped with the $\mathbb{Z}$-grading

$$\hat{\mathfrak{g}} = \coprod_{n \in \mathbb{Z}} \hat{\mathfrak{g}}_{(n)}, \tag{6.2.3}$$

where

$$\hat{\mathfrak{g}}_{(0)} = \mathfrak{g} \oplus \mathbb{C}\mathbf{k} \quad \text{and} \quad \hat{\mathfrak{g}}_{(n)} = \mathfrak{g} \otimes t^{-n} \quad \text{for } n \neq 0, \tag{6.2.4}$$

is a $\mathbb{Z}$-graded Lie algebra, and we have the following graded subalgebras of $\hat{\mathfrak{g}}$:

$$\hat{\mathfrak{g}}_{(\pm)} = \coprod_{n>0} \hat{\mathfrak{g}}_{(\pm n)} = \coprod_{n>0} \mathfrak{g} \otimes t^{\mp n}, \tag{6.2.5}$$

$$\hat{\mathfrak{g}}_{(\leq 0)} = \coprod_{n \leq 0} \hat{\mathfrak{g}}_{(n)} = \hat{\mathfrak{g}}_{(-)} \oplus \hat{\mathfrak{g}}_{(0)} = \hat{\mathfrak{g}}_{(-)} \oplus \mathfrak{g} \oplus \mathbb{C}\mathbf{k}. \tag{6.2.6}$$

Remark 6.2.1 As with the Virasoro algebra, the affine Lie algebra $\hat{\mathfrak{g}}$ is often graded by *degree* as $\hat{\mathfrak{g}} = \coprod_{n \in \mathbb{Z}} \hat{\mathfrak{g}}(n)$, where $\hat{\mathfrak{g}}(0) = \mathfrak{g} \oplus \mathbb{C}\mathbf{k}$ and $\hat{\mathfrak{g}}(n) = \mathfrak{g} \otimes t^n$ (rather than $\mathfrak{g} \otimes t^{-n}$) for $n \neq 0$. We shall realize a Virasoro algebra action on $\hat{\mathfrak{g}}$-modules W of a certain type so that our $\mathbb{Z}$-grading of the image of $\hat{\mathfrak{g}}$ in End W will coincide with its grading by eigenvalues of the operator ad $L(0)$. We correspondingly say that the affine Lie algebra $\hat{\mathfrak{g}}$ with the $\mathbb{Z}$-grading we are using is graded by *weight*.

For $a \in \mathfrak{g}$ we define the generating function

$$a(x) = \sum_{n \in \mathbb{Z}} (a \otimes t^n) x^{-n-1} \in \hat{\mathfrak{g}}[[x, x^{-1}]]. \tag{6.2.7}$$

In terms of generating functions, the defining relations (6.2.2) can be equivalently written as

$$\begin{aligned}
&[a(x_1), b(x_2)] \\
&= \sum_{m,n \in \mathbb{Z}} ([a, b] \otimes t^{m+n}) x_1^{-m-1} x_2^{-n-1} + \sum_{m \in \mathbb{Z}} m \langle a, b \rangle x_1^{-m-1} x_2^{m-1} \mathbf{k} \\
&= \sum_{m,n \in \mathbb{Z}} ([a, b] \otimes t^{m+n}) x_2^{-m-n-1} (x_1^{-m-1} x_2^m) + \sum_{m \in \mathbb{Z}} m \langle a, b \rangle x_1^{-m-1} x_2^{m-1} \mathbf{k} \\
&= [a, b](x_2) x_1^{-1} \delta\left(\frac{x_2}{x_1}\right) + \langle a, b \rangle \frac{\partial}{\partial x_2} x_1^{-1} \delta\left(\frac{x_2}{x_1}\right) \mathbf{k} \\
&= [a, b](x_2) x_2^{-1} \delta\left(\frac{x_1}{x_2}\right) - \langle a, b \rangle \frac{\partial}{\partial x_1} x_2^{-1} \delta\left(\frac{x_1}{x_2}\right) \mathbf{k};
\end{aligned} \tag{6.2.8}$$

here we use the fact that

$$\left(\frac{\partial}{\partial x_1} + \frac{\partial}{\partial x_2}\right) \left((x_1 - x_2)^{-1} + (x_2 - x_1)^{-1}\right) = 0$$

(recall Remark 2.3.10) for the last equality. In view of (6.2.8) and (2.3.13), we have

$$(x_1 - x_2)^2 [a(x_1), b(x_2)] = 0 \tag{6.2.9}$$

in $\hat{\mathfrak{g}}[[x_1, x_1^{-1}, x_2, x_2^{-1}]]$.

Remark 6.2.2 Given a $\hat{\mathfrak{g}}$-module W, we shall typically write $a(n)$ for the operator on W corresponding to $a \otimes t^n$, for $a \in \mathfrak{g}$ and $n \in \mathbb{Z}$, and we shall continue to use the notation $a(x)$, or $a_W(x)$, for the action of the expression (6.2.7) on W:

$$a_W(x) = a(x) = \sum_{n \in \mathbb{Z}} a(n)x^{-n-1} \in (\text{End } W)[[x, x^{-1}]]. \qquad (6.2.10)$$

This is analogous to our notational conventions for the Virasoro algebra (cf. Remark 6.1.2 and (6.1.19), for instance).

Remark 6.2.3 If W is a *restricted module* (cf. [K6]) in the sense that for every $a \in \mathfrak{g}$ and $w \in W$, $a(n)w = 0$ for n sufficiently large, then $\{a_W(x) \mid a \in \mathfrak{g}\}$ is a local subspace of $\mathcal{E}(W)$, from (6.2.9).

From Theorem 5.5.18 we immediately have:

Theorem 6.2.4 *Let W be a restricted $\hat{\mathfrak{g}}$-module. Set*

$$U_W = \text{span}\{a_W^{(1)}(x)_{n_1} \cdots a_W^{(r)}(x)_{n_r} 1_W \mid r \geq 0, \ a^{(i)} \in \mathfrak{g}, \ n_i \in \mathbb{Z}\} \subset \mathcal{E}(W).$$
$$(6.2.11)$$

Then $(U_W, Y_{\mathcal{E}}, 1_W)$ carries the structure of a vertex algebra with W as a natural faithful module, where the action Y_W of U_W on W is given by $Y_W(\alpha(x), x_0) = \alpha(x_0)$ for $\alpha(x) \in U_W$. $\square$

As with the Virasoro algebra, we shall be interested in restricted $\hat{\mathfrak{g}}$-modules on which the central element $\mathbf{k}$ acts as a scalar.

Definition 6.2.5 If W is a $\hat{\mathfrak{g}}$-module on which $\mathbf{k}$ acts as a scalar ℓ in $\mathbb{C}$, we say that W is of *level ℓ* (cf. [FL], [K6]).

We are going to show that if W is a restricted $\hat{\mathfrak{g}}$-module of level ℓ, then the vertex algebra U_W obtained in Theorem 6.2.4 is naturally a restricted $\hat{\mathfrak{g}}$-module of level ℓ. First we have the following analogue of Proposition 5.6.8 (cf. Remark 5.6.10):

Proposition 6.2.6 *Let V be a vertex algebra equipped with a linear map ϕ from our Lie algebra $\mathfrak{g}$ to V and let W be a V-module. Set $a_V(x) = Y(\phi(a), x)$ for $a \in \mathfrak{g}$, and suppose that V is a $\hat{\mathfrak{g}}$-module of level ℓ, in the sense that the outer equality in (6.2.8) holds for $a, b \in \mathfrak{g}$, with $\mathbf{k}$ replaced by ℓ (and with $\alpha(x)$ replaced by $\alpha_V(x)$ for any $\alpha \in \mathfrak{g}$). Then W is a restricted $\hat{\mathfrak{g}}$-module of level ℓ, with $a_W(x) = Y_W(\phi(a), x)$ and with $a(n)$ defined as in (6.2.10), for $a \in \mathfrak{g}$, in the analogous sense, for W in place of V. On the other hand, if W is a faithful V-module such that W is a $\hat{\mathfrak{g}}$-module of level ℓ, with $a_W(x) = Y_W(\phi(a), x)$ for $a \in \mathfrak{g}$, then V is a restricted $\hat{\mathfrak{g}}$-module of level ℓ, with $a_V(x) = Y(\phi(a), x)$ for $a \in \mathfrak{g}$.*

Proof. This follows immediately from Proposition 5.6.7 and (6.2.8); we have $\phi(a)_0\phi(b) = \phi([a,b])$, $\phi(a)_1\phi(b) = \ell\langle a,b\rangle 1$ and $\phi(a)_n\phi(b) = 0$ for $a, b \in \mathfrak{g}$ and $n \geq 2$. $\quad\square$

Let W be a restricted $\hat{\mathfrak{g}}$-module of level ℓ. Notice that in Theorem 6.2.4, W is a faithful module for the vertex algebra U_W. We would now like to apply the last assertion of Proposition 6.2.6 to $V = U_W$, acting on W, but first we need a linear map ϕ from $\mathfrak{g}$ to U_W. Note that for $a \in \mathfrak{g}$ and with $a_W(x)$ defined as usual in (6.2.10),

$$a_W(x) = a_W(x)_{-1}1_W \tag{6.2.12}$$

in U_W, so that in particular, $a_W(x)$ is indeed an element of U_W. This gives us a linear map

$$\begin{aligned} \phi : \mathfrak{g} &\to U_W \\ a &\mapsto a_W(x) = \sum_{n \in \mathbb{Z}} a(n)x^{-n-1}. \end{aligned} \tag{6.2.13}$$

We now verify that the last assertion of Proposition 6.2.6 applies to this situation: To see that $a_W(x) = Y_W(\phi(a), x)$ for $a \in \mathfrak{g}$, we note that

$$Y_W(\phi(a), x_0) = Y_W(a_W(x), x_0) = a_W(x_0), \tag{6.2.14}$$

as desired, and we know that W is a $\hat{\mathfrak{g}}$-module of level ℓ in the sense of the outer equality in (6.2.8) with $\mathbf{k}$ replaced by ℓ and with $a(x)$ taken to be $a_W(x)$ for $a \in \mathfrak{g}$. Thus from Proposition 6.2.6, U_W is a restricted $\hat{\mathfrak{g}}$-module of level ℓ, with $a_{U_W}(x_0) = Y_{\mathcal{E}}(\phi(a), x_0) = Y_{\mathcal{E}}(a_W(x), x_0)$ for $a \in \mathfrak{g}$; that is, the outer equality in (6.2.8) holds in the usual sense. In particular, for $a \in \mathfrak{g}$ and $n \in \mathbb{Z}$, $a \otimes t^n \in \hat{\mathfrak{g}}$ acts on U_W as $a_W(x)_n$. As a $\hat{\mathfrak{g}}$-module, U_W is clearly generated by 1_W, and by the creation property,

$$a_W(x)_n 1_W = 0 \quad \text{for } a \in \mathfrak{g}, \ n \geq 0. \tag{6.2.15}$$

Hence we have (cf. Theorems 5.6.11 and 6.1.4):

Theorem 6.2.7 *Let W be a restricted $\hat{\mathfrak{g}}$-module of level ℓ. Then the vertex algebra U_W obtained in Theorem 6.2.4 is naturally a restricted $\hat{\mathfrak{g}}$-module of level ℓ, with $a \otimes t^n$ acting as $a_W(x)_n$ for $a \in \mathfrak{g}$, $n \in \mathbb{Z}$. Furthermore, as a $\hat{\mathfrak{g}}$-module, U_W is generated by 1_W and we have the relations*

$$a(n)1_W = a_W(x)_n 1_W = 0 \quad \textit{for } a \in \mathfrak{g}, \ n \geq 0. \quad\square \tag{6.2.16}$$

Next, we shall construct a universal restricted $\hat{\mathfrak{g}}$-module of level ℓ, generated by a vector 1 satisfying the outer equality in (6.2.16), and endow it with a natural vertex algebra structure. Induced from a suitable "parabolic subalgebra" rather than from a "Borel subalgebra," this module is a "generalized Verma module" as in [Le1], [GarL] and [Le4]. Below (see (6.2.74)) we shall also consider more general such modules.

Let ℓ be a complex number as before. Let $\hat{\mathfrak{g}}_{(-)}$ and $\mathfrak{g}$ act trivially on $\mathbb{C}$ and let $\mathbf{k}$ act as the scalar ℓ, making $\mathbb{C}$ a $\hat{\mathfrak{g}}_{(\leq 0)}$-module (recall (6.2.6)), which we denote by $\mathbb{C}_\ell$. Form the induced module (cf. (6.1.12))

$$V_{\hat{\mathfrak{g}}}(\ell, 0) = U(\hat{\mathfrak{g}}) \otimes_{U(\hat{\mathfrak{g}}_{(\leq 0)})} \mathbb{C}_\ell. \tag{6.2.17}$$

Equip $\mathbb{C}_\ell$ ($= \mathbb{C}$) with the $\mathbb{Z}$-grading $\mathbb{C}_\ell = \coprod_{n \in \mathbb{Z}}(\mathbb{C}_\ell)_{(n)}$, where $(\mathbb{C}_\ell)_{(0)} = \mathbb{C}_\ell$ and $(\mathbb{C}_\ell)_{(n)} = 0$ for $n \neq 0$, making $\mathbb{C}_\ell$ a $\mathbb{Z}$-graded $\hat{\mathfrak{g}}_{(\leq 0)}$-module. Then $V_{\hat{\mathfrak{g}}}(\ell, 0)$ is naturally a $\mathbb{Z}$-graded $\hat{\mathfrak{g}}$-module of level ℓ. We shall call this grading the *weight*-grading; it will turn out to be compatible with a grading by $L(0)$-eigenvalues.

From the Poincaré–Birkhoff–Witt theorem we have

$$V_{\hat{\mathfrak{g}}}(\ell, 0) = U(\hat{\mathfrak{g}}_{(+)}) \simeq S(\hat{\mathfrak{g}}_{(+)}) \tag{6.2.18}$$

as a $\mathbb{Z}$-graded vector space. Set

$$\mathbf{1} = 1 \in \mathbb{C} \subset V_{\hat{\mathfrak{g}}}(\ell, 0). \tag{6.2.19}$$

Then

$$V_{\hat{\mathfrak{g}}}(\ell, 0) = \coprod_{n \geq 0} V_{\hat{\mathfrak{g}}}(\ell, 0)_{(n)}, \tag{6.2.20}$$

where $V_{\hat{\mathfrak{g}}}(\ell, 0)_{(n)}$ is spanned by the vectors

$$a^{(1)}(-m_1) \cdots a^{(r)}(-m_r)\mathbf{1} \tag{6.2.21}$$

for $r \geq 0$, $a^{(i)} \in \mathfrak{g}$, $m_i \geq 1$, with $n = m_1 + \cdots + m_r$. One can furthermore obtain a basis of $V_{\hat{\mathfrak{g}}}(\ell, 0)_{(n)}$ by using an ordered basis of $\hat{\mathfrak{g}}_+$. It follows from (6.2.20) that $V_{\hat{\mathfrak{g}}}(\ell, 0)$ is a restricted $\hat{\mathfrak{g}}$-module, since $a \otimes t^n$ is of weight $-n$ for $a \in \mathfrak{g}$ and $n \in \mathbb{Z}$. If $\mathfrak{g}$ is finite dimensional, then $V_{\hat{\mathfrak{g}}}(\ell, 0)$ with its weight grading satisfies the grading restrictions (3.1.38) and (3.1.39). We consider $\mathfrak{g}$ as a subspace of $V_{\hat{\mathfrak{g}}}(\ell, 0)$ through the map

$$\mathfrak{g} \to V_{\hat{\mathfrak{g}}}(\ell, 0)$$
$$a \mapsto a(-1)\mathbf{1}. \tag{6.2.22}$$

Then $\mathfrak{g}$ is in fact the full subspace of $V_{\hat{\mathfrak{g}}}(\ell, 0)$ of weight 1:

$$\mathfrak{g} = V_{\hat{\mathfrak{g}}}(\ell, 0)_{(1)}. \tag{6.2.23}$$

Remark 6.2.8 As a $\hat{\mathfrak{g}}$-module, $V_{\hat{\mathfrak{g}}}(\ell, 0)$ is generated by $\mathbf{1}$, with the relations $\mathbf{k} = \ell$ and $a(n)\mathbf{1} = 0$ for $a \in \mathfrak{g}$, $n \geq 0$, and in fact, $V_{\hat{\mathfrak{g}}}(\ell, 0)$ is universal in the sense that for any $\hat{\mathfrak{g}}$-module W of level ℓ equipped with a vector $w \in W$ such that $a(n)w = 0$ for $a \in \mathfrak{g}$, $n \geq 0$, there exists a unique $\hat{\mathfrak{g}}$-module homomorphism from $V_{\hat{\mathfrak{g}}}(\ell, 0)$ to W sending $\mathbf{1}$ to w.

Remark 6.2.9 We are going to put a natural vertex algebra structure onto $V = V_{\hat{\mathfrak{g}}}(\ell, 0)$ by applying Theorem 5.7.1. To do this, we need a linear operator d on $V_{\hat{\mathfrak{g}}}(\ell, 0)$ such that $d(\mathbf{1}) = 0$ and such that

$$[d, a(x)] = \frac{d}{dx} a(x) \quad \text{for } a \in \mathfrak{g}. \tag{6.2.24}$$

Here we construct such a (necessarily unique) linear operator. First we define a linear operator $d_{(1)}$ on $\hat{\mathfrak{g}}$ by

$$d_{(1)}(\mathbf{k}) = 0, \tag{6.2.25}$$

$$d_{(1)}(a \otimes t^n) = -n(a \otimes t^{n-1}) \ \left(= -\frac{d}{dt}(a \otimes t^n) \right) \quad \text{for } a \in \mathfrak{g}, \ n \in \mathbb{Z}. \tag{6.2.26}$$

Then $d_{(1)}\hat{\mathfrak{g}}_{(n)} \subset \hat{\mathfrak{g}}_{(n+1)}$ for $n \in \mathbb{Z}$; that is, $d_{(1)}$ is an operator of weight one. It is straightforward to check that $d_{(1)}$ is a derivation of the Lie algebra $\hat{\mathfrak{g}}$, so that $d_{(1)}$ naturally extends to a derivation of the associative algebra $U(\hat{\mathfrak{g}})$. Clearly, $d_{(1)}$ preserves the subspace $\hat{\mathfrak{g}}_{(-)} \oplus \mathfrak{g} \oplus \mathbb{C}(\mathbf{k} - \ell)$ of $U(\hat{\mathfrak{g}})$. Since as a (left) $U(\hat{\mathfrak{g}})$-module,

$$V_{\hat{\mathfrak{g}}}(\ell, 0) = U(\hat{\mathfrak{g}})/U(\hat{\mathfrak{g}})(\hat{\mathfrak{g}}_{(-)} \oplus \mathfrak{g} \oplus \mathbb{C}(\mathbf{k} - \ell)) \tag{6.2.27}$$

(from Remark 6.2.8), it follows that $d_{(1)}$ induces a linear operator on $V_{\hat{\mathfrak{g}}}(\ell, 0)$, which we denote by d. Then as operators on $V_{\hat{\mathfrak{g}}}(\ell, 0)$, we have

$$[d, a(n)] = -na(n-1) \quad \text{for } a \in \mathfrak{g}, \ n \in \mathbb{Z}. \tag{6.2.28}$$

Writing these relations in terms of generating functions we get (6.2.24), and we also have the relation $d(\mathbf{1}) = 0$.

Remark 6.2.10 On the affine Lie algebra $\hat{\mathfrak{g}}$, there is a different linear operator, namely, the "degree operator," of weight zero, that is also often denoted by d in the literature (cf. [Gar1], [GarL], [K6], [FLM6]), but it will not be explicitly needed much in this work; as soon as we construct a Virasoro algebra action on suitable restricted $\hat{\mathfrak{g}}$-modules, we will have an operator $L(0)$, which is closely related to this degree operator; see Remark 6.6.1.

Now we are prepared to present the following important result, which gives vertex algebra structure to $V_{\hat{\mathfrak{g}}}(\ell, 0)$ (cf. Theorem 6.1.5; see also [Lia]):

Theorem 6.2.11 *Let ℓ be any complex number. Then there exists a unique vertex algebra structure $(V_{\hat{\mathfrak{g}}}(\ell, 0), Y, \mathbf{1})$ on $V_{\hat{\mathfrak{g}}}(\ell, 0)$ such that $\mathbf{1} = 1 \in \mathbb{C}$ is the vacuum vector and such that*

$$Y(a, x) = a(x) \in (\text{End } V_{\hat{\mathfrak{g}}}(\ell, 0))[[x, x^{-1}]] \quad \text{for } a \in \mathfrak{g}. \tag{6.2.29}$$

The vertex operator map for this vertex algebra structure is given by

$$Y(a^{(1)}(n_1) \cdots a^{(r)}(n_r)\mathbf{1}, x) = a^{(1)}(x)_{n_1} \cdots a^{(r)}(x)_{n_r} 1 \tag{6.2.30}$$

for $r \geq 0$, $a^{(i)} \subset \mathfrak{g}$ and $n_i \subset \mathbb{Z}$, where 1 at the end is the identity operator on $V_{\hat{\mathfrak{g}}}(\ell, 0)$.

Proof. We know that $V_{\hat{\mathfrak{g}}}(\ell, 0)$ is a restricted $\hat{\mathfrak{g}}$-module. The relations (6.2.8) imply locality (recall (6.2.9)), and with the operator d constructed in Remark 6.2.9, the existence and uniqueness of vertex algebra structure on $V_{\hat{\mathfrak{g}}}(\ell, 0)$, as well as formula (6.2.30) for the action, follow immediately from Theorem 5.7.1 with $V = V_{\hat{\mathfrak{g}}}(\ell, 0)$, $T = \mathfrak{g} \subset V$ (recall (6.2.22)) and $Y_0(a, x) = Y_0(a(-1)\mathbf{1}, x) = a(x)$ for $a \in \mathfrak{g}$. $\square$

Furthermore, we have $V_{\hat{\mathfrak{g}}}(\ell, 0)$-module structure on any restricted $\hat{\mathfrak{g}}$-module of level ℓ (cf. Theorem 6.1.7):

Theorem 6.2.12 *Let ℓ be any complex number and let W be any restricted $\hat{\mathfrak{g}}$-module of level ℓ. Then there exists a unique $V_{\hat{\mathfrak{g}}}(\ell, 0)$-module structure on W such that for $a \in \mathfrak{g}$,*

$$Y_W(a, x) = a_W(x) \left(= \sum_{n \in \mathbb{Z}} a(n)x^{-n-1} \right). \tag{6.2.31}$$

The vertex operator map for this module structure is given by

$$Y_W(a^{(1)}(n_1) \cdots a^{(r)}(n_r)\mathbf{1}, x) = a_W^{(1)}(x)_{n_1} \cdots a_W^{(r)}(x)_{n_r} 1_W \tag{6.2.32}$$

for $r \geq 0$, $a^{(i)} \in \mathfrak{g}$, $n_i \in \mathbb{Z}$.

Proof. By Theorem 6.2.7, the vertex algebra U_W ($\subset \mathcal{E}(W)$) obtained in Theorem 6.2.4 is a restricted $\hat{\mathfrak{g}}$-module of level ℓ, with $a \otimes t^n$ acting as $a_W(x)_n$ for $a \in \mathfrak{g}$ and $n \in \mathbb{Z}$, and U_W is generated by 1_W and satisfies the relations $a(n)1_W = a_W(x)_n 1_W = 0$ for $a \in \mathfrak{g}$, $n \geq 0$. By Remark 6.2.8 there is a unique $\hat{\mathfrak{g}}$-module map ψ from $V_{\hat{\mathfrak{g}}}(\ell, 0)$ to U_W such that $\psi(\mathbf{1}) = 1_W$. Then

$$\psi(a^{(1)}(n_1) \cdots a^{(r)}(n_r)\mathbf{1}) = a_W^{(1)}(x)_{n_1} \cdots a_W^{(r)}(x)_{n_r} 1_W \tag{6.2.33}$$

for $r \geq 0$, $a^{(i)} \in \mathfrak{g}$, $n_i \in \mathbb{Z}$. It is clear that the vertex algebra $V_{\hat{\mathfrak{g}}}(\ell, 0)$ is generated by the subspace $\mathfrak{g}$. With the locality relation (6.2.9), the existence and uniqueness of $V_{\hat{\mathfrak{g}}}(\ell, 0)$-module structure on W, and formula (6.2.32), follow immediately from Theorem 5.7.6 applied to $T = \mathfrak{g} \subset V_{\hat{\mathfrak{g}}}(\ell, 0)$. $\square$

Next we summarize some facts related to Proposition 6.2.6 and Theorem 6.2.12, using Proposition 4.5.17 (cf. Theorem 6.1.8):

Theorem 6.2.13 *Let ℓ be any complex number. Any module W for the vertex algebra $V_{\hat{\mathfrak{g}}}(\ell, 0)$ is naturally a restricted $\hat{\mathfrak{g}}$-module of level ℓ, with $a_W(x) = Y_W(a, x)$ for $a \in \mathfrak{g}$. Conversely, any restricted $\hat{\mathfrak{g}}$-module W of level ℓ is naturally a $V_{\hat{\mathfrak{g}}}(\ell, 0)$-module with*

$$Y_W(a^{(1)}(n_1) \cdots a^{(r)}(n_r)\mathbf{1}, x) = a_W^{(1)}(x)_{n_1} \cdots a_W^{(r)}(x)_{n_r} 1_W \tag{6.2.34}$$

for $r \geq 0$, $a^{(i)} \in \mathfrak{g}$, $n_i \in \mathbb{Z}$. Furthermore, for any $V_{\hat{\mathfrak{g}}}(\ell, 0)$-module W, the $V_{\hat{\mathfrak{g}}}(\ell, 0)$-submodules of W coincide with the $\hat{\mathfrak{g}}$-submodules of W. $\square$

So far, each $V_{\hat{\mathfrak{g}}}(\ell, 0)$ is just a vertex algebra. In the following we shall show that $V_{\hat{\mathfrak{g}}}(\ell, 0)$ is in fact a vertex operator algebra under certain conditions. We refer the reader to [Lia] for a more general study.

We now assume that $\mathfrak{g}$ *is finite dimensional and that our symmetric invariant bilinear form $\langle \cdot, \cdot \rangle$ on $\mathfrak{g}$ is nondegenerate.* Let

$$d = \dim \mathfrak{g} \qquad (6.2.35)$$

and let $\{u^{(1)}, \ldots, u^{(d)}\}$ be an orthonormal basis of $\mathfrak{g}$ with respect to the form $\langle \cdot, \cdot \rangle$. (Here we re-use the letter d, but this should cause no confusion.) Then for $a \in \mathfrak{g}$,

$$a = \sum_{i=1}^{d} \langle a, u^{(i)} \rangle u^{(i)}. \qquad (6.2.36)$$

Set

$$\Omega = \sum_{i=1}^{d} u^{(i)} u^{(i)} \in U(\mathfrak{g}), \qquad (6.2.37)$$

which we call the *Casimir element* associated to $\mathfrak{g}$ and $\langle \cdot, \cdot \rangle$. It is a basic simple fact that the Casimir element is independent of the choice of the orthonormal basis of $\mathfrak{g}$. It is also a basic fact that the Casimir element is $\mathfrak{g}$-invariant:

$$[\mathfrak{g}, \Omega] = 0 \qquad (6.2.38)$$

in $U(\mathfrak{g})$; the argument in (6.2.54) below proves this. In particular, on any $\mathfrak{g}$-module, $\mathfrak{g}$ centralizes the action of Ω. (In the literature, the Casimir element is sometimes considered only for $\mathfrak{g}$ a semisimple Lie algebra. Also, various normalizations of the Casimir element are used.)

Remark 6.2.14 The Casimir element Ω can be written still more generally as follows. Let $\{w^{(1)}, \ldots, w^{(d)}\}$ be any basis (not necessarily orthonormal) of $\mathfrak{g}$, and let $\{w_{(1)}, \ldots, w_{(d)}\}$ be the corresponding dual basis of $\mathfrak{g}$ with respect to the form $\langle \cdot, \cdot \rangle$, that is, $\langle w^{(i)}, w_{(j)} \rangle = \delta_{ij}$ for $i, j = 1, \ldots, d$. Then

$$\Omega = \sum_{i=1}^{d} w^{(i)} w_{(i)} \in U(\mathfrak{g}). \qquad (6.2.39)$$

We assume that Ω acts on $\mathfrak{g}$ *(under the adjoint representation) as a scalar, say $2h$, where $h \in \mathbb{C}$:*

$$\Omega = 2h \quad \text{on } \mathfrak{g}, \qquad (6.2.40)$$

that is,

$$\sum_{i=1}^{d} [u^{(i)}, [u^{(i)}, a]] = 2ha \quad \text{for } a \in \mathfrak{g}. \qquad (6.2.41)$$

Remark 6.2.15 Later we shall be considering the cases in which $\mathfrak{g}$ is a finite-dimensional abelian or simple Lie algebra. If $\mathfrak{g}$ is abelian, Ω acts on $\mathfrak{g}$ as zero, so that $h = 0$ in (6.2.40). Suppose that $\mathfrak{g}$ is simple. Then our nondegenerate symmetric invariant bilinear form $\langle \cdot, \cdot \rangle$ on $\mathfrak{g}$ is a nonzero multiple of the Killing form of $\mathfrak{g}$ and must remain nondegenerate on any Cartan subalgebra $\mathfrak{h}$ of $\mathfrak{g}$. This gives us an identification of $\mathfrak{h}$ with its dual space $\mathfrak{h}^*$, allowing us to transport the form $\langle \cdot, \cdot \rangle$ from $\mathfrak{h}$ to $\mathfrak{h}^*$; we again denote the resulting nondegenerate symmetric bilinear form on $\mathfrak{h}^*$ by $\langle \cdot, \cdot \rangle$. Let us normalize the form $\langle \cdot, \cdot \rangle$ on $\mathfrak{g}$ so that $\langle \alpha, \alpha \rangle = 2$, where $\alpha \in \mathfrak{h}^* = \mathfrak{h}$ is any long root of $\mathfrak{g}$. Then Ω acts on $\mathfrak{g}$ as a scalar, and because of the normalization, h in (6.2.40) is the dual Coxeter number of $\mathfrak{g}$, a positive integer; cf. [Bou1].

Let ℓ be a complex number such that $\ell \neq -h$ (recall (6.2.40)). Set

$$\omega = \frac{1}{2(\ell + h)} \sum_{i=1}^{d} u^{(i)}(-1)u^{(i)}(-1)\mathbf{1} \in V_{\hat{\mathfrak{g}}}(\ell, 0)_{(2)}, \qquad (6.2.42)$$

and define operators $L(n)$ for $n \in \mathbb{Z}$ by:

$$Y(\omega, x) = \sum_{n \in \mathbb{Z}} \omega_n x^{-n-1} = \sum_{n \in \mathbb{Z}} L(n)x^{-n-2}. \qquad (6.2.43)$$

We shall show that these operators give a representation of the Virasoro algebra on any restricted $\hat{\mathfrak{g}}$-module of level ℓ, and compute the central charge. Since any restricted $\hat{\mathfrak{g}}$-module of level ℓ is naturally a module for the vertex algebra $V_{\hat{\mathfrak{g}}}(\ell, 0)$ (by Theorem 6.2.12), the $L(n)$ for $n \in \mathbb{Z}$ are naturally operators on any restricted $\hat{\mathfrak{g}}$-module of level ℓ. Furthermore, using (3.8.10) we get the following expression of $L(n)$ in terms of elements of $\hat{\mathfrak{g}}$:

$$L(n) = \mathrm{Res}_x x^{n+1} Y(\omega, x)$$

$$= \frac{1}{2(\ell + h)} \sum_{i=1}^{d} \mathrm{Res}_x x^{n+1} Y(u^{(i)}(-1)u^{(i)}, x)$$

$$= \frac{1}{2(\ell + h)} \sum_{i=1}^{d} \mathrm{Res}_x x^{n+1} {}^{\circ}_{\circ} Y(u^{(i)}, x)Y(u^{(i)}, x) {}^{\circ}_{\circ}$$

$$= \frac{1}{2(\ell + h)} \sum_{i=1}^{d} \sum_{m \in \mathbb{Z}} {}^{\circ}_{\circ} u^{(i)}(m)u^{(i)}(n - m) {}^{\circ}_{\circ}. \qquad (6.2.44)$$

For $\mathfrak{g}$ a simple Lie algebra, this is the well-known Segal–Sugawara construction of the Virasoro algebra action (on restricted $\hat{\mathfrak{g}}$-modules); for $\mathfrak{g}$ abelian, so that $h = 0$ and $\ell \neq 0$, this is the Virasoro construction. Next, following [DL3] we shall use the vertex algebra structure to establish the Virasoro algebra relations for these operators.

We have (cf. [DL3], Propositions 13.5 and 13.9):

Theorem 6.2.16 *Let $\mathfrak{g}$ be a finite-dimensional Lie algebra equipped with a nondegenerate symmetric invariant bilinear form $\langle \cdot, \cdot \rangle$. Assume that the Casimir operator Ω acts on $\mathfrak{g}$ as a scalar $2h \in \mathbb{C}$ and that $\ell \neq -h$. Then for $a \in \mathfrak{g}$ and $m, n \in \mathbb{Z}$,*

$$[L(m), a(n)] = -na(m+n), \tag{6.2.45}$$

$$[L(m), L(n)] = (m-n)L(m+n) + \frac{1}{12}(m^3 - m)\frac{d\ell}{\ell + h}\delta_{m+n,0} \tag{6.2.46}$$

on any restricted $\hat{\mathfrak{g}}$-module W of level ℓ. In particular, these relations hold on $V_{\hat{\mathfrak{g}}}(\ell, 0)$. Furthermore, on $V_{\hat{\mathfrak{g}}}(\ell, 0)$,

$$L(0)v = nv \quad \textit{for } v \in V_{\hat{\mathfrak{g}}}(\ell, 0)_{(n)}, \ n \geq 0, \tag{6.2.47}$$

$$L(-1) = \mathcal{D}, \tag{6.2.48}$$

where $\mathcal{D}$ is the $\mathcal{D}$-operator of the vertex algebra $V_{\hat{\mathfrak{g}}}(\ell, 0)$.

Proof. As we have mentioned, by Theorem 6.2.12 any restricted $\hat{\mathfrak{g}}$-module W of level ℓ is naturally a $V_{\hat{\mathfrak{g}}}(\ell, 0)$-module. The relations (6.2.45) can be written

$$[a(m), L(n)] = ma(m+n) \quad \text{for } a \in \mathfrak{g}, \ m, n \in \mathbb{Z}, \tag{6.2.49}$$

which can be equivalently written in terms of generating functions as

$$[Y(a, x_1), Y(\omega, x_2)] = -a(x_2)\frac{\partial}{\partial x_1}x_2^{-1}\delta\left(\frac{x_1}{x_2}\right). \tag{6.2.50}$$

For (6.2.45), in view of commutator formula (5.6.11) for vertex algebras and modules, it suffices to prove that

$$a_n \omega = a(n)\omega = \delta_{n,1}a \quad \text{for } n \geq 0. \tag{6.2.51}$$

Since $V_{\hat{\mathfrak{g}}}(\ell, 0)$ is a $\mathbb{Z}$-graded $\hat{\mathfrak{g}}$-module with $V_{\hat{\mathfrak{g}}}(\ell, 0)_{(n)} = 0$ for $n < 0$ and since $\omega \in V_{\hat{\mathfrak{g}}}(\ell, 0)_{(2)}$, we have $a(m)\omega = 0$ for $m > 2$. Now we compute $a(2)\omega$, using the affine Lie algebra relations (6.2.2), with **k** acting as ℓ:

$$\begin{aligned}
2(\ell + h)a(2)\omega &= \sum_{i=1}^{d} a(2)u^{(i)}(-1)^2 \mathbf{1} \\
&= \sum_{i=1}^{d}([a, u^{(i)}](1)u^{(i)}(-1)\mathbf{1} + u^{(i)}(-1)a(2)u^{(i)}(-1)\mathbf{1}) \\
&= \sum_{i=1}^{d} \ell\langle[a, u^{(i)}], u^{(i)}\rangle \mathbf{1} \\
&= \sum_{i=1}^{d} \ell\langle a, [u^{(i)}, u^{(i)}]\rangle \mathbf{1} \\
&= 0,
\end{aligned} \tag{6.2.52}$$

where we use the fact that $b(m)\mathbf{1} = 0$ for $b \in \mathfrak{g}$ and $m \geq 0$. Similarly,

$$
\begin{aligned}
2(\ell + h)a(1)\omega \\
= \sum_{i=1}^{d} \left([a, u^{(i)}](0)u^{(i)}(-1)\mathbf{1} + \ell\langle a, u^{(i)}\rangle u^{(i)}(-1)\mathbf{1} + u^{(i)}(-1)a(1)u^{(i)}(-1)\mathbf{1} \right) \\
= \sum_{i=1}^{d} \left([[a, u^{(i)}], u^{(i)}](-1)\mathbf{1} + 2\ell\langle a, u^{(i)}\rangle u^{(i)}(-1)\mathbf{1} \right) \\
= 2ha(-1)\mathbf{1} + 2\ell a(-1)\mathbf{1} \\
= 2(\ell + h)a(-1)\mathbf{1} \\
= 2(\ell + h)a,
\end{aligned}
\tag{6.2.53}
$$

since by assumption Ω acts as the scalar $2h$ on $\mathfrak{g}$, and

$$
\begin{aligned}
2(\ell + h)a(0)\omega &= \sum_{i=1}^{d} \left([a, u^{(i)}](-1)u^{(i)}(-1)\mathbf{1} + u^{(i)}(-1)[a, u^{(i)}](-1)\mathbf{1} \right) \\
&= \sum_{1 \leq i,j \leq d} \langle [a, u^{(i)}], u^{(j)}\rangle u^{(j)}(-1)u^{(i)}(-1)\mathbf{1} \\
&\quad + \sum_{1 \leq i,j \leq d} \langle [a, u^{(i)}], u^{(j)}\rangle u^{(i)}(-1)u^{(j)}(-1)\mathbf{1} \\
&= \sum_{1 \leq i,j \leq d} \left(\langle [a, u^{(i)}], u^{(j)}\rangle + \langle [a, u^{(j)}], u^{(i)}\rangle \right) u^{(i)}(-1)u^{(j)}(-1)\mathbf{1} \\
&= \sum_{1 \leq i,j \leq d} \left(\langle a, [u^{(i)}, u^{(j)}]\rangle + \langle a, [u^{(j)}, u^{(i)}]\rangle \right) u^{(i)}(-1)u^{(j)}(-1)\mathbf{1} \\
&= 0,
\end{aligned}
\tag{6.2.54}
$$

where we also use (6.2.36) for $[a, u^{(i)}]$ in place of a. This proves (6.2.51), so that (6.2.45) holds.

For $a \in \mathfrak{g}$, $n \in \mathbb{Z}$, from (6.2.45) and (3.1.32) we have

$$
[L(-1) - \mathcal{D}, a(n)] = 0
\tag{6.2.55}
$$

as operators on $V_{\hat{\mathfrak{g}}}(\ell, 0)$, and we also have

$$
(L(-1) - \mathcal{D})\mathbf{1} = \omega_0 \mathbf{1} - \mathbf{1}_{-2}\mathbf{1} = 0.
\tag{6.2.56}
$$

Since $V_{\hat{\mathfrak{g}}}(\ell, 0)$ is generated from $\mathbf{1}$ by $U(\hat{\mathfrak{g}})$, it follows that $L(-1) = \mathcal{D}$ on $V_{\hat{\mathfrak{g}}}(\ell, 0)$. Similarly, $L(0)$ coincides with the *weight operator* $d_{(0)}$ on $V_{\hat{\mathfrak{g}}}(\ell, 0)$ defined by $d_{(0)}u = nu$ for $u \in V_{\hat{\mathfrak{g}}}(\ell, 0)_{(n)}$ with $n \in \mathbb{Z}$. This proves (6.2.47) and (6.2.48).

For the Virasoro algebra relations (6.2.46), now that we know that $L(-1) = \mathcal{D}$ it suffices to prove (see the end of Section 3.1)

$$\omega_1 \omega = L(0)\omega = 2\omega, \tag{6.2.57}$$

$$\omega_3 \omega = L(2)\omega = \frac{d\ell}{2(\ell + h)}\mathbf{1}, \tag{6.2.58}$$

$$\omega_n \omega = L(n-1)\omega = 0 \quad \text{for } n = 2, \text{ or } n \geq 4. \tag{6.2.59}$$

But using (6.2.45) we get

$$L(0)\omega = \frac{1}{2(\ell + h)} \sum_{i=1}^{d} 2u^{(i)}(-1)^2 \mathbf{1} = 2\omega, \tag{6.2.60}$$

$$L(2)\omega = \frac{1}{2(\ell + h)} \sum_{i=1}^{d} \left([L(2), u^{(i)}(-1)]u^{(i)}(-1)\mathbf{1} + u^{(i)}(-1)[L(2), u^{(i)}(-1)]\mathbf{1}\right)$$

$$= \frac{1}{2(\ell + h)} \sum_{i=1}^{d} u^{(i)}(1)u^{(i)}(-1)\mathbf{1}$$

$$= \frac{1}{2(\ell + h)} \sum_{i=1}^{d} \ell\langle u^{(i)}, u^{(i)}\rangle \mathbf{1}$$

$$= \frac{d\ell}{2(\ell + h)}\mathbf{1}, \tag{6.2.61}$$

$$L(1)\omega = \frac{1}{2(\ell + h)} \sum_{i=1}^{d} (u^{(i)}(0)u^{(i)}(-1)\mathbf{1} + u^{(i)}(-1)u^{(i)}(0)\mathbf{1}) = 0, \tag{6.2.62}$$

$$L(n)\omega = \frac{1}{2(\ell + h)} \sum_{i=1}^{d} (u^{(i)}(n-1)u^{(i)}(-1)\mathbf{1} + u^{(i)}(-1)u^{(i)}(n-1)\mathbf{1})$$

$$= 0 \tag{6.2.63}$$

for $n \geq 3$, and so the proof is complete. $\square$

Remark 6.2.17 Note how the desired commutators were determined in the proof above, from the relations (6.2.51) and (6.2.57)–(6.2.59) (cf. (3.1.70)–(3.1.74)); these relations of course give $u_n v$ for certain pairs u and v in our vertex algebra, for $n \geq 0$, and as usual, this is precisely the information needed for commutators in general. As we mentioned in the proof of Proposition 6.2.6, the corresponding relations for $u = a \in \mathfrak{g}$ and $v = b \in \mathfrak{g}$ are

$$a_0 b = [a, b] \ (\in \mathfrak{g}) \tag{6.2.64}$$

$$a_1 b = \ell\langle a, b\rangle \mathbf{1} \tag{6.2.65}$$

$$a_n b = 0 \ \text{for } n \geq 2; \tag{6.2.66}$$

these relations are equivalent to the affine Lie algebra commutator relation (6.2.8) (with $\mathbf{k}$ replaced by ℓ). Recall from Remark 3.3.14 the interpretation of relations such as these in terms of operator product expansions.

Combining Theorem 6.2.11 with Theorem 6.2.16 we immediately have (cf. [DL3], [FZ], [Lia]):

Theorem 6.2.18 *Let $\mathfrak{g}$ be a finite-dimensional Lie algebra equipped with a nondegenerate symmetric invariant bilinear form $\langle \cdot, \cdot \rangle$. Assume that the Casimir operator Ω acts on $\mathfrak{g}$ as a scalar $2h$ in $\mathbb{C}$ and that $\ell \neq -h$. Then the vertex algebra $V_{\hat{\mathfrak{g}}}(\ell, 0)$ constructed in Theorem 6.2.11 is a vertex operator algebra of central charge $d\ell/(\ell + h)$ with ω defined in (6.2.42) as the conformal vector. The $\mathbb{Z}$-grading on $V_{\hat{\mathfrak{g}}}(\ell, 0)$ we use is given by $L(0)$-eigenvalues. Furthermore, $\mathfrak{g} = V_{\hat{\mathfrak{g}}}(\ell, 0)_{(1)}$, which generates $V_{\hat{\mathfrak{g}}}(\ell, 0)$ as a vertex algebra, and*

$$[L(m), a(n)] = -na(m + n) \quad \text{for } a \in \mathfrak{g}, \ m, n \in \mathbb{Z}. \quad \square \tag{6.2.67}$$

Remark 6.2.19 In this setting, we have

$$L(1)\mathfrak{g} = 0. \tag{6.2.68}$$

Indeed, by (6.2.67), for $a \in \mathfrak{g}$,

$$L(1)a = L(1)a(-1)\mathbf{1} = [L(1), a(-1)]\mathbf{1} = a(0)\mathbf{1} = 0. \tag{6.2.69}$$

Thus we have

$$L(n)\mathfrak{g} = 0 \quad \text{for } n \geq 1. \tag{6.2.70}$$

Remark 6.2.20 We have been assuming that $\ell \neq -h$ in order to equip the vertex algebra $V_{\hat{\mathfrak{g}}}(\ell, 0)$ with a natural vertex operator algebra structure (recall (6.2.42) and Theorems 6.2.16 and 6.2.18). On the other hand, when $\ell = -h$, the situation is very different and it is also very important and interesting, especially for $\mathfrak{g}$ a finite-dimensional simple Lie algebra. This situation has been investigated extensively by E. Frenkel and B. Feigin; see [FF8], [F1], [F6]. In the literature, the negative dual Coxeter number $-h$ is called the *critical level* of the affine Lie algebra $\hat{\mathfrak{g}}$. (Recall from Remark 6.2.15 that when $\mathfrak{g}$ is a finite-dimensional simple Lie algebra, h is the dual Coxeter number of $\mathfrak{g}$.) Now let $\ell = -h$. Set

$$Z = \sum_{i=1}^{d} u^{(i)}(-1)u^{(i)}(-1)\mathbf{1} \in V_{\hat{\mathfrak{g}}}(-h, 0)_{(2)}, \tag{6.2.71}$$

where $\{u^{(1)}, \ldots, u^{(d)}\}$ is an orthonormal basis of $\mathfrak{g}$ as before (cf. (6.2.42)). From (6.2.52)–(6.2.54) we immediately have

$$a(n)Z = 0 \quad \text{for } a \in \mathfrak{g}, \ n \geq 0, \tag{6.2.72}$$

which asserts that $\mathfrak{g}$ is contained in the centralizer of $\{Z\}$ in $V_{\hat{\mathfrak{g}}}(-h, 0)$ (recall (3.11.1), (3.11.3)). It then follows that the whole vertex algebra $V_{\hat{\mathfrak{g}}}(-h, 0)$ is contained in the centralizer of $\{Z\}$, since the centralizer of $\{Z\}$ is a vertex subalgebra and $\mathfrak{g}$ generates $V_{\hat{\mathfrak{g}}}(-h, 0)$ as a vertex algebra. This proves that Z lies in the center of $V_{\hat{\mathfrak{g}}}(-h, 0)$ (recall (3.11.2)), so that the vertex subalgebra $\langle Z \rangle$ generated by Z is contained in the center. The whole center of $V_{\hat{\mathfrak{g}}}(-h, 0)$ was completely determined in [F6] and certain other related results can also be found therein.

For the rest of this section we maintain all the assumptions in Theorem 6.2.18, so that in particular $\ell \neq -h$ and $V_{\hat{\mathfrak{g}}}(\ell, 0)$ is a vertex operator algebra. In the following we shall construct and classify the irreducible modules for $V_{\hat{\mathfrak{g}}}(\ell, 0)$ viewed as a vertex operator algebra.

We generalize the construction of the $\hat{\mathfrak{g}}$-module $V_{\hat{\mathfrak{g}}}(\ell, 0)$ to get a certain natural family of $\hat{\mathfrak{g}}$-modules, and then we endow them with natural $V_{\hat{\mathfrak{g}}}(\ell, 0)$-module structure.

Let U be a $\mathfrak{g}$-module on which the Casimir operator Ω acts as a scalar, denoted by h_U:

$$\Omega = h_U \quad \text{on} \quad U. \tag{6.2.73}$$

For example, if U is a finite-dimensional irreducible $\mathfrak{g}$-module, then by Schur's lemma the Casimir operator Ω acts as a scalar on U. Consider U as a $\hat{\mathfrak{g}}_{(\leq 0)}$-module with $\hat{\mathfrak{g}}_{(-)}$ acting trivially and with $\mathbf{k}$ acting as the scalar ℓ (recall (6.2.6)). Form the induced module

$$\text{Ind}_{\mathfrak{g}}^{\hat{\mathfrak{g}}}(U) = U(\hat{\mathfrak{g}}) \otimes_{U(\hat{\mathfrak{g}}_{(\leq 0)})} U, \tag{6.2.74}$$

a "generalized Verma module" as in (6.2.17). In fact, if $U = \mathbb{C}$ is the trivial $\mathfrak{g}$-module, then $\text{Ind}_{\mathfrak{g}}^{\hat{\mathfrak{g}}}(\mathbb{C}) = V_{\hat{\mathfrak{g}}}(\ell, 0)$.

Consider U as a $\mathbb{C}$-graded $\hat{\mathfrak{g}}_{(\leq 0)}$-module $U = \coprod_{r \in \mathbb{C}} U_{(r)}$ with

$$U_{(r)} = U \quad \text{for} \quad r = \frac{1}{2(\ell + h)} h_U \tag{6.2.75}$$

(recall that we are assuming that $\ell \neq -h$) and $U_{(r)} = 0$ otherwise, and correspondingly make $\text{Ind}_{\mathfrak{g}}^{\hat{\mathfrak{g}}}(U)$ a $\mathbb{C}$-graded $\hat{\mathfrak{g}}$-module. (The assignment of any complex number as the weight of U would make $\text{Ind}_{\mathfrak{g}}^{\hat{\mathfrak{g}}}(U)$ a $\mathbb{C}$-graded $\hat{\mathfrak{g}}$-module.) Again, from the Poincaré–Birkhoff–Witt theorem, as a $\mathbb{C}$-graded vector space,

$$\text{Ind}_{\mathfrak{g}}^{\hat{\mathfrak{g}}}(U) = U(\hat{\mathfrak{g}}_{(+)}) \otimes_{\mathbb{C}} U, \tag{6.2.76}$$

and in particular, $\text{Ind}_{\mathfrak{g}}^{\hat{\mathfrak{g}}}(U)$ is a restricted $\hat{\mathfrak{g}}$-module of level ℓ. We naturally consider U as a subspace of $\text{Ind}_{\mathfrak{g}}^{\hat{\mathfrak{g}}}(U)$. Using the expression (6.2.44) of $L(n)$ with $n = 0$ and the fact that $\hat{\mathfrak{g}}_{(-)}U = 0$, we find that

$$L(0) = \frac{1}{2(\ell + h)} \Omega \quad \text{on } U, \tag{6.2.77}$$

so that our $\mathbb{C}$-grading on $\text{Ind}_{\mathfrak{g}}^{\hat{\mathfrak{g}}}(U)$ is actually given by $L(0)$-eigenvalues. Since $\text{Ind}_{\mathfrak{g}}^{\hat{\mathfrak{g}}}(U)$, being a restricted $\hat{\mathfrak{g}}$-module of level ℓ, is already a module for $V_{\hat{\mathfrak{g}}}(\ell, 0)$ viewed as a vertex algebra (by Theorem 6.2.12), $\text{Ind}_{\mathfrak{g}}^{\hat{\mathfrak{g}}}(U)$ is a module for $V_{\hat{\mathfrak{g}}}(\ell, 0)$ viewed as a vertex operator algebra except that the homogeneous subspaces (the $L(0)$-eigenspaces) may be infinite dimensional. Moreover, any $\hat{\mathfrak{g}}$-submodule of the $\hat{\mathfrak{g}}$-module $\text{Ind}_{\mathfrak{g}}^{\hat{\mathfrak{g}}}(U)$ is $L(0)$-stable and hence is $\mathbb{C}$-graded by $L(0)$-eigenvalues in view of the last assertion of Theorem 6.2.13. If we assume in addition that U is finite dimensional, then from (6.2.76) all the homogeneous subspaces are finite dimensional. Summarizing, we have (see also [FZ]):

Theorem 6.2.21 *Let U be a finite-dimensional $\mathfrak{g}$-module on which the Casimir operator Ω acts as a scalar h_U. Then $W = \mathrm{Ind}_{\mathfrak{g}}^{\hat{\mathfrak{g}}}(U)$ has a unique module structure for the vertex operator algebra $V_{\hat{\mathfrak{g}}}(\ell, 0)$ such that $Y_W(a, x) = a_W(x)$ for $a \in \mathfrak{g}$. Furthermore,*

$$\mathrm{Ind}_{\mathfrak{g}}^{\hat{\mathfrak{g}}}(U) = \coprod_{n \in \mathbb{N}} \left(\mathrm{Ind}_{\mathfrak{g}}^{\hat{\mathfrak{g}}}(U) \right)_{(r+n)} \tag{6.2.78}$$

with $\left(\mathrm{Ind}_{\mathfrak{g}}^{\hat{\mathfrak{g}}}(U) \right)_{(r)} = U$, *where* $r = \frac{1}{2(\ell+h)} h_U$. *Moreover, any $\hat{\mathfrak{g}}$-submodule of the $\hat{\mathfrak{g}}$-module W is $\mathbb{C}$-graded by $L(0)$-eigenvalues. If U is infinite-dimensional, all the assertions still hold except that the homogeneous subspaces of $W = \mathrm{Ind}_{\mathfrak{g}}^{\hat{\mathfrak{g}}}(U)$ are infinite dimensional.* $\square$

Remark 6.2.22 As a $\hat{\mathfrak{g}}$-module, $\mathrm{Ind}_{\mathfrak{g}}^{\hat{\mathfrak{g}}}(U)$ is generated by the $\mathfrak{g}$-submodule U and satisfies the relations $\mathbf{k} = \ell$ and $\hat{\mathfrak{g}}_- U = 0$, and furthermore, it is universal in the sense that for any $\hat{\mathfrak{g}}$-module W of level ℓ and any $\mathfrak{g}$-module map $\rho : U \to W$ such that $\hat{\mathfrak{g}}_- \rho(U) = 0$, there is a unique $\hat{\mathfrak{g}}$-module homomorphism from $\mathrm{Ind}_{\mathfrak{g}}^{\hat{\mathfrak{g}}}(U)$ to W extending the map ρ.

We now assume that U is a finite-dimensional irreducible $\mathfrak{g}$-module. As with the Virasoro algebra module $M_{Vir}(\ell, h)$ in Section 6.1, the sum, say N, of all the proper $\hat{\mathfrak{g}}$-submodules of $\mathrm{Ind}_{\mathfrak{g}}^{\hat{\mathfrak{g}}}(U)$ is also proper, so that we have a unique irreducible quotient $\hat{\mathfrak{g}}$-module of $\mathrm{Ind}_{\mathfrak{g}}^{\hat{\mathfrak{g}}}(U)$, which we denote by $L_{\hat{\mathfrak{g}}}(\ell, U)$:

$$L_{\hat{\mathfrak{g}}}(\ell, U) = \mathrm{Ind}_{\mathfrak{g}}^{\hat{\mathfrak{g}}}(U)/N. \tag{6.2.79}$$

In view of Theorem 6.2.13, $L_{\hat{\mathfrak{g}}}(\ell, U)$ is also an irreducible $V_{\hat{\mathfrak{g}}}(\ell, 0)$-module. This proves the first assertion of the following analogue of Theorem 6.1.12:

Theorem 6.2.23 *For any finite-dimensional irreducible $\mathfrak{g}$-module U, the irreducible $\hat{\mathfrak{g}}$-module $L_{\hat{\mathfrak{g}}}(\ell, U)$ is naturally an irreducible module for the vertex operator algebra $V_{\hat{\mathfrak{g}}}(\ell, 0)$. Furthermore, the modules $L_{\hat{\mathfrak{g}}}(\ell, U)$ for finite-dimensional irreducible $\mathfrak{g}$-modules U exhaust the irreducible $V_{\hat{\mathfrak{g}}}(\ell, 0)$-modules up to equivalence.*

Proof. This proof is similar to the proof of Theorem 6.1.12. Let W be an irreducible $V_{\hat{\mathfrak{g}}}(\ell, 0)$-module. Then $W = \coprod_{n \in \mathbb{N}} W_{(n+r)}$ for some complex number r, with $W_{(r)} \neq 0$ and $\dim W_{(n+r)} < \infty$ for $n \in \mathbb{N}$. From Theorem 6.2.13, W is an irreducible $\hat{\mathfrak{g}}$-module. In view of (6.2.45), each homogeneous subspace $W_{(n+r)}$ is a $\mathfrak{g}$-module, and in particular, $W_{(r)}$ is a $\mathfrak{g}$-module. Using the fact that $\hat{\mathfrak{g}}_- W_{(r)} = 0$ together with the Poincaré–Birkhoff–Witt theorem we see that any proper $\mathfrak{g}$-submodule of $W_{(r)}$ generates a proper $\hat{\mathfrak{g}}$-submodule of W. Thus $W_{(r)}$ must be a (finite-dimensional) irreducible $\mathfrak{g}$-module. Set $U = W_{(r)}$, a $\mathfrak{g}$-module. By Remark 6.2.22, there is a unique $\hat{\mathfrak{g}}$-module homomorphism ψ from $\mathrm{Ind}_{\mathfrak{g}}^{\hat{\mathfrak{g}}}(U)$ onto W such that $\psi|_U = 1$, and hence by Proposition 4.5.1 ψ is a $V_{\hat{\mathfrak{g}}}(\ell, 0)$-module homomorphism. It follows that ψ induces a $V_{\hat{\mathfrak{g}}}(\ell, 0)$-module isomorphism from $L_{\hat{\mathfrak{g}}}(\ell, U)$ onto W. This completes the proof. $\square$

Remark 6.2.24 In certain cases, $V_{\hat{\mathfrak{g}}}(\ell, 0)$ is a simple vertex operator algebra. (One example is the case in which $\mathfrak{g}$ is abelian and ℓ is nonzero; see Section 6.3.) But $V_{\hat{\mathfrak{g}}}(\ell, 0)$ is not simple in general. Since $V_{\hat{\mathfrak{g}}}(\ell, 0) = \coprod_{n \geq 0} V_{\hat{\mathfrak{g}}}(\ell, 0)_{(n)}$ with $V_{\hat{\mathfrak{g}}}(\ell, 0)_{(0)} = \mathbb{C}\mathbf{1}$, from Remark 3.9.11 the sum $I_{\hat{\mathfrak{g}}}(\ell, 0)$ of all the proper ideals of $V_{\hat{\mathfrak{g}}}(\ell, 0)$ is the (unique) largest proper ideal, so that $V_{\hat{\mathfrak{g}}}(\ell, 0)/I_{\hat{\mathfrak{g}}}(\ell, 0)$ is simple vertex operator algebra, which we denote by $L_{\hat{\mathfrak{g}}}(\ell, 0)$:

$$L_{\hat{\mathfrak{g}}}(\ell, 0) = V_{\hat{\mathfrak{g}}}(\ell, 0)/I_{\hat{\mathfrak{g}}}(\ell, 0). \tag{6.2.80}$$

In view of Theorem 6.2.13, $I_{\hat{\mathfrak{g}}}(\ell, 0)$ is the unique largest proper $\hat{\mathfrak{g}}$-submodule of $V_{\hat{\mathfrak{g}}}(\ell, 0)$, and $L_{\hat{\mathfrak{g}}}(\ell, 0)$ is an irreducible $\hat{\mathfrak{g}}$-module.

Remark 6.2.25 Concerning our notation, note that

$$L_{\hat{\mathfrak{g}}}(\ell, 0) = L_{\hat{\mathfrak{g}}}(\ell, \mathbb{C}), \tag{6.2.81}$$

where $U = \mathbb{C}$ is taken to be the trivial one-dimensional $\mathfrak{g}$-module. We trust that these two notations for $L_{\hat{\mathfrak{g}}}(\ell, 0)$ will not cause any confusion.

Remark 6.2.26 It is important and interesting to classify the irreducible modules for the simple vertex operator algebra $L_{\hat{\mathfrak{g}}}(\ell, 0)$. If $L_{\hat{\mathfrak{g}}}(\ell, 0) = V_{\hat{\mathfrak{g}}}(\ell, 0)$, that is, $I_{\hat{\mathfrak{g}}}(\ell, 0) = 0$, then by Theorem 6.2.23 all the irreducible $L_{\hat{\mathfrak{g}}}(\ell, 0)$-modules are classified. In general, by Theorem 6.2.23, any irreducible $L_{\hat{\mathfrak{g}}}(\ell, 0)$-module, which is naturally a $V_{\hat{\mathfrak{g}}}(\ell, 0)$-module, must be isomorphic to $L_{\hat{\mathfrak{g}}}(\ell, U)$ for some (finite-dimensional) irreducible $\mathfrak{g}$-module U. However, as we shall see in Section 6.6, for certain important and interesting cases, *only for certain finitely many* irreducible $\mathfrak{g}$-modules U (up to equivalence) is it true that $L_{\hat{\mathfrak{g}}}(\ell, U)$ carries the structure of an (irreducible) $L_{\hat{\mathfrak{g}}}(\ell, 0)$-module.

6.3 Vertex operator algebras and modules associated to Heisenberg algebras

Here we shall continue with the discussion in Section 6.2 to study the vertex operator algebras $V_{\hat{\mathfrak{g}}}(\ell, 0)$ with $\mathfrak{g}$ taken to be an abelian Lie algebra, which we shall now call $\mathfrak{h}$. In this section, we prove that the vertex operator algebras $V_{\hat{\mathfrak{h}}}(\ell, 0)$ with $\ell \neq 0$ are simple, we classify their irreducible modules and we show that all the vertex operator algebras $V_{\hat{\mathfrak{h}}}(\ell, 0)$ are isomorphic for $\ell \neq 0$.

In this special case, the subalgebra $\mathfrak{h} = \mathfrak{h} \otimes t^0$ of the affine Lie algebra $\hat{\mathfrak{h}}$ is central and $\hat{\mathfrak{h}}$ is the direct sum of $\mathfrak{h}$ and the ideal $\mathfrak{h} \otimes t\mathbb{C}[t] \oplus \mathfrak{h} \otimes t^{-1}\mathbb{C}[t^{-1}] \oplus \mathbb{C}\mathbf{k}$, which we shall denote by $\hat{\mathfrak{h}}_*$ and which is a "Heisenberg (Lie) algebra" in the precise sense defined below. Furthermore, $\mathfrak{h}$ acts trivially on the vertex operator algebra $V_{\hat{\mathfrak{h}}}(\ell, 0)$ and the quotient Lie algebra $\hat{\mathfrak{h}}/\mathfrak{h}$ is naturally isomorphic to the Heisenberg subalgebra $\hat{\mathfrak{h}}_*$. In view of this, the vertex operator algebra $V_{\hat{\mathfrak{h}}}(\ell, 0)$ is often called "the vertex operator algebra associated to the Heisenberg algebra $\hat{\mathfrak{h}}_*$ of level ℓ."

In Sections 6.4 and 6.5 we shall be studying vertex operator algebras and modules associated with lattices, where Heisenberg algebras will play an important role (see [FLM6]). This section also serves as a preparation for that study. For this purpose we also review certain exponential operators and their basic properties, in [LW1], [FK], [Se1], [FLM6].

Let $\mathfrak{h}$ be a finite-dimensional vector space equipped with a nondegenerate symmetric bilinear form $\langle \cdot, \cdot \rangle$. Consider $\mathfrak{h}$ as an abelian Lie algebra with $\langle \cdot, \cdot \rangle$ as an invariant symmetric bilinear form. Specializing the considerations in Section 6.2, we have the affine Lie algebra

$$\hat{\mathfrak{h}} = \mathfrak{h} \otimes \mathbb{C}[t, t^{-1}] \oplus \mathbb{C}\mathbf{k}, \tag{6.3.1}$$

with the Lie bracket relations

$$[\mathbf{k}, \hat{\mathfrak{h}}] = 0 \tag{6.3.2}$$

$$[\alpha \otimes t^m, \beta \otimes t^n] = \langle \alpha, \beta \rangle m \delta_{m+n,0} \mathbf{k} \tag{6.3.3}$$

for $\alpha, \beta \in \mathfrak{h}$ and $m, n \in \mathbb{Z}$. The affine Lie algebra $\hat{\mathfrak{h}}$ is a $\mathbb{Z}$-graded Lie algebra with

$$\hat{\mathfrak{h}} = \coprod_{n \in \mathbb{Z}} \hat{\mathfrak{h}}_{(n)},$$

where

$$\hat{\mathfrak{h}}_{(0)} = \mathfrak{h} \oplus \mathbb{C}\mathbf{k} \quad \text{and} \quad \hat{\mathfrak{h}}_{(n)} = \mathfrak{h} \otimes t^{-n} \quad \text{for } n \neq 0.$$

It has graded subalgebras

$$\hat{\mathfrak{h}}_+ = \mathfrak{h} \otimes t^{-1}\mathbb{C}[t^{-1}] \quad \text{and} \quad \hat{\mathfrak{h}}_- = \mathfrak{h} \otimes t\mathbb{C}[t]. \tag{6.3.4}$$

Remark 6.3.1 In the literature (cf. [FLM6]), the following notation is often used:

$$\hat{\mathfrak{h}}^+ = \mathfrak{h} \otimes t\mathbb{C}[t] \left(= \hat{\mathfrak{h}}_-\right) \quad \text{and} \quad \hat{\mathfrak{h}}^- = \mathfrak{h} \otimes t^{-1}\mathbb{C}[t^{-1}] \left(= \hat{\mathfrak{h}}_+\right), \tag{6.3.5}$$

corresponding to the $\mathbb{Z}$-grading on $\hat{\mathfrak{h}}$ given by degree rather than weight (recall Remark 6.2.1).

Set

$$\hat{\mathfrak{h}}_* = \hat{\mathfrak{h}}_- \oplus \hat{\mathfrak{h}}_+ \oplus \mathbb{C}\mathbf{k} = \coprod_{n \neq 0} (\mathfrak{h} \otimes t^n) \oplus \mathbb{C}\mathbf{k}, \tag{6.3.6}$$

a graded subalgebra of $\hat{\mathfrak{h}}$. Then $\hat{\mathfrak{h}}_*$ is a Heisenberg (Lie) algebra in the precise sense that its commutator subalgebra coincides with its center, which is one-dimensional. Observe that $\mathfrak{h}$ and $\hat{\mathfrak{h}}_*$ are ideals of $\hat{\mathfrak{h}}$ and that

$$\hat{\mathfrak{h}} = \hat{\mathfrak{h}}_* \oplus \mathfrak{h}. \tag{6.3.7}$$

We shall continue to use the notational conventions for module actions given in Remark 6.2.2, so that for an $\hat{\mathfrak{h}}$-module W and $\alpha \in \mathfrak{h}$, $n \in \mathbb{Z}$, we write $\alpha(n)$ for the action of $\alpha \otimes t^n$ on W and write

$$\alpha_W(x) = \alpha(x) = \sum_{n \in \mathbb{Z}} \alpha(n) x^{-n-1}. \tag{6.3.8}$$

Noticing that the Casimir operator Ω associated to $\mathfrak{h}$ and $\langle \cdot, \cdot \rangle$ (recall (6.2.37)) acts trivially on $\mathfrak{h}$ so that $h = 0$ (recall (6.2.40) and (6.2.41)), as a specialization of Theorem 6.2.18 we have:

Theorem 6.3.2 *Let ℓ be any nonzero complex number. Then the vertex algebra $V_{\hat{\mathfrak{h}}}(\ell, 0)$ constructed in Theorem 6.2.11 is a vertex operator algebra of central charge $d = \dim \mathfrak{h}$ with the conformal vector given by*

$$\omega = \frac{1}{2\ell} \sum_{i=1}^{d} u^{(i)}(-1) u^{(i)}(-1) \mathbf{1}, \tag{6.3.9}$$

where $\{u^{(1)}, \ldots, u^{(d)}\}$ is any orthonormal basis of $\mathfrak{h}$. The weight grading on $V_{\hat{\mathfrak{h}}}(\ell, 0)$ is given by $L(0)$-eigenvalues, $\mathfrak{h} = V_{\hat{\mathfrak{h}}}(\ell, 0)_{(1)}$, which generates $V_{\hat{\mathfrak{h}}}(\ell, 0)$ as a vertex algebra, and

$$[L(m), a(n)] = -na(m + n) \quad for\ a \in \mathfrak{h},\ m, n \in \mathbb{Z}. \quad \square \tag{6.3.10}$$

Since $\hat{\mathfrak{h}}_+$ is abelian, as a graded vector space,

$$V_{\hat{\mathfrak{h}}}(\ell, 0) = U(\hat{\mathfrak{h}}_+) = S(\hat{\mathfrak{h}}_+); \tag{6.3.11}$$

that is, $V_{\hat{\mathfrak{h}}}(\ell, 0)$ can be identified with the symmetric algebra $S(\hat{\mathfrak{h}}_+)$. The vertex operator algebra $V_{\hat{\mathfrak{h}}}(\ell, 0)$ is commonly referred to as *the vertex operator algebra associated to the Heisenberg algebra $\hat{\mathfrak{h}}_*$ of level ℓ.*

The form $\langle \cdot, \cdot \rangle$ being nondegenerate, we may and do identify $\mathfrak{h}$ with its dual space $\mathfrak{h}^*$. Let $\alpha \in \mathfrak{h} = \mathfrak{h}^*$. Denote by $\mathbb{C}_\alpha$ the one-dimensional $\mathfrak{h}$-module with $h \in \mathfrak{h}$ acting as the scalar $\langle h, \alpha \rangle$. Then the Casimir operator Ω acts on $\mathbb{C}_\alpha$ as the scalar $\langle \alpha, \alpha \rangle$:

$$\Omega = \langle \alpha, \alpha \rangle \quad \text{on } \mathbb{C}_\alpha, \tag{6.3.12}$$

since

$$\Omega \cdot 1 = \sum_{i=1}^{d} u^{(i)} u^{(i)} 1 = \sum_{i=1}^{d} \langle u^{(i)}, \alpha \rangle^2 1 = \langle \alpha, \alpha \rangle 1. \tag{6.3.13}$$

Set

$$M(\ell, \alpha) = \mathrm{Ind}_{\hat{\mathfrak{h}}}^{\hat{\mathfrak{h}}}(\mathbb{C}_\alpha) = U(\hat{\mathfrak{h}}_+) \otimes_{\mathbb{C}} \mathbb{C}_\alpha = S(\hat{\mathfrak{h}}_+) \otimes_{\mathbb{C}} \mathbb{C}_\alpha, \tag{6.3.14}$$

an $\hat{\mathfrak{h}}$-module of level ℓ. Taking $\alpha = 0$ we have in particular that $M(\ell, 0) = V_{\hat{\mathfrak{h}}}(\ell, 0)$. Following [FLM6] we alternatively set

$$M(\ell) = M(\ell, 0) = V_{\hat{\mathfrak{h}}}(\ell, 0), \tag{6.3.15}$$

as an $\hat{\mathfrak{h}}$-module. As a specialization of Theorem 6.2.21 we have:

Theorem 6.3.3 *For any $\alpha \in \mathfrak{h} = \mathfrak{h}^*$, the $\hat{\mathfrak{h}}$-module $M(\ell, \alpha)$ is naturally a module for the vertex operator algebra $V_{\hat{\mathfrak{h}}}(\ell, 0)$, with $Y_W(a, x) = a_W(x)$ for $a \in \mathfrak{h}$, and*

$$M(\ell, \alpha) = \coprod_{n \in \mathbb{N}} M(\ell, \alpha)_{(n + \frac{1}{2\ell}\langle \alpha, \alpha \rangle)}, \tag{6.3.16}$$

with $M(\ell, \alpha)_{(\frac{1}{2\ell}\langle \alpha, \alpha \rangle)} = \mathbb{C}\alpha$. $\square$

In the following we shall show that for any nonzero complex number ℓ, the vertex operator algebra $V_{\hat{\mathfrak{h}}}(\ell, 0)$ is simple and that for any $\alpha \in \mathfrak{h}$, the $V_{\hat{\mathfrak{h}}}(\ell, 0)$-module $M(\ell, \alpha)$ is irreducible.

First we present the following well-known canonical realization of the affine Lie algebra $\hat{\mathfrak{h}}$ (and the Heisenberg algebra $\hat{\mathfrak{h}}_*$):

Proposition 6.3.4 *Let ℓ be any nonzero complex number and let $\alpha \in \mathfrak{h} = \mathfrak{h}^*$. Let $\{u^{(1)}, \ldots, u^{(d)}\}$ be an orthonormal basis of $\mathfrak{h}$. Set*

$$P(\ell, \alpha) = \mathbb{C}[x_{ij} \mid i = 1, \ldots, d, \; j = 1, 2, \ldots], \tag{6.3.17}$$

as a vector space, where x_{ij} are mutually commuting independent formal variables. Let $\hat{\mathfrak{h}}$ act on $P(\ell, \alpha)$ by

$$\mathbf{k} \mapsto \ell, \tag{6.3.18}$$

$$u^{(i)}(0) = \langle u^{(i)}, \alpha \rangle, \tag{6.3.19}$$

$$u^{(i)}(n) = n\ell \frac{\partial}{\partial x_{in}}, \tag{6.3.20}$$

$$u^{(i)}(-n) = x_{in} \quad (\text{the multiplication operator}) \tag{6.3.21}$$

for $i = 1, \ldots, d$ and $n = 1, 2, \ldots$. Then $P(\ell, \alpha)$ is an irreducible $\hat{\mathfrak{h}}$-module and an irreducible $\hat{\mathfrak{h}}_*$-module.

Proof. It is routine to check that (6.3.18)–(6.3.21) define an $\hat{\mathfrak{h}}$-module action on $P(\ell, \alpha)$. As an $\hat{\mathfrak{h}}_*$-module, $P(\ell, \alpha)$ is generated by 1. Let U be any nonzero $\hat{\mathfrak{h}}_*$-submodule. Let f be a nonzero polynomial in U of the least possible total degree (where each variable is assigned degree one); f is not necessarily homogeneous. Since $n\ell \frac{\partial}{\partial x_{in}} f = u^{(i)}(n)f \in U$ we must have $\frac{\partial}{\partial x_{in}} f = 0$ for all i and all $n \geq 1$, so that $f \in \mathbb{C}$. Since 1 generates the whole module, U must be the whole module, and so $P(\ell, \alpha)$ is an irreducible $\hat{\mathfrak{h}}_*$-module and hence an irreducible $\hat{\mathfrak{h}}$-module. $\square$

Definition 6.3.5 A nonzero vector w in an $\hat{\mathfrak{h}}_*$-module W is called a *vacuum vector* if $\hat{\mathfrak{h}}_- w = 0$. The *vacuum space* of W, denoted by Ω_W, is the subspace consisting of its vacuum vectors and 0.

Remark 6.3.6 In the context of Proposition 6.3.4, the multiplication operators $u^{(i)}(-n) = x_{in}$ are called "creation operators" in that they "create" vectors from the vacuum space $\Omega_{P(\ell,\alpha)}$, which is one-dimensional and consists of the scalars, and the operators $\frac{\partial}{\partial x_{in}}$, or their rescalings $u^{(i)}(n)$, are called "annihilation operators." This terminology of annihilation and creation operators, which comes from quantum mechanics and quantum field theory, is naturally extended to all $\hat{\mathfrak{h}}$-modules W generated by vacuum vectors (see below); for $h \in \mathfrak{h}$ and $n > 0$, the operators $h(-n)$ on W are typically called *creation operators* and the operators $h(n)$, *annihilation operators*.

Remark 6.3.7 Again in the context of Proposition 6.3.4, we observe that any operator on $P(\ell, \alpha)$ that commutes with the action of the Heisenberg algebra $\hat{\mathfrak{h}}_*$ must be a scalar multiplication operator. Indeed, such an operator must preserve the vacuum space $\Omega_{P(\ell,\alpha)}$, which is one-dimensional, and so must act as a scalar on it; thus the operator must act as a scalar on all of $P(\ell, \alpha)$. This is of course consistent with the fact that for $h \in \mathfrak{h}$, the operator $h(0)$ acts as a scalar on $P(\ell, \alpha)$ (namely, the scalar $\langle h, \alpha \rangle$).

The following well-known result expresses the "uniqueness of the Heisenberg commutation relations" in our present context:

Proposition 6.3.8 *For any $\alpha \in \mathfrak{h}$, $M(\ell, \alpha)$ is an irreducible $\hat{\mathfrak{h}}$-module and an irreducible $\hat{\mathfrak{h}}_*$-module. Moreover, for any $\hat{\mathfrak{h}}$-module of level ℓ, its submodule generated by any vacuum vector w satisfying the relations $hw = \langle h, \alpha \rangle w$ for $h \in \mathfrak{h}$ is equivalent to $M(\ell, \alpha)$. For any $\hat{\mathfrak{h}}_*$-module of level ℓ, its submodule generated by any vacuum vector is equivalent to $M(\ell, 0)$ viewed as an $\hat{\mathfrak{h}}_*$-module. In particular, $M(\ell, 0)$ is the unique (up to equivalence) irreducible $\hat{\mathfrak{h}}_*$-module that contains a vacuum vector.*

Proof. From the construction of the $\hat{\mathfrak{h}}$-module $P(\ell, \alpha)$ in Proposition 6.3.4, $P(\ell, \alpha)$ is generated by 1, and we have the relations $\mathbf{k} = \ell$, $\hat{\mathfrak{h}}_- 1 = 0$ and $h \cdot 1 = \langle h, \alpha \rangle 1$ for $h \in \mathfrak{h}$, so that the $\mathfrak{h}$-submodule $\mathbb{C}$ of $P(\ell, \alpha)$ is equivalent to $\mathbb{C}_\alpha$. By Remark 6.2.22, there is a (unique) $\hat{\mathfrak{h}}$-module homomorphism ψ from $M(\ell, \alpha)$ to $P(\ell, \alpha)$ such that $\psi(1) = 1$. Then

$$\psi\left(u^{(i_1)}(-n_1) \cdots u^{(i_r)}(-n_r)1\right) = x_{i_1 n_1} \cdots x_{i_r n_r} \tag{6.3.22}$$

for $r \geq 0$, $1 \leq i_1, \ldots, i_r \leq d$ and $n_i \geq 1$. From (6.3.14) ψ is a linear isomorphism, so that ψ is an $\hat{\mathfrak{h}}$-module isomorphism. In view of Proposition 6.3.4, $M(\ell, \alpha)$ is an irreducible $\hat{\mathfrak{h}}$-module and an irreducible $\hat{\mathfrak{h}}_*$-module.

Let W be an $\hat{\mathfrak{h}}$-module of level ℓ generated by a vacuum vector w such that $hw = \langle h, \alpha \rangle w$ for $h \in \mathfrak{h}$. Then $\mathbb{C}w = \mathbb{C}_\alpha$ as an $\mathfrak{h}$-module. From Remark 6.2.22, there is a (unique) $\hat{\mathfrak{h}}$-module homomorphism ϕ from $M(\ell, \alpha)$ to W such that $\phi(1) = w$, and since $M(\ell, \alpha)$ is irreducible, ϕ must be an isomorphism.

If W is an $\hat{\mathfrak{h}}_*$-module generated by a vacuum vector w, we let $\mathfrak{h}$ act trivially on W, making W an $\hat{\mathfrak{h}}$-module (recall (6.3.7)). By the result just proved, there is an $\hat{\mathfrak{h}}$-module isomorphism ϕ from $M(\ell, 0)$ onto W such that $\phi(1) = w$. Of course, ϕ is also an $\hat{\mathfrak{h}}_*$-module isomorphism, completing the proof. $\square$

As an application of Proposition 6.3.8 we have:

Theorem 6.3.9 *Let ℓ be any nonzero complex number. Then the vertex operator algebra $V_{\hat{\mathfrak{h}}}(\ell, 0)$ is simple and for every $\alpha \in \mathfrak{h}$, $M(\ell, \alpha)$ is naturally an irreducible $V_{\hat{\mathfrak{h}}}(\ell, 0)$-module with lowest weight $\frac{1}{2\ell}\langle \alpha, \alpha \rangle$. Furthermore, the modules $M(\ell, \alpha)$ for $\alpha \in \mathfrak{h}$ exhaust the irreducible $V_{\hat{\mathfrak{h}}}(\ell, 0)$-modules up to equivalence.*

Proof. From Proposition 6.3.8 and Theorem 6.2.13, $M(\ell, \alpha)$ is naturally an irreducible $V_{\hat{\mathfrak{h}}}(\ell, 0)$-module. In particular, $V_{\hat{\mathfrak{h}}}(\ell, 0) = M(\ell, 0)$ is an irreducible $V_{\hat{\mathfrak{h}}}(\ell, 0)$-module, and hence from Remark 4.5.4 $V_{\hat{\mathfrak{h}}}(\ell, 0)$ is a simple vertex operator algebra. By Theorem 6.3.3, the lowest $L(0)$-weight of $M(\ell, \alpha)$ is $\frac{1}{2\ell}\langle \alpha, \alpha \rangle$.

Since $\mathfrak{h}$ is abelian, any finite-dimensional irreducible $\mathfrak{h}$-module U is one-dimensional, and each $h \in \mathfrak{h}$ acts as a scalar, so that U is equivalent to $\mathbb{C}_\alpha$ for some $\alpha \in \mathfrak{h}$. Since $M(\ell, \alpha)$ is irreducible, $M(\ell, \alpha) = \text{Ind}_{\mathfrak{h}}^{\hat{\mathfrak{h}}} \mathbb{C}_\alpha = L_{\hat{\mathfrak{h}}}(\ell, \mathbb{C}_\alpha)$ (recall Theorem 6.2.23 for the notation). The last assertion follows immediately from Theorem 6.2.23. $\square$

Since the central charge of the vertex operator algebra $V_{\hat{\mathfrak{h}}}(\ell, 0)$, namely, $d = \dim \mathfrak{h}$ (recall Theorem 6.3.2), is independent of ℓ, it is natural to expect that all the vertex operator algebras $V_{\hat{\mathfrak{h}}}(\ell, 0)$ for nonzero complex numbers ℓ are isomorphic. This is indeed true, as we now show:

Proposition 6.3.10 *Let ℓ be any nonzero complex number and let $\sqrt{\ell}$ be one of the two square roots of ℓ. Then the (well-defined) linear map*

$$\psi : V_{\hat{\mathfrak{h}}}(\ell, 0) \to V_{\hat{\mathfrak{h}}}(1, 0)$$
$$\alpha^{(1)}(-n_1) \cdots \alpha^{(r)}(-n_r)\mathbf{1} \mapsto (\sqrt{\ell})^r \alpha^{(1)}(-n_1) \cdots \alpha^{(r)}(-n_r)\mathbf{1} \qquad (6.3.23)$$

for $r \geq 0$, $\alpha^{(i)} \in \mathfrak{h}$, $n_i \geq 1$ is a vertex operator algebra isomorphism.

Proof. Since $V_{\hat{\mathfrak{h}}}(\lambda, 0) = S(\hat{\mathfrak{h}}_+)$ as a vector space for any nonzero number λ, ψ is well defined and is a linear isomorphism. By definition we have $\psi(\mathbf{1}) = \mathbf{1}$ and

$$\psi(\alpha) = \psi(\alpha(-1)\mathbf{1}) = \sqrt{\ell}\alpha(-1)\mathbf{1} = \sqrt{\ell}\alpha \qquad (6.3.24)$$

for $\alpha \in \mathfrak{h} \subset V_{\hat{\mathfrak{h}}}(\ell, 0)$. We also have

$$\psi(\omega) = \psi\left(\frac{1}{2\ell} \sum_{i=1}^{d} u^{(i)}(-1)u^{(i)}(-1)\mathbf{1}\right)$$
$$= \frac{1}{2} \sum_{i=1}^{d} u^{(i)}(-1)u^{(i)}(-1)\mathbf{1} = \omega \ (\in V_{\hat{\mathfrak{h}}}(1, 0)). \qquad (6.3.25)$$

What we must prove is that ψ is a vertex algebra homomorphism. Since $\mathfrak{h}$ generates $V_{\hat{\mathfrak{h}}}(\ell, 0)$ as a vertex algebra, in view of Proposition 5.7.9 it suffices to show that

$$\psi(h(n)u) = \sqrt{\ell}\,h(n)\psi(u) \quad \text{for } h \in \mathfrak{h},\ n \in \mathbb{Z},\ u \in V_{\hat{\mathfrak{h}}}(\ell, 0). \qquad (6.3.26)$$

From the definition of ψ, (6.3.26) holds if $n < 0$, and it also holds if $n = 0$ because $h(0)$ acts trivially on both $V_{\hat{\mathfrak{h}}}(\ell, 0)$ and $V_{\hat{\mathfrak{h}}}(1, 0)$. Assume $n > 0$. Let $\alpha \in \mathfrak{h}$, $k \geq 1$, $v \in V_{\hat{\mathfrak{h}}}(\ell, 0)$. Using the commutator relations (6.3.3) and the property (6.3.26) for $n < 0$ we have

$$\begin{aligned}
\psi(h(n)\alpha(-k)v) &= n\ell\langle h, \alpha\rangle \delta_{n,k}\psi(v) + \psi(\alpha(-k)h(n)v) \\
&= n\ell\langle h, \alpha\rangle \delta_{n,k}\psi(v) + \sqrt{\ell}\,\alpha(-k)\psi(h(n)v), \qquad (6.3.27)
\end{aligned}$$

while

$$\begin{aligned}
\sqrt{\ell}\,h(n)\psi(\alpha(-k)v) &= \ell h(n)\alpha(-k)\psi(v) \\
&= n\ell\langle h, \alpha\rangle \delta_{n,k}\psi(v) + \ell\alpha(-k)h(n)\psi(v). \qquad (6.3.28)
\end{aligned}$$

Thus we see that if $\psi(h(n)v) = \sqrt{\ell}\,h(n)\psi(v)$, then

$$\psi(h(n)\alpha(-k)v) = \sqrt{\ell}\,h(n)\psi(\alpha(-k)v),$$

and it follows immediately (by induction) that (6.3.26) holds for $n > 0$, as desired. $\square$

Remark 6.3.11 Recall from Theorem 6.2.18 that the central charge of the vertex operator algebra $V_{\hat{\mathfrak{a}}}(\ell, 0)$ is $d\ell/(\ell + h)$. If $h \neq 0$, we see that for complex numbers ℓ_1, ℓ_2 not equal to $-h$, $d\ell_1/(\ell_1 + h) = d\ell_2/(\ell_2 + h)$ if and only if $\ell_1 = \ell_2$. Since isomorphic vertex operator algebras necessarily have the same central charge, if $h \neq 0$, the vertex operator algebras $V_{\hat{\mathfrak{a}}}(\ell_1, 0)$ and $V_{\hat{\mathfrak{a}}}(\ell_2, 0)$ are isomorphic if and only if $\ell_1 = \ell_2$.

We next discuss certain exponential operators and some basic properties, in preparation for the construction of the vertex algebras associated to even lattices in the next section.

For the rest of this section, we assume that W *is an $\hat{\mathfrak{h}}$-module of level $\ell = 1$*. The reason for the assumption $\ell = 1$ is that it will be natural to use $\ell = 1$ in the next section.

For $\alpha \in \mathfrak{h}$, following [FLM6], using the formal exponential series $\exp(\cdot)$ we define

$$E^{\pm}(\alpha, x) = \exp\left(\sum_{n \in \pm \mathbb{Z}_+} \frac{\alpha(n)}{n} x^{-n}\right), \qquad (6.3.29)$$

viewed as elements of $(\text{End } W)[[x^{\mp 1}]]$. Note that the elements $E^{\pm}(\alpha, x)$ are indeed well defined, just as the element e^S in (2.2.12) is well defined.

The following three results were treated in [FLM6]:

Proposition 6.3.12 *On our level-1 $\hat{\mathfrak{h}}$-module W we have*

$$E^{\pm}(0, x) = 1 \tag{6.3.30}$$

$$E^{\pm}(\alpha + \beta, x) = E^{\pm}(\alpha, x) E^{\pm}(\beta, x) \tag{6.3.31}$$

for $\alpha, \beta \in \mathfrak{h}$. For $h, \alpha \in \mathfrak{h}$ and $m \in \mathbb{Z}$,

$$[h(m), E^{+}(\alpha, x)] = 0 \qquad\qquad\qquad if\, m \geq 0 \tag{6.3.32}$$

$$[h(m), E^{-}(\alpha, x)] = -\langle h, \alpha\rangle x^{m} E^{-}(\alpha, x) \quad if\, m > 0 \tag{6.3.33}$$

$$[h(m), E^{+}(\alpha, x)] = -\langle h, \alpha\rangle x^{m} E^{+}(\alpha, x) \quad if\, m < 0 \tag{6.3.34}$$

$$[h(m), E^{-}(\alpha, x)] = 0 \qquad\qquad\qquad if\, m \leq 0. \tag{6.3.35}$$

Proof. Property (6.3.30) is clear, and properties (6.3.31), (6.3.32) and (6.3.35) follow from the fact that the operators $\alpha(n)$ for $\alpha \in \mathfrak{h}$, $n \geq 0$ commute with one another, as do the operators $\alpha(n)$ for $\alpha \in \mathfrak{h}$, $n \leq 0$.

To prove (6.3.33) and (6.3.34), set

$$A_{\pm} = \sum_{n \in \pm \mathbb{Z}_{+}} \frac{\alpha(n)}{n} x^{-n}. \tag{6.3.36}$$

From the relations $[h(m), \alpha(n)] = m\langle h, \alpha\rangle \delta_{m+n,0}$ for $m, n \in \mathbb{Z}$, we have

$$[h(m), A_{-}] = -\langle h, \alpha\rangle x^{m} \quad if\, m > 0 \tag{6.3.37}$$

$$[h(m), A_{+}] = -\langle h, \alpha\rangle x^{m} \quad if\, m < 0. \tag{6.3.38}$$

Formulas (6.3.33) and (6.3.34) now follow from the formal principle

$$\Delta(e^{A}) = \Delta(A)e^{A} \tag{6.3.39}$$

for a derivation Δ such that $\Delta(A)$ commutes with A. $\square$

Proposition 6.3.13 *Suppose that our $\hat{\mathfrak{h}}$-module W is equipped with a linear operator d such that*

$$[d, \alpha(n)] = -n\alpha(n - 1) \quad for\, \alpha \in \mathfrak{h},\, n \in \mathbb{Z}. \tag{6.3.40}$$

Then for $\alpha \in \mathfrak{h}$,

$$d, E^{+}(\alpha, x)] = -\left(\sum_{n \geq 0} \alpha(n) x^{-n-1}\right) E^{+}(\alpha, x)$$

$$= \left(-\alpha(0) x^{-1} + \frac{d}{dx}\right) E^{+}(\alpha, x) \tag{6.3.41}$$

$$[d, E^{-}(\alpha, x)] = -\left(\sum_{n \leq -2} \alpha(n) x^{-n-1}\right) E^{-}(\alpha, x)$$

$$= \left(\alpha(-1) + \frac{d}{dx}\right) E^{-}(\alpha, x). \tag{6.3.42}$$

Proof. With $A_{\pm}$ as defined as in (6.3.36), we have

$$[d, A_{\pm}] = - \sum_{n \in \pm \mathbb{Z}_+} \alpha(n-1)x^{-n}, \tag{6.3.43}$$

which commutes with $A_{\pm}$, and all the desired equalities follow from the principle (6.3.39). $\square$

Proposition 6.3.14 *For* $\alpha, \beta \in \mathfrak{h}$,

$$E^+(\alpha, x_1)E^-(\beta, x_2) = \left(1 - \frac{x_2}{x_1}\right)^{\langle \alpha, \beta \rangle} E^-(\beta, x_2)E^+(\alpha, x_1) \tag{6.3.44}$$

in $(\text{End } W)[[x_1^{-1}, x_2]]$.

Proof. We have

$$\left[\sum_{m \in \mathbb{Z}_+} \frac{\alpha(m)x_1^{-m}}{m}, \sum_{n \in -\mathbb{Z}_+} \frac{\beta(n)x_2^{-n}}{n} \right]$$
$$= -\langle \alpha, \beta \rangle \sum_{m \in \mathbb{Z}_+} \frac{1}{m}\left(\frac{x_2}{x_1}\right)^m$$
$$= \langle \alpha, \beta \rangle \log\left(1 - \frac{x_2}{x_1}\right). \tag{6.3.45}$$

The result now follows from the formal rule

$$e^A e^B = e^B e^A e^{[A,B]} \quad \text{if } [A, B] \text{ commutes with } A \text{ and } B. \tag{6.3.46}$$

This rule is easily established by the following formal argument:

$$Ae^B = e^B A + [A, e^B] = e^B A + e^B[A, B] = e^B(A + [A, B]).$$

Iterating, we obtain

$$A^k e^B = e^B(A + [A, B])^k \quad \text{for } k \geq 0.$$

Now divide by $k!$ and sum over k to get

$$e^A e^B = e^B e^{A+[A,B]} = e^B e^A e^{[A,B]}. \quad \square$$

Remark 6.3.15 There is a useful complete reducibility theorem for $\hat{\mathfrak{h}}_*$-modules satisfying natural hypotheses; see [LW3] (and [FLM6], Theorem 1.7.3), [K6]. Correspondingly, one can formulate a complete reducibility theorem for suitable $V_{\hat{\mathfrak{h}}}(\ell, 0)$-modules.

6.4 Vertex operator algebras and modules associated to even lattices—the setting

In this section and the next we shall apply Theorems 5.7.1 and 5.7.6 to construct a family of vertex (operator) algebras V_{L_0} associated to even lattices L_0, along with their irreducible modules, and we classify the irreducible V_{L_0}-modules. The existence of vertex algebra structure on V_{L_0} was announced in [B1] and the vertex algebras V_{L_0}, along with a natural set of irreducible modules, were explicitly constructed in [FLM6], Chapter 8; the proof of the vertex algebra properties in [FLM6] is different from the proof here. It was proved in [D2] (see also [DLM4]) that the irreducible V_{L_0}-modules constructed in [FLM6] in fact exhaust the irreducible V_{L_0}-modules up to equivalence. To apply Theorem 5.7.6 we introduce a certain associative algebra $A(L_0)$ associated with an even lattice L_0 and classify the irreducible $A(L_0)$-modules of a certain type. The associative algebra $A(L_0)$ replaces, and can be thought of as analogous to, the Lie algebras used in Sections 6.1–6.3. The idea of introducing and exploiting algebras like $A(L_0)$ is expected to be useful in studying the modules for still more general classes of vertex algebras. In the present section we lay out the setting and the basic constructions, and in the next section formulate and prove the main theorems about these structures.

We closely follow [FLM6], Section 7.1 and Chapter 8, in setting up the basic constructions. Let L be a *nondegenerate (rational) lattice of rank* $d \in \mathbb{N}$, in the sense that L is a free abelian group of rank d provided with a nondegenerate rational-valued symmetric $\mathbb{Z}$-bilinear form $\langle \cdot, \cdot \rangle$. The nondegenerateness property is the condition

$$\langle \alpha, L \rangle = 0 \quad \text{implies } \alpha = 0. \tag{6.4.1}$$

Form the vector space

$$\mathfrak{h} = L \otimes_{\mathbb{Z}} \mathbb{C}, \tag{6.4.2}$$

so that

$$\dim \mathfrak{h} = d, \tag{6.4.3}$$

and $\mathfrak{h}$ is naturally equipped with a nondegenerate symmetric bilinear form $\langle \cdot, \cdot \rangle$, the natural extension of the form on L. We identify L with $L \otimes 1 \subset \mathfrak{h}$.

As in Section 6.3, we view $\mathfrak{h}$ as an abelian Lie algebra and form the affine Lie algebra

$$\hat{\mathfrak{h}} = \coprod_{n \in \mathbb{Z}} \mathfrak{h} \otimes t^n \oplus \mathbb{C}\mathbf{k} \tag{6.4.4}$$

and its Heisenberg subalgebra

$$\hat{\mathfrak{h}}_* = \coprod_{n \neq 0} \mathfrak{h} \otimes t^n \oplus \mathbb{C}\mathbf{k} = \hat{\mathfrak{h}}_+ \oplus \hat{\mathfrak{h}}_- \oplus \mathbb{C}\mathbf{k}, \tag{6.4.5}$$

where

$$\hat{\mathfrak{h}}_- = \mathfrak{h} \otimes t\mathbb{C}[t] \quad \text{and} \quad \hat{\mathfrak{h}}_+ = \mathfrak{h} \otimes t^{-1}\mathbb{C}[t^{-1}]. \tag{6.4.6}$$

In the setting of Section 6.3, *we take the level ℓ equal to* 1. We have the $\mathbb{Z}$-graded $\hat{\mathfrak{h}}_*$-irreducible $\hat{\mathfrak{h}}$-module

$$M(1) = M(1,0) = V_{\hat{\mathfrak{h}}}(1,0) = S(\hat{\mathfrak{h}}_+). \tag{6.4.7}$$

As in Section 6.3, we use the notation $\alpha(n)$ for the action of $\alpha \otimes t^n$ on $M(1)$, for $\alpha \in \mathfrak{h}$ and $n \in \mathbb{Z}$, so that

$$[\alpha(m), \beta(n)] = \langle \alpha, \beta \rangle m \delta_{m+n,0} \tag{6.4.8}$$

for $\alpha, \beta \in \mathfrak{h}$ and $m, n \in \mathbb{Z}$. The module $M(1)$ is endowed with a natural vertex operator algebra structure $(M(1), Y, \mathbf{1}, \omega)$ of central charge d ($= \operatorname{rank} L$), where

$$\omega = \frac{1}{2} \sum_{i=1}^{d} u^{(i)}(-1)u^{(i)}(-1)\mathbf{1} \tag{6.4.9}$$

(recall Theorem 6.3.2). Roughly speaking, the vertex (operator) algebra that we shall construct extends the simple vertex operator algebra $M(1)$ by adjoining a certain set of (irreducible) $M(1)$-modules $M(1, \alpha)$ for $\alpha \in \mathfrak{h} = \mathfrak{h}^*$.

Remark 6.4.1 The fact that the central charge of the vertex operator algebra $M(1)$ is the rank of the lattice L and that this same number will be the central charge of the larger vertex operator algebra we shall construct, motivated the terminology "rank V" for the central charge of a vertex operator algebra V in [FLM6] and [FHL]; recall Remark 3.1.23.

Let $(\hat{L}, ^-)$ be a central extension of L by a finite cyclic group

$$\langle \kappa \rangle = \langle \kappa \mid \kappa^s = 1 \rangle \tag{6.4.10}$$

of order $s > 0$; that is, we have an exact sequence

$$1 \to \langle \kappa \rangle \to \hat{L} \xrightarrow{^-} L \to 1. \tag{6.4.11}$$

(We are using the notation $\to 1$ rather than $\to 0$ at the end of the exact sequence, even though the abelian group L is written additively.) Denote by

$$c_0 : L \times L \to \mathbb{Z}/s\mathbb{Z} \tag{6.4.12}$$

the *associated commutator map*, which is defined, and indeed well defined, by the condition

$$aba^{-1}b^{-1} = \kappa^{c_0(\bar{a},\bar{b})} \quad \text{for } a, b \in \hat{L}. \tag{6.4.13}$$

The central extension $(\hat{L}, ^-)$ is *equivalent* to a central extension (M, ϕ) of L by the same group $\langle \kappa \rangle$ if there is an isomorphism $\psi : \hat{L} \to M$ such that the diagram

$$1 \to \langle \kappa \rangle \to \quad \hat{L} \xrightarrow{\bar{\ }} L \to 1$$
$$\| \qquad \psi \downarrow \qquad \|$$
$$1 \to \langle \kappa \rangle \to \quad M \xrightarrow{\phi} L \to 1 \tag{6.4.14}$$

commutes. Clearly, equivalent central extensions of L by $\langle \kappa \rangle$ determine the same commutator map.

Remark 6.4.2 The commutator map c_0 is an alternating $\mathbb{Z}$-bilinear form ("alternating" means that $c_0(\alpha, \alpha) = 0$ for $\alpha \in L$) and it characterizes the central extension uniquely up to equivalence; moreover, every alternating $\mathbb{Z}$-bilinear map $L \times L \to \mathbb{Z}/s\mathbb{Z}$ arises in this way from a central extension of L by $\langle \kappa \rangle$. For these elementary facts, see [FLM6], Sections 5.1 and 5.2; this bijection between the set of alternating $\mathbb{Z}$-bilinear maps and the set of equivalence classes of central extensions (as described) is given in Proposition 5.2.3 of [FLM6]. Note from (6.4.13) that the alternating property of c_0 corresponds to the fact that the commutator of an element $a \in \hat{L}$ with itself is 1.

Let

$$e : L \to \hat{L}$$
$$\alpha \mapsto e_\alpha \tag{6.4.15}$$

be a section of $\hat{L}$, that is, a map e such that $\bar{\ } \circ e = 1$, such that

$$e_0 = 1, \tag{6.4.16}$$

and denote by

$$\epsilon_0 : L \times L \to \mathbb{Z}/s\mathbb{Z} \tag{6.4.17}$$

the *corresponding* 2-*cocycle*, which is defined by the condition

$$e_\alpha e_\beta = \kappa^{\epsilon_0(\alpha, \beta)} e_{\alpha + \beta} \quad \text{for } \alpha, \beta \in L. \tag{6.4.18}$$

Then

$$\epsilon_0(\alpha, \beta) + \epsilon_0(\alpha + \beta, \gamma) = \epsilon_0(\beta, \gamma) + \epsilon_0(\alpha, \beta + \gamma) \tag{6.4.19}$$
$$\epsilon_0(\alpha, \beta) - \epsilon_0(\beta, \alpha) = c_0(\alpha, \beta) \tag{6.4.20}$$
$$\epsilon_0(\alpha, 0) = \epsilon_0(0, \alpha) = 0 \tag{6.4.21}$$

for $\alpha, \beta, \gamma \in L$; again see [FLM6], Sections 5.1 and 5.2, for this elementary material. Formula (6.4.19) is the 2-cocycle condition; (6.4.20) expresses the compatibility of ϵ_0 with the commutator map, and (6.4.21), which corresponds to the condition (6.4.16), is the normalization condition for the 2-cocycle ϵ_0.

Let $\omega_s \in \mathbb{C}$ be a primitive s^{th} root of unity and define the faithful character

$$\chi : \langle \kappa \rangle \to \mathbb{C}^\times \tag{6.4.22}$$

(an injective homomorphism from the s-element cyclic group to $\mathbb{C}^\times$) by the condition

$$\chi(\kappa) = \omega_s. \tag{6.4.23}$$

Denote by $\mathbb{C}_\chi$ the one-dimensional space $\mathbb{C}$ viewed as a $\langle\kappa\rangle$-module on which $\langle\kappa\rangle$ acts according to χ:

$$\kappa \cdot 1 = \omega_s. \tag{6.4.24}$$

(The reason we want to make a distinction between the abstract s-element cyclic group and the "concrete" s-element cyclic group of sth roots of unity in $\mathbb{C}^\times$ is that there are situations where it is appropriate to vary the primitive s^{th} root of unity by which κ acts.) Denote by $\mathbb{C}\{L\}$ the induced $\hat{L}$-module

$$\mathbb{C}\{L\} = \mathrm{Ind}_{\langle\kappa\rangle}^{\hat{L}} \mathbb{C}_\chi = \mathbb{C}[\hat{L}] \otimes_{\mathbb{C}[\langle\kappa\rangle]} \mathbb{C}_\chi = \mathbb{C}[\hat{L}]/(\kappa - \omega_s)\mathbb{C}[\hat{L}]. \tag{6.4.25}$$

For $a \in \hat{L}$, set

$$\iota(a) = a \otimes 1 \in \mathbb{C}\{L\}. \tag{6.4.26}$$

These elements span $\mathbb{C}\{L\}$ and

$$a \cdot \iota(b) = \iota(ab) \tag{6.4.27}$$
$$\kappa \cdot \iota(b) = \iota(\kappa b) = \omega_s \iota(b) \tag{6.4.28}$$

for $a, b \in \hat{L}$. In particular, κ acts as multiplication by ω_s on $\mathbb{C}\{L\}$.

Remark 6.4.3 The map from $\hat{L}$ to $\mathbb{C}\{L\}$ given by $a \mapsto \iota(a)$ is an injection. The relation $\iota(\kappa b) = \omega_s \iota(b)$ in (6.4.28) gives rise to the only possible linear relations among the elements $\iota(b)$ for $b \in \hat{L}$, since the elements $\iota(e_\alpha)$ of $\mathbb{C}\{L\}$ (recall our section (6.4.15)) form a basis of $\mathbb{C}\{L\}$.

Define a map

$$c : L \times L \to \mathbb{C}^\times$$
$$(\alpha, \beta) \mapsto \omega_s^{c_0(\alpha,\beta)}. \tag{6.4.29}$$

Then as operators on $\mathbb{C}\{L\}$,

$$ab = c(\bar{a}, \bar{b})ba \quad \text{for } a, b \in \hat{L}. \tag{6.4.30}$$

The choice of section (recall (6.4.15)) allows us to identify $\mathbb{C}\{L\}$ with the group algebra $\mathbb{C}[L]$, viewed as a vector space, by the linear isomorphism

$$\mathbb{C}[L] \to \mathbb{C}\{L\},$$
$$e^\alpha \mapsto \iota(e_\alpha) \quad \text{for } \alpha \in L. \tag{6.4.31}$$

Define

$$\epsilon : L \times L \to \mathbb{C}^{\times}$$
$$(\alpha, \beta) \mapsto \omega_s^{\epsilon_0(\alpha,\beta)}. \tag{6.4.32}$$

Then we have the multiplicative analogues of (6.4.19)–(6.4.21):

$$\epsilon(\alpha, \beta)\epsilon(\alpha + \beta, \gamma) = \epsilon(\beta, \gamma)\epsilon(\alpha, \beta + \gamma) \tag{6.4.33}$$
$$\epsilon(\alpha, \beta)/\epsilon(\beta, \alpha) = c(\alpha, \beta) \tag{6.4.34}$$
$$\epsilon(\alpha, 0) = \epsilon(0, \alpha) = 1 \tag{6.4.35}$$

for $\alpha, \beta, \gamma \in L$. As operators on $\mathbb{C}\{L\}$, we have

$$e_\alpha e_\beta = \epsilon(\alpha, \beta)e_{\alpha+\beta} \quad \text{for } \alpha, \beta \in L, \tag{6.4.36}$$

and the action of $\hat{L}$ on $\mathbb{C}[L]$ is given by

$$e_\alpha \cdot e^\beta = \epsilon(\alpha, \beta)e^{\alpha+\beta} \tag{6.4.37}$$
$$\kappa \cdot e^\beta = \omega_s e^\beta \qquad \text{for } \alpha, \beta \in L. \tag{6.4.38}$$

In particular,

$$e_\alpha \cdot 1 = e_\alpha \cdot e^0 = e^\alpha. \tag{6.4.39}$$

Remark 6.4.4 In addition to being an induced $\hat{L}$-module (recall (6.4.25)), $\mathbb{C}\{L\}$ is an associative algebra, since $(\kappa - \omega_s)\mathbb{C}[\hat{L}]$ is a two-sided ideal of the group algebra $\mathbb{C}[\hat{L}]$. This algebra is isomorphic to the "twisted group algebra of L," twisted by the $\mathbb{C}^{\times}$-valued 2-cocycle ϵ; this twisted group algebra is defined by the formula (6.4.36) with e_γ ($\gamma \in L$) replaced by e^γ, under the identification (6.4.31):

$$e^\alpha e^\beta = \epsilon(\alpha, \beta)e^{\alpha+\beta} \quad \text{for } \alpha, \beta \in L. \tag{6.4.40}$$

(Recall that $\mathbb{C}\{L\}$ is linearly isomorphic to $\mathbb{C}[L]$ by (6.4.31) and recall (6.4.18) and (6.4.32)–(6.4.33).)

Now we are ready to form the main space we will consider. Set

$$V_L = M(1) \otimes \mathbb{C}\{L\} = S(\hat{\mathfrak{h}}_+) \otimes \mathbb{C}\{L\} = S(\hat{\mathfrak{h}}^-) \otimes \mathbb{C}\{L\} \tag{6.4.41}$$

(recall the notation discussed in Remark 6.3.1), and set

$$\mathbf{1} = 1 \otimes \iota(1) \in V_L. \tag{6.4.42}$$

Next we shall equip V_L with more structures. Regard $S(\hat{\mathfrak{h}}_+)$ as a trivial $\hat{L}$-module and V_L as the corresponding tensor product $\hat{L}$-module, so that $a \in \hat{L}$ acts as the operator

$$a = 1 \otimes a \in \text{End } V_L. \tag{6.4.43}$$

In particular, for $\alpha \in L$, e_α acts as the operator

$$e_\alpha = 1 \otimes e_\alpha \in \text{End } V_L. \qquad (6.4.44)$$

View $\mathbb{C}\{L\}$ as a trivial $\hat{\mathfrak{h}}_*$-module and for $h \in \mathfrak{h}$, define

$$h(0) : \mathbb{C}\{L\} \to \mathbb{C}\{L\}$$
$$\iota(a) \mapsto \langle h, \bar{a}\rangle \iota(a) \qquad (6.4.45)$$

for $a \in \hat{L}$. Note that this is indeed a well-defined linear operator, since the linear relation $\iota(\kappa a) = \omega_s \iota(a)$ is also satisfied by the image elements:

$$\langle h, \bar{\kappa a}\rangle \iota(\kappa a) = \omega_s \langle h, \bar{a}\rangle \iota(a). \qquad (6.4.46)$$

This makes $\mathbb{C}\{L\}$ an $\hat{\mathfrak{h}}$-module (recall (6.3.7)). We view V_L as the tensor product $\hat{\mathfrak{h}}$-module, so that $\mathbf{k} \in \hat{\mathfrak{h}}$ acts as the identity operator, and

$$h(0)(v \otimes \iota(e_\alpha)) = \langle h, \alpha\rangle (v \otimes \iota(e_\alpha)) \qquad (6.4.47)$$
$$h(n)(v \otimes \iota(e_\alpha)) = h(n)v \otimes \iota(e_\alpha) \qquad (6.4.48)$$

for $h \in \mathfrak{h}$, $n \neq 0$, $v \in S(\hat{\mathfrak{h}}_+)$ and $\alpha \in L$.

Now, V_L is a restricted $\hat{\mathfrak{h}}$-module of level 1. Furthermore, V_L is the direct sum of $\hat{\mathfrak{h}}$-submodules $M(1) \otimes \mathbb{C}\iota(e_\alpha)$ for $\alpha \in L$, where the submodule $M(1) \otimes \mathbb{C}\iota(e_\alpha)$ is irreducible and is equivalent to the irreducible $\hat{\mathfrak{h}}$-module $M(1, \alpha)$ by Proposition 6.3.8. By Theorem 6.3.3, each $M(1, \alpha)$ is naturally a module for $M(1)$ viewed as a vertex operator algebra. In particular, V_L is naturally a Virasoro algebra module, and is an (infinite) direct sum of vertex operator algebra modules; V_L has a natural grading (by the rational numbers) compatible with the action of $L(0)$ on V_L, so that in particular,

$$\text{wt } \iota(e_\alpha) = \frac{1}{2}\langle \alpha, \alpha\rangle \qquad (6.4.49)$$

for $\alpha \in L$ (recall Theorem 6.3.3). The explicit expression of $L(n)$ $(n \in \mathbb{Z})$ as an operator on V_L is given in (6.2.44) (with $\ell = 1$ and $h = 0$), and for $j \in \mathbb{Z}$, the operator $u^{(i)}(j)$ appearing in (6.2.44) acts only on the tensor factor $M(1)$ of V_L if $j \neq 0$ and it acts only on the tensor factor $\mathbb{C}\{L\}$ of V_L if $j = 0$.

For any vector space U, we denote by $U\{x\}$ the space of all formal series with *complex* (rather than integral) powers of x and with coefficients in U:

$$U\{x\} = \left\{\sum_{r \in \mathbb{C}} u_r x^r \mid u_r \in U\right\}. \qquad (6.4.50)$$

For $h \in \mathfrak{h}$, define $x^h \in (\text{End } V_L)\{x\}$ (thought of as $x^{h(0)}$) by

$$x^h(v \otimes \iota(a)) = x^{\langle h, \bar{a}\rangle}(v \otimes \iota(a)) \qquad (6.4.51)$$

for $v \in S(\hat{\mathfrak{h}}_+)$, $a \in \hat{L}$. The following are the basic commutation relations among our operators e_α, $h(m)$, $L(m)$ and x^h on V_L for $\alpha \in L$, $h \in \mathfrak{h}$ and $m \in \mathbb{Z}$:

Proposition 6.4.5 *On V_L we have*

$$[h(m), x^\alpha] = 0 \tag{6.4.52}$$

$$x^\alpha e_\beta = x^{\langle \alpha, \beta \rangle} e_\beta x^\alpha = e_\beta x^{\alpha + \langle \alpha, \beta \rangle} \tag{6.4.53}$$

$$[h(m), e_\alpha] = \delta_{m,0} \langle h, \alpha \rangle e_\alpha \tag{6.4.54}$$

$$[L(m), x^\alpha] = 0 \tag{6.4.55}$$

for $h \in \mathfrak{h}$, $\alpha, \beta \in L$ and $m \in \mathbb{Z}$. We also have

$$[L(m), h(n)] = -nh(m+n) \qquad \text{for } h \in \mathfrak{h}, \ m, n \in \mathbb{Z} \tag{6.4.56}$$

$$[L(m), e_\alpha] = \alpha(m)e_\alpha \qquad \text{for } \alpha \in L, \ 0 \neq m \in \mathbb{Z} \tag{6.4.57}$$

$$[L(0), e_\alpha] = e_\alpha \left(\alpha(0) + \frac{1}{2}\langle \alpha, \alpha \rangle \right). \tag{6.4.58}$$

Proof. The relations (6.4.52)–(6.4.54) are straightforward and the relation (6.4.55) follows from (6.2.44) (the explicit expression of $L(n)$) and (6.4.52). The relation (6.4.56) follows from (6.2.45). What we have to prove is (6.4.57) and (6.4.58). We start with (6.4.57) for $m = -1$, a particularly important case. From (6.2.44) with $\ell = 1$ and $h = 0$, we have

$$L(-1) = \frac{1}{2} \sum_{i=1}^{d} \sum_{m \in \mathbb{Z}} {}^{\circ}_{\circ} u^{(i)}(m)u^{(i)}(-1-m) {}^{\circ}_{\circ} = \sum_{i=1}^{d} \sum_{m \geq 1} u^{(i)}(-m)u^{(i)}(m-1).$$

$$\tag{6.4.59}$$

Using (6.4.54) we obtain

$$[L(-1), e_\alpha] = \left[\sum_{i=1}^{d} u^{(i)}(-1)u^{(i)}(0), e_\alpha \right]$$

$$= \sum_{i=1}^{d} u^{(i)}(-1)[u^{(i)}(0), e_\alpha]$$

$$= \sum_{i=1}^{d} \langle \alpha, u^{(i)} \rangle u^{(i)}(-1)e_\alpha$$

$$= \alpha(-1)e_\alpha, \tag{6.4.60}$$

proving (6.4.57) for $m = -1$. A similar calculation proves (6.4.57) in general. Finally,

$$[L(0), e_\alpha] = \frac{1}{2} \left[\sum_{i=1}^{d} u^{(i)}(0)u^{(i)}(0), e_\alpha \right]$$

$$= \frac{1}{2} \sum_{i=1}^{d} [u^{(i)}(0), e_\alpha]u^{(i)}(0) + \frac{1}{2} \sum_{i=1}^{d} u^{(i)}(0)[u^{(i)}(0), e_\alpha]$$

$$= \frac{1}{2}\sum_{i=1}^{d} \langle \alpha, u^{(i)}\rangle e_\alpha u^{(i)}(0) + \frac{1}{2}\sum_{i=1}^{d} u^{(i)}(0)\langle \alpha, u^{(i)}\rangle e_\alpha$$

$$= \frac{1}{2}e_\alpha \alpha(0) + \frac{1}{2}\alpha(0)e_\alpha$$

$$= e_\alpha \left(\alpha(0) + \frac{1}{2}\langle \alpha, \alpha\rangle \right). \quad \Box \tag{6.4.61}$$

For $a \in \hat{L}$, we define the fundamental operator (cf. [FLM6], Section 8.4)

$$Y(\iota(a), x) = E^-(-\bar{a}, x)E^+(-\bar{a}, x)ax^{\bar{a}} \in (\text{End } V_L)\{x\}, \tag{6.4.62}$$

recalling from Section 6.3 that for $h \in \mathfrak{h}$,

$$E^\pm(-h, x) = \exp\left(\sum_{n \in \pm \mathbb{Z}_+} \frac{-h(n)}{n}x^{-n}\right) \in (\text{End } V_L)[[x, x^{-1}]].$$

(When we determine the properties of the operator (6.4.62) in Section 6.5, we shall see how natural this operator is, and how, for example, the use of $-\bar{a}$ rather than, say, $\bar{a}$ in (6.4.62) makes the calculations come out right.) We need to verify that this operator $Y(\iota(a), x)$ is well defined. First, the parametrization of this operator by $\iota(a)$ rather than by a is acceptable because, as we have observed in Remark 6.4.3, the correspondence $a \mapsto \iota(a)$ is injective. (As we shall see momentarily, it is in fact very natural to parametrize this operator by an element of V_L.) Next, the product

$$E^-(-\bar{a}, x)E^+(-\bar{a}, x) \tag{6.4.63}$$

is a well-defined element of $(\text{End } M(1))[[x, x^{-1}]]$ and hence of $(\text{End } V_L)[[x, x^{-1}]]$ (recall (6.4.41)). Indeed, when $E^+(-\bar{a}, x)$ is applied to any element of $M(1)$, the result is a polynomial in x^{-1} with coefficients in $M(1)$, so that the product (6.4.63) is a well-defined operator from $M(1)$ to $M(1)((x))$; thus the product (6.4.63) is not only well defined as an element of $(\text{End } V_L)[[x, x^{-1}]]$, but it also satisfies the truncation condition. In particular, the operator (6.4.62) is well defined.

We observe that the definition (6.4.62) in fact extends (uniquely) to a well-defined linear map

$$Y(\cdot, x) : \mathbb{C}\{L\} \to (\text{End } V_L)\{x\}$$
$$v \mapsto Y(v, x). \tag{6.4.64}$$

Indeed, in view of Remark 6.4.3, (6.4.27) and (6.4.28), the only linear relations among the elements $\iota(a)$ are also satisfied by the operators $Y(\iota(a), x)$, since if we replace a by κa on the right-hand side of (6.4.62), then the operator is multiplied by ω_s. (This is very similar to the justification that the map (6.4.45) is well defined; recall (6.4.46).)

Remark 6.4.6 We might also, as in [FLM6], Chapter 8, define the operator $Y(a, x)$ for $a \in \hat{L}$ to be exactly the expression (6.4.62), but this will not be necessary, and we would like instead to keep our focus on the point of view that vertex operators should be defined so as to depend on elements (such as $\iota(a)$) of our space (in this case, V_L) that will be a vertex algebra. More precisely, our space V_L will contain vertex algebras as natural substructures, as we shall explain.

For $\alpha \in L$, using our section e (6.4.15) we have the operator

$$Y(\iota(e_\alpha), x) = E^-(-\alpha, x) E^+(-\alpha, x) e_\alpha x^\alpha \in (\text{End } V_L)\{x\}. \tag{6.4.65}$$

Taking $a = 1$ (or $\alpha = 0$) we have in particular

$$Y(1, x) = Y(\iota(1), x) = 1 \ (= 1_{V_L}). \tag{6.4.66}$$

Now we need to extend the definition of the operators (6.4.62) from the elements $\iota(a)$ of V_L, or rather, from the subspace $\mathbb{C}\{L\}$ of V_L (recall (6.4.64)), to the whole space V_L. For this, we shall use the constructions and concepts we have discussed concerning normal-ordered products of operators.

Recall that in Section 3.8 we defined and discussed normal-ordered products of vertex operators in a vertex algebra, including multiple normal-ordered products (Remark 3.8.4), and that in Section 4.4 we commented that all of these considerations extend to modules for a vertex algebra (Remark 4.4.8); indeed, we have been using such normal ordering in this chapter. Also recall from Section 5.2, starting with Remark 5.2.6, that we have extended these normal-ordering notions to weak vertex operators, since normal ordering is based on "iterate vertex operators," and the action of one weak vertex operator on another is expressed in terms of the iterate formula. Next we shall (easily) extend this notion of normal ordering in a natural way to our current situation.

For $h \in \mathfrak{h}$, we set

$$h(x) = \sum_{n \in \mathbb{Z}} h(n) x^{-n-1} \in (\text{End } V_L)[[x, x^{-1}]] \tag{6.4.67}$$

as usual. We extend the map $Y(\cdot, x)$ from $\mathbb{C}\{L\}$ to all of V_L as follows. For $r \geq 0$, $\alpha^{(i)} \in \mathfrak{h}$, $n_i \geq 1$ and $a \in \hat{L}$, we define, using (6.4.62) and (6.4.67),

$$Y(\alpha^{(1)}(-n_1) \cdots \alpha^{(r)}(-n_r) \iota(a), x) =$$

$$\left. {}_\circ^\circ \left(\frac{1}{(n_1 - 1)!} \left(\frac{d}{dx} \right)^{n_1 - 1} \alpha^{(1)}(x) \right) \cdots \left(\frac{1}{(n_r - 1)!} \left(\frac{d}{dx} \right)^{n_r - 1} \alpha^{(r)}(x) \right) Y(\iota(a), x) {}_\circ^\circ \right. .$$

$$\tag{6.4.68}$$

Remark 6.4.7 This expression requires some discussion. In this expression, the multiple normal-ordered product has the usual meaning, as in Remark 3.8.4, even though we are not in the usual situation, because the operator $Y(\iota(a), x)$ in general involves nonintegral powers of x. The reason for this nonintegrality is that L is not necessarily an integral lattice; that is, we do not necessarily have $\langle \alpha, \beta \rangle \in \mathbb{Z}$ for $\alpha, \beta \in L$. For this reason, the operator x^α, which is a factor of the operator $Y(\iota(a), x)$ (or of the operator (6.4.65)), may involve nonintegral rational powers of x (recall (6.4.51)), and so $Y(\iota(a), x)$ may involve nonintegral rational powers of x, while by definition, vertex operators for vertex algebras and for modules, and also, weak vertex operators as discussed in Chapter 5, involve only integral powers of x. With the normal-ordering procedure taken to be the usual one, the operator (6.4.68) is still well defined. Indeed, the factors

$$\frac{1}{(n-1)!} \left(\frac{d}{dx} \right)^{n-1} h(x) \qquad (6.4.69)$$

on the right-hand side of (6.4.68) are familiar from (3.8.8) and (5.2.25), and in our Heisenberg algebra situation the regular part of each of these operators (6.4.69) involves only creation operators $h(-n)$ for $h \in \mathfrak{h}$ and $n > 0$, and the singular part involves only annihilation operators $h(n)$ for $h \in \mathfrak{h}$ and $n > 0$, as well as operators $h(0)$, where we are using the terminology of Remark 6.3.6 (also recall Remark 3.8.2). This makes it clear that the operator (6.4.68) is well defined. Allowing ourselves to extend the notion and notation of the action of weak vertex operators on the space of weak vertex operators (Section 5.2) to our current situation in which $Y(\iota(a), x)$ in general involves nonintegral powers of x, we can write (6.4.68) as

$$Y(\alpha^{(1)}(-n_1) \cdots \alpha^{(r)}(-n_r) \iota(a), x)$$
$$= \alpha^{(1)}(x)_{-n_1} \cdots \alpha^{(r)}(x)_{-n_r} Y(\iota(a), x). \qquad (6.4.70)$$

Now we are ready to extend the map $Y(\cdot, x)$ from the subspace $\mathbb{C}\{L\}$ of V_L to V_L itself. Formula (6.4.68) indeed uniquely defines a linear map

$$Y(\cdot, x) : V_L \rightarrow (\text{End } V_L)\{x\}$$
$$v \mapsto Y(v, x). \qquad (6.4.71)$$

Remark 6.4.8 The fact that this uniquely determined linear map from the vector space $V_L = S(\hat{\mathfrak{h}}_+) \otimes \mathbb{C}\{L\}$ to $(\text{End } V_L)\{x\}$ is in fact well defined follows from the fact that the right-hand side of (6.4.68) is invariant under permutation of the factors $\alpha^{(1)}(-n_1), \cdots, \alpha^{(r)}(-n_r)$, and this is because the creation operators all commute with one another, and so do the annihilation operators and the operators $h(0)$ for $h \in \mathfrak{h}$. (Even though normal ordering is not in general commutative, as we commented in Section 3.8, it is indeed commutative in our present Heisenberg algebra situation.)

For the special case $a = 1$ in (6.4.68) and (6.4.70), we have

$$Y(\alpha^{(1)}(-n_1) \cdots \alpha^{(r)}(-n_r)\mathbf{1}, x)$$
$$= {}^{\circ}_{\circ} \left(\frac{1}{(n_1 - 1)!} \left(\frac{d}{dx} \right)^{n_1 - 1} \alpha^{(1)}(x) \right) \cdots \left(\frac{1}{(n_r - 1)!} \left(\frac{d}{dx} \right)^{n_r - 1} \alpha^{(r)}(x) \right) {}^{\circ}_{\circ}$$
$$\qquad (6.4.72)$$
$$= \alpha^{(1)}(x)_{-n_1} \cdots \alpha^{(r)}(x)_{-n_r} \mathbf{1} \qquad (6.4.73)$$

(cf. (6.2.30) and (6.2.32)). In particular,

$$Y(h(-1)\mathbf{1}, x) = h(x) \quad \text{for } h \in \mathfrak{h}, \tag{6.4.74}$$

$$Y(\alpha^{(1)}(-1) \cdots \alpha^{(r)}(-1)\mathbf{1}, x) = {}^{\circ}_{\circ}\alpha^{(1)}(x) \cdots a^{(r)}(x){}^{\circ}_{\circ} = \alpha^{(1)}(x)_{-1} \cdots \alpha^{(r)}(x)_{-1}\mathbf{1}, \tag{6.4.75}$$

and

$$Y(\omega, x) = \frac{1}{2} \sum_{i=1}^{d} Y(u^{(i)}(-1)u^{(i)}(-1)\mathbf{1}, x) = \frac{1}{2} \sum_{i=1}^{d} {}^{\circ}_{\circ}u^{(i)}(x)u^{(i)}(x){}^{\circ}_{\circ}. \tag{6.4.76}$$

Remark 6.4.9 In view of Theorems 6.2.11 and 6.2.12, the restriction of our map $Y(\cdot, x)$ to the tensor factor $M(1)$ of V_L indeed agrees, as it should, with the vertex operator action $Y(\cdot, x)$ of the vertex operator algebra $M(1) = V_{\hat{\mathfrak{h}}}(1, 0)$ on $M(1)$ itself, and also agrees with the vertex operator action $Y_W(\cdot, x)$ of $M(1)$ on $W = V_L$ (an infinite direct sum of modules for the vertex operator algebra $M(1)$); to see this, we observe that the case in which the integers in (6.2.30) and (6.2.32) are negative agrees with (6.4.72). Our goal is to extend this algebra and module structure to all of V_L, under certain natural conditions, and in a certain natural sense; the next Remark discusses what we can expect in general.

Remark 6.4.10 We mentioned in Remark 6.4.7 that because of the nonintegrality of the rational lattice L (in general), the space V_L is not a candidate as a vertex algebra. It turns out that for many reasons it is important to generalize the notion of vertex algebra and to equip V_L with such a generalized structure. In fact, in [DL3], a precise notion of "generalized vertex algebra" was formulated and studied, and it was proved that for any nondegenerate rational lattice L, V_L equipped with the linear map $Y(\cdot, x)$ defined above, together with certain other natural structure, is a generalized vertex algebra. A similar notion was also formulated and studied in [FFR]; see also [Mos1], [Mos2]. In this terminology, a "vertex superalgebra" is a particularly simple kind of generalized vertex algebra. In [DL3], a still further generalization—the notion of "abelian intertwining algebra"—was introduced and studied. And the natural "nonabelian" generalization of this notion, a natural notion of "intertwining operator algebra," was introduced and studied in [Hua11], [Hua12], [Hua16] (see also [HL9], [HM1], [HM2]). In general, the nonintegrality of the powers of the formal variables in these generalized structures is deeply related to the monodromy of intertwining operators among modules for an underlying vertex algebra.

In order to get vertex algebras and modules for them, we shall consider certain subspaces of V_L, namely, structures of the form V_{L_0}, where L_0 is an even sublattice of L.

Remark 6.4.11 In what follows, the sublattice L_0 of L is certainly allowed to equal L itself; the case in which L is itself an even lattice is very important.

A lattice L_0 with $\mathbb{Z}$-bilinear form $\langle \cdot, \cdot \rangle$ is said to be *even* if

$$\langle \alpha, \alpha \rangle \in 2\mathbb{Z} \quad \text{for } \alpha \in L_0. \tag{6.4.77}$$

An even lattice L_0 is automatically integral, that is, $\langle \alpha, \beta \rangle \in \mathbb{Z}$ for $\alpha, \beta \in L_0$, since

$$\langle \alpha, \beta \rangle = \frac{1}{2} \left(\langle \alpha + \beta, \alpha + \beta \rangle - \langle \alpha, \alpha \rangle - \langle \beta, \beta \rangle \right). \tag{6.4.78}$$

We now make our basic assumptions on L and a sublattice L_0: Let

$$L_0 \subset L \tag{6.4.79}$$

be a (free abelian) subgroup of our nondegenerate lattice L such that

$$\text{rank } L_0 = \text{rank } L \quad (= d), \tag{6.4.80}$$

so that L_0 is itself a nondegenerate (rational) lattice with respect to the form $\langle \cdot, \cdot \rangle$. Define the *dual lattice* of L_0 to be

$$(L_0)^\circ = \{ \beta \in \mathfrak{h} \; (= L \otimes_\mathbb{Z} \mathbb{C}) \mid \langle \alpha, \beta \rangle \in \mathbb{Z} \quad \text{for } \alpha \in L_0 \}; \tag{6.4.81}$$

then $(L_0)^\circ$ is indeed again a (nondegenerate rational) lattice. We assume that

$$L_0 \text{ is even,} \tag{6.4.82}$$

so that in particular, $L_0 \subset (L_0)^\circ$, and that

$$L \subset (L_0)^\circ, \tag{6.4.83}$$

that is,

$$\langle \alpha, \beta \rangle \in \mathbb{Z} \quad \text{for } \alpha \in L_0, \; \beta \in L; \tag{6.4.84}$$

then we have the inclusions

$$L_0 \subset L \subset (L_0)^\circ. \tag{6.4.85}$$

In view of (6.4.83) or (6.4.84),

$$x^\alpha \in \text{Hom}(V_L, V_L[x, x^{-1}]) \quad \text{for } \alpha \in L_0, \tag{6.4.86}$$

that is, only integral powers of x occur in the expansion of the operator x^α on V_L. We also assume that

$$c(\alpha, \beta) = \omega_s^{c_0(\alpha, \beta)} = (-1)^{\langle \alpha, \beta \rangle} \quad \text{for } \alpha, \beta \in L_0 \tag{6.4.87}$$

(recall (6.4.13) and (6.4.29)); note that the evenness property (6.4.82) of L_0 ensures that $(-1)^{\langle \alpha, \alpha \rangle} = 1$, and hence that $c_0(\alpha, \alpha) = 0$ in $\mathbb{Z}/s\mathbb{Z}$, which agrees with the fact that the commutator map c_0 is alternating. From (6.4.34), we also have

$$\epsilon(\alpha, \beta)/\epsilon(\beta, \alpha) \ (= c(\alpha, \beta)) = (-1)^{\langle\alpha,\beta\rangle} \quad \text{for } \alpha, \beta \in L_0. \tag{6.4.88}$$

With these assumptions, we shall prove in the next section the basic results that V_{L_0} is a vertex operator algebra, possibly without the two grading restrictions (3.1.38) and (3.1.39), and that V_L is a V_{L_0}-module, possibly without the grading restrictions (4.1.16) and (4.1.17). (Of course, in the case $L_0 = L$ (recall Remark 6.4.11), this second assertion is vacuous.)

A (nondegenerate rational) lattice K is said to be *positive definite* if

$$\langle\alpha, \alpha\rangle > 0 \quad \text{for any nonzero } \alpha \in K, \tag{6.4.89}$$

or equivalently, for any nonzero α in the real vector space $K \otimes_{\mathbb{Z}} \mathbb{R}$. Clearly, our lattice L is positive definite if and only if L_0 is positive definite.

Remark 6.4.12 Here we show that given a nondegenerate rational lattice L and an even sublattice L_0 as in (6.4.79) and (6.4.80), there always exists $s > 0$ and a central extension $\hat{L}$ of L by the finite cyclic group $\langle \kappa \mid \kappa^s = 1 \rangle$ such that the associated commutator map c_0 satisfies (6.4.87), or equivalently, there exists an alternating $\mathbb{Z}$-bilinear form c_0 on L whose restriction to L_0 satisfies (6.4.87). (Recall from Remark 6.4.2 that such a form c_0 gives rise to a central extension as desired.) First choose a $\mathbb{Z}$-base $\{\alpha_1, \ldots, \alpha_d\}$ of L such that $\{m_1\alpha_1, \ldots, m_d\alpha_d\}$ is a $\mathbb{Z}$-base of L_0, where $m_1, \ldots, m_d$ are suitable positive integers. Consideration of the base of L shows that there exists a positive integer s, which we take to be even, such that

$$\frac{s}{2}\langle\alpha, \beta\rangle \in \mathbb{Z} \quad \text{for } \alpha, \beta \in L. \tag{6.4.90}$$

Now define the $\mathbb{Z}$-bilinear form $c_0 : L \times L \to \mathbb{Z}/s\mathbb{Z}$ on L by the conditions

$$c_0(\alpha_i, \alpha_i) = 0 + s\mathbb{Z} \quad \text{for } 1 \le i \le d, \tag{6.4.91}$$

$$c_0(\alpha_i, \alpha_j) = \frac{s}{2}\langle\alpha_i, \alpha_j\rangle + s\mathbb{Z} \quad \text{for } 1 \le i < j \le d, \tag{6.4.92}$$

$$c_0(\alpha_j, \alpha_i) = -c_0(\alpha_i, \alpha_j) \quad \text{for } 1 \le i < j \le d, \tag{6.4.93}$$

where we use the property (6.4.90). The form c_0 is clearly alternating. Since L_0 is even, the $\mathbb{Z}$-bilinear form $(\alpha, \beta) \mapsto \frac{s}{2}\langle\alpha, \beta\rangle + s\mathbb{Z}$ on L_0 is also alternating, and for $1 \le i < j \le d$, we have

$$c_0(m_i\alpha_i, m_j\alpha_j) = \frac{s}{2}\langle m_i\alpha_i, m_j\alpha_j\rangle + s\mathbb{Z}. \tag{6.4.94}$$

Thus for $\alpha, \beta \in L_0$,

$$c_0(\alpha, \beta) = \frac{s}{2}\langle\alpha, \beta\rangle + s\mathbb{Z}, \tag{6.4.95}$$

and we have indeed constructed an alternating form c_0 on L such that

$$c(\alpha, \beta) = \omega_s^{c_0(\alpha,\beta)} = \omega_s^{s\langle\alpha,\beta\rangle/2} = (-1)^{\langle\alpha,\beta\rangle} \tag{6.4.96}$$

for $\alpha, \beta \in L_0$.

Remark 6.4.13 In particular situations, the mere existence of a form c_0 on L whose restriction to L_0 satisfies the conditions achieved in Remark 6.4.12 is often less important than the actual construction of a form on L satisfying these conditions together with additional conditions. Notably, in the construction of the moonshine module vertex operator algebra $V^\natural$ [FLM6], one takes the lattice L_0 to be the rank-24 Leech lattice and L to be a suitable larger rational lattice. Then one constructs a form c_0 on L satisfying a number of conditions needed for the construction of $V^\natural$ and of the action of the Monster group on it; see in particular [FLM6], Section 12.1.

Before embarking on the proof of the main results about V_L and V_{L_0}, we need to give a precise definition of the space V_{L_0} as a substructure of V_L. For any subset E of L (not necessarily a sublattice), we write

$$\hat{E} = \{a \in \hat{L} \mid \bar{a} \in E\} \tag{6.4.97}$$

(recall (6.4.11)) and

$$\mathbb{C}\{E\} = \mathrm{span}\{\iota(a) \mid a \in \hat{E}\} \subset \mathbb{C}\{L\}, \tag{6.4.98}$$
$$V_E = M(1) \otimes \mathbb{C}\{E\} \subset V_L \tag{6.4.99}$$

(recall (6.4.25), (6.4.26), (6.4.41)). In particular, we have

$$\mathbb{C}\{L_0\} = \mathbb{C}[\hat{L}_0]/(\kappa - \omega_s)\mathbb{C}[\hat{L}_0], \tag{6.4.100}$$
$$V_{L_0} = M(1) \otimes \mathbb{C}\{L_0\}. \tag{6.4.101}$$

In view of (6.4.86), from the definition (6.4.68) of the operators $Y(\cdot, x)$ we have

$$Y(u, x) \in (\mathrm{End}\ V_L)[[x, x^{-1}]] \quad \text{for } u \in V_{L_0}, \tag{6.4.102}$$

that is, the operators $Y(u, x)$, acting on V_L, involve only integral powers of x.

With the setting in place, we now turn to the formulation and proof of the basic results concerning V_L and V_{L_0}.

6.5 Vertex operator algebras and modules associated to even lattices—the main results

In the last section we have laid the groundwork for the next theorem, due to Borcherds [B1] and Frenkel–Lepowsky–Meurman [FLM6], Chapter 8 (where the proof of the theorem is different from the present proof). The previous section is devoted to setting up the structure V_{L_0} and V_L based on an even lattice L_0 and a larger lattice L. In this theorem, we are continuing to use the notation, setting and assumptions of Section 6.4, including (6.4.79)–(6.4.88), which we just discussed, so that L_0 is an even lattice, L is a lattice between L_0 and its dual lattice ($L_0 \subset L \subset (L_0)^\circ$), and $c(\alpha, \beta) = (-1)^{\langle \alpha, \beta \rangle}$ for $\alpha, \beta \in L_0$. The theorem states:

Theorem 6.5.1 *The quadruple $(V_{L_0}, Y, \mathbf{1}, \omega)$ carries the structure of a vertex operator algebra, possibly without the grading restrictions (3.1.38) and (3.1.39). The pair (V_L, Y) carries the structure of a V_{L_0}-module, possibly without the grading restrictions (4.1.16) and (4.1.17). Furthermore, if L_0 is positive definite, then the grading restrictions (3.1.38) and (3.1.39) hold for V_{L_0} and the grading restrictions (4.1.16) and (4.1.17) hold for V_L.*

In the following we shall prove the two assertions of this theorem separately by using Theorems 5.7.1 and 5.7.6. The "algebra" part will be quite straightforward, and the "module" part will occupy much of this section.

First we prove the following relations:

Proposition 6.5.2 *For $h^{(1)}, h^{(2)}, h \in \mathfrak{h}$, $\alpha, \beta \in L_0$, we have, as operators on V_L:*

$$[L(-1), Y(h, x)] = \frac{d}{dx} Y(h, x), \tag{6.5.1}$$

$$[L(-1), Y(\iota(e_\alpha), x)] = \frac{d}{dx} Y(\iota(e_\alpha), x), \tag{6.5.2}$$

$$[Y(h^{(1)}, x_1), Y(h^{(2)}, x_2)] = -\langle h^{(1)}, h^{(2)}\rangle \frac{\partial}{\partial x_1} x_2^{-1} \delta\left(\frac{x_1}{x_2}\right), \tag{6.5.3}$$

$$[Y(h, x_1), Y(\iota(e_\alpha), x_2)] = \langle h, \alpha\rangle x_2^{-1}\delta\left(\frac{x_1}{x_2}\right) Y(\iota(e_\alpha), x_2), \tag{6.5.4}$$

and

$$
x_0^{-1}\delta\left(\frac{x_1 - x_2}{x_0}\right) Y(\iota(e_\alpha), x_1) Y(\iota(e_\beta), x_2)
$$
$$
\quad - x_0^{-1}\delta\left(\frac{x_2 - x_1}{-x_0}\right) Y(\iota(e_\beta), x_2) Y(\iota(e_\alpha), x_1)
$$
$$
= x_2^{-1}\delta\left(\frac{x_1 - x_0}{x_2}\right) \epsilon(\alpha, \beta) x_0^{\langle\alpha, \beta\rangle} \cdot
$$
$$
E^-(-\alpha, x_1) E^-(\alpha, x_2) Y(\iota(e_{\alpha+\beta}), x_2) E^+(\alpha, x_2) E^+(-\alpha, x_1) \left(\frac{x_1}{x_2}\right)^\alpha. \tag{6.5.5}
$$

Proof. The relation (6.5.1) amounts to the relations (6.4.56) with $m = -1$, in terms of generating functions. For $\alpha \in L_0$, using Proposition 6.3.13 and the relations (6.4.57) and (6.4.55) we get

$$[L(-1), E^-(-\alpha, x)e_\alpha] = \frac{d}{dx} E^-(-\alpha, x)e_\alpha, \tag{6.5.6}$$

$$[L(-1), E^+(-\alpha, x)x^\alpha] = \frac{d}{dx}\left(E^+(-\alpha, x)x^\alpha\right), \tag{6.5.7}$$

so that

$$[L(-1), Y(\iota(e_\alpha), x)]$$

$$= [L(-1), E^-(-\alpha, x)e_\alpha]E^+(-\alpha, x)x^\alpha + E^-(-\alpha, x)e_\alpha[L(-1), E^+(-\alpha, x)x^\alpha]$$

$$= \left(\frac{d}{dx}E^-(-\alpha, x)e_\alpha\right)E^+(-\alpha, x)x^\alpha + E^-(-\alpha, x)e_\alpha\frac{d}{dx}\left(E^+(-\alpha, x)x^\alpha\right)$$

$$= \frac{d}{dx}Y(\iota(e_\alpha), x), \tag{6.5.8}$$

proving (6.5.2) (notice that from (6.4.54), e_α commutes with $E^+(-\alpha, x)$).

The relation (6.5.3) immediately follows from (6.2.8), and using (6.4.52), (6.4.54) and Proposition 6.3.12 we obtain (6.5.4) as follows:

$$[Y(h, x_1), Y(\iota(e_\alpha), x_2)]$$

$$= x_1^{-1}E^-(-\alpha, x_2)E^+(-\alpha, x_2)[h(0), e_\alpha]x_2^\alpha$$

$$+ \sum_{m\geq 1}x_1^{-m-1}[h(m), E^-(-\alpha, x_2)]E^+(-\alpha, x_2)e_\alpha x_2^\alpha$$

$$+ \sum_{m\leq -1}x_1^{-m-1}E^-(-\alpha, x_2)[h(m), E^+(-\alpha, x_2)]e_\alpha x_2^\alpha$$

$$= \langle h, \alpha\rangle x_1^{-1}Y(\iota(e_\alpha), x_2)$$

$$+ \langle h, \alpha\rangle \sum_{m\geq 1}x_1^{-m-1}x_2^m Y(\iota(e_\alpha), x_2)$$

$$+ \langle h, \alpha\rangle \sum_{m\leq -1}x_1^{-m-1}x_2^m Y(\iota(e_\alpha), x_2)$$

$$= \langle h, \alpha\rangle x_1^{-1}\delta\left(\frac{x_2}{x_1}\right)Y(\iota(e_\alpha), x_2)$$

$$= \langle h, \alpha\rangle x_2^{-1}\delta\left(\frac{x_1}{x_2}\right)Y(\iota(e_\alpha), x_2). \tag{6.5.9}$$

We finally prove (6.5.5). For $\alpha, \beta \in L_0$, using Proposition 6.3.14, (6.4.53) and (6.4.36) we get

$$Y(\iota(e_\alpha), x_1)Y(\iota(e_\beta), x_2)$$

$$= x_1^{\langle\alpha,\beta\rangle}\left(1 - \frac{x_2}{x_1}\right)^{\langle\alpha,\beta\rangle}E^-(-\alpha, x_1)E^-(-\beta, x_2)E^+(-\alpha, x_1)E^+(-\beta, x_2)e_\alpha e_\beta x_1^\alpha x_2^\beta$$

$$= \epsilon(\alpha, \beta)(x_1 - x_2)^{\langle\alpha,\beta\rangle}E^-(-\alpha, x_1)E^-(-\beta, x_2)E^+(-\alpha, x_1)E^+(-\beta, x_2)e_{\alpha+\beta}x_1^\alpha x_2^\beta. \tag{6.5.10}$$

Symmetrically, we have

$$Y(\iota(e_\beta), x_2)Y(\iota(e_\alpha), x_1)$$

$$= \epsilon(\beta, \alpha)(x_2 - x_1)^{\langle\alpha,\beta\rangle}E^-(-\alpha, x_1)E^-(-\beta, x_2)E^+(-\alpha, x_1)E^+(-\beta, x_2)e_{\alpha+\beta}x_1^\alpha x_2^\beta$$

$$= (-1)^{\langle\alpha,\beta\rangle}\epsilon(\alpha, \beta)(x_2 - x_1)^{\langle\alpha,\beta\rangle}E^-(-\alpha, x_1)E^-(-\beta, x_2)E^+(-\alpha, x_1)$$

$$\cdot E^+(-\beta, x_2)e_{\alpha+\beta}x_1^\alpha x_2^\beta, \tag{6.5.11}$$

using the relation $\epsilon(\beta, \alpha) = \epsilon(\alpha, \beta)(-1)^{\langle \alpha, \beta \rangle}$ (recall (6.4.88)). Then using the delta function substitution formula (2.3.56) and the delta function identity (2.3.18) we get

$$x_0^{-1}\delta\left(\frac{x_1 - x_2}{x_0}\right) Y(\iota(e_\alpha), x_1)Y(\iota(e_\beta), x_2) - x_0^{-1}\delta\left(\frac{x_2 - x_1}{-x_0}\right) Y(\iota(e_\beta), x_2)Y(\iota(e_\alpha), x_1)$$

$$= x_0^{-1}\delta\left(\frac{x_1 - x_2}{x_0}\right) \epsilon(\alpha, \beta)x_0^{\langle \alpha, \beta \rangle} E^-(-\alpha, x_1)E^-(-\beta, x_2)E^+(-\alpha, x_1)$$

$$\cdot E^+(-\beta, x_2)e_{\alpha+\beta}x_1^\alpha x_2^\beta$$

$$- x_0^{-1}\delta\left(\frac{x_2 - x_1}{-x_0}\right) \epsilon(\alpha, \beta)x_0^{\langle \alpha, \beta \rangle} E^-(-\alpha, x_1)E^-(-\beta, x_2)E^+(-\alpha, x_1)$$

$$\cdot E^+(-\beta, x_2)e_{\alpha+\beta}x_1^\alpha x_2^\beta$$

$$= x_2^{-1}\delta\left(\frac{x_1 - x_0}{x_2}\right) \epsilon(\alpha, \beta)x_0^{\langle \alpha, \beta \rangle} E^-(-\alpha, x_1)E^-(-\beta, x_2)E^+(-\alpha, x_1)$$

$$\cdot E^+(-\beta, x_2)e_{\alpha+\beta}x_1^\alpha x_2^\beta$$

$$= x_2^{-1}\delta\left(\frac{x_1 - x_0}{x_2}\right) \epsilon(\alpha, \beta)x_0^{\langle \alpha, \beta \rangle} E^-(-\alpha, x_1)E^-(\alpha, x_2)Y(\iota(e_{\alpha+\beta}), x_2)E^+(\alpha, x_2)$$

$$\cdot E^+(-\alpha, x_1)\left(\frac{x_1}{x_2}\right)^\alpha, \tag{6.5.12}$$

where we have also used (6.3.31). This completes the proof. $\square$

Now we are ready to prove the "algebra" half of Theorem 6.5.1, in the following precise form:

Theorem 6.5.3 *There exists a unique vertex algebra structure $(V_{L_0}, Y, \mathbf{1})$ on V_{L_0} with $\mathbf{1}$ as in (6.4.42) such that*

$$Y(h, x) = h(x) \quad \text{for } h \in \mathfrak{h} \tag{6.5.13}$$

and

$$Y(\iota(e_\alpha), x) = E^-(-\alpha, x)E^+(-\alpha, x)e_\alpha x^\alpha \quad \text{for } \alpha \in L_0. \tag{6.5.14}$$

The vertex operator map Y on V_{L_0} is given by (6.4.68). Furthermore, with ω as in (6.4.9), $(V_{L_0}, Y, \mathbf{1}, \omega)$ carries the structure of a vertex operator algebra of central charge (or rank) $\ell = \dim \mathfrak{h} = \operatorname{rank} L_0$, possibly without the grading restrictions (3.1.38) and (3.1.39), and the grading is given by the following action of $L(0)$:

$$L(0)(h^{(1)}(-n_1) \cdots h^{(r)}(-n_r)\iota(e_\alpha))$$

$$= \left(n_1 + \cdots + n_r + \frac{1}{2}\langle \alpha, \alpha \rangle\right) (h^{(1)}(-n_1) \cdots h^{(r)}(-n_r)\iota(e_\alpha)) \tag{6.5.15}$$

for $r \geq 0$, $h^{(i)} \in \mathfrak{h}$, $n_i \geq 1$, $\alpha \in L_0$. If our lattice L_0 is positive definite, then the grading restrictions (3.1.38) and (3.1.39) hold.

Proof. We shall apply Theorem 5.7.1, taking the generating set T to be

$$T = \mathfrak{h} \cup \{\iota(e_\alpha) \mid \alpha \in L_0\} \subset V_{L_0}. \tag{6.5.16}$$

Define

$$Y_0(u, x) = Y(u, x) \quad \text{for } u \in T. \tag{6.5.17}$$

We now verify that Theorem 5.7.1 applies to our situation with $V = V_{L_0}$ and $d = L(-1)$. For $h \in \mathfrak{h}$, we know from Section 6.2 that $Y_0(h, x)\mathbf{1} = h(x)\mathbf{1} \in M(1)[[x]]$ and $\lim_{x\to 0} h(x)\mathbf{1} = h$, and for $\alpha \in L_0$,

$$Y(\iota(e_\alpha), x)\mathbf{1} = E^-(-\alpha, x)e_\alpha\mathbf{1} = E^-(-\alpha, x)e^\alpha \in V_{L_0}[[x]]$$

and $\lim_{x\to 0} Y(\iota(e_\alpha), x)\mathbf{1} = e^\alpha = \iota(e_\alpha)$. From Proposition 6.5.2, using (2.3.13) we get

$$(x_1 - x_2)^2[Y_0(h^{(1)}, x_1), Y_0(h^{(2)}, x_2)] = 0, \tag{6.5.18}$$

$$(x_1 - x_2)[Y_0(h, x_1), Y_0(\iota(e_\alpha), x_2)] = 0 \tag{6.5.19}$$

for $h^{(1)}, h^{(2)}, h \in \mathfrak{h}$, $\alpha \in L_0$. Let $\alpha, \beta \in L_0$ and recall that $\langle \alpha, \beta \rangle \in \mathbb{Z}$. If $\langle \alpha, \beta \rangle \geq 0$, then applying Res_{x_0} to (6.5.5) we see immediately that

$$[Y_0(\iota(e_\alpha), x_1), Y_0(\iota(e_\beta), x_2)] = 0. \tag{6.5.20}$$

If $\langle \alpha, \beta \rangle < 0$, we apply $\mathrm{Res}_{x_0} x_0^{-\langle \alpha, \beta \rangle}$ to (6.5.5) (extracting the coefficient of $x_0^{\langle \alpha, \beta \rangle - 1}$) to get

$$(x_1 - x_2)^{-\langle \alpha, \beta \rangle}[Y_0(\iota(e_\alpha), x_1), Y_0(\iota(e_\beta), x_2)] = 0. \tag{6.5.21}$$

By Proposition 6.5.2, we also have

$$[L(-1), Y_0(a, x)] = \frac{d}{dx} Y_0(a, x) \quad \text{for } a \in T. \tag{6.5.22}$$

It is clear that $L(-1)\mathbf{1} = 0$, and in addition, $V_{L_0} = \sum_{\alpha \in L_0} S(\hat{\mathfrak{h}}_+)\iota(e_\alpha)$. Hence by Theorem 5.7.1, Y_0 extends uniquely to a linear map $\tilde{Y}_0$ from V_{L_0} to $(\mathrm{End}\, V_{L_0})[[x, x^{-1}]]$ such that $(V_{L_0}, \tilde{Y}_0, \mathbf{1})$ carries the structure of a vertex algebra, and we have $\mathcal{D} = L(-1)$. For $h \in \mathfrak{h}$, $n \geq 1$, $v \in V_{L_0}$, by Remark 5.2.5 and (5.2.25) we have

$$\tilde{Y}_0(h(-n)v, x) = h(x)_{-n}\tilde{Y}_0(v, x) = {}^{\circ}_{\circ}\left(\frac{1}{(n-1)!}\left(\frac{d}{dx}\right)^{n-1} h(x)\right)\tilde{Y}_0(v, x){}^{\circ}_{\circ}.$$

$$\tag{6.5.23}$$

It follows that $\tilde{Y}_0 = Y$ on V_{L_0}, so $(V_{L_0}, Y, \mathbf{1})$ carries the structure of a vertex algebra with $\mathcal{D} = L(-1)$.

For $\alpha \in L_0$, since $h(n)\iota(e_\alpha) = 0$ for $h \in \mathfrak{h}$ and $n \geq 1$ we have

$$L(0)\iota(e_\alpha) = \frac{1}{2}\sum_{i=1}^{d} u^{(i)}(0)u^{(i)}(0)\iota(e_\alpha) = \frac{1}{2}\sum_{i=1}^{d}\langle u^{(i)}, \alpha\rangle^2\iota(e_\alpha) = \frac{1}{2}\langle\alpha, \alpha\rangle\iota(e_\alpha)$$

$$(6.5.24)$$

(cf. (6.4.49)). Then (6.5.15) follows from (6.4.56).

Finally, if L_0 is positive definite, in view of (6.5.15) we have that $(V_{L_0})_{(n)} = 0$ for $n < 0$ and $\dim(V_{L_0})_{(n)} < \infty$ for $n \geq 0$ because for any integer k, the set $\{\alpha \in L_0 \mid \langle\alpha, \alpha\rangle = k\}$ is finite. Thus, the grading restrictions (3.1.38) and (3.1.39) hold for V_{L_0}, and the proof is complete. $\square$

Remark 6.5.4 Since the construction of the vertex algebra V_{L_0} (where we take $L = L_0$ in the construction of V_L) starts with a central extension of L_0 by a finite cyclic group $\langle\kappa\rangle$ of order s such that the associated commutator map satisfies (6.4.87), some obvious questions are: To what extent does the vertex algebra V_{L_0} depend on this choice of central extension of L_0 and also on the choice of the primitive s^{th} root of unity by which κ acts (recall (6.4.23))? (Recall that our definitions of V_{L_0} (6.4.41) and $Y(\cdot, x)$ ((6.4.62) and (6.4.68)) use $\mathbb{C}\{L_0\}$, and therefore also use the positive integer s and the primitive root of unity ω_s.) Note that one reason for raising these questions is that in Remark 6.4.12, where L_0 is embedded in a given lattice L, we had to find a suitable positive integer s and construct a corresponding central extension $\hat{L}$ of L containing a central extension of L_0 of the desired type, and we want to know that the resulting vertex algebras arising from L_0 are all isomorphic, independently of the choice of the structure associated with L. Next we show that in fact the vertex algebra V_{L_0} is, as suggested by our notation V_{L_0}, unique up to isomorphism.

Proposition 6.5.5 *The vertex algebra V_{L_0} is independent, up to isomorphism of vertex algebras, of the choices of the positive integer s, of the central extension of our even lattice L_0 (6.4.11) by the cyclic group $\langle\kappa \mid \kappa^s = 1\rangle$, and of the primitive s^{th} root of unity ω_s such that (6.4.87) holds. The vertex algebra isomorphism between any two such structures can be chosen to be an $\hat{\mathfrak{h}}$-module isomorphism.*

Proof. First we consider the case in which

$$\langle\alpha, \beta\rangle \in 2\mathbb{Z} \quad \text{for } \alpha, \beta \in L_0. \tag{6.5.25}$$

For example, this happens when our lattice L_0 is taken to be the root lattice of the three-dimensional simple Lie algebra $\mathfrak{sl}(2, \mathbb{C})$, that is, $L_0 = \mathbb{Z}\alpha$ with the root α normalized so that $\langle\alpha, \alpha\rangle = 2$. For any (odd or even) positive integer s (possibly equal to 1), set

$$M_s = L_0 \times \langle\kappa\rangle, \tag{6.5.26}$$

the product group (where as usual κ has order s). Then M_s gives an obvious central extension of L_0 by $\langle\kappa\rangle$, with the maps being those compatible with the product group. The associated commutator map, which is trivial, satisfies (6.4.87), since $(-1)^{\langle\alpha,\beta\rangle} = 1$ for $\alpha, \beta \in L_0$. Conversely, for any central extension $\hat{L}_0$ of L_0 by $\langle\kappa\rangle$ such that the associated commutator map satisfies (6.4.87), the associated commutator map is trivial, so that by Proposition 5.2.3 of [FLM6], the central extension $\hat{L}_0$ is *trivial* in the sense that $\hat{L}_0 = M_s = L_0 \times \langle \kappa \rangle$. Notice that $M_1 = L_0$. We shall use M_1 to build a vertex algebra $\mathcal{V}_{L_0}$ and show that $\mathcal{V}_{L_0}$ is isomorphic to any V_{L_0} for which (6.5.25) holds.

Let ω_s be *any* primitive s^{th} root of unity. The natural injection of L_0 into $M_s = L_0 \times \langle \kappa \rangle$ naturally extends to an algebra map from $\mathbb{C}[L_0]$ to $\mathbb{C}[M_s]$, which in turn induces an algebra map

$$\theta : \mathbb{C}[L_0] \to \mathbb{C}[M_s]/(\kappa - \omega_s)\mathbb{C}[M_s]$$
$$e^\alpha \mapsto \iota(e_\alpha) = \iota(e^\alpha), \tag{6.5.27}$$

where the injection map of L_0 into M_s is used as the section $\{e_\alpha = e^\alpha \mid \alpha \in L_0\}$ for the central extension M_s (recall (6.4.15), (6.4.26) and (6.4.31)). In particular, the map (6.4.31) is an algebra isomorphism as well as a linear isomorphism in this case. Notice that in view of (6.4.31) and (6.4.37) we have

$$e_\alpha \cdot \iota(e_\beta) = e_\alpha \cdot e^\beta = \epsilon(\alpha, \beta)e^{\alpha+\beta} = \epsilon(\alpha, \beta)\iota(e_{\alpha+\beta}) = \iota(e_\alpha)\iota(e_\beta) \tag{6.5.28}$$

for $\alpha, \beta \in L_0$, recalling Remark 6.4.4. (For our trivial extension M_s and our choice of section, for any primitive s^{th} root of unity ω_s, $\epsilon(\alpha, \beta) = 1$ for $\alpha, \beta \in L_0$.) Set

$$\mathcal{V}_{L_0} = M(1) \otimes \mathbb{C}[L_0] \tag{6.5.29}$$

(cf. (6.4.25) and (6.4.41)), which is the vertex algebra associated to the central extension $M_1 = L_0$ with $s = 1$ (so that $\mathbb{C}\{L_0\} = \mathbb{C}[L_0]$), and let V_{L_0} be the usual vertex algebra associated to our central extension M_s and with the primitive root of unity ω_s (for our general positive integer s), so that

$$V_{L_0} = M(1) \otimes (\mathbb{C}[M_s]/(\kappa - \omega_s)\mathbb{C}[M_s]) \tag{6.5.30}$$

(again recall (6.4.25) and (6.4.41)). Then $\mathcal{V}_{L_0}$ can be identified as a vector space with V_{L_0} through the map $1 \otimes \theta$. Clearly, $1 \otimes \theta$ preserves both the vacuum vector $\mathbf{1}$ and the conformal vector ω. It is also clear that $1 \otimes \theta$ is an $\hat{\mathfrak{h}}$-module isomorphism. Using this together with the fact that θ is an algebra isomorphism and (6.5.28) we get

$$(1 \otimes \theta)Y(e^\alpha, x) = Y(\iota(e^\alpha), x)(1 \otimes \theta) \tag{6.5.31}$$

for $\alpha \in L_0$, recalling (6.4.62) (the definition of $Y(\iota(a), x)$). Since $\mathfrak{h}$ and e^α for $\alpha \in L_0$ generate $\mathcal{V}_{L_0}$ as a vertex algebra, in view of Proposition 5.7.9, $1 \otimes \theta$ is a vertex algebra isomorphism.

It remains to consider the (more interesting) case in which

$$\langle \alpha, \beta \rangle \notin 2\mathbb{Z} \quad \text{for some } \alpha, \beta \in L_0. \tag{6.5.32}$$

We first show that for any central extension of L_0 by a cyclic group $\langle \kappa \rangle$ of order s such that the associated commutator map satisfies (6.4.87), s must be even. From (6.4.87), we have

$$\omega_s^{2c_0(\alpha,\beta)} = (-1)^{2\langle\alpha,\beta\rangle} = 1,$$

which implies that if s is odd, we have $\omega_s^{c_0(\alpha,\beta)} = 1$, and then by (6.4.87) again,

$$(-1)^{\langle\alpha,\beta\rangle} = \omega_s^{c_0(\alpha,\beta)} = 1.$$

But $(-1)^{\langle\alpha,\beta\rangle} = -1$, and so s must be even.

The rest of the proof follows the outline of the proof for the first case. By Proposition 5.2.3 of [FLM6], there exists a central extension $\hat{L}_0$ of L_0 by the group $\langle\kappa_2\rangle$ (of order 2) such that the associated commutator map satisfies (6.4.87). Denote by $\mathcal{V}_{L_0}$ the associated vertex algebra, so that

$$\mathcal{V}_{L_0} = M(1) \otimes \left(\mathbb{C}[\hat{L}_0]/(\kappa_2 + 1)\mathbb{C}[\hat{L}_0]\right), \tag{6.5.33}$$

since -1 is the unique primitive square root of unity in $\mathbb{C}^\times$.

Let s be an arbitrary positive even integer. We shall construct a particular central extension N_s of L_0 by $\langle\kappa \mid \kappa^s = 1\rangle$ with the desired properties and we shall embed $\hat{L}_0$ into N_s; N_s will be analogous to M_s in the first case. Let N_s be the quotient group of the product group $\hat{L}_0 \times \langle\kappa\rangle$ by the (central two-element) subgroup generated by the element $\kappa_2 \cdot \kappa^{s/2}$. Denote by $f : \langle\kappa\rangle \to N_s$ the composition of the natural injection of $\langle\kappa\rangle$ into $\hat{L}_0 \times \langle\kappa\rangle$ with the natural map from $\hat{L}_0 \times \langle\kappa\rangle$ to N_s. Since $\langle\kappa\rangle \cap \langle\kappa_2\kappa^{s/2}\rangle = 1$ in $\hat{L}_0 \times \langle\kappa\rangle$, f is injective. Let $\tilde{\phi} : \hat{L}_0 \times \langle\kappa\rangle \to L_0$ be the composition of the projection map of $\hat{L}_0 \times \langle\kappa\rangle$ onto $\hat{L}_0$ with the map $^-$ from $\hat{L}_0$ onto L_0. Since $\tilde{\phi}(\kappa_2) = \tilde{\phi}(\kappa) = 1$, $\tilde{\phi}$ gives rise to a group surjection $\phi : N_s \to L_0$. Clearly, Ker $\tilde{\phi} = \langle\kappa_2\rangle \times \langle\kappa\rangle$, so that

$$\mathrm{Ker}\,\phi = (\langle\kappa_2\rangle \times \langle\kappa\rangle)/\langle\kappa_2\kappa^{s/2}\rangle = f(\langle\kappa\rangle). \tag{6.5.34}$$

Thus N_s equipped with the maps f and ϕ is a central extension of L_0 by $\langle\kappa\rangle$. Since $\hat{L}_0 \cap \langle\kappa_2\kappa^{s/2}\rangle = 1$, the natural group map ψ from $\hat{L}_0$ to N_s ($\hat{L}_0 \to \hat{L}_0 \times \langle\kappa\rangle \to N_s$) is injective. Furthermore, the diagram

$$\begin{array}{ccccccccc}
1 & \to & \langle\kappa_2\rangle & \to & \hat{L}_0 & \overset{-}{\to} & L_0 & \to & 1 \\
 & & \downarrow & & \psi\downarrow & & \| & & \\
1 & \to & \langle\kappa\rangle & \overset{f}{\to} & N_s & \overset{\phi}{\to} & L_0 & \to & 1
\end{array} \tag{6.5.35}$$

commutes, where the map from $\langle\kappa_2\rangle$ to $\langle\kappa\rangle$ is the injection map $\kappa_2 \mapsto \kappa^{s/2}$. Thus we have constructed the analogue of the injection $M_1 = L_0 \to M_s$ in the trivial case above.

Let $e : L_0 \to \hat{L}_0$ ($\alpha \mapsto e_\alpha$) be a section of the extension $\hat{L}_0$. Then $\psi \circ e$ (that is, $\alpha \mapsto \psi(e_\alpha)$) is a section of the extension N_s. Using this section, we now compute the associated commutator map $c_0 : L_0 \times L_0 \to \mathbb{Z}/s\mathbb{Z}$ for N_s (recall (6.4.12), (6.4.13)): For $\alpha, \beta \in L_0$,

$$\begin{aligned}
\psi(e_\alpha)\psi(e_\beta)(\psi(e_\alpha))^{-1}(\psi(e_\beta))^{-1} &= \psi(e_\alpha e_\beta e_\alpha^{-1} e_\beta^{-1}) \\
&= \psi(\kappa_2^{c_0(\alpha,\beta)}) = \psi(\kappa_2^{\langle\alpha,\beta\rangle}) = \kappa^{s\langle\alpha,\beta\rangle/2}
\end{aligned} \tag{6.5.36}$$

(since the commutator map for $\hat{L}_0$ takes the pair (α, β) to $\langle \alpha, \beta \rangle + 2\mathbb{Z}$ in $\mathbb{Z}/2\mathbb{Z}$), and so

$$c_0(\alpha, \beta) = \frac{s}{2} \langle \alpha, \beta \rangle + s\mathbb{Z} \tag{6.5.37}$$

in $\mathbb{Z}/s\mathbb{Z}$. In particular, the values of c_0 all lie in the 2-element subgroup of $\mathbb{Z}/s\mathbb{Z}$ generated by $s/2 + s\mathbb{Z}$. Moreover, by Proposition 5.2.3 of [FLM6], *any* central extension of L_0 by $\langle \kappa \rangle$ (with our fixed even s) with the commutator map given by the (alternating) $\mathbb{Z}$-bilinear form (6.5.37) on $L_0 \times L_0$ is equivalent to the central extension N_s that we have just constructed. But since any central extension of L_0 by $\langle \kappa \rangle$ (of even order s) satisfying (6.4.87) (for *any* choice of ω_s) must have commutator map c_0 given by (6.5.37), we see that any central extension as indicated in the statement of the proposition (in our present case (6.5.32)) is equivalent to our extension N_s. Thus, to complete the proof, we may, without loss of generality, replace any extension in the statement of the proposition by the extension N_s in (6.5.35), along with the embedding of $\hat{L}_0$ into it.

Let ω_s be any primitive s^{th} root of unity. As we have already commented (recall (6.5.33)), the primitive root of unity ω_2 is of course unique and equals -1. Let V_{L_0} be the vertex algebra associated with the central extension N_s and the primitive s^{th} root of unity ω_s, so that

$$V_{L_0} = M(1) \otimes (\mathbb{C}[N_s]/(\kappa - \omega_s)\mathbb{C}[N_s]) . \tag{6.5.38}$$

The group injection ψ from $\hat{L}_0$ into N_s naturally extends to an algebra injection from $\mathbb{C}[\hat{L}_0]$ to $\mathbb{C}[N_s]$. Using this map we obtain an algebra map

$$\psi_1 : \mathbb{C}[\hat{L}_0] \to \mathbb{C}[N_s]/(\kappa - \omega_s)\mathbb{C}[N_s]. \tag{6.5.39}$$

This algebra map ψ_1 in turn induces an algebra map

$$\psi_2 : \mathbb{C}[\hat{L}_0]/(\kappa_2 + 1)\mathbb{C}[\hat{L}_0] \to \mathbb{C}[N_s]/(\kappa - \omega_s)\mathbb{C}[N_s], \tag{6.5.40}$$

since $\psi(\kappa_2) = \kappa^{s/2}$ and $\omega_s^{s/2} = -1$. Using sections e for $\hat{L}_0$ and $\psi \circ e$ for N_s we see that ψ_2 is an isomorphism (recall (6.4.31)). Now, $1 \otimes \psi_2$ is a linear isomorphism from the vertex algebra V_{L_0} to the vertex algebra V_{L_0} (recall (6.5.33) and (6.5.38)), and it follows just as in the proof of the first case that $1 \otimes \psi_2$ is a vertex algebra isomorphism. $\square$

Remark 6.5.6 In the proof of the second case of Proposition 6.5.5, one of the key points was to prove that for any central extension, say N, of L_0 by a cyclic group $\langle \kappa \rangle$ of order s (a positive even integer) such that the associated commutator map satisfies (6.4.87), there exists a section for N such that the associated 2-cocycle $\epsilon_0^{(s)}$ (with values in $\mathbb{Z}/s\mathbb{Z}$) is induced from a $\mathbb{Z}/2\mathbb{Z}$-valued 2-cocycle ϵ_0 for the central extension $1 \to \langle \kappa_2 \rangle \to \hat{L}_0 \to L_0 \to 1$, in the sense that

$$\epsilon_0^{(s)}(\alpha, \beta) = \frac{s}{2} \epsilon_0(\alpha, \beta) + s\mathbb{Z} \tag{6.5.41}$$

for $\alpha, \beta \in L_0$. (The 2-cocycle associated with the section $\psi \circ e$ constructed in the proof has this property, where we take ϵ_0 to be the 2-cocycle associated to the section

e for $\hat{L}_0$.) This fact can be proved alternatively by using the proof (rather than the statement) of Proposition 5.2.3 of [FLM6], which, as we have mentioned, gives the bijection between the set of alternating $\mathbb{Z}$-bilinear maps $c_0 : L_0 \times L_0 \to \mathbb{Z}/s\mathbb{Z}$ and the set of equivalence classes of central extensions of L_0 by $\langle \kappa \mid \kappa^s = 1 \rangle$. Indeed, given an alternating $\mathbb{Z}$-bilinear map c_0, that proof exhibits a 2-cocycle ϵ_0, all of whose values lie in the subgroup of $\mathbb{Z}/s\mathbb{Z}$ generated by the values of c_0. In our present situation, since the hypothesis (6.4.87) implies that our commutator map c_0 for our central extension N is given by (6.5.37), we have that the values of c_0 all lie in the 2-element subgroup of $\mathbb{Z}/s\mathbb{Z}$ generated by $s/2 + s\mathbb{Z}$, and thus there exists a section for N whose cocycle takes only the values $s/2 + s\mathbb{Z}$ and 0 in $\mathbb{Z}/s\mathbb{Z}$. This section for N exhibits the central extension $\hat{L}_0$ (by $\langle \kappa_2 \rangle$) as a subgroup of N, and we have bypassed the portion of the proof above in which an embedding of $\hat{L}_0$ into N_s was constructed. The algebra isomorphism (6.5.40), with N_s replaced by N, is then constructed as in the proof above, and the desired isomorphism of the vertex algebras (6.5.33) and (6.5.38) (with N_s replaced by N) is also constructed as above.

Next, we shall apply Theorem 5.7.6 to prove that the pair (V_L, Y) carries the structure of a V_{L_0}-module. (Of course, in the case $L_0 = L$, there is nothing more to prove.) In attempting to apply Theorem 5.7.6, we want to take T as in (6.5.16), and we certainly know how $\iota_{V_L}^0 : T \to \mathcal{E}(V_L)$ should be extended to $\iota_{V_L} : V_{L_0} \to \mathcal{E}(V_L)$, namely, we should use the action (6.4.68) for $a = e_\alpha$ with $\alpha \in L_0$. However, it would be difficult to verify (5.7.28) for the element a in (5.7.28) of the form e_α, $\alpha \in L_0$. Recall that in Sections 6.1–6.3 we used the corresponding Lie algebras and certain induced modules. In the present case, in order to bypass our difficulty, and also prove additional results, we shall introduce and exploit a certain associative algebra, which will play a role analogous to that of the Lie algebras in Sections 6.1–6.3.

Form the vector space

$$U = \mathfrak{h} \oplus \mathbb{C}[L_0]. \tag{6.5.42}$$

Define $A(L_0)$ to be the quotient algebra of the tensor algebra $T(U \otimes \mathbb{C}[t, t^{-1}])$ over the space $U \otimes \mathbb{C}[t, t^{-1}]$, modulo the two-sided ideal generated by the elements

$$e^0 \otimes t^n - \delta_{n,-1}, \tag{6.5.43}$$

$$[h^{(1)} \otimes t^m, h^{(2)} \otimes t^n] - m \langle h^{(1)}, h^{(2)} \rangle \delta_{m+n,0}, \tag{6.5.44}$$

$$[h \otimes t^m, e^\alpha \otimes t^n] - \langle h, \alpha \rangle (e^\alpha \otimes t^{m+n}), \tag{6.5.45}$$

$$[e^\alpha \otimes t^m, e^\beta \otimes t^n] \quad \text{if } \langle \alpha, \beta \rangle \geq 0, \tag{6.5.46}$$

$$\sum_{i=0}^{-\langle \alpha, \beta \rangle} (-1)^i \binom{-\langle \alpha, \beta \rangle}{i} [e^\alpha \otimes t^{m-\langle \alpha, \beta \rangle - i}, e^\beta \otimes t^{n+i}] \quad \text{if } \langle \alpha, \beta \rangle < 0, \tag{6.5.47}$$

for $h^{(1)}, h^{(2)}, h \in \mathfrak{h}$, $\alpha, \beta \in L_0$ and $m, n \in \mathbb{Z}$, where $[a, b] = ab - ba$ for $a, b \in T(U \otimes \mathbb{C}[t, t^{-1}])$.

Remark 6.5.7 For $u \in U$, set

$$u(x) = \sum_{n \in \mathbb{Z}} (u \otimes t^n) x^{-n-1} \in T(U \otimes \mathbb{C}[t, t^{-1}])[[x, x^{-1}]]. \qquad (6.5.48)$$

As with the affine Lie algebra $\hat{\mathfrak{g}}$ in Section 6.2 (Remark 6.2.2), given a $T(U \otimes \mathbb{C}[t, t^{-1}])$-module W, we shall typically write $u(n)$ for the operator on W corresponding to $u \otimes t^n$, for $u \in U$, $n \in \mathbb{Z}$, and we shall use the notation $u(x)$, or $u_W(x)$, for the action of the expression (6.5.48) on W:

$$u(x) = u_W(x) = \sum_{n \in \mathbb{Z}} u(n) x^{-n-1} \in (\text{End } W)[[x, x^{-1}]] \qquad (6.5.49)$$

for $u \in U$. Since any $A(L_0)$-module is a natural $T(U \otimes \mathbb{C}[t, t^{-1}])$-module, we shall use the same notation for an $A(L_0)$-module W.

In terms of these notations, we observe that a $T(U \otimes \mathbb{C}[t, t^{-1}])$-module W is an $A(L_0)$-module if and only if the following relations hold:

$$e^0(n) = \delta_{n,-1}, \qquad (6.5.50)$$

$$[h^{(1)}(m), h^{(2)}(n)] = m \langle h^{(1)}, h^{(2)} \rangle \delta_{m+n,0}, \qquad (6.5.51)$$

$$[h(m), e^\alpha(n)] = \langle h, \alpha \rangle e^\alpha(m+n), \qquad (6.5.52)$$

$$[e^\alpha(m), e^\beta(n)] = 0 \quad \text{if } \langle \alpha, \beta \rangle \geq 0, \qquad (6.5.53)$$

$$\sum_{i=0}^{-\langle \alpha, \beta \rangle} (-1)^i \binom{-\langle \alpha, \beta \rangle}{i} [e^\alpha(m - \langle \alpha, \beta \rangle - i), e^\beta(n+i)] = 0 \quad \text{if } \langle \alpha, \beta \rangle < 0,$$

$$(6.5.54)$$

for $h^{(1)}, h^{(2)}, h \in \mathfrak{h}$, $\alpha, \beta \in L_0$ and $m, n \in \mathbb{Z}$. These relations can be equivalently written in terms of generating functions as

$$e^0(x) = 1, \qquad (6.5.55)$$

$$[h^{(1)}(x_1), h^{(2)}(x_2)] = -\langle h^{(1)}, h^{(2)} \rangle \frac{\partial}{\partial x_1} x_2^{-1} \delta \left(\frac{x_1}{x_2} \right), \qquad (6.5.56)$$

$$[h(x_1), e^\alpha(x_2)] = \langle h, \alpha \rangle x_2^{-1} \delta \left(\frac{x_1}{x_2} \right) e^\alpha(x_2), \qquad (6.5.57)$$

$$[e^\alpha(x_1), e^\beta(x_2)] = 0 \quad \text{if } \langle \alpha, \beta \rangle \geq 0, \qquad (6.5.58)$$

$$(x_1 - x_2)^{-\langle \alpha, \beta \rangle} [e^\alpha(x_1), e^\beta(x_2)] = 0 \quad \text{if } \langle \alpha, \beta \rangle < 0. \qquad (6.5.59)$$

Remark 6.5.8 Because of the relations (6.5.51), any $A(L_0)$-module is naturally an $\hat{\mathfrak{h}}$-module of level 1. Correspondingly, for any $A(L_0)$-module W we set

$$\Omega_W = \{ w \in W \mid h(n)w = 0 \quad \text{for } h \in \mathfrak{h}, \ n \geq 1 \}, \qquad (6.5.60)$$

as in Definition 6.3.5.

Remark 6.5.9 We define a notion of restricted $A(L_0)$-module in the obvious way. An $A(L_0)$-module W is said to be *restricted* if for $u \in U$ and $w \in W$, $u(n)w = 0$ for n sufficiently large. Then for an $A(L_0)$-module W, W is restricted if and only if

$$u(x) = u_W(x) \in \mathcal{E}(W) \quad \text{for } u \in U. \tag{6.5.61}$$

Let W be a restricted $A(L_0)$-module. Set

$$U(x) = \{u(x) = u_W(x) \mid u \in U\} \subset \mathcal{E}(W). \tag{6.5.62}$$

From (6.5.56)–(6.5.59), $U(x)$ is a local subspace of $\mathcal{E}(W)$.

Define a linear map

$$\phi : U \to V_{L_0} \tag{6.5.63}$$

by $\phi(h) = h$ for $h \in \mathfrak{h}$ and $\phi(e^\alpha) = \iota(e_\alpha)$ for $\alpha \in L_0$. We shall sometimes identify U with its image in V_{L_0} under ϕ. In terms of this map and the notations $a(x)^{\pm}$ defined in (5.2.18) (cf. (3.8.1)) we have:

Proposition 6.5.10 *The space V_L is a restricted $A(L_0)$-module with $u(x) = Y(\phi(u), x)$ for $u \in U$, and for any $\gamma \in L$, $V_{L_0+\gamma}$ is an irreducible $A(L_0)$-submodule of V_L. Moreover, the following relations hold for $\alpha, \beta \in L_0$:*

$$\frac{d}{dx}e^\alpha(x) = \alpha(x)^+ e^\alpha(x) + e^\alpha(x)\alpha(x)^-, \tag{6.5.64}$$

$$\mathrm{Res}_{x_1}\left((x_1 - x_2)^{-\langle\alpha,\beta\rangle-1}e^\alpha(x_1)e^\beta(x_2) - (-x_2 + x_1)^{-\langle\alpha,\beta\rangle-1}e^\beta(x_2)e^\alpha(x_1)\right)$$
$$= \epsilon(\alpha, \beta)e^{\alpha+\beta}(x_2). \tag{6.5.65}$$

Proof. The first assertion follows from Proposition 6.5.2, (6.5.20) and (6.5.21). Clearly $V_{L_0+\gamma}$ is an $A(L_0)$-submodule of V_L. With $e^\alpha(x) = E^-(-\alpha, x)E^+(-\alpha, x)e_\alpha x^{\alpha(0)}$ and therefore

$$e_\alpha = E^-(\alpha, x)e^\alpha(x)x^{-\alpha(0)}E^+(\alpha, x) \tag{6.5.66}$$

for $\alpha \in L_0$, we see that the subspaces of V_L invariant under $A(L_0)$ are precisely the subspaces invariant under the operators $h(m)$ and e_α for $h \in \mathfrak{h}$, $m \in \mathbb{Z}$ and $\alpha \in L_0$. Hence for the irreducibility of the $A(L_0)$-module $V_{L_0+\gamma}$ it suffices to prove that $V_{L_0+\gamma}$ is irreducible under the action of the operators $h(m)$ and e_α for $h \in \mathfrak{h}$, $m \in \mathbb{Z}$, $\alpha \in L_0$. Let W be a nonzero $A(L_0)$-submodule of $V_{L_0+\gamma}$. Since $V_{L_0+\gamma}$ is the direct sum of the irreducible $\hat{\mathfrak{h}}$-modules $S(\hat{\mathfrak{h}}_+) \otimes \mathbb{C}\iota(e_{\alpha+\gamma})$ for $\alpha \in L_0$ and since these $\hat{\mathfrak{h}}$-modules are inequivalent (because $\mathfrak{h}$ acts differently on them), W is a sum of certain irreducible $\hat{\mathfrak{h}}$-modules $S(\hat{\mathfrak{h}}_+) \otimes \mathbb{C}\iota(e_{\alpha+\gamma})$ for $\alpha \in L_0$. Clearly, for any $\alpha \in L_0$, $S(\hat{\mathfrak{h}}_+) \otimes \iota(e_{\alpha+\gamma})$ generates $V_{L_0+\gamma}$ as an $\hat{L}_0$-module, so that W must be the whole module $V_{L_0+\gamma}$. Thus $V_{L_0+\gamma}$ is irreducible.

Since, by (6.3.41) and (6.3.42),

$$\frac{d}{dx}E^{-}(-\alpha, x) = \left(\sum_{n \le -1} \alpha(n)x^{-n-1}\right) E^{-}(-\alpha, x) = \alpha(x)^{+}E^{-}(-\alpha, x),$$

$$(6.5.67)$$

$$\frac{d}{dx}(E^{+}(-\alpha, x)x^{\alpha}) = \left(\sum_{n \ge 1} \alpha(n)x^{-n-1} + \alpha(0)x^{-1}\right) E^{+}(-\alpha, x)x^{\alpha}$$

$$= \alpha(x)^{-}E^{+}(-\alpha, x)x^{\alpha}, \qquad (6.5.68)$$

using the definition of $Y(\iota(e_{\alpha}), x)$ we get

$$\frac{d}{dx}Y(\iota(e_{\alpha}), x) = \alpha(x)^{+}Y(\iota(e_{\alpha}), x) + Y(\iota(e_{\alpha}), x)\alpha(x)^{-}. \qquad (6.5.69)$$

The relation (6.5.65) also holds because applying $\mathrm{Res}_{x_1} \mathrm{Res}_{x_0} x_0^{-\langle \alpha, \beta \rangle - 1}$ to the left-hand side of (6.5.5) gives the left-hand side of (6.5.65) and applying this operation to the right-hand side of (6.5.5) gives

$$\mathrm{Res}_{x_1} x_2^{-1}\delta\left(\frac{x_1}{x_2}\right) \epsilon(\alpha, \beta)E^{-}(-\alpha, x_1)E^{-}(\alpha, x_2)Y(\iota(e_{\alpha+\beta}), x_2)E^{+}(\alpha, x_2)$$

$$\cdot E^{+}(-\alpha, x_1)\left(\frac{x_1}{x_2}\right)^{\alpha} = \epsilon(\alpha, \beta)Y(\iota(e_{\alpha+\beta}), x_2), \qquad (6.5.70)$$

where we use the fundamental delta-function properties, in particular, (2.3.36). $\square$

From Remark 6.5.9 and Theorem 5.5.18 we immediately have (cf. Theorem 6.2.4):

Theorem 6.5.11 *Let W be a restricted $A(L_0)$-module. Set*

$$V_W = \mathrm{span}\{u_W^{(1)}(x)_{n_1} \cdots u_W^{(r)}(x)_{n_r} 1_W \mid r \ge 0, \ u^{(i)} \in U, \ n_i \in \mathbb{Z}\} \subset \mathcal{E}(W).$$

$$(6.5.71)$$

Then $(V_W, Y_{\mathcal{E}}, 1_W)$ carries the structure of a vertex algebra with W as a natural faithful module, where the action Y_W of V_W on W is given by $Y_W(a(x), x_0) = a(x_0)$ for any $a(x) \in V_W$. $\square$

Our central goal is to show that any restricted $A(L_0)$-module W that satisfies the relations (6.5.64) and (6.5.65) is naturally a V_{L_0}-module. We first prove that the vertex algebra V_W obtained in Theorem 6.5.11 is naturally a restricted $A(L_0)$-module that is isomorphic to V_{L_0} if $W \ne 0$, and we then apply Theorem 5.7.6 to prove that W is naturally a V_{L_0}-module. In fact, we prove that V_W is naturally isomorphic to V_{L_0} as a vertex algebra if $W \ne 0$.

As the first step we prove the following analogue of Proposition 6.2.6 (cf. Proposition 5.6.8 and Remark 5.6.10):

Proposition 6.5.12 *Let V be a vertex algebra equipped with a linear map ϕ from $U = \mathfrak{h} \oplus \mathbb{C}[L_0]$ to V and let (W, Y_W) be a faithful V-module such that W is an $A(L_0)$-module (necessarily restricted) with $u_W(x) = Y_W(\phi(u), x)$ for $u \in U$. Then V is a*

*restricted $A(L_0)$ module with $u_V(x) = Y(\phi(u), x)$ for $u \in U$. Furthermore, if (6.5.64)
and (6.5.65) hold on W, then they also hold on V.*

Proof. Since W is a faithful V-module, the relations (6.5.56) and (6.5.57) with
$u(x) = u_V(x) = Y(\phi(u), x)$ for $u \in U$ hold, using Proposition 5.6.7 as in the proof of
Proposition 6.2.6. It is clear that (6.5.55) holds on V. For $a, b \in V$, $n \geq 0$, we have
(cf. (3.8.14))

$$\mathrm{Res}_{x_1}(x_1 - x_2)^n[Y_W(a, x_1), Y_W(b, x_2)] = Y_W(a_n b, x_2). \tag{6.5.72}$$

Since from our assumption we have

$$\mathrm{Res}_{x_1}(x_1 - x_2)^n[Y_W(\phi(e^\alpha), x_1), Y_W(\phi(e^\beta), x_2)] = 0 \tag{6.5.73}$$

for all $n \geq 0$ if $\langle \alpha, \beta \rangle \geq 0$ and for all $n \geq -\langle \alpha, \beta \rangle$ if $\langle \alpha, \beta \rangle < 0$, we must have

$$\phi(e^\alpha)_n \phi(e^\beta) = 0 \tag{6.5.74}$$

for the same n, since W is faithful. In view of Remark 3.2.2 we have

$$(x_1 - x_2)^n[Y(\phi(e^\alpha), x_1), Y(\phi(e^\beta), x_2)] = 0, \tag{6.5.75}$$

again for the same n. Thus the relations (6.5.58) and (6.5.59) hold with $e^\alpha(x) = Y(\phi(e^\alpha), x)$ for $\alpha \in L_0$. Hence V is an $A(L_0)$-module with $u_V(x) = Y(\phi(u), x)$
for $u \in U$, and it is restricted because of the truncation property of the vertex operator
$Y(\phi(u), x)$ for $u \in U$.

Assume that (6.5.64) holds on W. Since W is a V-module we have

$$Y_W(\mathcal{D}\phi(e^\alpha), x) = \frac{d}{dx}Y_W(\phi(e^\alpha), x), \tag{6.5.76}$$

$$\begin{aligned}
Y_W(\phi(\alpha)_{-1}\phi(e^\alpha), x) &= {}^\circ_\circ Y_W(\phi(\alpha), x)Y_W(\phi(e^\alpha), x){}^\circ_\circ \\
&= Y_W(\phi(\alpha), x)^+ Y_W(\phi(e^\alpha), x) + Y_W(\phi(e^\alpha), x)Y_W(\phi(\alpha), x)^-
\end{aligned}$$
$$\tag{6.5.77}$$

for $\alpha \in L_0 \subset \mathfrak{h} \subset U$, where $\mathcal{D}$ is the $\mathcal{D}$-operator of V, recalling Proposition 4.1.3 and
(3.8.10). Combining these relations with (6.5.64) we obtain

$$Y_W(\mathcal{D}\phi(e^\alpha), x) = Y_W(\phi(\alpha)_{-1}\phi(e^\alpha), x), \tag{6.5.78}$$

which implies

$$\mathcal{D}\phi(e^\alpha) = \phi(\alpha)_{-1}\phi(e^\alpha), \tag{6.5.79}$$

since W is faithful. Then we have

$$\begin{aligned}
\frac{d}{dx}Y(\phi(e^\alpha), x) &= Y(\mathcal{D}\phi(e^\alpha), x) = Y(\phi(\alpha)_{-1}\phi(e^\alpha), x) \\
&= Y(\phi(\alpha), x)^+ Y(\phi(e^\alpha), x) + Y(\phi(e^\alpha), x)Y(\phi(\alpha), x)^-
\end{aligned}$$
$$\tag{6.5.80}$$

for $\alpha \in L_0 \subset \mathfrak{h}$, and (6.5.64) holds on V.

Finally, suppose that (6.5.65) holds on W. This relation amounts to

$$\operatorname{Res}_{x_1}(x_1 - x_2)^{-\langle \alpha, \beta \rangle - 1} Y_W(\phi(e^\alpha), x_1) Y_W(\phi(e^\beta), x_2)$$
$$-\operatorname{Res}_{x_1}(-x_2 + x_1)^{-\langle \alpha, \beta \rangle - 1} Y_W(\phi(e^\beta), x_2) Y_W(\phi(e^\alpha), x_1)$$
$$= \epsilon(\alpha, \beta) Y_W(\phi(e^{\alpha+\beta}), x_2). \tag{6.5.81}$$

On the other hand, we have

$$\operatorname{Res}_{x_1}(x_1 - x_2)^{-\langle \alpha, \beta \rangle - 1} Y_W(\phi(e^\alpha), x_1) Y_W(\phi(e^\beta), x_2)$$
$$-\operatorname{Res}_{x_1}(-x_2 + x_1)^{-\langle \alpha, \beta \rangle - 1} Y_W(\phi(e^\beta), x_2) Y_W(\phi(e^\alpha), x_1)$$
$$= Y_W(\phi(e^\alpha)_{-\langle \alpha, \beta \rangle - 1} \phi(e^\beta), x_2). \tag{6.5.82}$$

The module W being faithful, we must have

$$\phi(e^\alpha)_{-\langle \alpha, \beta \rangle - 1} \phi(e^\beta) = \epsilon(\alpha, \beta) \phi(e^{\alpha+\beta}) \tag{6.5.83}$$

in V. Then

$$\operatorname{Res}_{x_1}(x_1 - x_2)^{-\langle \alpha, \beta \rangle - 1} Y(\phi(e^\alpha), x_1) Y(\phi(e^\beta), x_2)$$
$$-\operatorname{Res}_{x_1}(-x_2 + x_1)^{-\langle \alpha, \beta \rangle - 1} Y(\phi(e^\beta), x_2) Y(\phi(e^\alpha), x_1)$$
$$= Y(\phi(e^\alpha)_{-\langle \alpha, \beta \rangle - 1} \phi(e^\beta), x_2)$$
$$= \epsilon(\alpha, \beta) Y(\phi(e^{\alpha+\beta}), x_2), \tag{6.5.84}$$

proving that (6.5.65) holds on V. $\square$

As a corollary of Proposition 6.5.12, as in the proof of Theorem 6.2.7 we immediately have (cf. Theorem 6.2.7):

Theorem 6.5.13 *Let W be a restricted $A(L_0)$-module such that (6.5.64) and (6.5.65) hold. Then the vertex algebra V_W constructed in Theorem 6.5.11 is naturally a restricted $A(L_0)$-module such that (6.5.64) and (6.5.65) hold, with $u_V(x_0) = Y_{\mathcal{E}}(u_W(x), x_0)$ for $u \in U = \mathfrak{h} \oplus \mathbb{C}[L_0]$. Moreover, V_W is generated by 1_W, and we have the relations*

$$u(n)1_W = u_W(x)_n 1_W = 0 \quad \text{for } u \in U,\ n \geq 0. \quad \square \tag{6.5.85}$$

Having proved Theorem 6.5.13, we shall prove that if $W \neq 0$, then the vertex algebra V_W as an $A(L_0)$-module is isomorphic to V_{L_0}. To achieve this we use the theory of Z-operators developed in [LW4], [LW5], [LW6], [LP1], [LP2] and [FLM6]. First we prove:

Proposition 6.5.14 *Let W be a restricted $A(L_0)$-module such that (6.5.64) and (6.5.65) hold. Assume that W is equivalent to a (direct) sum of some irreducible $\hat{\mathfrak{h}}$-submodules $M(1, \lambda)$ for $\lambda \in \mathfrak{h} = \mathfrak{h}^*$. For $\alpha \in L_0$, set*

$$Z_\alpha(x) - E^-(\alpha, x)e^\alpha(x)E^+(\alpha, x)x^{-\alpha(0)}, \tag{6.5.86}$$

a well-defined operator. Then

$$Z_\alpha(x) \in \text{End } W, \tag{6.5.87}$$

that is, $Z_\alpha(x)$ does not depend on x, so that we can write Z_α for $Z_\alpha(x)$. Furthermore, for $h \in \mathfrak{h}$, $m \in \mathbb{Z}$, $\alpha, \beta \in L_0$,

$$[h(m), Z_\alpha] = \delta_{m,0}\langle h, \alpha\rangle Z_\alpha \tag{6.5.88}$$

$$x^{h(0)} Z_\alpha = x^{\langle h, \alpha\rangle} Z_\alpha x^{h(0)} \tag{6.5.89}$$

$$Z_\alpha Z_\beta = (-1)^{\langle \alpha, \beta\rangle} Z_\beta Z_\alpha \tag{6.5.90}$$

$$Z_\alpha Z_\beta = \epsilon(\alpha, \beta) Z_{\alpha+\beta}. \tag{6.5.91}$$

In particular (from (6.5.88)), each Z-operator Z_α commutes with the action of the Heisenberg algebra $\hat{\mathfrak{h}}_*$.

Proof. Observe that under our complete reducibility assumption, for $h \in \mathfrak{h}$, $h(0)$ acts semisimply on W (as the scalar $\langle h, \lambda\rangle$) on $M(1, \lambda)$) and $E^+(\alpha, x)w \in W[x^{-1}]$ for $\alpha \in L_0$, $w \in W$. Since W is a restricted $A(L_0)$-module, $Z_\alpha(x)$ is well defined for $\alpha \in L_0$. Since

$$\frac{d}{dx} E^-(\alpha, x) = -\alpha(x)^+ E^-(\alpha, x), \tag{6.5.92}$$

$$\frac{d}{dx}\left(E^+(\alpha, x)x^{-\alpha(0)}\right) = -\alpha(x)^- E^+(\alpha, x)x^{-\alpha(0)} \tag{6.5.93}$$

(cf. (6.5.67), (6.5.68)) it follows immediately from the assumption (6.5.64) that $\frac{d}{dx}Z_\alpha(x) = 0$. This proves the first assertion. The relations (6.5.88) follow from the relations (6.3.32)–(6.3.35) and (6.5.57). Since $\mathfrak{h}$ $(= \mathfrak{h} \otimes t^0)$ acts semisimply on W, it suffices to prove (6.5.89) acting on $w \in W$ such that $\beta w = \langle \lambda, \beta\rangle w$ for $\beta \in \mathfrak{h}$, where λ is some element of $\mathfrak{h} = \mathfrak{h}^*$. But this follows immediately from (6.5.88) for $m = 0$.

From our complete reducibility assumption, we have $W = S(\hat{\mathfrak{h}}_+)\Omega_W$. Since $[h(m), Z_\alpha] = 0$ for $h \in \mathfrak{h}$, $m < 0$, $\alpha \in L_0$, it suffices to prove (6.5.90) and (6.5.91) applied to each $u \in \Omega_W$. Again since $\mathfrak{h}$ acts semisimply on W and since Ω_W is $\mathfrak{h}$-stable, it suffices to consider $u \in \Omega_W$ such that $hu = \langle \lambda, h\rangle u$ for $h \in \mathfrak{h}$, where $\lambda \in \mathfrak{h}$. From (6.5.86) we see that

$$e^\alpha(x) = E^-(-\alpha, x)Z_\alpha E^+(-\alpha, x)x^{\alpha(0)} \tag{6.5.94}$$

and using Proposition 6.3.14 and (6.5.89) we find that

$$e^\alpha(x_1)e^\beta(x_2)u = (x_1 - x_2)^{\langle \alpha, \beta\rangle} E^-(-\alpha, x_1)E^-(-\beta, x_2)Z_\alpha Z_\beta x_1^{\langle \lambda, \alpha\rangle} x_2^{\langle \lambda, \beta\rangle} u. \tag{6.5.95}$$

Symmetrically we have

$$e^\beta(x_2)e^\alpha(x_1)u = (x_2 - x_1)^{\langle \alpha, \beta\rangle} E^-(-\alpha, x_1)E^-(-\beta, x_2)Z_\beta Z_\alpha x_1^{\langle \lambda, \alpha\rangle} x_2^{\langle \lambda, \beta\rangle} u. \tag{6.5.96}$$

Since W is an $A(L_0)$-module, we have (6.5.58) and (6.5.59), which together with (6.5.95) and (6.5.96) imply that

$$Z_\alpha Z_\beta u = (-1)^{\langle \alpha, \beta \rangle} Z_\beta Z_\alpha u, \tag{6.5.97}$$

proving (6.5.90). From the assumption (6.5.65), using (6.5.95), (6.5.96), (6.5.90) and (6.5.94) we get

$$\mathrm{Res}_{x_1} \left((x_1 - x_2)^{-1} - (-x_2 + x_1)^{-1} \right) E^-(-\alpha, x_1) E^-(-\beta, x_2) Z_\alpha Z_\beta x_1^{\langle \lambda, \alpha \rangle} x_2^{\langle \lambda, \beta \rangle} u$$
$$= \epsilon(\alpha, \beta) E^-(-\alpha - \beta, x_2) Z_{\alpha+\beta} x_2^{\langle \lambda, \alpha+\beta \rangle} u, \tag{6.5.98}$$

while

$$\mathrm{Res}_{x_1} \left((x_1 - x_2)^{-1} - (-x_2 + x_1)^{-1} \right) E^-(-\alpha, x_1) E^-(-\beta, x_2) Z_\alpha Z_\beta x_1^{\langle \lambda, \alpha \rangle} x_2^{\langle \lambda, \beta \rangle} u$$
$$= \mathrm{Res}_{x_1} x_2^{-1} \delta\left(\frac{x_1}{x_2} \right) E^-(-\alpha, x_1) E^-(-\beta, x_2) Z_\alpha Z_\beta x_1^{\langle \lambda, \alpha \rangle} x_2^{\langle \lambda, \beta \rangle} u$$
$$= E^-(-\alpha, x_2) E^-(-\beta, x_2) Z_\alpha Z_\beta x_2^{\langle \lambda, \alpha+\beta \rangle} u, \tag{6.5.99}$$

where we use (2.3.36). Thus $Z_\alpha Z_\beta u = \epsilon(\alpha, \beta) Z_{\alpha+\beta} u$, and (6.5.91) is proved. $\square$

Remark 6.5.15 The reason why the operators (6.5.86), and similar, much more general, operators, were called "Z-operators" in [LW4]-[LW6], [LP1], [LP2] and [FLM6] is that they centralize the action of the relevant Heisenberg algebra, as stated in the last assertion of Proposition 6.5.14.

We need the next two results to prove that the vertex algebra V_W constructed in Theorem 6.5.11 is a completely reducible $\hat{\mathfrak{h}}$-module:

Lemma 6.5.16 *Let W be any restricted $A(L_0)$-module such that the relation (6.5.64) holds. For $\alpha \in L_0$, set*

$$A(\alpha)(x) = E^-(\alpha, x) e^\alpha(x) \in (\mathrm{End}\ W)[[x, x^{-1}]] \tag{6.5.100}$$

and write

$$A(\alpha)(x) = \sum_{n \in \mathbb{Z}} A(\alpha)_n x^{-n-1}. \tag{6.5.101}$$

Then for $h \in \mathfrak{h}$, $\alpha \in L_0$, $m, n \in \mathbb{Z}$,

$$[h(m), A(\alpha)_n] = \langle h, \alpha \rangle A(\alpha)_{m+n} \qquad if\ m \leq 0 \tag{6.5.102}$$
$$[h(m), A(\alpha)_n] = 0 \qquad if\ m \geq 1 \tag{6.5.103}$$
$$(-n - 1) A(\alpha)_n = \sum_{i \geq 0} A(\alpha)_{n-i} \alpha(i). \tag{6.5.104}$$

Proof. First notice that since W is a restricted $A(L_0)$-module, for every $\alpha \in L_0$, $A(\alpha)(x)$ is a well-defined element of $(\text{End } W)[[x, x^{-1}]]$. The properties (6.5.102) and (6.5.103) follow directly from the relations (6.3.33), (6.3.35) and (6.5.57). Since

$$\frac{d}{dx}E^-(\alpha, x) = \left(-\sum_{n \leq -1} \alpha(n)x^{-n-1}\right)E^-(\alpha, x) = -\alpha(x)^+ E^-(\alpha, x),$$

using the assumption (6.5.64) we get

$$\frac{d}{dx}A(\alpha)(x) = A(\alpha)(x)\alpha(x)^-, \tag{6.5.105}$$

which is equivalent to (6.5.104). $\square$

Recall that our lattice L is a sublattice of the full dual lattice $(L_0)^\circ$ of L_0 (recall (6.4.79)–(6.4.85)), and that all our considerations of course apply to the case $L = (L_0)^\circ$. Let $\{\gamma_1, \ldots, \gamma_k\}$ be a complete set of representatives of equivalence classes of L_0 in its dual lattice $(L_0)^\circ$. Then using the notation in (6.4.98) and (6.4.99), we have

$$\mathbb{C}\{(L_0)^\circ\} = \mathbb{C}\{L_0 + \gamma_1\} \oplus \cdots \oplus \mathbb{C}\{L_0 + \gamma_k\}, \tag{6.5.106}$$

$$V_{(L_0)^\circ} = V_{L_0+\gamma_1} \oplus \cdots \oplus V_{L_0+\gamma_k}, \tag{6.5.107}$$

and by Proposition 6.5.10 each $V_{L_0+\gamma_i}$ is an irreducible $A(L_0)$-submodule of $V_{(L_0)^\circ}$.

Now we can formulate and prove the central results:

Proposition 6.5.17 *Let W be a restricted $A(L_0)$-module such that (6.5.64) and (6.5.65) hold. Suppose that there exist $0 \neq e \in W$ and $\gamma \in \mathfrak{h}$ such that $W = A(L_0)e$ and $h(n)e = \delta_{n,0}\langle h, \gamma \rangle e$ for $h \in \mathfrak{h}$, $n \geq 0$. Then $\gamma \in (L_0)^\circ$ and there exists a (unique) $A(L_0)$-module isomorphism ψ from the irreducible $A(L_0)$-submodule $V_{L_0+\gamma}$ of $V_{(L_0)^\circ}$ onto W such that $\psi(\iota(e_\gamma)) = e$.*

Proof. We take $L = (L_0)^\circ$. First, since $h(0)e = \langle h, \gamma \rangle e$ for $h \in \mathfrak{h}$ and $W = A(L_0)e$, it follows from the relations (6.5.51) and (6.5.52) that $\mathfrak{h}$ acts semisimply on W, and hence on Ω_W. From (6.5.103), we have

$$A(\alpha)_n \Omega_W \subset \Omega_W \quad \text{for } \alpha \in L_0, \ n \in \mathbb{Z}. \tag{6.5.108}$$

Thus using the commutation relations (6.5.102) and (6.5.103) (and induction) we have

$$A(\alpha)_n U(\hat{\mathfrak{h}})\Omega_W \subset U(\hat{\mathfrak{h}})\Omega_W \quad \text{for } \alpha \in L_0, \ n \in \mathbb{Z}. \tag{6.5.109}$$

Since

$$e^\alpha(x) = E^-(-\alpha, x)A(\alpha)(x) \tag{6.5.110}$$

by (6.5.100), we see from (6.5.109) that

$$e^{\alpha}(n)U(\hat{\mathfrak{h}})\Omega_W \subset U(\hat{\mathfrak{h}})\Omega_W \quad \text{for } \alpha \in L_0, \; n \in \mathbb{Z}. \tag{6.5.111}$$

Thus $U(\hat{\mathfrak{h}})\Omega_W$ is an $A(L_0)$-submodule of W, and since $e \in \Omega_W$ and $W = A(L_0)e$, we must have $W = U(\hat{\mathfrak{h}})\Omega_W$. It follows from Proposition 6.3.8 that W is isomorphic to a sum of some irreducible $\hat{\mathfrak{h}}$-submodules $M(1, \lambda)$ for $\lambda \in \mathfrak{h} = \mathfrak{h}^*$. Thus Proposition 6.5.14 applies to W.

We next show that $\gamma \in (L_0)^\circ$. Applying (6.5.104) to the vector e, using the property that $\alpha(i)e = 0$ for $i \geq 1$ we get

$$(-n-1)A(\alpha)_n e = A(\alpha)_n \alpha(0)e = \langle \alpha, \gamma \rangle A(\alpha)_n e \tag{6.5.112}$$

for $\alpha \in L_0, \; n \in \mathbb{Z}$. But by (6.5.91) we have

$$Z_{-\alpha}Z_\alpha e = \epsilon(\alpha, -\alpha)Z_0 e = \epsilon(\alpha, -\alpha)e,$$

so that $Z_\alpha e \neq 0$ because we have assumed that $e \neq 0$. (Notice that from the definition and (6.5.55), $Z_0 = 1$.) Thus from (6.5.86) and (6.5.88),

$$A(\alpha)(x)e = Z_\alpha E^+(-\alpha, x)x^{\alpha(0)}e = x^{\langle \gamma, \alpha \rangle}E^+(-\alpha, x)Z_\alpha e \neq 0. \tag{6.5.113}$$

That is, $A(\alpha)_n e \neq 0$ for some $n \in \mathbb{Z}$, so from (6.5.112) we get $\langle \alpha, \gamma \rangle = -n - 1 \in \mathbb{Z}$. This proves that $\gamma \in (L_0)^\circ = L$.

We finally prove that $W \simeq V_{L_0+\gamma}$ as an $A(L_0)$-module. Define a linear map

$$\begin{aligned}
\psi : V_{L_0+\gamma} = S(\hat{\mathfrak{h}}_+) \otimes \mathbb{C}\{L_0 + \gamma\} &\to W \\
v \otimes \iota(e_{\alpha+\gamma}) &\mapsto \epsilon(\alpha, \gamma)^{-1}vZ_\alpha \cdot e
\end{aligned} \tag{6.5.114}$$

for $v \in S(\hat{\mathfrak{h}}_+), \; \alpha \in L_0$, and note that $\psi(\iota(e_\gamma)) = e$. Clearly,

$$\psi h(n) = h(n)\psi \quad \text{for } h \in \mathfrak{h}, \; n \neq 0. \tag{6.5.115}$$

For $h \in \mathfrak{h}, \; v \in S(\hat{\mathfrak{h}}_+), \; \alpha \in L_0$ we have

$$h(0)(v \otimes \iota(e_{\alpha+\gamma})) = \langle \alpha + \gamma, h \rangle(v \otimes \iota(e_{\alpha+\gamma})) \tag{6.5.116}$$

and

$$h(0)vZ_\alpha e = vh(0)Z_\alpha e = \langle \alpha + \gamma, h \rangle vZ_\alpha e, \tag{6.5.117}$$

using (6.5.88), so that

$$\psi h(0) = h(0)\psi \quad \text{for } h \in \mathfrak{h}. \tag{6.5.118}$$

Similarly, we have

$$\psi x^{h(0)} = x^{h(0)}\psi \quad \text{for } h \in \mathfrak{h}. \tag{6.5.119}$$

For $\alpha, \beta \in L_0, \; v \in S(\hat{\mathfrak{h}}_+)$, using (6.4.31), (6.4.33), (6.4.37) and (6.5.91) we have

$$\begin{aligned}
\psi(e_\alpha(v \otimes \iota(e_{\beta+\gamma}))) &= \epsilon(\alpha, \beta+\gamma)\psi(v \otimes \iota(e_{\alpha+\beta+\gamma})) \\
&= \epsilon(\alpha, \beta+\gamma)\epsilon(\alpha+\beta, \gamma)^{-1}vZ_{\alpha+\beta} \cdot e \\
&= \epsilon(\alpha, \beta)\epsilon(\beta, \gamma)^{-1}vZ_{\alpha+\beta} \cdot e \\
&= \epsilon(\beta, \gamma)^{-1}vZ_\alpha Z_\beta \cdot e \\
&= Z_\alpha \psi(v \otimes \iota(e_{\beta+\gamma})).
\end{aligned} \tag{6.5.120}$$

Noticing that on W,

$$e^\alpha(x) = E^-(-\alpha, x)Z_\alpha E^+(-\alpha, x)x^{\alpha(0)},$$

while on $V_{L_0+\gamma}$,

$$e^\alpha(x) = E^-(-\alpha, x)e_\alpha E^+(-\alpha, x)x^{\alpha(0)},$$

we have $\psi(e^\alpha(x)u) = e^\alpha(x)\psi(u)$ for $\alpha \in L_0$, $u \in V_{L_0+\gamma}$, and we have proved that ψ is an $A(L_0)$-module homomorphism. Since e generates W as an $A(L_0)$-module, ψ is surjective, and since $V_{L_0+\gamma}$ is an irreducible $A(L_0)$-module (by Proposition 6.5.10), ψ is injective. Thus ψ is an isomorphism. $\square$

We are now in a position to prove the following canonical analogue of Theorems 6.1.7 and 6.2.12:

Theorem 6.5.18 *Let W be any restricted $A(L_0)$-module such that the relations (6.5.64) and (6.5.65) hold. Then W admits a unique module structure for the vertex algebra V_{L_0} such that $Y_W(h, x) = h(x)$ and $Y_W(\iota(e_\alpha), x) = e^\alpha(x)$ for $h \in \mathfrak{h}$, $\alpha \in L_0$. The module structure is given by*

$$Y_W(h^{(1)}(-n_1)\cdots h^{(r)}(-n_r)\iota(e_\alpha), x)$$
$$= {}^\circ_\circ \left(\frac{1}{(n_1-1)!}\left(\frac{d}{dx}\right)^{n_1-1}h^{(1)}(x)\right)\cdots\left(\frac{1}{(n_r-1)!}\left(\frac{d}{dx}\right)^{n_r-1}h^{(r)}(x)\right)e^\alpha(x){}^\circ_\circ$$

$$\tag{6.5.121}$$

for $r \geq 0$, $h^{(i)} \in \mathfrak{h}$, $n_i \geq 1$, $\alpha \in L_0$.

Proof. Let V_W be the vertex algebra obtained in Theorem 6.5.11. By Theorem 6.5.13 V_W is a restricted $A(L_0)$-module such that (6.5.64) and (6.5.65) hold; it is generated by 1_W; and it satisfies the relations $u(n)1_W = 0$ for $u \in U$, $n \geq 0$. If $W = 0$ there is nothing to prove, so we assume that $W \neq 0$. Applying Proposition 6.5.17 to V_W as the $A(L_0)$-module, 1_W ($\neq 0$) as the element e and $\gamma = 0 \in \mathfrak{h}$, we see that there exists a (unique) linear isomorphism (of $A(L_0)$-modules) ψ from V_{L_0} to V_W such that

$$\psi(\mathbf{1}) = 1_W, \tag{6.5.122}$$

$$\psi(u(n)v) = u(n)\psi(v) \quad \text{for } u \in U = \mathfrak{h} \oplus \mathbb{C}[L_0],\ v \in V_{L_0},\ n \in \mathbb{Z}. \tag{6.5.123}$$

In particular, for $u \in U$,

$$\psi(u) = \psi(u(-1)\mathbf{1}) = u(-1)1_W = u_W(x)_{-1}1_W = u_W(x). \qquad (6.5.124)$$

Now we apply Theorem 5.7.6 to $V = V_{L_0}$ and W, taking $T = U = \mathfrak{h} \oplus \mathbb{C}[L_0]$, $Y_W^0(u, x) = u_W(x)$ (the given $A(L_0)$-module action) for $u \in U$, and $Y(\cdot, x) = \iota_W = \psi$, and from Theorem 5.7.6 we conclude that there is a unique V_{L_0}-module structure on W satisfying the desired conditions. By Theorem 6.5.11 and the fact that ψ is an isomorphism, the V_{L_0}-module W is faithful. Thus Remark 5.2.5 applies, and by (5.2.14) and the fact that

$$h(x)_{-n}a(x) = {}^{\circ}_{\circ}\left(\frac{1}{(n-1)!}\left(\frac{d}{dx}\right)^{n-1}h(x)\right)a(x){}^{\circ}_{\circ} \qquad (6.5.125)$$

for $h \in \mathfrak{h}$, $n \geq 1$, $a(x) \in \mathcal{E}(W)$, the action (6.5.121) follows. $\square$

Remark 6.5.19 Recall from the last assertion of Theorem 5.7.6 that $Y(\cdot, x)$ $(=\psi)$ is a homomorphism of weak vertex algebras. Thus if $W \neq 0$ then $\psi : V_{L_0} \to V_W$ is in fact an isomorphism of vertex algebras, and in particular, V_W is isomorphic to V_{L_0} as a vertex algebra.

Now we can prove the "module" half of Theorem 6.5.1.

Theorem 6.5.20 *With L and L_0 as in Theorem 6.5.1, the pair (V_L, Y) carries the structure of a module for the vertex operator algebra V_{L_0} (possibly without grading restrictions on V_{L_0} or V_L), and $V_{L_0+\gamma}$ is an irreducible submodule for any $\gamma \in L$. In particular, V_{L_0} is a simple vertex algebra. If L_0 is positive definite, then the grading restrictions (4.1.16) and (4.1.17) for V_L also hold, and in particular, V_L is a module for the vertex operator algebra V_{L_0} (with grading restrictions).*

Proof. It follows from Proposition 6.5.10 and Theorem 6.5.18 that the pair (V_L, Y) carries the structure of a module for V_{L_0} viewed as a vertex algebra, with $V_{L_0+\gamma}$ irreducible submodules for $\gamma \in L$. The fact that V_L is also a module for V_{L_0} viewed as a vertex operator algebra (possibly without grading restrictions) is simply the statement that the action of $L(0)$ on V_L is compatible with the grading of V_L (recall (6.4.49)). It is clear that (6.5.15) also holds for $\alpha \in L$. If L_0 is positive definite, we know that L is positive definite, so that from the proof of the last assertion of Theorem 6.5.3 the grading restrictions on V_L hold. $\square$

Now we shall use our structures and methods to classify the irreducible V_{L_0}-modules ([FLM6], [D2]). First we prove the following converse of Theorem 6.5.18 (cf. Proposition 6.5.10):

Proposition 6.5.21 *Any module W for the vertex algebra V_{L_0} is a restricted $A(L_0)$-module with $h(x) = Y_W(h, x)$ and $e^{\alpha}(x) = Y_W(\iota(e_\alpha), x)$ for $h \in \mathfrak{h}$, $\alpha \in L_0$, and (6.5.64) and (6.5.65) hold.*

Proof. Since

$$[Y(h^{(1)}, x_1), Y(h^{(2)}, x_2)] = -\langle h^{(1)}, h^{(2)}\rangle \frac{\partial}{\partial x_1} x_2^{-1} \delta\left(\frac{x_1}{x_2}\right),$$

$$[Y(h, x_1), Y(\iota(e_\alpha), x_2)] = \langle h, \alpha\rangle x_2^{-1} \delta\left(\frac{x_1}{x_2}\right) Y(\iota(e_\alpha), x_2)$$

on V_{L_0}, in view of Proposition 5.6.7 we have

$$[Y_W(h^{(1)}, x_1), Y_W(h^{(2)}, x_2)] = -\langle h^{(1)}, h^{(2)}\rangle \frac{\partial}{\partial x_1} x_2^{-1} \delta\left(\frac{x_1}{x_2}\right) \qquad (6.5.126)$$

$$[Y_W(h, x_1), Y_W(\iota(e_\alpha), x_2)] = \langle h, \alpha\rangle x_2^{-1} \delta\left(\frac{x_1}{x_2}\right) Y_W(\iota(e_\alpha), x_2)$$

$$(6.5.127)$$

on W, for $h^{(1)}, h^{(2)}, h \in \mathfrak{h}$, $\alpha \in L_0$.

Let $\alpha, \beta \in L_0$. Since

$$Y(\iota(e_\alpha), x)\iota(e_\beta) = \epsilon(\alpha, \beta) x^{\langle\alpha,\beta\rangle} E^-(-\alpha, x)\iota(e_{\alpha+\beta}),$$

we have

$$\iota(e_\alpha)_i \iota(e_\beta) = 0 \quad \text{for } i \geq -\langle\alpha, \beta\rangle$$

and

$$\iota(e_\alpha)_{-\langle\alpha,\beta\rangle-1}\iota(e_\beta) = \epsilon(\alpha, \beta)\iota(e_{\alpha+\beta}). \qquad (6.5.128)$$

Thus

$$[Y_W(\iota(e_\alpha), x_1), Y_W(\iota(e_\beta), x_2)] = 0 \quad \text{if } \langle\alpha, \beta\rangle \geq 0, \qquad (6.5.129)$$

$$(x_1 - x_2)^{-\langle\alpha,\beta\rangle}[Y_W(\iota(e_\alpha), x_1), Y_W(\iota(e_\beta), x_2)] = 0 \quad \text{if } \langle\alpha, \beta\rangle < 0,$$

$$(6.5.130)$$

and

$$\begin{aligned}
&\mathrm{Res}_{x_1}(x_1 - x_2)^{-\langle\alpha,\beta\rangle-1} Y_W(\iota(e_\alpha), x_1) Y_W(\iota(e_\beta), x_2)\\
&\quad -\mathrm{Res}_{x_1}(-x_2 + x_1)^{-\langle\alpha,\beta\rangle-1} Y_W(\iota(e_\beta), x_2) Y_W(\iota(e_\alpha), x_1)\\
&= Y_W(\iota(e_\alpha)_{-\langle\alpha,\beta\rangle-1}\iota(e_\beta), x_2)\\
&= \epsilon(\alpha, \beta) Y_W(\iota(e_{\alpha+\beta}), x_2). \qquad (6.5.131)
\end{aligned}$$

Finally,

$$\begin{aligned}
\frac{d}{dx} Y_W(\iota(e_\alpha), x) &= Y_W(L(-1)\iota(e_\alpha), x)\\
&= Y_W(\alpha(-1)\iota(e_\alpha), x) = {}^\circ_\circ Y_W(\alpha, x) Y_W(\iota(e_\alpha), x){}^\circ_\circ, \qquad (6.5.132)
\end{aligned}$$

completing the proof. $\square$

Remark 6.5.22 From Theorem 6.5.18 and Proposition 6.5.21 we know that the modules for V_{L_0}, viewed as a vertex algebra, are exactly the restricted $A(L_0)$-modules such that (6.5.64) and (6.5.65) hold. By Proposition 4.5.17, for such a module W the V_{L_0}-submodules of W are exactly the $A(L_0)$-submodules. (Recall the analogous results in Theorems 6.1.8 and 6.2.13.)

The following result was proved in [DLM4]:

Proposition 6.5.23 *Let W be any nonzero module for V_{L_0} viewed as a vertex algebra. Then $\Omega_W \neq 0$ and $\mathfrak{h}$ has a simultaneous eigenvector in Ω_W.*

Proof. By definition, $w \in \Omega_W$ if and only if $\hat{\mathfrak{h}}_- w = 0$. With W a V_{L_0}-module, for any $h \in \mathfrak{h}$, $w \in W$, we have $h(n)w = 0$ for n sufficiently large. Since $\dim \mathfrak{h} < \infty$, this holds uniformly for $h \in \mathfrak{h}$, and so $\dim \hat{\mathfrak{h}}_- w < \infty$ for any $w \in W$. Let $0 \neq w \in W$ be such that $\dim \hat{\mathfrak{h}}_- w$ is minimal. We shall show that $\hat{\mathfrak{h}}_- w = 0$, so that $w \in \Omega_W \neq 0$.

Assume that $\hat{\mathfrak{h}}_- w \neq 0$. Then there is a positive integer k such that $h(n)w = 0$ for $h \in \mathfrak{h}$, $n > k$ and $\alpha(k)w \neq 0$ for some $\alpha \in \mathfrak{h}$. Since L_0 spans $\mathfrak{h}$, $\alpha(k)w \neq 0$ for some $\alpha \in L_0 \subset \mathfrak{h}$.

In view of Proposition 6.5.21, W is naturally a restricted $A(L_0)$-module such that (6.5.64) and (6.5.65) hold. Since V_{L_0} is a simple vertex algebra, by Corollary 4.5.15 $Y(\iota(e_\alpha), x)w \neq 0$. Thus using the notation (6.5.100) we have

$$A(\alpha)(x)w = E^-(\alpha, x)e^\alpha(x)w = E^-(\alpha, x)Y(\iota(e_\alpha), x)w \neq 0.$$

Since $A(\alpha)(x)w \in W((x))$, there is an integer m such that $A(\alpha)_m w \neq 0$ and $A(\alpha)_n w = 0$ for all $n > m$. Using the relation (6.5.104) with $n = m + k$ and the relations (6.5.102) and (6.5.103) we get

$$0 = (\,m \quad k \quad 1)A(\alpha)_{m+k}w$$
$$= \sum_{i=0}^{k} A(\alpha)_{m+k-i}\alpha(i)w$$
$$= A(\alpha)_{m+k}\alpha(0)w + \sum_{i=1}^{k} \alpha(i)A(\alpha)_{m+k-i}w$$
$$= (\alpha(0) - \langle \alpha, \alpha \rangle)A(\alpha)_{m+k}w + \sum_{i=1}^{k} \alpha(i)A(\alpha)_{m+k-i}w$$
$$= \alpha(k)A(\alpha)_m w. \tag{6.5.133}$$

For any $f \in \hat{\mathfrak{h}}_-$, if $fw = 0$, then from (6.5.103), $fA(\alpha)_m w = A(\alpha)_m fw = 0$. Also, $\alpha(k)A(\alpha)_m w = 0$ (from (6.5.133)), but $\alpha(k)w \neq 0$. Thus the annihilator of w in $\hat{\mathfrak{h}}_-$ is strictly smaller than the annihilator of $A(\alpha)_m w$ in $\hat{\mathfrak{h}}_-$, and we have a natural surjection $\hat{\mathfrak{h}}_- w \rightarrow \hat{\mathfrak{h}}_- A(\alpha)_m w$ whose kernel contains $\alpha(k)w \neq 0$. Hence $\dim \hat{\mathfrak{h}}_- A(\alpha)_m w < \dim \hat{\mathfrak{h}}_- w$. This contradicts the choice of w. Thus $\hat{\mathfrak{h}}_- w = 0$, and so $w \in \Omega_W \neq 0$.

Let $0 \neq w \in \Omega_W$. Assume that for a given subset S of $\mathfrak{h}$, w is a simultaneous eigenvector for $h(0)$ for each $h \in S$, and let $\alpha \in L_0$. As above there is an integer m such that $A(\alpha)_m w \neq 0$ and $A(\alpha)_n w = 0$ for $n > m$. Using the relation (6.5.104) with $n = m$ together with (6.5.102) we get

$$(-m - 1)A(\alpha)_m w = \alpha(0)A(\alpha)_m w - \langle \alpha, \alpha \rangle A(\alpha)_m w.$$

Thus

$$\alpha(0)A(\alpha)_m w = (-m - 1 + \langle \alpha, \alpha \rangle)A(\alpha)_m w, \tag{6.5.134}$$

so that $A(\alpha)_m w$ is an $\alpha(0)$-eigenvector and from (6.5.103) is in Ω_W. From our assumption on S, using (6.5.102) again, we see that $A(\alpha)_m w$ is an eigenvector for $h(0)$ for each $h \in S$ as well as for $\alpha(0)$. Thus from a simultaneous eigenvector in Ω_W for $h(0)$ ($h \in S$), we have constructed a simultaneous eigenvector in Ω_W for all these operators together with $\alpha(0)$. Applying this process to α ranging through a finite subset of L_0 spanning $\mathfrak{h}$, we see that Ω_W contains a simultaneous $\mathfrak{h}$-eigenvector, as desired. $\square$

Now we have the following classification theorem due to Dong [D2]:

Theorem 6.5.24 *Any irreducible module for V_{L_0} viewed as a vertex algebra is equivalent to a V_{L_0}-module of the form $V_{L_0 + \gamma} \subset V_{(L_0)^\circ}$ for some $\gamma \in (L_0)^\circ$, and in particular, to one of the V_{L_0}-modules $V_{L_0 + \gamma_i}$ in (6.5.107). Moreover, the V_{L_0}-modules $V_{L_0 + \gamma_1}, \ldots, V_{L_0 + \gamma_k}$ are inequivalent.*

Proof. From Propositions 6.5.21 and 6.5.23, such a V_{L_0}-module W is naturally a restricted $A(L_0)$-module satisfying the relations (6.5.64) and (6.5.65), and there exist $0 \neq e \in W$ and $\gamma \in \mathfrak{h}$ such that $h(n)e = \delta_{n,0}\langle \gamma, h \rangle e$ for $h \in \mathfrak{h}$, $n \geq 0$. From Remark 6.5.22, $W = A(L_0)e$. Then by Proposition 6.5.17, $\gamma \in (L_0)^\circ$ and there is an $A(L_0)$-module isomorphism ψ from $V_{L_0 + \gamma} \subset V_{(L_0)^\circ}$ onto W such that $\psi(\iota(e_\gamma)) = e$. In view of Proposition 4.5.1, ψ is a V_{L_0}-module isomorphism. The inequivalence follows from the fact that the modules $V_{L_0 + \gamma_i}$ have distinct simultaneous eigenspaces for the action of $\mathfrak{h}$. $\square$

Remark 6.5.25 Recall from Remark 6.4.12 and Theorem 6.5.20 that V_L was constructed by using a suitable choice of positive integer s, of central extension $\hat{L}$ of L by the cyclic group $\langle \kappa \mid \kappa^s = 1 \rangle$ and of primitive s^{th} root of unity ω_s, and that V_{L_0} is a substructure of V_L. Recall also from Proposition 6.5.5 that for two such sets of choices the two vertex algebras V_{L_0} are isomorphic by an isomorphism that is also an $\hat{\mathfrak{h}}$-module isomorphism. Now, Theorem 6.5.24 shows that for any such set of choices, the inequivalent irreducible V_{L_0}-modules are classified as indicated. Thus, for two sets of choices (with $L = (L_0)^\circ$), the two V_{L_0}-modules $V_{L_0 + \gamma_i}$ are equivalent for each $i = 1, \ldots, k$, where we identify the two copies of V_{L_0} using our isomorphism (of vertex algebras and $\hat{\mathfrak{h}}$-modules).

Remark 6.5.26 In [GaoL], Theorems 5.5.14 and 5.7.1 were generalized in order to construct generalized vertex algebras in the sense of [DL3]. One can apply the results of [GaoL] to show that V_L equipped with the linear map Y and the vacuum vector $\mathbf{1}$ together with certain other data is a generalized vertex algebra, in the same way that Theorem 6.5.3 was proved. Then using this generalized vertex algebra structure, one can obtain an alternate proof of the first assertion of Theorem 6.5.20, that is, the second assertion of Theorem 6.5.1—that V_L is a V_{L_0}-module. Other proofs of this result have been given in [Xu2] (cf. [Xu12]), [BDT], and, as we have mentioned, in [FLM6], Chapter 8.

Remark 6.5.27 The results of this section do not simplify the construction of the moonshine module vertex operator algebra $V^\natural$ in [FLM6], even though the construction of the vertex operator algebra V_{L_0}, with L_0 taken as the Leech lattice Λ, is a fundamental part of the construction of $V^\natural$. The reason is that the part of the construction of $V^\natural$ in which V_Λ is constructed entails situations still more general than the situation in Theorem 6.5.1; while Theorem 6.5.1 was indeed proved in Chapter 8 of [FLM6], more general results such as Theorem 8.8.23 in [FLM6] are also needed in [FLM6] in the construction of $V^\natural$. As we have mentioned, the proof of Theorem 6.5.1 in [FLM6] is different from the present proof, and that proof also proved the more general results necessary for the construction of $V^\natural$. On the other hand, the proof of Theorem 6.5.1 that we have presented here, based as it is on the theory developed in this work, reveals conceptually interesting structure and methods that we have of course been exploiting in many different contexts. (Incidentally, another aspect of the construction of the moonshine module vertex operator algebra $V^\natural$ in [FLM6] is that even after such results as Theorem 6.5.1 (above) and Theorem 8.8.23 in [FLM6] are proved, a great deal more needs to be done; see Chapters 9 to 13 of [FLM6].)

Remark 6.5.28 Besides the case in which the even lattice L_0 is taken to be the Leech lattice Λ in the construction of the moonshine module vertex operator algebra $V^\natural$, another fundamental and interesting case is that in which L_0 is taken to be the root lattice of a finite-dimensional simple Lie algebra $\mathfrak{g}$ of type A, D or E (that is, of type A_ℓ, D_ℓ, E_6, E_7 or E_8), with $\langle \alpha, \alpha \rangle$ normalized to equal 2 for each root α. In this case, using a natural generating substructure of the vertex operator algebra V_{L_0} one recovers the well-known "untwisted" vertex operator construction ([FK], [Se1]; cf. [Ha], [BHN]) of the corresponding affine Lie algebra $\hat{\mathfrak{g}}$ on V_{L_0} and also on V_L with $L = (L_0)^\circ$; the dual lattice L of L_0 is the weight lattice of $\mathfrak{g}$. See [FLM6] for an extensive treatment of this, and also of a "twisted" analogue, in which the "twisted" vertex operator constructions of affine Lie algebras in [LW1] and [FLM1] are placed in an analogous context centering around a "twisted Jacobi identity"; both the "untwisted" and "twisted" theories are needed in the construction of the vertex operator algebra $V^\natural$. The vertex operator algebra V_{L_0}, which is naturally an irreducible $\hat{\mathfrak{g}}$-module of level one, is actually isomorphic to the vertex operator algebra $L_{\hat{\mathfrak{g}}}(1, 0)$ constructed in Section 6.2 above and discussed in Remarks 6.2.24 and 6.2.26; we are taking $\ell = 1$ in Remarks 6.2.24 and 6.2.26. That is, the vertex operator algebra V_{L_0} is the vertex operator algebra associated with a certain distinguished level-one standard module ($=$

integrable highest weight module) for the affine Lie algebra $\hat{\mathfrak{g}}$; this $\hat{\mathfrak{g}}$-module is the irreducible $\hat{\mathfrak{g}}$-module $L_{\hat{\mathfrak{g}}}(1, 0)$ (viewed as a $\hat{\mathfrak{g}}$-module; recall Remark 6.2.24 above). In particular, the Virasoro algebras and the conformal vectors ω in these two vertex operator algebras V_{L_0} and $L_{\hat{\mathfrak{g}}}(1, 0)$ are identified. This is a nontrivial fact, since the constructions of the two conformal vectors are different. For a proof of this identification of the two conformal vectors, see for example Proposition 13.10 in [DL3]. In Section 6.6 we shall discuss and treat the standard $\hat{\mathfrak{g}}$-modules of arbitrary positive integral level, for $\mathfrak{g}$ a finite-dimensional simple Lie algebra not necessarily of type A, D or E (see [K6]), from the point of view of the theory developed in the present work. The treatment of the vertex operator algebras $L_{\hat{\mathfrak{g}}}(1, 0)$ associated with level-one standard modules in Sections 6.2 and 6.6 is very different from the treatment of these vertex operator algebras in the form V_{L_0} (in the cases of type A, D or E). But it happens that in Section 6.6 we shall work out in detail the important special case $\widehat{\mathfrak{sl}(2, \mathbb{C})}$ of the lattice constructions treated in Sections 6.4 and 6.5 in the course of proving Proposition 6.6.19, a result about $\widehat{\mathfrak{sl}(2, \mathbb{C})}$ used in the study of the higher-level standard modules for general $\mathfrak{g}$. The theory of lattice vertex algebras presented in Sections 6.4 and 6.5 generalizes to very general not necessarily integral lattices and to twisted constructions, including the theory needed in the construction of the moonshine module vertex operator algebra. On the other hand, the theory presented in Sections 6.2 and 6.6 (and the alternative approaches to vertex operator algebras for affine Lie algebras presented in [FZ] and [DL3]) generalize in other directions, including the directions treated in the present chapter of this work.

6.6 Classification of the irreducible $L_{\hat{\mathfrak{g}}}(\ell, 0)$-modules for $\mathfrak{g}$ finite-dimensional simple and ℓ a positive integer

We continue here from discussions at the end of Section 6.2 to study the simple vertex operator algebra $L_{\hat{\mathfrak{g}}}(\ell, 0)$ with $\mathfrak{g}$ now taken to be a finite-dimensional simple Lie algebra equipped with the (suitably normalized) Killing form. Our main goal is to focus on the case in which ℓ is a positive integer and to classify the irreducible $L_{\hat{\mathfrak{g}}}(\ell, 0)$-modules in this case. The main tool is the weak nilpotency property, in the sense of Section 3.10, of certain elements of $L_{\hat{\mathfrak{g}}}(\ell, 0)$. We show that in fact the irreducible $L_{\hat{\mathfrak{g}}}(\ell, 0)$-modules, including the simple vertex operator algebra $L_{\hat{\mathfrak{g}}}(\ell, 0)$ itself, are precisely the standard modules (the integrable highest weight modules) of level ℓ (see [K6]; cf. [Le2]) for the untwisted affine Kac–Moody Lie algebra associated with $\mathfrak{g}$ (see [K6], [MoP]). In the course of this proof, we will find it convenient at a certain point to use the construction of vertex operator algebras and modules from lattices carried out in Sections 6.4 and 6.5. Then we give some details about the Goddard–Kent–Olive "coset construction" of certain "unitary" irreducible lowest weight modules for the Virasoro algebra and show how this is essentially a special case of I. Frenkel–Zhu's results on centralizers treated in Section 3.11.

The main results in this section were originally obtained in [FZ] and [DL3] (see also [Li3], [MP2], [MP4]), by means of methods closely related to but different from those used here. Our approach here, based on [Li3], uses the theory developed in Chapter 5 and exploits the idea of [DL3] to employ the weak nilpotency of certain elements. As we have been doing, we make our entire exposition quite detailed and self-contained, other than in Remark 6.6.16 where we collect certain results about standard modules that can be found in or follow easily from results in [K6], and we use these results at the end.

Let $\mathfrak{g}$ be a finite-dimensional simple Lie algebra with nondegenerate symmetric invariant bilinear form $\langle \cdot, \cdot \rangle$, and let $\mathfrak{h}$ be a Cartan subalgebra of $\mathfrak{g}$. As in Remark 6.2.15, we identify $\mathfrak{h}$ with $\mathfrak{h}^*$ by means of the form $\langle \cdot, \cdot \rangle$. Let Δ be the root system of $\mathfrak{g}$ with respect to $\mathfrak{h}$, viewed as a subset of $\mathfrak{h}$ $(= \mathfrak{h}^*)$, and assume that the form is normalized so that

$$\langle \alpha, \alpha \rangle = 2 \quad \text{for long roots} \quad \alpha \in \Delta. \tag{6.6.1}$$

A basic fact is that the Casimir operator Ω (recall (6.2.37)) associated to $\mathfrak{g}$ and $\langle \cdot, \cdot \rangle$ acts on $\mathfrak{g}$ as follows:

$$\Omega = 2h \quad \text{on} \quad \mathfrak{g}, \tag{6.6.2}$$

where h is the dual Coxeter number of $\mathfrak{g}$ (cf. [Bou1]; recall (6.2.40) and Remark 6.2.15). We have the root space decomposition

$$\mathfrak{g} = \mathfrak{h} \oplus \sum_{\alpha \in \Delta} \mathfrak{g}_\alpha, \tag{6.6.3}$$

where $\mathfrak{g}_\alpha = \{ a \in \mathfrak{g} \mid [h, a] = \alpha(h)a \text{ for } h \in \mathfrak{h} \}$, for $\alpha \in \Delta$.

We continue reviewing basic facts about finite-dimensional simple Lie algebras (over $\mathbb{C}$). Such background can be found in [Bou1] or [Hum]; in this section we shall sometimes invoke standard facts about semisimple Lie algebras and their modules without giving explicit references for these facts. We have $\dim \mathfrak{g}_\alpha = 1$ and $\dim [\mathfrak{g}_\alpha, \mathfrak{g}_{-\alpha}] = 1$ for $\alpha \in \Delta$. For $\alpha \in \Delta$, let h_α be the unique vector in $\mathfrak{h}$ such that $[\mathfrak{g}_\alpha, \mathfrak{g}_{-\alpha}] = \mathbb{C}h_\alpha$ and $\alpha(h_\alpha) = 2$. Then under the identification between $\mathfrak{h}$ and $\mathfrak{h}^*$ we have

$$h_\alpha = \frac{2\alpha}{\langle \alpha, \alpha \rangle}, \tag{6.6.4}$$

and for $a \in \mathfrak{g}_\alpha$ and $b \in \mathfrak{g}_{-\alpha}$,

$$[h_\alpha, a] = 2a, \tag{6.6.5}$$

$$[h_\alpha, b] = -2b, \tag{6.6.6}$$

$$[a, b] = \langle a, b \rangle \alpha = \langle a, b \rangle \frac{2}{\langle h_\alpha, h_\alpha \rangle} h_\alpha; \tag{6.6.7}$$

we also have

$$\frac{1}{2}\langle \alpha, \alpha \rangle = \frac{2}{\langle h_\alpha, h_\alpha \rangle}. \tag{6.6.8}$$

Set

$$n_\alpha = \frac{1}{2}\langle h_\alpha, h_\alpha \rangle = \frac{2}{\langle \alpha, \alpha \rangle}. \qquad (6.6.9)$$

Then n_α is always a positive integer, and in fact $n_\alpha = 1, 2$ or 3, according to whether α is a long root, a short root if $\mathfrak{g}$ is of type B_n, C_n or F_4, or a short root if $\mathfrak{g}$ is of type G_2. The case in which α is a long root (i.e., $n_\alpha = 1$) is the case

$$\langle \alpha, \alpha \rangle = 2, \quad \text{that is,} \quad h_\alpha = \alpha \qquad (6.6.10)$$

(recall (6.6.1)). For $\alpha \in \Delta$, set

$$\mathfrak{g}^\alpha = \mathfrak{g}_\alpha \oplus \mathbb{C}h_\alpha \oplus \mathfrak{g}_{-\alpha} \subset \mathfrak{g}. \qquad (6.6.11)$$

In view of (6.6.5)–(6.6.7), $\mathfrak{g}^\alpha$ is a subalgebra isomorphic to the three-dimensional simple Lie algebra $\mathfrak{sl}(2, \mathbb{C})$.

Fix a system $\Delta_+ \subset \Delta$ of positive roots. We have the triangular decomposition

$$\mathfrak{g} = \mathfrak{g}_+ \oplus \mathfrak{h} \oplus \mathfrak{g}_-, \qquad (6.6.12)$$

where

$$\mathfrak{g}_\pm = \sum_{\alpha \in \Delta_+} \mathfrak{g}_{\pm\alpha}. \qquad (6.6.13)$$

For $\lambda \in \mathfrak{h}^*$, we denote by $V(\lambda)$ the Verma $\mathfrak{g}$-module with highest weight λ:

$$V(\lambda) = U(\mathfrak{g}) \otimes_{U(\mathfrak{h}\oplus\mathfrak{g}_+)} \mathbb{C}_\lambda, \qquad (6.6.14)$$

where $\mathbb{C}_\lambda$ is the $\mathfrak{h} \oplus \mathfrak{g}_+$-module on which $\mathfrak{h}$ acts according to the character λ and $\mathfrak{g}_+$ acts trivially. We also denote by $L(\lambda)$ the irreducible highest weight $\mathfrak{g}$-module, the quotient of $V(\lambda)$ by its (unique) maximal proper submodule. The $\mathfrak{g}$-module $L(\lambda)$ is finite dimensional if and only if λ is *dominant integral* in the sense that

$$\lambda(h_\alpha) = \frac{2\langle \lambda, \alpha \rangle}{\langle \alpha, \alpha \rangle} \in \mathbb{N} \quad \text{for } \alpha \in \Delta_+. \qquad (6.6.15)$$

Let θ be the highest root, which is also the highest weight of the adjoint $\mathfrak{g}$-module. Notice that

$$\langle \theta, \theta \rangle = 2. \qquad (6.6.16)$$

Fix nonzero vectors $e_\theta \in \mathfrak{g}_\theta$ and $f_\theta \in \mathfrak{g}_{-\theta}$ such that $\langle e_\theta, f_\theta \rangle = 1$. We have

$$[h_\theta, e_\theta] = 2e_\theta, \quad [h_\theta, f_\theta] = -2f_\theta, \quad [e_\theta, f_\theta] = h_\theta. \qquad (6.6.17)$$

As in Section 6.2, associated to the pair $(\mathfrak{g}, \langle \cdot, \cdot \rangle)$ we have the affine Lie algebra

$$\hat{\mathfrak{g}} = \mathfrak{g} \otimes \mathbb{C}[t, t^{-1}] \oplus \mathbb{C}\mathbf{k}, \tag{6.6.18}$$

where for $a, b \in \mathfrak{g}$, $m, n \in \mathbb{Z}$,

$$[a \otimes t^m, b \otimes t^n] = [a, b] \otimes t^{m+n} + m\langle a, b\rangle \delta_{m+n,0}\mathbf{k} \tag{6.6.19}$$

and the nonzero element $\mathbf{k}$ is central. Then $\hat{\mathfrak{g}}$ has the triangular decomposition

$$\hat{\mathfrak{g}} = (\mathfrak{g} \otimes t\mathbb{C}[t]) \oplus (\mathfrak{g} \oplus \mathbb{C}\mathbf{k}) \oplus (\mathfrak{g} \otimes t^{-1}\mathbb{C}[t^{-1}]). \tag{6.6.20}$$

Remark 6.6.1 It is often very useful to enlarge the Lie algebra $\hat{\mathfrak{g}}$ to the following usual version of the untwisted affine Kac–Moody Lie algebra associated to $\mathfrak{g}$:

$$\tilde{\mathfrak{g}} = \mathbb{C}\mathbf{d} \oplus \hat{\mathfrak{g}}, \tag{6.6.21}$$

where $\mathbf{d}$ is the "degree derivation," or "degree operator," of $\mathfrak{g}$, which acts on $\mathfrak{g}$ according to the following bracket relations in $\tilde{\mathfrak{g}}$:

$$[\mathbf{d}, a \otimes t^n] = na \otimes t^n \quad \text{for } a \in \mathfrak{g}, \ n \in \mathbb{Z}, \tag{6.6.22}$$

$$[\mathbf{d}, \mathbf{k}] = 0 \tag{6.6.23}$$

(see [Gar1], [GarL], [K6]). From Theorem 6.2.16 we recall that for any complex number ℓ not equal to $-h$ (recall (6.6.2)), we have

$$[L(0), a(n)] = -na(n) \quad \text{for } a \in \mathfrak{g}, \ n \in \mathbb{Z} \tag{6.6.24}$$

on any restricted $\hat{\mathfrak{g}}$-module W of level ℓ (see (6.2.45)). Comparing (6.6.22) and (6.6.24) we see that $L(0)$ plays the role of $-\mathbf{d}$ under these conditions, and in fact W canonically becomes a $\tilde{\mathfrak{g}}$-module, with $\mathbf{d}$ acting as $-L(0)$. However, without further assumptions, W might not be graded by $L(0)$-eigenvalues ($\mathbf{d}$ might not act semisimply on W). (Recall Remark 6.2.10.)

Assume that $\ell \neq -h$ (cf. Remark 6.2.20). From Theorem 6.2.18, we have a vertex operator algebra $V_{\hat{\mathfrak{g}}}(\ell, 0)$, which by construction is precisely the $\hat{\mathfrak{g}}$-module generated by the vacuum vector $\mathbf{1}$ subject to the defining relations $\mathbf{k}\mathbf{1} = \ell\mathbf{1}$ and $(\mathfrak{g} \otimes \mathbb{C}[t])\mathbf{1} = 0$. (In particular, $V_{\hat{\mathfrak{g}}}(\ell, 0)$ is a restricted $\hat{\mathfrak{g}}$-module of level ℓ.) Furthermore, as a vertex algebra, $V_{\hat{\mathfrak{g}}}(\ell, 0)$ is generated by its weight–1 subspace, which is canonically identified with $\mathfrak{g}$ through the map $a \mapsto a(-1)\mathbf{1}$ (recall (6.2.22)). The relationship between the restricted $\hat{\mathfrak{g}}$-modules of level ℓ and the modules for $V_{\hat{\mathfrak{g}}}(\ell, 0)$ viewed as a vertex algebra was summarized in Theorem 6.2.13.

The vertex operator algebra $L_{\hat{\mathfrak{g}}}(\ell, 0)$, which was defined in Remark 6.2.24 (recall (6.2.80)), is the quotient algebra of $V_{\hat{\mathfrak{g}}}(\ell, 0)$ by its largest proper ideal $I_{\hat{\mathfrak{g}}}(\ell, 0)$, which is also its largest proper $\hat{\mathfrak{g}}$-submodule; thus $L_{\hat{\mathfrak{g}}}(\ell, 0)$ is a simple vertex operator algebra and is naturally an irreducible $\hat{\mathfrak{g}}$-module of level ℓ.

Remark 6.6.2 Assume that $\ell \neq 0$ (and $\ell \neq -h$). For any nonzero element $a \in \mathfrak{g}$, there exists $b \in \mathfrak{g}$ such that $\langle b, a \rangle \neq 0$, so that in the vertex operator algebra $V_{\hat{\mathfrak{g}}}(\ell, 0)$,

$$b(1)a(-1)\mathbf{1} = a(-1)b(1)\mathbf{1} + [b, a](0)\mathbf{1} + \ell\langle b, a\rangle\mathbf{1} = \ell\langle b, a\rangle\mathbf{1} \neq 0 \quad (6.6.25)$$

(recall (6.6.19)). Thus the linear map

$$\begin{aligned}
\mathfrak{g} &\to L_{\hat{\mathfrak{g}}}(\ell, 0) \\
a &\mapsto a(-1)\mathbf{1}
\end{aligned} \quad (6.6.26)$$

(cf. (6.2.22)) is injective. Through this map we consider $\mathfrak{g}$ as a subspace of $L_{\hat{\mathfrak{g}}}(\ell, 0)$; then $\mathfrak{g}$ is the full subspace of $L_{\hat{\mathfrak{g}}}(\ell, 0)$ of weight 1:

$$\mathfrak{g} = L_{\hat{\mathfrak{g}}}(\ell, 0)_{(1)}. \quad (6.6.27)$$

The same argument in fact shows that for *any* nonzero quotient vertex operator algebra V of $V_{\hat{\mathfrak{g}}}(\ell, 0)$, the same assertions hold with V in place of $L_{\hat{\mathfrak{g}}}(\ell, 0)$. Just as in Remark 6.2.19 we have

$$L(1)\mathfrak{g} = 0 \quad (6.6.28)$$

and more generally,

$$L(n)\mathfrak{g} = 0 \quad \text{for } n \geq 1. \quad (6.6.29)$$

Remark 6.6.3 Now take $\ell = 0$. Using a calculation similar to (6.6.25) we get

$$b(n)a(-1)\mathbf{1} = 0 \quad \text{for } a, b \in \mathfrak{g}, \ n \geq 1. \quad (6.6.30)$$

It follows from (6.6.20) that $V_{\hat{\mathfrak{g}}}(0, 0)_{(1)}$ $(= \mathfrak{g})$ generates a proper $\hat{\mathfrak{g}}$-submodule, so that

$$a(-1)\mathbf{1} = 0 \quad \text{in } L_{\hat{\mathfrak{g}}}(\ell, 0) \quad (6.6.31)$$

for $a \in \mathfrak{g}$. That is, the linear map $\mathfrak{g} \to L_{\hat{\mathfrak{g}}}(0, 0)$, $a \mapsto a(-1)\mathbf{1}$ is zero. Since $\mathfrak{g}$ generates $V_{\hat{\mathfrak{g}}}(0, 0)$ as a vertex algebra and $L_{\hat{\mathfrak{g}}}(0, 0)$ is a homomorphic image of $V_{\hat{\mathfrak{g}}}(0, 0)$, we have

$$L_{\hat{\mathfrak{g}}}(0, 0) = \mathbb{C}, \quad (6.6.32)$$

the one-dimensional vertex operator algebra (recall Remark 3.1.26).

Our main goal in this section is to determine all the irreducible $L_{\hat{\mathfrak{g}}}(\ell, 0)$-modules when ℓ is a nonnegative integer. First, for the trivial case $\ell = 0$, we have $L_{\hat{\mathfrak{g}}}(\ell, 0) = \mathbb{C}$, so that $\mathbb{C}$ is the only irreducible $L_{\hat{\mathfrak{g}}}(0, 0)$-module up to equivalence.

Now assume that ℓ is an arbitrary nonzero complex number (not equal to $-h$). From Theorem 6.2.23 (cf. Remark 6.2.26), any irreducible $L_{\hat{\mathfrak{g}}}(\ell, 0)$-module is equivalent to $L_{\hat{\mathfrak{g}}}(\ell, U)$ for some finite-dimensional irreducible $\mathfrak{g}$-module U. So we must determine for which finite-dimensional irreducible $\mathfrak{g}$-modules U there exists an $L_{\hat{\mathfrak{g}}}(\ell, 0)$-module structure on $W = L_{\hat{\mathfrak{g}}}(\ell, U)$ such that $Y_W(a, x) = a_W(x)$ for $a \in \mathfrak{g} \subset L_{\hat{\mathfrak{g}}}(\ell, 0)$. By Theorem 6.2.13, for any restricted $\hat{\mathfrak{g}}$-module W of level ℓ, W is naturally a module for $V_{\hat{\mathfrak{g}}}(\ell, 0)$ viewed as a vertex algebra, with $Y_W(a, x) = a_W(x)$ for $a \in \mathfrak{g}$. Thus a restricted $\hat{\mathfrak{g}}$-module W of level ℓ is naturally a module for $L_{\hat{\mathfrak{g}}}(\ell, 0)$ viewed as a vertex algebra with $Y_W(a, x) = a_W(x)$ for $a \in \mathfrak{g}$ if and only if

$$Y_W(u, x) = 0 \quad \text{on} \quad W \quad \text{for all } u \in I_{\hat{\mathfrak{g}}}(\ell, 0). \tag{6.6.33}$$

In view of this, the structure of the $\hat{\mathfrak{g}}$-module $I_{\hat{\mathfrak{g}}}(\ell, 0)$ is important for achieving our goal.

A fact from Kac–Moody algebra theory (see [K6]) is that when ℓ is a nonnegative integer, the largest proper $\hat{\mathfrak{g}}$-submodule $I_{\hat{\mathfrak{g}}}(\ell, 0)$ of $V_{\hat{\mathfrak{g}}}(\ell, 0)$ has a very simple structure, as we shall see below.

Until further notice, we assume that ℓ is a positive integer. (When we discuss highest weight $\tilde{\mathfrak{g}}$-modules below, ℓ will be allowed to be arbitrary. Sometimes ℓ will be assumed to be distinct from $-h$.)

The next result (cf. [K6], [Li3], [MP4]) is central to our method. It establishes the weak nilpotency, in the sense of Section 3.10, of the root vectors $e \in \mathfrak{g}$ associated to the long roots α, where $\mathfrak{g}$ is viewed as embedded in our vertex operator algebra $L_{\hat{\mathfrak{g}}}(\ell, 0)$ via (6.6.26).

Proposition 6.6.4 *Let $\alpha \in \Delta$ be a long root, i.e., $\langle \alpha, \alpha \rangle = 2$, and let $e \in \mathfrak{g}_\alpha$. Then*

$$e(-1)^{\ell+1}\mathbf{1} \in I_{\hat{\mathfrak{g}}}(\ell, 0) \subset V_{\hat{\mathfrak{g}}}(\ell, 0), \tag{6.6.34}$$

or equivalently,

$$e(-1)^{\ell+1}\mathbf{1} = 0 \quad \text{in} \quad L_{\hat{\mathfrak{g}}}(\ell, 0). \tag{6.6.35}$$

In particular (recall (6.6.17)),

$$e_\theta(-1)^{\ell+1}\mathbf{1}, \ f_\theta(-1)^{\ell+1}\mathbf{1} = 0 \quad \text{in} \quad L_{\hat{\mathfrak{g}}}(\ell, 0). \tag{6.6.36}$$

Proof. Set

$$v = e(-1)^{\ell+1}\mathbf{1} \in V_{\hat{\mathfrak{g}}}(\ell, 0). \tag{6.6.37}$$

What we must prove is that $U(\hat{\mathfrak{g}})v$ is a proper $\hat{\mathfrak{g}}$-submodule of $V_{\hat{\mathfrak{g}}}(\ell, 0)$. For this we are going to prove that if $v \neq 0$, the lowest $L(0)$-weight of the $V_{\hat{\mathfrak{g}}}(\ell, 0)$-submodule $U(\hat{\mathfrak{g}})v$ is the $L(0)$-weight of v, which is $\ell + 1$, from (6.2.45). In view of (6.2.45) and the triangular decomposition (6.6.20), this will follow from the relation $(\mathfrak{g} \otimes t\mathbb{C}[t])v = 0$, which we proceed to prove.

Writing $C_{\mathfrak{g}}(S)$ for the centralizer in $\mathfrak{g}$ of a subset S of $\mathfrak{g}$, we see from the commutation relations (6.6.19) that

$$a(0)v = e(-1)^{\ell+1}a(0)\mathbf{1} = 0 \quad \text{for} \quad a \in C_{\mathfrak{g}}(\mathfrak{g}_\alpha). \tag{6.6.38}$$

Let $f \in \mathfrak{g}_{-\alpha}$ be such that $\langle e, f \rangle = 1$. (Note that v is assumed to be nonzero, so $e \neq 0$.) Then

$$[h_\alpha, e] = 2e, \quad [h_\alpha, f] = -2f, \quad [e, f] = h_\alpha \tag{6.6.39}$$

(recall (6.6.5)–(6.6.10)) and notice that $\langle h_\alpha, h_\alpha \rangle = \langle \alpha, \alpha \rangle = 2$). Set

$$E = f \otimes t, \quad F = e \otimes t^{-1}, \quad H = \mathbf{k} - h_\alpha \in \hat{\mathfrak{g}}, \tag{6.6.40}$$

so that

$$[H, E] = 2E, \quad [H, F] = -2F, \quad [E, F] = H \tag{6.6.41}$$

(recall (6.6.19)). We have the standard commutation relation

$$[E, F^{r+1}] = -(r + 1)F^r(r - H) \tag{6.6.42}$$

in $U(\hat{\mathfrak{g}})$ for $r \in \mathbb{N}$ (cf. [Hum]). Since $E \cdot \mathbf{1} = f(1)\mathbf{1} = 0$ and $H \cdot \mathbf{1} = (\mathbf{k} - h_\alpha)\mathbf{1} = \ell\mathbf{1}$, we get

$$E \cdot v = EF^{\ell+1}\mathbf{1} = -(\ell + 1)F^\ell(\ell - H)\mathbf{1} = 0, \tag{6.6.43}$$

that is,

$$f(1)v = 0. \tag{6.6.44}$$

In view of (6.6.38) and (6.6.44), to prove $(\mathfrak{g} \otimes t\mathbb{C}[t])v = 0$ it suffices to prove that $\mathfrak{g} \otimes t\mathbb{C}[t]$ is contained in the subalgebra of $\hat{\mathfrak{g}}$ generated by $C_{\mathfrak{g}}(\mathfrak{g}_\alpha)$ and $f \otimes t$.

For $\beta \in \Delta$, if $\langle \alpha, \beta \rangle \geq 0$, then

$$\langle \alpha + \beta, \alpha + \beta \rangle \geq \langle \alpha, \alpha \rangle + \langle \beta, \beta \rangle > \langle \alpha, \alpha \rangle = 2, \tag{6.6.45}$$

which implies that $\alpha + \beta \notin \Delta$. Thus $[\mathfrak{g}_\alpha, \mathfrak{g}_\beta] = 0$, that is, $\mathfrak{g}_\beta \subset C_{\mathfrak{g}}(\mathfrak{g}_\alpha)$. Thus

$$\sum_{\langle \alpha, \beta \rangle \geq 0} \mathfrak{g}_\beta \subset C_{\mathfrak{g}}(\mathfrak{g}_\alpha), \tag{6.6.46}$$

and of course we also have

$$\sum_{\langle \alpha, \beta \rangle \leq 0} \mathfrak{g}_\beta \subset C_{\mathfrak{g}}(\mathfrak{g}_{-\alpha}). \tag{6.6.47}$$

Thus

$$\mathfrak{g} = C_{\mathfrak{g}}(\mathfrak{g}_\alpha) + \mathfrak{h} + C_{\mathfrak{g}}(\mathfrak{g}_{-\alpha}). \tag{6.6.48}$$

Since $\mathfrak{g}$ is an irreducible $\mathfrak{g}$-module, we have

$$\mathfrak{g} = U(\mathfrak{g}) \cdot f = U(C_{\mathfrak{g}}(\mathfrak{g}_\alpha))U(\mathfrak{h})U(C_{\mathfrak{g}}(\mathfrak{g}_{-\alpha})) \cdot f = U(C_{\mathfrak{g}}(\mathfrak{g}_\alpha)) \cdot f. \tag{6.6.49}$$

From this we get

$$U(C_{\mathfrak{g}}(\mathfrak{g}_\alpha)) \cdot (f \otimes t) = (U(C_{\mathfrak{g}}(\mathfrak{g}_\alpha)) \cdot f) \otimes t = \mathfrak{g} \otimes t, \qquad (6.6.50)$$

where we use the adjoint action of $\hat{\mathfrak{g}}$, and since $\mathfrak{g} \otimes t$ generates $\mathfrak{g} \otimes t\mathbb{C}[t]$, we see that $\mathfrak{g} \otimes t\mathbb{C}[t]$ is contained in the subalgebra of $\hat{\mathfrak{g}}$ generated by $C_{\mathfrak{g}}(\mathfrak{g}_\alpha)$ and $f \otimes t$, completing the proof. $\square$

Now we exploit the weak nilpotency property, which we have just established, of the element e of our vertex operator algebra $L_{\hat{\mathfrak{g}}}(\ell, 0)$: $e(-1)^{\ell+1}\mathbf{1} = 0$ in $L_{\hat{\mathfrak{g}}}(\ell, 0)$. Since $[e, e] = 0$ and $\langle e, e \rangle = 0$, we have

$$[e \otimes t^m, e \otimes t^n] = 0 \quad \text{for } m, n \in \mathbb{Z}. \qquad (6.6.51)$$

Thus in the vertex algebra $L_{\hat{\mathfrak{g}}}(\ell, 0)$,

$$e_n e = e(n)e(-1)\mathbf{1} = e(-1)e(n)\mathbf{1} = 0 \quad \text{for } n \geq 0. \qquad (6.6.52)$$

In view of Proposition 4.5.19 and Remark 4.5.20 we immediately have (cf. [DL3], [Li3], [MP2], [MP4]):

Proposition 6.6.5 *Let $\alpha \in \Delta$ be a long root and let $e \in \mathfrak{g}_\alpha$. Then $Y(e, x)^{\ell+1}$ is defined and*

$$Y(e, x)^{\ell+1} = 0 \quad on \quad L_{\hat{\mathfrak{g}}}(\ell, 0). \qquad (6.6.53)$$

Furthermore, for any module W for $L_{\hat{\mathfrak{g}}}(\ell, 0)$ viewed as a vertex algebra,

$$Y_W(e, x)^{\ell+1} = 0. \qquad (6.6.54)$$

In particular,

$$Y_W(e_\theta, x)^{\ell+1} = Y_W(f_\theta, x)^{\ell+1} = 0. \quad \square \qquad (6.6.55)$$

We shall next apply Proposition 6.6.5 to obtain a necessary condition on a finite-dimensional irreducible $\mathfrak{g}$-module U such that $L_{\hat{\mathfrak{g}}}(\ell, U)$ is naturally an $L_{\hat{\mathfrak{g}}}(\ell, 0)$-module. We shall in fact work more generally with possibly infinite-dimensional highest weight irreducible $\mathfrak{g}$-modules U.

Remark 6.6.6 Observe that in the context of Section 6.2, for any $\ell \neq -h$ the structure $L_{\hat{\mathfrak{g}}}(\ell, U)$ can certainly be defined, as in (6.2.79), but more generally for any irreducible $\mathfrak{g}$-module U on which the Casimir operator Ω (recall (6.2.37)) acts as a scalar (recall (6.2.73)–(6.2.79)). For example, for any $\lambda \in \mathfrak{h}^*$, consider the unique (up to equivalence) irreducible $\mathfrak{g}$-module $L(\lambda)$ with highest weight λ; recall that $L(\lambda)$ is finite–dimensional if and only if λ is dominant integral. Then $U = L(\lambda)$ is an irreducible $\mathfrak{g}$-module on which Ω acts as a scalar. In fact, in the context of Remark 6.2.14, let us choose a basis of $\mathfrak{g}$ consisting of an orthonormal basis $\{w^{(i)}\}$ of $\mathfrak{h}$ (with respect to the form $\langle \cdot, \cdot \rangle$) together with a set of root vectors $\{x_\alpha \in \mathfrak{g}_\alpha \mid \alpha \in \Delta\}$ such that $\langle x_\alpha, x_{-\alpha} \rangle = 1$ for all $\alpha \in \Delta$. Using Remark 6.2.14 we find that

$$
\begin{aligned}
\Omega &= \sum_{i=1}^{\dim \mathfrak{h}} w^{(i)} w^{(i)} + \sum_{\alpha \in \Delta_+} (2 x_{-\alpha} x_\alpha + [x_\alpha, x_{-\alpha}]) \\
&= \sum_{i=1}^{\dim \mathfrak{h}} w^{(i)} w^{(i)} + \sum_{\alpha \in \Delta_+} (2 x_{-\alpha} x_\alpha + \alpha),
\end{aligned}
\tag{6.6.56}
$$

from (6.6.7) (where as usual we are identifying $\mathfrak{h}$ with $\mathfrak{h}^*$ using $\langle \cdot, \cdot \rangle$). Writing

$$
\rho = \frac{1}{2} \sum_{\alpha \in \Delta_+} \alpha
\tag{6.6.57}
$$

for the half-sum of the positive roots, and applying Ω to a highest weight vector v of $L(\lambda)$, we have

$$
\begin{aligned}
\Omega v &= \langle \lambda, \lambda \rangle v + 2 \langle \rho, \lambda \rangle v \\
&= \langle \lambda, \lambda + 2\rho \rangle v \\
&= (\langle \lambda + \rho, \lambda + \rho \rangle - \langle \rho, \rho \rangle) v,
\end{aligned}
\tag{6.6.58}
$$

and Ω must act by this same scalar on all of $L(\lambda)$. (This argument shows that Ω acts as this same scalar on *any* highest weight $\mathfrak{g}$-module with highest weight λ in place of $L(\lambda)$.) Thus $L_{\hat{\mathfrak{g}}}(\ell, L(\lambda))$ is defined. In general, $L_{\hat{\mathfrak{g}}}(\ell, U)$ is an irreducible module for $V_{\hat{\mathfrak{g}}}(\ell, 0)$ viewed as a vertex operator algebra, except possibly with infinite-dimensional homogeneous subspaces (cf. Theorems 6.2.23 and 6.2.21). Note, incidentally, that in case $\lambda = 0 \in \mathfrak{h}^*$, $L(0)$ (which should not be confused with the Virasoro algebra operator $L(0)$!) is the trivial one-dimensional $\mathfrak{g}$-module, and the $V_{\hat{\mathfrak{g}}}(\ell, 0)$-module we are calling $L_{\hat{\mathfrak{g}}}(\ell, L(0))$ is the same as the structure that we have also been calling $L_{\hat{\mathfrak{g}}}(\ell, 0)$; recall Remark 6.2.25. In addition to being a $V_{\hat{\mathfrak{g}}}(\ell, 0)$-module, $L_{\hat{\mathfrak{g}}}(\ell, 0)$ is also a simple vertex operator algebra (recall Remark 6.2.24). Note that the Casimir element Ω acts on $L(0)$ by the scalar 0.

We now have (cf. [DL3], Proposition 13.17):

Theorem 6.6.7 *Assume that ℓ is a positive integer and let $\lambda \in \mathfrak{h}^*$. If $L_{\hat{\mathfrak{g}}}(\ell, L(\lambda))$ is naturally a module for $L_{\hat{\mathfrak{g}}}(\ell, 0)$ viewed as a vertex algebra, then $L(\lambda)$ is finite dimensional, or equivalently, λ is dominant integral, and*

$$
\lambda(h_\theta) \le \ell.
\tag{6.6.59}
$$

In particular, $L_{\hat{\mathfrak{g}}}(\ell, L(\lambda))$ is in fact a module for $L_{\hat{\mathfrak{g}}}(\ell, 0)$ viewed as a vertex operator algebra. Furthermore,

$$
e_\theta(-1)^{\ell+1} v = 0,
\tag{6.6.60}
$$

where v is a highest weight vector of the $\mathfrak{g}$-module $L(\lambda)$.

Proof. Let α be a long root and let $e \in \mathfrak{g}_\alpha$. Since $L(\lambda)$ is the lowest $L(0)$-weight subspace of $L_{\hat{\mathfrak{g}}}(\ell, L(\lambda))$ (from Theorem 6.2.21), we have

$$e(n)u = 0 \quad \text{for } n \geq 1, \ u \in L(\lambda) \subset L_{\hat{\mathfrak{g}}}(\ell, L(\lambda)). \tag{6.6.61}$$

Thus the coefficient of $x^{-\ell-1}$ in $Y(e, x)^{\ell+1}u$ is $e(0)^{\ell+1}u$, which must vanish by Proposition 6.6.5. Hence

$$e^{\ell+1}\left(= e(0)^{\ell+1}\right) = 0 \quad \text{on} \quad L(\lambda). \tag{6.6.62}$$

In particular,

$$e_\theta^{\ell+1} = f_\theta^{\ell+1} = 0 \quad \text{on} \quad L(\lambda). \tag{6.6.63}$$

For a highest weight vector v of $L(\lambda)$, since $e_\theta(n)v = 0$ for $n \geq 0$, the constant term of $Y(e_\theta, x)^{\ell+1}v$ is $e_\theta(-1)^{\ell+1}v$, which must vanish by Proposition 6.6.5.

The finite dimensionality of $L(\lambda)$ will follow immediately from (6.6.63) and the following lemma. It then follows that $U(\mathfrak{g}^\theta)v$ is an irreducible module for the three-dimensional simple Lie algebra $\mathfrak{g}^\theta$ (recall (6.6.11)) with highest weight $\lambda(h_\theta)$, so that $\dim U(\mathfrak{g}^\theta)v = \lambda(h_\theta) + 1$. Since $e_\theta^{\ell+1}L(\lambda) = 0$, we must have $\lambda(h_\theta) + 1 \leq \ell + 1$. That is, $\lambda(h_\theta) \leq \ell$. $\square$

Lemma 6.6.8 *Let W be a $\mathfrak{g}$-module on which $\mathfrak{h}$ acts semisimply. Suppose that $\mathfrak{g}_\alpha$ and $\mathfrak{g}_{-\alpha}$ act nilpotently on W for some $\alpha \in \Delta$. Then for each $\beta \in \Delta$, $\mathfrak{g}_\beta$ acts nilpotently on W and W is a direct sum of finite-dimensional irreducible $\mathfrak{g}$-modules.*

Proof. For $\beta \in \Delta$, $n \in \mathbb{C}$, set

$$W_\beta(n) = \{w \in W \mid h_\beta w = nw\}. \tag{6.6.64}$$

Since $\mathfrak{h}$ acts semisimply on W,

$$W = \coprod_{n \in \mathbb{C}} W_\beta(n). \tag{6.6.65}$$

It follows from the triangular decomposition (6.6.11) that any $\mathfrak{h}$-weight vector in W, and hence any vector in W, generates a finite-dimensional $\mathfrak{g}^\alpha$-submodule. Consequently, W is a direct sum of finite-dimensional irreducible $\mathfrak{g}^\alpha$-modules. Assume that $\mathfrak{g}_\alpha^{k+1} = \mathfrak{g}_{-\alpha}^{k+1} = 0$ for some $k \in \mathbb{N}$. Then W is a direct sum of irreducible $\mathfrak{g}^\alpha$-modules of dimension $\leq k + 1$. Thus

$$W = W_\alpha(k) \oplus W_\alpha(k - 1) \oplus \cdots \oplus W_\alpha(-k). \tag{6.6.66}$$

Let $\beta \in \Delta$ be such that $\langle \alpha, \beta \rangle \neq 0$, so that $\beta(h_\alpha) \neq 0$. Since

$$\mathfrak{g}_\beta W_\alpha(n) \subset W_\alpha(n + \beta(h_\alpha)) \tag{6.6.67}$$

for $n \in \mathbb{C}$, $\mathfrak{g}_\beta$ acts nilpotently on W. The same argument also shows that $\mathfrak{g}_{-\beta}$ acts nilpotently on W. Since $\mathfrak{g}$ is simple, for any $\gamma \in \Delta$, there exist $\beta_1, \ldots, \beta_m \in \Delta$ such that $\langle \beta_i, \beta_{i+1} \rangle \neq 0$ for $i = 0, 1, \ldots, m$, with $\beta_0 = \alpha$ and $\beta_{m+1} = \gamma$. It follows that $\mathfrak{g}_\gamma$ acts nilpotently on W for every $\gamma \in \Delta$. This proves the first assertion, which in turn implies that each weight vector in W, and hence each vector in W, generates a finite-dimensional $\mathfrak{g}$-module. Consequently, W is a direct sum of finite-dimensional irreducible $\mathfrak{g}$-modules. $\square$

In Theorem 6.6.7 we allowed the irreducible $\mathfrak{g}$-module $L(\lambda)$ to be infinite-dimensional. If we assume instead that $L_{\hat{\mathfrak{g}}}(\ell, U)$ is an irreducible module for $L_{\hat{\mathfrak{g}}}(\ell, 0)$ viewed as a vertex operator algebra, then U is finite dimensional by assumption, and we have, using Theorems 6.2.23 and 6.6.7:

Corollary 6.6.9 *Assume that ℓ is a positive integer. Every irreducible module for $L_{\hat{\mathfrak{g}}}(\ell, 0)$, viewed as a vertex operator algebra, is equivalent to $L_{\hat{\mathfrak{g}}}(\ell, L(\lambda))$ for some dominant integral $\lambda \in \mathfrak{h}^*$, and we have $\lambda(h_\theta) \leq \ell$ and $e_\theta(-1)^{\ell+1} v = 0$, where v is a highest weight vector of the $\mathfrak{g}$-module $L(\lambda)$.* $\square$

Remark 6.6.10 Note that there are only finitely many dominant integral λ such that $\lambda(h_\theta) \leq \ell$.

We are going to use Corollary 6.6.9 to identify the irreducible $L_{\hat{\mathfrak{g}}}(\ell, 0)$-modules with integrable highest weight modules for $\tilde{\mathfrak{g}}$ (recall Remark 6.6.1), or, as they are also known, standard $\tilde{\mathfrak{g}}$-modules, of level ℓ. Let us recall the definition of this notion and related notions (see [K6]).

First, in addition to the triangular decomposition (6.6.20), the affine algebra $\hat{\mathfrak{g}}$ has another important triangular decomposition,

$$\hat{\mathfrak{g}} = \hat{\mathfrak{g}}^+ \oplus (\mathfrak{h} \oplus \mathbb{C}\mathbf{k}) \oplus \hat{\mathfrak{g}}^-, \tag{6.6.68}$$

where

$$\hat{\mathfrak{g}}^\pm = \mathfrak{g}_\pm \oplus \left(\mathfrak{g} \otimes t^{\pm 1} \mathbb{C}[t^{\pm 1}] \right) \tag{6.6.69}$$

(recall (6.6.13)). Recall from Remark 6.6.1 that $\tilde{\mathfrak{g}}$ has the degree derivation $\mathbf{d}$ adjoined to $\hat{\mathfrak{g}}$. The abelian subalgebra

$$\mathcal{H} = \mathbb{C}\mathbf{d} \oplus \mathbb{C}\mathbf{k} \oplus \mathfrak{h} \tag{6.6.70}$$

plays the role of the Cartan subalgebra of the Kac–Moody algebra $\tilde{\mathfrak{g}}$. A nonzero vector v in a $\tilde{\mathfrak{g}}$-module is called a *weight vector*, with *weight* $\mu \in \mathcal{H}^*$, if

$$a \cdot v = \mu(a)v \quad \text{for } a \in \mathcal{H}. \tag{6.6.71}$$

A weight vector v is called a *highest weight vector* if

$$\hat{\mathfrak{g}}^+ v = 0. \tag{6.6.72}$$

A *highest weight $\tilde{\mathfrak{g}}$-module* is a $\tilde{\mathfrak{g}}$-module generated by a highest weight vector. It is clear that a generating highest weight vector in a highest weight $\tilde{\mathfrak{g}}$-module is unique up to scalar multiple and that its weight μ has the property that the space of all vectors of weight μ (together with the zero vector) is one-dimensional. This element μ is the *highest weight* of the highest weight module. Equivalent highest weight $\tilde{\mathfrak{g}}$-modules clearly have the same highest weight.

Let W be a highest weight $\tilde{\mathfrak{g}}$-module with highest weight vector v and highest weight $\mu \in \mathcal{H}^*$. We may view μ as a triple

$$\mu = (\ell, m, \lambda) \tag{6.6.73}$$

where $\ell \in \mathbb{C}$ is the scalar by which $\mathbf{k}$ acts on v and hence on all of W (ℓ is the *level* of W), $m \in \mathbb{C}$ is the scalar by which $\mathbf{d}$ acts on v and $\lambda \in \mathfrak{h}^*$ is the $\mathfrak{h}$-weight of v. Note that the $\mathbf{d}$-eigenspace of W with eigenvalue m is a highest weight module for the finite-dimensional simple Lie algebra $\mathfrak{g}$ with highest weight λ and highest weight vector v, and that W has a $\mathbf{d}$-eigenspace decomposition, with all the eigenvalues other than m of the form $m - n$ where n is a positive integer.

A highest weight $\tilde{\mathfrak{g}}$-*module* W is called a *standard $\tilde{\mathfrak{g}}$-module*, or an *integrable highest weight $\tilde{\mathfrak{g}}$-module*, if there is a nonnegative integer k such that

$$(\mathfrak{g}_{-\alpha})^{k+1}v = 0 \quad \text{and} \quad e_\theta(-1)^{k+1}v = 0 \tag{6.6.74}$$

for every simple root α of $\mathfrak{g}$ with respect to $\mathfrak{h}$, where v is a generating highest weight vector of W. The *level* of a standard $\tilde{\mathfrak{g}}$-module (the scalar by which $\mathbf{k}$ acts) is a nonnegative integer (see [K6]).

We shall next exhibit the irreducible $\hat{\mathfrak{g}}$-modules $L_{\hat{\mathfrak{g}}}(\ell, L(\lambda))$ in Corollary 6.6.9 as standard $\tilde{\mathfrak{g}}$-modules (with $-L(0)$ giving rise to the action of the element $\mathbf{d}$ of $\tilde{\mathfrak{g}}$), and for this, it is convenient to recall the construction of the irreducible highest weight $\tilde{\mathfrak{g}}$-modules using Verma modules for $\tilde{\mathfrak{g}}$ (cf. [K6]). For any $\ell, m \in \mathbb{C}$ and $\lambda \in \mathfrak{h}^*$, let $\mathbb{C}_{(\ell,m,\lambda)}$ be the one-dimensional $(\hat{\mathfrak{g}}^+ \oplus \mathcal{H})$-module $\mathbb{C}$ on which $\hat{\mathfrak{g}}^+$ acts trivially, $\mathfrak{h}$ acts according to the character λ, $\mathbf{k}$ acts as the scalar ℓ and $\mathbf{d}$ acts as the scalar m (recall (6.6.21), (6.6.68) and (6.6.70)). Form the induced $\tilde{\mathfrak{g}}$-module

$$M_{\tilde{\mathfrak{g}}}(\ell, m, \lambda) = U(\tilde{\mathfrak{g}}) \otimes_{U(\hat{\mathfrak{g}}^+ \oplus \mathcal{H})} \mathbb{C}_{(\ell,m,\lambda)}. \tag{6.6.75}$$

In view of the Poincaré–Birkhoff–Witt theorem, we have

$$M_{\tilde{\mathfrak{g}}}(\ell, m, \lambda) = U(\hat{\mathfrak{g}}^-) \otimes \mathbb{C}_{(\ell,m,\lambda)} \tag{6.6.76}$$

as a vector space, so that we may and do consider $\mathbb{C}_{(\ell,m,\lambda)}$ as a subspace of $M_{\tilde{\mathfrak{g}}}(\ell, m, \lambda)$. Set

$$1_{(\ell,m,\lambda)} = 1 \in \mathbb{C}_{(\ell,m,\lambda)} \subset M_{\tilde{\mathfrak{g}}}(\ell, m, \lambda). \tag{6.6.77}$$

The $\tilde{\mathfrak{g}}$-module $M_{\tilde{\mathfrak{g}}}(\ell, m, \lambda)$ is the *Verma module* for $\tilde{\mathfrak{g}}$ with highest weight (ℓ, m, λ), where this triple is viewed as an element of $\mathcal{H}^*$ (recall (6.6.73)); it is clear that

$M_{\tilde{\mathfrak{g}}}(\ell, m, \lambda)$ is in fact a highest weight $\tilde{\mathfrak{g}}$-module with highest weight (ℓ, m, λ) and with 1 as a generating highest weight vector, and that $M_{\tilde{\mathfrak{g}}}(\ell, m, \lambda)$ is universal in the sense that for any highest weight $\tilde{\mathfrak{g}}$-module W with highest weight (ℓ, m, λ) and generating highest weight vector v, there is a unique $\tilde{\mathfrak{g}}$-module homomorphism from $M_{\tilde{\mathfrak{g}}}(\ell, m, \lambda)$ onto W taking 1 to v.

The Verma module $M_{\tilde{\mathfrak{g}}}(\ell, m, \lambda)$ clearly has a (unique) largest proper $\tilde{\mathfrak{g}}$-submodule. Calling it J, we form the irreducible quotient module

$$L_{\tilde{\mathfrak{g}}}(\ell, m, \lambda) = M_{\tilde{\mathfrak{g}}}(\ell, m, \lambda)/J. \tag{6.6.78}$$

Remark 6.6.11 The $\tilde{\mathfrak{g}}$-module $L_{\tilde{\mathfrak{g}}}(\ell, m, \lambda)$ is clearly the unique (up to equivalence) irreducible highest weight $\tilde{\mathfrak{g}}$-module with highest weight (ℓ, m, λ). Furthermore, equivalent irreducible highest weight $\tilde{\mathfrak{g}}$-modules have the same highest weight.

Remark 6.6.12 Now let ℓ be any complex number not equal to $-h$ and let $\lambda \in \mathfrak{h}^*$, and recall from Remark 6.6.6 the irreducible $\hat{\mathfrak{g}}$-module and $V_{\hat{\mathfrak{g}}}(\ell, 0)$-module $L_{\hat{\mathfrak{g}}}(\ell, L(\lambda))$. Also recall from Remark 6.6.6 that the Casimir element Ω of $\mathfrak{g}$ acts as the scalar

$$h_{L(\lambda)} = \langle \lambda, \lambda + 2\rho \rangle = \langle \lambda + \rho, \lambda + \rho \rangle - \langle \rho, \rho \rangle \tag{6.6.79}$$

on $L(\lambda)$. The canonical action of $L(0)$ on $L_{\hat{\mathfrak{g}}}(\ell, L(\lambda))$ makes this $\hat{\mathfrak{g}}$-module a $\tilde{\mathfrak{g}}$-module, with $\mathbf{d}$ acting as $-L(0)$ (recall Remark 6.6.1), and as such, $L_{\hat{\mathfrak{g}}}(\ell, L(\lambda))$ is clearly equivalent to the irreducible highest weight $\tilde{\mathfrak{g}}$-module with highest weight $(\ell, -\langle \lambda, \lambda + 2\rho \rangle/2(\ell + h), \lambda)$ (recall Theorem 6.2.21; cf. Theorem 6.2.23):

$$L_{\hat{\mathfrak{g}}}(\ell, L(\lambda)) \simeq L_{\tilde{\mathfrak{g}}}(\ell, -\langle \lambda, \lambda + 2\rho \rangle/2(\ell + h), \lambda); \tag{6.6.80}$$

the highest weight vector of the irreducible $\mathfrak{g}$-module $L(\lambda)$ (viewed as a subspace of $L_{\hat{\mathfrak{g}}}(\ell, L(\lambda))$) is also a highest weight vector of the $\tilde{\mathfrak{g}}$-module $L_{\hat{\mathfrak{g}}}(\ell, L(\lambda))$. In particular, this highest weight $\tilde{\mathfrak{g}}$-module carries a natural structure of irreducible module for the vertex operator algebra $V_{\hat{\mathfrak{g}}}(\ell, 0)$, except that its homogeneous subspaces need not be finite-dimensional. Note that for the case $\lambda = 0$ we have (recalling Remark 6.6.6)

$$L_{\hat{\mathfrak{g}}}(\ell, 0) = L_{\hat{\mathfrak{g}}}(\ell, L(0)) \simeq L_{\tilde{\mathfrak{g}}}(\ell, 0, 0). \tag{6.6.81}$$

Remark 6.6.13 With ℓ and λ as in Remark 6.6.12, let $V(\lambda)$ be the Verma module for $\mathfrak{g}$ with highest weight λ. Again from Remark 6.6.6, the Casimir element Ω of $\mathfrak{g}$ acts as the scalar

$$h_{V(\lambda)} = \langle \lambda, \lambda + 2\rho \rangle = \langle \lambda + \rho, \lambda + \rho \rangle - \langle \rho, \rho \rangle \tag{6.6.82}$$

on $V(\lambda)$, and so the Verma module $M_{\tilde{\mathfrak{g}}}(\ell, -\langle \lambda, \lambda + 2\rho \rangle/2(\ell + h), \lambda)$ for $\tilde{\mathfrak{g}}$ also carries a natural module structure for the vertex operator algebra $V_{\hat{\mathfrak{g}}}(\ell, 0)$ (except that it has infinite-dimensional $L(0)$-eigenspaces), since we have the natural identification

$$\mathrm{Ind}_{\mathfrak{g}}^{\hat{\mathfrak{g}}} V(\lambda) \simeq M_{\tilde{\mathfrak{g}}}(\ell, -\langle \lambda, \lambda + 2\rho \rangle/2(\ell + h), \lambda), \tag{6.6.83}$$

from the Poincaré–Birkhoff–Witt theorem. Of course, we also have that $\mathrm{Ind}_{\mathfrak{g}}^{\hat{\mathfrak{g}}} V(\lambda)$ is a highest weight $\tilde{\mathfrak{g}}$-module. For the case $\lambda = 0$ we have

$$\mathrm{Ind}_{\mathfrak{g}}^{\hat{\mathfrak{g}}} V(0) \simeq M_{\tilde{\mathfrak{g}}}(\ell, 0, 0). \tag{6.6.84}$$

Now that we have identified the structures $L_{\hat{\mathfrak{g}}}(\ell, 0)$ $(= L_{\hat{\mathfrak{g}}}(\ell, L(0))$ and $L_{\hat{\mathfrak{g}}}(\ell, L(\lambda))$ as highest weight $\tilde{\mathfrak{g}}$-modules (with $\mathbf{d}$ acting as $-L(0)$), Proposition 6.6.4, Theorem 6.6.7 and Corollary 6.6.9 immediately give (cf. [FZ], [DL3], [Li3], [MP4]):

Theorem 6.6.14 *Let ℓ be a positive integer. Then the simple vertex operator algebra $L_{\hat{\mathfrak{g}}}(\ell, 0)$ is a standard $\tilde{\mathfrak{g}}$-module of level ℓ, namely,*

$$L_{\hat{\mathfrak{g}}}(\ell, 0) \simeq L_{\tilde{\mathfrak{g}}}(\ell, 0, 0), \tag{6.6.85}$$

with $\mathbf{d}$ acting as $-L(0)$. Moreover, every irreducible module W for the vertex operator algebra $L_{\hat{\mathfrak{g}}}(\ell, 0)$ is a standard $\tilde{\mathfrak{g}}$-module of level ℓ, with $\mathbf{d}$ acting as $-L(0)$:

$$W \simeq L_{\hat{\mathfrak{g}}}(\ell, L(\lambda)) \simeq L_{\tilde{\mathfrak{g}}}(\ell, -\langle \lambda, \lambda + 2\rho \rangle / 2(\ell + h), \lambda) \tag{6.6.86}$$

for some dominant integral $\lambda \in \mathfrak{h}^*$, with $\lambda(h_\theta) \leq \ell$. Furthermore, for $\lambda \in \mathfrak{h}^*$, if $L_{\hat{\mathfrak{g}}}(\ell, L(\lambda))$ is an (irreducible) module for $L_{\hat{\mathfrak{g}}}(\ell, 0)$ viewed as a vertex algebra, then λ is dominant integral and $L_{\hat{\mathfrak{g}}}(\ell, L(\lambda))$ is as in (6.6.86). $\square$

Remark 6.6.15 The assertions of Theorem 6.6.14 remain valid for $\ell = 0$, for trivial reasons; recall from Remark 6.6.3 that $L_{\hat{\mathfrak{g}}}(0, 0)$ is the trivial one-dimensional vertex operator algebra.

Next we are going to prove that the converse of the second assertion of Theorem 6.6.14 also holds, that is, every standard $\tilde{\mathfrak{g}}$-module of level ℓ is naturally an (irreducible) $L_{\hat{\mathfrak{g}}}(\ell, 0)$-module, except that, as we expect, the grading associated with the action of $\mathbf{d}$ on the given standard module may have to be shifted in order to provide the natural $L(0)$-action.

Remark 6.6.16 Here we collect some basic facts about standard $\tilde{\mathfrak{g}}$-modules, which can be found in [K6] or follow immediately from results in [K6]. First, every standard module is irreducible and remains irreducible under $\hat{\mathfrak{g}}$. Given a standard $\tilde{\mathfrak{g}}$-module W, consider its $\mathbb{C}$-grading determined by the eigenspace decomposition of W with respect to the given action of $\mathbf{d}$ on W. (The eigenvalues of $\mathbf{d}$ are of the form $b - n$ for some fixed $b \in \mathbb{C}$ and for nonnegative integers n.) The operator $L(0)$, defined as in (6.2.44) for $n = 0$, acts naturally on W and the operator $\mathbf{d} + L(0)$ is a complex scalar operator on W; that is, up to a grading shift, the grading on W is defined by the eigenvalues of the operator $-L(0)$. *It is convenient for us to assume that $\mathbf{d}$ acts as $-L(0)$ on a given standard $\tilde{\mathfrak{g}}$-module.* With this assumption, every standard $\tilde{\mathfrak{g}}$-module is equivalent to a $\tilde{\mathfrak{g}}$-module of the form $L_{\hat{\mathfrak{g}}}(\ell, L(\lambda))$ (recall (6.2.79)), where ℓ is a nonnegative integer and λ is a dominant integral element of $\mathfrak{h}^*$, with $-L(0)$ providing the action of $\mathbf{d}$. Among these $\tilde{\mathfrak{g}}$-modules, $L_{\hat{\mathfrak{g}}}(\ell, L(\lambda))$ is standard if and only if $\lambda(h_\theta) \leq \ell$. The only standard

$\tilde{\mathfrak{g}}$-module of level 0 is the trivial one-dimensional module. It is convenient to define a $\hat{\mathfrak{g}}$-module to be a *standard $\hat{\mathfrak{g}}$-module* if it is a standard $\tilde{\mathfrak{g}}$-module but with the action of $\mathbf{d}\ (= -L(0))$ *ignored*. With this notion, the standard $\hat{\mathfrak{g}}$-modules are exactly those equivalent to the $\hat{\mathfrak{g}}$-modules of the form $L_{\hat{\mathfrak{g}}}(\ell, L(\lambda))$ with ℓ a nonnegative integer and λ a dominant integral element of $\mathfrak{h}^*$ with $\lambda(h_\theta) \leq \ell$. Consider $\mathfrak{sl}(2, \mathbb{C})$ as a subalgebra of $\mathfrak{g}$ through the embedding

$$\phi : h \mapsto h_\theta, \quad e \mapsto e_\theta, \quad f \mapsto f_\theta, \tag{6.6.87}$$

where $\{h, e, f\}$ is a standard Chevalley basis of $\mathfrak{sl}(2, \mathbb{C})$. Since the embedding map ϕ preserves the bilinear forms, we naturally consider $\widehat{\mathfrak{sl}(2, \mathbb{C})}$ as a subalgebra of $\hat{\mathfrak{g}}$ (with $\mathbf{k}$ for $\widehat{\mathfrak{sl}(2, \mathbb{C})}$ identified with $\mathbf{k}$ for $\hat{\mathfrak{g}}$), so that a $\hat{\mathfrak{g}}$-module of level ℓ is naturally an $\widehat{\mathfrak{sl}(2, \mathbb{C})}$-module of level ℓ. We shall need the fact that every standard $\hat{\mathfrak{g}}$-module W of level ℓ, viewed as an $\widehat{\mathfrak{sl}(2, \mathbb{C})}$-module, is a direct sum of standard $\widehat{\mathfrak{sl}(2, \mathbb{C})}$-modules of level ℓ.

Assume that ℓ is a positive integer. Notice that Proposition 6.6.4 implies that

$$U(\hat{\mathfrak{g}})e_\theta(-1)^{\ell+1}\mathbf{1} \subset I_{\hat{\mathfrak{g}}}(\ell, 0). \tag{6.6.88}$$

In fact, equality holds. Indeed, let $\mathbf{d}$ act on $V_{\hat{\mathfrak{g}}}(\ell, 0)$ as $-L(0)$. Since $\mathbf{d} \cdot \mathbf{1} = -L(0)\mathbf{1} = 0$, $\mathfrak{g}_\alpha \cdot \mathbf{1} = 0$ for $\alpha \in \Delta$, and $e_\theta(-1)^{\ell+1}\mathbf{1} = 0$ in the quotient $\tilde{\mathfrak{g}}$-module $V_{\hat{\mathfrak{g}}}(\ell, 0)/U(\hat{\mathfrak{g}})e_\theta(-1)^{\ell+1}\mathbf{1}$, we have that this quotient module is a standard $\tilde{\mathfrak{g}}$-module, which is irreducible in view of Remark 6.6.16. Consequently, $I_{\hat{\mathfrak{g}}}(\ell, 0) = U(\hat{\mathfrak{g}})e_\theta(-1)^{\ell+1}\mathbf{1}$. Therefore, we have (cf. [K6]):

Proposition 6.6.17 *Let ℓ be a positive integer. As a $\hat{\mathfrak{g}}$-submodule of $V_{\hat{\mathfrak{g}}}(\ell, 0)$,*

$$I_{\hat{\mathfrak{g}}}(\ell, 0) = U(\hat{\mathfrak{g}})e_\theta(-1)^{\ell+1}\mathbf{1}. \quad \square \tag{6.6.89}$$

With this result, we are able to prove the following central result, to a certain extent strengthening Proposition 6.6.5:

Theorem 6.6.18 *Let ℓ be a positive integer and let W be a restricted $\hat{\mathfrak{g}}$-module of level ℓ. Then W is a module for $L_{\hat{\mathfrak{g}}}(\ell, 0)$ viewed as a vertex algebra, with $Y_W(a, x) = a_W(x) = a(x)$ for $a \in \mathfrak{g} \subset L_{\hat{\mathfrak{g}}}(\ell, 0)$, if and only if*

$$e_\theta(x)^{\ell+1} = 0 \quad on \quad W. \tag{6.6.90}$$

(Notice that $e_\theta(x)^{\ell+1}$ is defined on W.)

Proof. We need only prove the "if" part. By Theorem 6.2.12, W is a module for $V_{\hat{\mathfrak{g}}}(\ell, 0)$ viewed as a vertex algebra, with $Y_W(a, x) = a_W(x)$ for $a \in \mathfrak{g} \subset V_{\hat{\mathfrak{g}}}(\ell, 0)$. From our assumption we have

$$Y_W(e_\theta, x)^{\ell+1} = e_\theta(x)^{\ell+1} = 0. \tag{6.6.91}$$

Set

$$V = V_{\hat{\mathfrak{g}}}(\ell, 0)/I_W, \tag{6.6.92}$$

where $I_W = \{v \in V_{\hat{\mathfrak{g}}}(\ell, 0) \mid Y_W(v, x) = 0\}$. By Proposition 4.5.11, I_W is an ideal of $V_{\hat{\mathfrak{g}}}(\ell, 0)$, so that V is a vertex algebra with W as a faithful module. Since the zero space is an $L_{\hat{\mathfrak{g}}}(\ell, 0)$-module, we now assume that $W \neq 0$. We have

$$I_W \subset I_{\hat{\mathfrak{g}}}(\ell, 0), \tag{6.6.93}$$

since $1 \notin I_W$ and $I_{\hat{\mathfrak{g}}}(\ell, 0)$ is the largest proper ideal of $V_{\hat{\mathfrak{g}}}(\ell, 0)$. With W being a faithful V-module and with the nilpotence property (6.6.91), by Proposition 4.5.19 and Remark 4.5.20 we have

$$e_\theta(-1)^{\ell+1}\mathbf{1} = 0 \quad \text{in} \quad V. \tag{6.6.94}$$

(Recall from Remark 6.6.2 that $\mathfrak{g}$ is naturally embedded in V.) That is, $e_\theta(-1)^{\ell+1}\mathbf{1} \in I_W$. Combining this with (6.6.93) we have

$$U(\hat{\mathfrak{g}})e_\theta(-1)^{\ell+1}\mathbf{1} \subset I_W \subset I_{\hat{\mathfrak{g}}}(\ell, 0), \tag{6.6.95}$$

which together with Proposition 6.6.17 immediately implies that $I_W = I_{\hat{\mathfrak{g}}}(\ell, 0)$. Now we have $V = L_{\hat{\mathfrak{g}}}(\ell, 0)$, as a vertex algebra, proving that W is a module for $L_{\hat{\mathfrak{g}}}(\ell, 0)$ viewed as a vertex algebra. $\square$

In view of Theorem 6.6.18, to prove that every standard $\hat{\mathfrak{g}}$-module W of level ℓ (a positive integer) is naturally an $L_{\hat{\mathfrak{g}}}(\ell, 0)$-module, all we need to prove is that $e_\theta(x)^{\ell+1} = 0$ on W. To do this, first we consider the special case $\mathfrak{g} = \mathfrak{sl}(2, \mathbb{C})$. Taking α to be the positive root, we have

$$\Delta = \{\pm\alpha\}, \quad \theta = \alpha, \quad \rho = \frac{1}{2}\alpha, \quad h = 2. \tag{6.6.96}$$

The dominant integral weights of $\mathfrak{sl}(2, \mathbb{C})$ are

$$\lambda = m\alpha/2 \quad \text{for } m \in \mathbb{N} \tag{6.6.97}$$

and the Casimir element acts on the $\mathfrak{g}$-modules $L(m\alpha/2)$ and $V(m\alpha/2)$ as the scalar

$$h_{L(m\alpha/2)} = \langle m\alpha/2, m\alpha/2 + 2\rho \rangle = \langle m\alpha/2, m\alpha/2 + \alpha \rangle = m(m+2)/2 \tag{6.6.98}$$

(recall Remarks 6.6.12 and 6.6.13). Then all the standard $\widehat{\mathfrak{sl}(2, \mathbb{C})}$-modules of level ℓ (up to equivalence) are

$$L_{\widehat{\mathfrak{sl}(2,\mathbb{C})}}(\ell, L(m\alpha/2)) \quad \text{for } m \in \mathbb{N}, \ 0 \le m \le \ell, \tag{6.6.99}$$

and as an $\widehat{\mathfrak{sl}(2, \mathbb{C})}$-module with $\mathbf{d}$ acting as $-L(0)$, we have

$$L_{\widehat{\mathfrak{sl}(2,\mathbb{C})}}\left(\ell, L(m\alpha/2)\right) \simeq L_{\widehat{\mathfrak{sl}(2,\mathbb{C})}}\left(\ell, -m(m+2)/4(\ell+2), m\alpha/2\right) \quad (6.6.100)$$

(again recall Remark 6.6.12). For brevity let us write $L(\ell, m)$ for $L_{\widehat{\mathfrak{sl}(2,\mathbb{C})}}(\ell, L(m\alpha/2))$ viewed as an $\widehat{\mathfrak{sl}(2,\mathbb{C})}$-module:

$$L(\ell, m) = L_{\widehat{\mathfrak{sl}(2,\mathbb{C})}}\left(\ell, L(m\alpha/2)\right). \quad (6.6.101)$$

Now, using the lattice constructions developed in Sections 6.4 and 6.5 in the special case of the rank-one root lattice and weight lattice of $\mathfrak{g} = \mathfrak{sl}(2, \mathbb{C})$ (recall Remark 6.5.28), we have as in [LP2] (cf. [Li3], [MP2], [MP4]):

Proposition 6.6.19 *Let* $\mathfrak{g} = \mathfrak{sl}(2, \mathbb{C})$ *equipped with a standard Chevalley basis* $\{h, e, f\}$, *let* ℓ *be a positive integer and let* $m \in \mathbb{N}$ *with* $0 \le m \le \ell$. *Then* $e(x)^{\ell+1}$ *and* $f(x)^{\ell+1}$ *are defined on* $L(\ell, m)$ *and*

$$e(x)^{\ell+1} = 0, \quad f(x)^{\ell+1} = 0 \quad \text{on } L(\ell, m). \quad (6.6.102)$$

Proof. We shall first prove the assertion for $\ell = 1$, using the vertex operator algebra V_{L_0} and its module V_L constructed in Section 6.5 with L_0 taken to be the root lattice of $\mathfrak{sl}(2, \mathbb{C})$ and L to be the weight lattice; then we shall prove the assertion for general (higher) ℓ, using a tensor product technique. Notice that by Proposition 6.6.5, we already have that $e(x)^2 = f(x)^2 = 0$ on $L(1, 0)$. But here we shall prove that $e(x)^2 = f(x)^2 = 0$ on both $L(1, 0)$ and $L(1, 1)$ simultaneously.

Let L_0 be the root lattice of $\mathfrak{sl}(2, \mathbb{C})$, i.e., $L_0 = \mathbb{Z}\alpha$ is a lattice of rank 1 with $\langle \alpha, \alpha \rangle = 2$. Then $L = \frac{1}{2}\mathbb{Z}\alpha$ is the dual lattice of L_0 (recall (6.4.81)); L is the weight lattice of $\mathfrak{sl}(2, \mathbb{C})$. Take $\hat{L} = L$ to be the trivial extension of L with $s = 1$, so that all the associated forms c_0, ϵ_0, c, and ϵ (recall (6.4.10)–(6.4.13), (6.4.17), (6.4.18), (6.4.29), (6.4.32)) are trivial; we are allowed to do this because $\langle \alpha, \beta \rangle \in 2\mathbb{Z}$ for all $\alpha, \beta \in L_0$ and so the basic condition $c(\alpha, \beta) = \omega_s^{c_0(\alpha,\beta)} = (-1)^{\langle \alpha,\beta \rangle}$ for $\alpha, \beta \in L_0$ (recall (6.4.87)) tells us that c_0 and c are trivial on L_0 and hence can be taken to be trivial on L. (Recall that in Remark 6.4.12 we showed in general that c_0 on L_0 satisfying (6.4.87) can always be extended to an alternating $\mathbb{Z}$-bilinear form on L with suitable s, but in our present situation we can certainly simply take c_0 to be trivial on L, with $s = 1$.) The conditions for Theorems 6.5.1 and 6.5.20 clearly hold, so V_{L_0} is a vertex operator algebra with V_L as a V_{L_0}-module. Furthermore, the subspaces V_{L_0} and $V_{L_0+\alpha/2}$ are irreducible V_{L_0}-submodules of V_L, which is in fact the direct sum of these two spaces.

We next show that V_L is naturally an $\widehat{\mathfrak{sl}(2, \mathbb{C})}$-module of level 1 with V_{L_0} and $V_{L_0+\alpha/2}$ as irreducible submodules such that $V_{L_0} \simeq L(1, 0)$ and $V_{L_0+\alpha/2} \simeq L(1, 1)$. We first claim that V_L is an $\widehat{\mathfrak{sl}(2, \mathbb{C})}$-module of level 1 with the Lie algebra element $h \otimes t^n$ acting as $\alpha(n)$, the element $e \otimes t^n$ as $(e^\alpha)_n$ and the element $f \otimes t^n$ as $(e^{-\alpha})_n$ for $n \in \mathbb{Z}$, and $\mathbf{k}$ acting as 1; recall that

$$Y(\alpha, x) = Y(\alpha(-1)\mathbf{1}, x) = \sum_{n \in \mathbb{Z}} \alpha(n) x^{-n-1} \quad (6.6.103)$$

and that

$$Y(e^{\pm\alpha}, x) = Y(\iota(e_{\pm\alpha}), x) = \sum_{n\in\mathbb{Z}}(e^{\pm\alpha})_n x^{-n-1} \qquad (6.6.104)$$

acting on V_L, where we are identifying $e^{\pm\alpha}$ with $\iota(e_{\pm\alpha})$ (recall (6.4.31)). In view of the defining relations (6.2.2) and (6.2.8), this amounts to proving the following commutator relations on V_L:

$$[Y(\alpha, x_1), Y(\alpha, x_2)] = -2\frac{\partial}{\partial x_1}x_2^{-1}\delta\left(\frac{x_1}{x_2}\right), \qquad (6.6.105)$$

$$[Y(\alpha, x_1), Y(e^{\pm\alpha}, x_2)] = \pm 2x_2^{-1}\delta\left(\frac{x_1}{x_2}\right)Y(e^{\pm\alpha}, x_2), \qquad (6.6.106)$$

$$[Y(e^{\pm\alpha}, x_1), Y(e^{\pm\alpha}, x_2)] = 0, \qquad (6.6.107)$$

$$[Y(e^{\alpha}, x_1), Y(e^{-\alpha}, x_2)] = x_2^{-1}\delta\left(\frac{x_1}{x_2}\right)Y(\alpha, x_2) - \frac{\partial}{\partial x_1}x_2^{-1}\delta\left(\frac{x_1}{x_2}\right).$$

$$(6.6.108)$$

The first two of these relations were established in Proposition 6.5.2, in the course of the proof of Theorem 6.5.1. For (6.6.107) and (6.6.108), in view of the commutator formula (cf. (5.6.11)) it suffices to prove the following relations in V_{L_0} for $n \geq 0$:

$$(e^{\pm\alpha})_n e^{\pm\alpha} = 0, \qquad (6.6.109)$$

$$(e^{\alpha})_0 e^{-\alpha} = \alpha, \qquad (6.6.110)$$

$$(e^{\alpha})_1 e^{-\alpha} = \mathbf{1}, \qquad (6.6.111)$$

$$(e^{\alpha})_n e^{-\alpha} = 0 \quad \text{if } n \geq 2. \qquad (6.6.112)$$

These relations hold because for $\beta \in L_0$ and $\gamma \in L$ we have

$$Y(e^{\beta}, x)e^{\gamma} = x^{\langle\beta,\gamma\rangle}E^-(-\beta, x)e^{\beta+\gamma}, \qquad (6.6.113)$$

recalling that

$$Y(e^{\beta}, x)\left(= \sum_{n\in\mathbb{Z}}(e^{\beta})_n x^{-n-1}\right) = E^-(-\beta, x)E^+(-\beta, x)e^{\beta}x^{\beta}. \qquad (6.6.114)$$

This proves that V_L is naturally an $\widehat{\mathfrak{sl}(2, \mathbb{C})}$-module of level 1. It is clearly a restricted $\widehat{\mathfrak{sl}(2, \mathbb{C})}$-module, so by Remark 6.6.1, V_L is naturally an $\widetilde{\mathfrak{sl}(2, \mathbb{C})}$-module of level 1 with $\mathbf{d}$ acting as $-L(0)$.

Now we shall identify the $\widetilde{\mathfrak{sl}(2, \mathbb{C})}$-modules V_0 and $V_{L_0+\alpha/2}$ with standard modules. Notice that as an $\hat{\mathfrak{h}}$-module $V_L = \coprod_{\beta\in L} M(1)\otimes\mathbb{C}e^{\beta}$, where $M(1)\otimes\mathbb{C}e^{\beta} \simeq M(1, \beta)$ for $\beta \in L$ and that these are inequivalent irreducible $\hat{\mathfrak{h}}$-modules (recall Proposition 6.3.8). Any $\widetilde{\mathfrak{sl}(2, \mathbb{C})}$-submodule of V_L is automatically an $\hat{\mathfrak{h}}$-submodule, and must also be stable under the action of e^α and $e^{-\alpha}$ from the explicit construction (6.6.114) of $Y(e^{\pm\alpha}, x)$. It follows immediately that V_{L_0} and $V_{L_0+\alpha/2}$ are irreducible $\widehat{\mathfrak{sl}(2,\mathbb{C})}$-modules. (This is essentially the same as the irreducibility argument in the proof of Proposition 6.5.10, which gave the irreducibility in Theorem 6.5.20.) Since $v_n\mathbf{1} = 0$ for $v \in V_{L_0}$, $n \geq 0$, the vacuum vector $\mathbf{1}$ is a highest $\mathcal{H}$-weight vector of $\mathcal{H}$-weight $(1, 0, 0)$ in V_{L_0} (recall (6.6.70)–(6.6.73)). In view of Remark 6.6.11,

$$V_{L_0} \simeq L_{\widehat{\mathfrak{sl}(2,\mathbb{C})}}(1, 0, 0) = L(1, 0) \tag{6.6.115}$$

as an $\widehat{\mathfrak{sl}(2,\mathbb{C})}$-module. The vector $e^{\alpha/2}$ is a highest $\mathcal{H}$-weight vector of $\mathcal{H}$-weight $(1, -1/4, \alpha/2)$ in $V_{L_0+\alpha/2}$, because for $n \geq 0$,

$$h(n)e^{\alpha/2} = \alpha(n)e^{\alpha/2} = \delta_{n,0}e^{\alpha/2}, \tag{6.6.116}$$

$$e(n)e^{\alpha/2} = (e^\alpha)_n e^{\alpha/2} = 0, \tag{6.6.117}$$

$$f(n+1)e^{\alpha/2} = (e^{-\alpha})_{n+1}e^{\alpha/2} = 0, \tag{6.6.118}$$

$$\mathbf{d}e^{\alpha/2} = -L(0)e^{\alpha/2} = -\frac{1}{2}\langle \alpha/2, \alpha/2 \rangle e^{\alpha/2} = -\frac{1}{4}e^{\alpha/2}, \tag{6.6.119}$$

where we are using (6.6.113). Thus as an $\widehat{\mathfrak{sl}(2,\mathbb{C})}$-module,

$$V_{L_0+\alpha/2} \simeq L_{\widehat{\mathfrak{sl}(2,\mathbb{C})}}(1, -1/4, \alpha/2) = L(1, 1). \tag{6.6.120}$$

Using (6.6.113) we also have

$$(e^{\pm\alpha})_{-1}e^{\pm\alpha} = 0. \tag{6.6.121}$$

That is, $(e^{\pm\alpha})^2_{-1}\mathbf{1} = 0$. In view of Proposition 4.5.19 and Remark 4.5.20 we have $Y(e^{\pm\alpha}, x)^2 = 0$ on V_L, so that

$$e(x)^2 = Y(e^\alpha, x)^2 = 0, \quad f(x)^2 = Y(e^{-\alpha}, x)^2 = 0 \tag{6.6.122}$$

on both $L(1, 0)$ and $L(1, 1)$, proving (6.6.102) for $\ell = 1$.

Now we consider the general case with $\ell \geq 1$, $0 \leq m \leq \ell$. Consider the tensor product $\widehat{\mathfrak{sl}(2,\mathbb{C})}$-module

$$W = (V_{L_0})^{\otimes(\ell-m)} \otimes (V_{L_0+\alpha/2})^{\otimes m}. \tag{6.6.123}$$

Since $e(x)^2 = 0$ and $f(x)^2 = 0$ on V_{L_0} and $V_{L_0+\alpha/2}$, we clearly have

$$e(x)^{\ell+1} = f(x)^{\ell+1} = 0 \quad \text{on} \quad W. \tag{6.6.124}$$

We now identify the $\widehat{\mathfrak{sl}(2,\mathbb{C})}$-module $L(\ell, m)$ with a submodule of W, so that (6.6.102) will follow immediately. Set

$$v = \mathbf{1}^{\otimes(\ell-m)} \otimes (e^{\alpha/2})^{\otimes m} \in W. \tag{6.6.125}$$

With $\mathbf{1}$ and $e^{\alpha/2}$ highest weight vectors with $\mathcal{H}$-weights $(1, 0, 0)$ and $(1, -1/4, \alpha/2)$ in V_{L_0} and $V_{L_0+\alpha/2}$, respectively, we see that v is a highest weight vector with $\mathcal{H}$-weight $(\ell, -m/4, m\alpha/2)$. Denote by W^0 the highest weight $\widetilde{\mathfrak{sl}(2, \mathbb{C})}$-submodule of W generated by v. Since $(e^{-\alpha})_n e^{\alpha/2} = 0$ for $n \geq 1$ (recall (6.6.118)), the coefficient of x^{-2} in $Y(e^{-\alpha}, x)^2 e^{\alpha/2}$ is $((e^{-\alpha})_0)^2 e^{\alpha/2}$. Since $Y(e^{-\alpha}, x)^2 = 0$ on $V_{L_0+\alpha/2}$ we have

$$f^2 e^{\alpha/2} = \left((e^{-\alpha})_0\right)^2 e^{\alpha/2} = 0. \tag{6.6.126}$$

We also have $f \cdot \mathbf{1} = f(0)\mathbf{1} = 0$, so

$$f^{m+1} v \left(= f^{m+1}(\mathbf{1}^{\otimes(\ell-m)} \otimes (e^{\alpha/2})^{\otimes m})\right) = 0. \tag{6.6.127}$$

Similarly, since $e(-1)e^{\alpha/2} = (e^{\alpha})_{-1}e^{\alpha/2} = 0$ and $e(-1)^2\mathbf{1} = ((e^{\alpha})_{-1})^2\mathbf{1} = (e^{\alpha})_{-1}e^{\alpha} = 0$ (from (6.6.113)) we have

$$e(-1)^{\ell-m+1} v \left(= e(-1)^{\ell-m+1}(\mathbf{1}^{\otimes(\ell-m)} \otimes (e^{\alpha/2})^{\otimes m})\right) = 0. \tag{6.6.128}$$

Thus W^0 is a standard $\widetilde{\mathfrak{sl}(2, \mathbb{C})}$-module with highest weight $(\ell, -m/4, m\alpha/2)$, and so it is irreducible, from Remark 6.6.16. In view of Remark 6.6.11, we have

$$W^0 \simeq L_{\widetilde{\mathfrak{sl}(2,\mathbb{C})}}(\ell, -m/4, m\alpha/2) \tag{6.6.129}$$

(as an $\widetilde{\mathfrak{sl}(2, \mathbb{C})}$-module). Since as an $\widetilde{\mathfrak{sl}(2, \mathbb{C})}$-module,

$$L_{\widetilde{\mathfrak{sl}(2,\mathbb{C})}}(\ell, -m/4, m\alpha/2) \simeq L_{\widetilde{\mathfrak{sl}(2,\mathbb{C})}}(\ell, -m(m + 2)/4(\ell + 2), m\alpha/2)$$
$$\simeq L_{\widetilde{\mathfrak{sl}(2,\mathbb{C})}}(\ell, L(m\alpha/2)) \tag{6.6.130}$$

(recall (6.6.100)), we have that $L(\ell, m) \simeq W^0 \subset W$ as an $\widetilde{\mathfrak{sl}(2, \mathbb{C})}$-module, proving (6.6.102). $\square$

Remark 6.6.20 This proof of Proposition 6.6.19 is the same as the one in [LP2], except that in [LP2] the vertex operator construction of the level 1 standard $\widetilde{\mathfrak{sl}(2, \mathbb{C})}$-modules from [FK], [Se1] was directly used, in place of the vertex operator algebra considerations; vertex (operator) algebras had not yet been discovered. In [LP2] this result was used in the course of a construction of all the *higher level* standard $\widetilde{\mathfrak{sl}(2, \mathbb{C})}$-modules. This construction was essentially based on Z-operators and Z-algebras; Z-operators were used in Section 6.5 above.

Now we use Proposition 6.6.19 to prove the following generalization for arbitrary $\mathfrak{g}$, as in [Li3] and [MP4] (cf. [DL3]):

Proposition 6.6.21 *Let ℓ be a nonnegative integer and let W be a standard $\hat{\mathfrak{g}}$-module of level ℓ. Then*

$$e_\theta(x)^{\ell+1} = 0 \quad on \quad W. \tag{6.6.131}$$

Proof. If $\ell = 0$, W is the trivial one-dimensional $\hat{\mathfrak{g}}$-module, so the assertion holds. Assume that ℓ is a positive integer. We now embed $\widehat{\mathfrak{sl}(2, \mathbb{C})}$ into $\hat{\mathfrak{g}}$ as described in Remark 6.6.16 (with $\mathbf{k} \mapsto \mathbf{k}$), making W an $\widehat{\mathfrak{sl}(2, \mathbb{C})}$-module of level ℓ. From Remark 6.6.16, W is a direct sum of standard $\widehat{\mathfrak{sl}(2, \mathbb{C})}$-modules of level ℓ, so from Proposition 6.6.19 we get

$$e_\theta(x)^{\ell+1} = 0 \quad \text{on} \quad W. \quad \square \tag{6.6.132}$$

Combining Theorem 6.6.18 with Proposition 6.6.21 we have that every standard $\hat{\mathfrak{g}}$-module of level ℓ, equipped with its natural action of $L(0)$, is naturally an irreducible module for $L_{\hat{\mathfrak{g}}}(\ell, 0)$ viewed as a vertex operator algebra; since for $\ell = 0$ we have $L_{\hat{\mathfrak{g}}}(0, 0) = \mathbb{C}$ (recall Remark 6.6.3), this is true even for $\ell = 0$. Meanwhile, Theorem 6.6.14 and Remark 6.6.15 give the converse. Summarizing, we have:

Theorem 6.6.22 *Let ℓ be a nonnegative integer. Every irreducible module W for the vertex operator algebra $L_{\hat{\mathfrak{g}}}(\ell, 0)$ is naturally a standard $\hat{\mathfrak{g}}$-module of level ℓ, and conversely, every standard $\hat{\mathfrak{g}}$-module W of level ℓ, equipped with its natural action of $L(0)$, is naturally an irreducible module for the vertex operator algebra $L_{\hat{\mathfrak{g}}}(\ell, 0)$.* $\quad \square$

Remark 6.6.23 It was proved in [FZ] and [Li3] that for any positive integer ℓ, every module for $L_{\hat{\mathfrak{g}}}(\ell, 0)$ viewed as a vertex operator algebra is completely reducible. Furthermore, it was proved in [DLM4] that every module for $L_{\hat{\mathfrak{g}}}(\ell, 0)$ viewed as a vertex algebra is a direct sum of irreducible modules for $L_{\hat{\mathfrak{g}}}(\ell, 0)$ viewed as a vertex operator algebra.

Remark 6.6.24 In Section 3.11 we presented I. Frenkel–Zhu's theorem (Theorem 3.11.12) on the centralizers of a pair of vertex operator subalgebras with possibly different conformal vectors from that of the large vertex operator algebra, and we mentioned at the beginning of Section 3.11 that this theorem generalizes the "coset construction" of [GKO1], [GKO2]. Now that we know about the vertex operator algebras $L_{\hat{\mathfrak{g}}}(\ell, 0)$, we give some details. The Goddard–Kent–Olive coset construction is a construction of certain irreducible lowest weight "unitary" modules for the Virasoro algebra using standard modules for affine Lie algebras. With $\mathfrak{g}$ as above, let ℓ_1 and ℓ_2 be positive integers. Set

$$V = L_{\hat{\mathfrak{g}}}(\ell_1, 0) \otimes L_{\hat{\mathfrak{g}}}(\ell_2, 0), \tag{6.6.133}$$

the tensor product vertex operator algebra (recall Section 3.12). Recall from Remark 6.6.2 that $\mathfrak{g}$ is naturally identified with the weight–1 subspaces of $L_{\hat{\mathfrak{g}}}(\ell_1, 0)$ and of $L_{\hat{\mathfrak{g}}}(\ell_2, 0)$ (with respect to $L(0)$). We now embed $\mathfrak{g}$ into the weight 1 subspace of V:

$$\mathfrak{g} \to V_{(1)} \subset V = L_{\hat{\mathfrak{g}}}(\ell_1, 0) \otimes L_{\hat{\mathfrak{g}}}(\ell_2, 0)$$
$$a \mapsto a \otimes \mathbf{1} + \mathbf{1} \otimes a. \tag{6.6.134}$$

We know that V is naturally a $\hat{\mathfrak{g}}$-module of level $\ell_1 + \ell_2$. Let U be the vertex subalgebra of V generated by the subspace $\mathfrak{g}$. In view of Proposition 3.9.3, U is exactly the $\hat{\mathfrak{g}}$-submodule of V generated by $\mathbf{1}$ ($= \mathbf{1} \otimes \mathbf{1}$), so U is naturally a highest weight $\tilde{\mathfrak{g}}$-module of highest ($\mathcal{H}$)-weight $(\ell_1 + \ell_2, 0, 0)$. Since $e_\theta(-1)^{\ell_1+1}\mathbf{1} = 0$ in $L_{\hat{\mathfrak{g}}}(\ell_1, 0)$ and $e_\theta(-1)^{\ell_2+1}\mathbf{1} = 0$ in $L_{\hat{\mathfrak{g}}}(\ell_2, 0)$, we have

$$e_\theta(-1)^{\ell_1+\ell_2+1}\mathbf{1} = 0 \quad \text{in} \quad V. \tag{6.6.135}$$

We also have that $a(0)\mathbf{1} = a_0\mathbf{1} = 0$ for $a \in \mathfrak{g}$. Thus U is a standard $\tilde{\mathfrak{g}}$-module, which must be isomorphic to $L_{\tilde{\mathfrak{g}}}(\ell_1 + \ell_2, 0, 0)$ in view of Remark 6.6.11, since it is irreducible from Remark 6.6.16. Now, U is isomorphic to $L_{\hat{\mathfrak{g}}}(\ell_1 + \ell_2, 0)$ as a $\hat{\mathfrak{g}}$-module, so that there is a (unique) $\hat{\mathfrak{g}}$-module embedding π of $L_{\hat{\mathfrak{g}}}(\ell_1 + \ell_2, 0)$ into V such that $\pi(\mathbf{1}) = \mathbf{1}$ ($= \mathbf{1} \otimes \mathbf{1}$). Since $\mathfrak{g}$ generates $L_{\hat{\mathfrak{g}}}(\ell_1 + \ell_2, 0)$ as a vertex algebra, in view of Proposition 5.7.9, π is an (injective) vertex algebra homomorphism. Thus we may and do consider $L_{\hat{\mathfrak{g}}}(\ell_1 + \ell_2, 0)$ a vertex subalgebra of V:

$$U = L_{\hat{\mathfrak{g}}}(\ell_1 + \ell_2, 0) \subset V = L_{\hat{\mathfrak{g}}}(\ell_1, 0) \otimes L_{\hat{\mathfrak{g}}}(\ell_2, 0). \tag{6.6.136}$$

Clearly, $V_{(n)} = 0$ for $n < 0$ and $V_{(0)} = \mathbb{C}\mathbf{1}$. The conformal vector ω_U of U (the image in V under our $\hat{\mathfrak{g}}$-module embedding of the element of $L_{\hat{\mathfrak{g}}}(\ell_1 + \ell_2, 0)$ given by the expression (6.2.42)) lies in $V_{(2)}$. To apply Theorem 3.11.12 to V and U, with ω_U playing the role of ω', we need to verify that $L(1)\omega_U = 0$; recall from (3.12.23) that $L(1)$ on V is given by $L(1) \otimes 1 + 1 \otimes L(1)$. Since $L(1)\mathfrak{g} = 0$ in $L_{\hat{\mathfrak{g}}}(\ell_i, 0)$ for $i = 1, 2$ (recall Remark 6.2.19), we see that $L(1)\mathfrak{g} = 0$ in V. Now $\mathfrak{g} \subset V_{(1)}$, $L(1)\mathfrak{g} = 0$ and $L(n)\mathfrak{g} = 0$ for $n \geq 2$. Thus in view of the commutator formula (5.6.11), for $a \in \mathfrak{g}$ ($\subset U$) $\subset V$ we have on V (cf. (6.2.49) and (6.2.50))

$$[Y(\omega, x_1), Y(a, x_2)]$$
$$= Y(L(-1)a, x_2)x_2^{-1}\delta\left(\frac{x_1}{x_2}\right) - Y(L(0)a, x_2)\frac{\partial}{\partial x_1}x_2^{-1}\delta\left(\frac{x_1}{x_2}\right)$$
$$= a'(x_2)x_2^{-1}\delta\left(\frac{x_1}{x_2}\right) - a(x_2)\frac{\partial}{\partial x_1}x_2^{-1}\delta\left(\frac{x_1}{x_2}\right)$$
$$= a'(x_2)x_2^{-1}\delta\left(\frac{x_1}{x_2}\right) + a(x_2)\frac{\partial}{\partial x_2}x_2^{-1}\delta\left(\frac{x_1}{x_2}\right)$$
$$= \frac{\partial}{\partial x_2}\left(a(x_2)x_2^{-1}\delta\left(\frac{x_1}{x_2}\right)\right)$$
$$= a(x_1)\frac{\partial}{\partial x_2}x_2^{-1}\delta\left(\frac{x_1}{x_2}\right) \tag{6.6.137}$$

(where we are writing $Y(a, x)$ as $a(x)$, as usual), which is just (6.2.50) with x_1 and x_2 playing opposite roles. Thus, on V,

$$[L(m), a(n)] = -na(m+n) \quad \text{for } a \in \mathfrak{g}, \ m, n \in \mathbb{Z}. \tag{6.6.138}$$

The same calculation as in (6.2.62) with ω_U in place of ω now yields

$$L(1)\omega_U = 0. \qquad\qquad (6.6.139)$$

Hence Theorem 3.11.12 applies to our present situation, and it asserts that the centralizer $C_V(U)$ of the vertex algebra U in V is a vertex operator algebra with $\omega - \omega_U$ as the conformal vector and with central charge c given by

$$c = c_V - c_U. \qquad\qquad (6.6.140)$$

From Theorem 6.2.18 we have

$$c = \frac{d\ell_1}{\ell_1 + h} + \frac{d\ell_2}{\ell_2 + h} - \frac{d(\ell_1 + \ell_2)}{\ell_1 + \ell_2 + h}, \qquad\qquad (6.6.141)$$

where $d = \dim \mathfrak{g}$ and h is the dual Coxeter number of $\mathfrak{g}$. For example, taking $\mathfrak{g} = \mathfrak{sl}(2, \mathbb{C})$, $\ell_1 = 1$ and $\ell_2 = \ell$, we have that $d = 3$ and $h = 2$, so that

$$c = 1 + \frac{3\ell}{\ell + 2} - \frac{3(\ell + 1)}{\ell + 3} = 1 - \frac{6}{(\ell + 2)(\ell + 3)}. \qquad\qquad (6.6.142)$$

In particular, $C_V(U)$ is a module for the Virasoro algebra of central charge c, which we now rewrite as c_ℓ to show its dependence on ℓ. It turns out that the Virasoro algebra submodule of $C_V(U)$ generated by $\mathbf{1}$ is irreducible and so is equivalent to $L_{Vir}(c_\ell, 0)$ as defined in Section 6.1, and that the modules $L_{Vir}(c_\ell, 0)$ for $\ell \geq 1$ (up to equivalence) are exactly those "unitary" (irreducible) lowest weight Virasoro algebra modules of lowest weight 0 with nonnegative real central charge less than 1. (Note that $c_\ell = c_{p,q}$ as defined in Remark 6.1.14 with $p = \ell+2$ and $q = \ell+3$.) Such Virasoro algebra modules had been classified by Friedan, Qiu and Shenker in [FQS1]–[FQS4], and in fact more generally for possibly nonzero lowest weights, but without a proof of their existence. Later, Goddard, Kent and Olive ([GKO1], [GKO2]) proved their existence, by the method we have just described (including the case of nonzero lowest weight), or rather, the method just described carried out "directly," without the theory of vertex operator algebras. This Goddard-Kent-Olive construction of the "unitary" Virasoro algebra modules of the central charges c_ℓ is called the GKO "coset construction."

Remark 6.6.25 In the theory of Kac–Moody algebras, there is another interesting and important class of irreducible highest weight $\hat{\mathfrak{g}}$-modules, more general than the standard $\hat{\mathfrak{g}}$-modules, associated to what Kac and Wakimoto called "admissible levels" and "admissible weights." The admissible levels for $\hat{\mathfrak{g}}$ are certain rational numbers related to the dual Coxeter number h of $\mathfrak{g}$. For each admissible level ℓ, there are only finitely many "admissible $\mathfrak{h}$-weights" λ of level ℓ (see [KW1], [KW2]) and the corresponding irreducible $\tilde{\mathfrak{g}}$-modules $L_{\tilde{\mathfrak{g}}}(\ell, m, \lambda)$ are called the admissible modules. It was proved by Kac and Wakimoto that the "characters" of the admissible $\tilde{\mathfrak{g}}$-modules $L_{\tilde{\mathfrak{g}}}(\ell, m, \lambda)$ satisfy a certain modular invariance property. For an admissible level ℓ and an admissible weight λ of level ℓ, the structure of the maximal proper $\tilde{\mathfrak{g}}$-submodule of the Verma module $M_{\tilde{\mathfrak{g}}}(\ell, m, \lambda)$ was determined by Malikov, Feigin and Fuchs ([MFF], [Fu2]). Later, using the results in [Fu2], the vertex operator algebras $L_{\hat{\mathfrak{g}}}(\ell, 0)$ for $\mathfrak{g} = \mathfrak{sl}(2, \mathbb{C})$ with ℓ

an admissible level, and their modules, were studied in [AM] and [DLM5], and it was proved that an irreducible highest weight $\hat{\mathfrak{g}}$-module $L_{\hat{\mathfrak{g}}}(\ell, L(\lambda))$ is naturally a module for $L_{\hat{\mathfrak{g}}}(\ell, 0)$ as a vertex algebra if and only if λ is an admissible weight of level ℓ (cf. Theorems 6.6.14 and 6.6.22).

References

[Ab1] T. Abe, Fusion rules for the free bosonic orbifold vertex operator algebra, *J. Algebra* **229** (2000), 333–374.

[Ab2] T. Abe, Fusion rules for the charge conjugation orbifold, *J. Algebra* **242** (2001), 624–655.

[Ab3] T. Abe, The charge conjugation orbifold $V_{\mathbb{Z}\alpha}^{+}$ is rational when $\langle \alpha, \alpha \rangle/2$ is prime, *Internat. Math. Res. Notices* **12** (2002), 647–665.

[ABD] T. Abe, G. Buhl and C. Dong, Rationality, regularity and C_2-cofiniteness, preprint, arXiv:math.QA/0204021.

[AD] T. Abe and C. Dong, Classification of irreducible modules for the vertex operator algebra V_L^{+}, preprint, arXiv:math.QA/0210274.

[AN] T. Abe and K. Nagatomo, Finiteness of conformal blocks over the projective line, in: *Vertex Operator Algebras in Mathematics and Physics, Proc. of Workshop at Fields Institute for Research in Mathematical Sciences, 2000*, ed. by S. Berman, Y. Billig, Y.-Z. Huang and J. Lepowsky, Fields Institute Communications **39**, Amer. Math. Soc., 2003.

[Ad1] D. Adamović, Some rational vertex algebras, *Glasnik Matematički* **29** (1994), 25–40.

[Ad2] D. Adamović, Vertex operator algebras and irreducibility of certain modules for affine Lie algebras, *Math. Res. Lett.* **4** (1997), 809–821.

[Ad3] D. Adamović, Rationality of Neveu–Schwarz vertex operator superalgebras, *Internat. Math. Res. Notices* **17** (1997), 865–874.

[Ad4] D. Adamović, Representations of vertex algebras, *Math. Commun.* **3** (1998), 109–114.

[Ad5] D. Adamović, Representations of the $N = 2$ superconformal vertex algebra, *Internat. Math. Res. Notices* **2** (1999), 61–79.

[Ad6] D. Adamović, Representations of the vertex algebra $W_{1+\infty}$ with a negative integer central charge, *Commun. Algebra* **29** (2001), 3153–3166.

[Ad7] D. Adamović, Vertex algebra approach to fusion rules for $N = 2$ superconformal minimal models, *J. Algebra* **239** (2001), 549–572.

[Ad8] D. Adamović, A construction of some ideals in affine vertex algebras, *Internat. J. Math. Math. Sci.* **15** (2003), 971–980.

[Ad9] D. Adamović, Regularity of certain vertex operator algebras with two generators, preprint, arXiv:math.QA/0111055.

[Ad10] D. Adamović, Classification of irreducible modules of certain subalgebras of free boson vertex algebra, preprint, arXiv:math.QA/0207155.

[AM] D. Adamović and A. Milas, Vertex operator algebras associated to modular invariant representations for $A_1^{(1)}$, *Math. Res. Lett.* **2** (1995), 563–575.

[A1] F. Akman, The semi-infinite Weil complex of a graded Lie algebra, Ph.D. thesis, Yale University, 1993.

[A2] F. Akman, Chicken or egg? A hierarchy of homotopy algebras, preprint, arXiv:math.QA/0306145.

290   References

[AABGP] B. N. Allison, S. Azam, S. Berman, Y. Gao and A. Pianzola, Extended affine Lie algebras and their root systems, *Memoirs Amer. Math. Soc.* **126**, 1997.

[AG]   S. Arkhipov and D. Gaitsgory, Differential operators on the loop group via chiral algebras, *Internat. Math. Res. Notices* **4** (2002), 165–210.

[BK]   B. Bakalov and V. Kac, Field algebras, *Internat. Math. Res. Notices* **3** (2003), 123–159.

[BKi]   B. Bakalov and A. Kirillov Jr., *Lectures on Tensor Categories and Modular Functors*, University Lecture Series, Vol. 21, Amer. Math. Soc., Providence, 2001.

[BHN]   T. Banks, D. Horn and H. Neuberger, Bosonization of the $SU(N)$ Thirring models, *Nucl. Phys.* **B108** (1976), 119.

[Ban]   P. Bantay, Permutation orbifolds and their applications, in: *Vertex Operator Algebras in Mathematics and Physics, Proc. of Workshop at Fields Institute for Research in Mathematical Sciences, 2000*, ed. by S. Berman, Y. Billig, Y.-Z. Huang and J. Lepowsky, Fields Institute Communications **39**, Amer. Math. Soc., 2003.

[BH]   K. Bardakci and M. B. Halpern, New dual quark models, *Phys. Rev.* **D3** (1971), 2493–2506.

[Ba1]   K. Barron, The supergeometric interpretation of vertex operator superalgebras, Ph.D. thesis, Rutgers University, 1996.

[Ba2]   K. Barron, The supergeometric interpretation of vertex operator superalgebras, *Internat. Math. Res. Notices* **9** (1996), 409–430.

[Ba3]   K. Barron, $N = 1$ Neveu–Schwarz vertex operator superalgebras over Grassmann algebras and with odd formal variables, in: *Representations and Quantizations (Shanghai, 1998)*, China High. Educ. Press, Beijing, 2000, 9–35.

[Ba4]   K. Barron, The moduli space of $N = 1$ superspheres with tubes and the sewing operation, *Memoirs Amer. Math. Soc.* **162**, 2003.

[Ba5]   K. Barron, The notion of $N = 1$ supergeometric vertex operator superalgebra and the isomorphism theorem, *Commun. Contemp. Math.*, to appear.

[Ba6]   K. Barron, Superconformal change of variables for $N = 1$ Neveu–Schwarz vertex operator superalgebras, preprint, arXiv:math.QA/0307038.

[BDM]   K. Barron, C. Dong and G. Mason, Twisted sectors for tensor product vertex operator algebras associated to permutation groups, *Commun. Math. Phys.* **227** (2002), 349–384.

[BHL]   K. Barron, Y.-Z. Huang and J. Lepowsky, Factorization of formal exponentials and uniformization *J. Algebra* **228** (2000), 551–579.

[BD]   A. Beilinson and V. Drinfeld, Chiral algebras, preprint.

[BFM]   A. Beilinson, B. Feigin and B. Mazur, Introduction to algebraic field theory on curves, preprint.

[BMS]   A. Beilinson, Y. Manin and V. Schechtman, Sheaves of the Virasoro and Neveu–Schwarz algebras, in: *K-theory, Arithmetic and Geometry (Moscow, 1984–1986)*, Lecture Notes in Math. **1289**, Springer-Verlag, 1987, 52–66.

[BS]   A. Beilinson and V. Schechtman, Determinant bundles and Virasoro algebra, *Commun. Math. Phys.* **118** (1988), 651–701.

[BPZ]   A. Belavin, A. M. Polyakov, A. B. Zamolodchikov, Infinite conformal symmetries in two-dimensional quantum field theory, *Nucl. Phys.* **B241** (1984), 333–380.

[BF1]   D. Ben-Zvi and E. Frenkel, Spectral curves, opers and integrable systems, *Publ. Math. Inst. Hautes Études Sci.* **94** (2001), 87–159.

[BF2]   D. Ben-Zvi and E. Frenkel, Geometric realization of the Segal–Sugawara construction, preprint, arXiv:math.AG/0301206.

[BB]   S. Berman, Y. Billig, Irreducible representations for toroidal Lie algebras, *J. Algebra* **221** (1999), 188–231.

[BBS]   S. Berman, Y. Billig and J. Szmigielski, Vertex operator algebras and the representation theory of toroidal algebras, in: *Recent Developments in Infinite-Dimensional Lie Algebras and Conformal Field Theory (Charlottesville, VA, 2000)*, Contemporary Math. **297**, Amer. Math. Soc., Providence, 2002, 1–26.

[BBHL] S. Berman, Y. Billig, Y.-Z. Huang and J. Lepowsky, eds., *Vertex Operator Algebras in Mathematics and Physics, Proc. of Workshop at Fields Institute for Research in Mathematical Sciences, 2000*, Fields Institute Communications **39**, Amer. Math. Soc., 2003.

[BDT]   S. Berman, C. Dong and S. Tan, Representations of a class of lattice type vertex algebras, *J. Pure Applied Algebra* **176** (2002), 27–47.

[BGT]   S. Berman, Y. Gao and S. Tan, A unified view of some vertex operator constructions, *Israel J. Math.* **134** (2003), 29–60.

[BJT]   S. Berman, E. Jurisich and S. Tan, Beyond Borcherds Lie algebras and inside, *Trans. Amer. Math. Soc.* **353** (2001), 1183–1219.

[BT]   D. Bernard and J. Thierry-Mieg, Level one representations of the simple affine Kac–Moody algebras in their homogeneous gradations, *Commun. Math. Phys.* **111** (1987), 181–246.

[Bi1]   Y. Billig, Principal vertex operator representations for toroidal Lie algebras, *J. Math. Phys.* **39** (1998), 3844–3864.

[Bi2]   Y. Billig, Energy-momentum tensor for the toroidal Lie algebras, preprint, arXiv:math.RT/0201313.

[Bi3]   Y. Billig, Representations of the twisted Heisenberg–Virasoro algebra at level zero, preprint, arXiv:math.RT/0201314.

[Blo]   S. Bloch, Zeta values and differential operators on the circle, *J. Algebra* **182** (1996), 476–500.

[Bl]   R. Block, On the Mills–Seligman axioms for Lie algebras of classical type, *Trans. Amer. Math. Soc.* **121** (1966), 378–392.

[Bo1]   J. Borcea, Annihilating fields of standard modules for affine Lie algebras, *Math. Z.* **237** (2001), 301–319.

[Bo2]   J. Borcea, Dualities and vertex operator algebras of affine type, *J. Algebra* **258** (2002), 389–441.

[B1]   R. E. Borcherds, Vertex algebras, Kac–Moody algebras, and the Monster, *Proc. Natl. Acad. Sci. USA* **83** (1986), 3068–3071.

[B2]   R. E. Borcherds, Generalized Kac–Moody algebras, *J. Algebra* **115** (1988), 501–512.

[B3]   R. E. Borcherds, Central extensions of generalized Kac–Moody algebras, *J. Algebra* **140** (1991), 330–335.

[B4]   R. E. Borcherds, A characterization of generalized Kac–Moody algebras, *J. Algebra* **174** (1995), 1073–1079.

[B5]   R. E. Borcherds, The monster Lie algebra, *Advances in Math.* **83** (1990), 3–47.

[B6]   R. E. Borcherds, Monstrous moonshine and monstrous Lie superalgebras, *Invent. Math.* **109** (1992), 405–444.

[B7]   R. E. Borcherds, Vertex algebras, in *"Topological Field Theory, Primitive Forms and Related Topics"* (Kyoto, 1996), edited by M. Kashiwara, A. Matsuo, K. Saito and I. Satake, Progress in Math., Vol. 160, Birkhäuser, Boston, 1998, 35–77.

[B8]   R. E. Borcherds, What is moonshine?, in: *Proc. Internat. Congr. Math., Vol. I, Berlin, 1998*, Doc. Math. 1998, Extra Vol. I, 607–615.

[B9]   R. E. Borcherds, Quantum vertex algebras, *Taniguchi Conference on Mathematics Nara '98*, Adv. Stud. Pure Math., 31, Math. Soc. Japan, Tokyo, 2001, 51–74.

[BCQS]   R. E. Borcherds, J. H. Conway, L. Queen and N. J. A. Sloane, A monster Lie algebra?, *Advances in Math.* **53** (1984), 75–79.

[Bor1]   L. Borisov, Vertex algebras and mirror symmetry, *Commun. Math. Phys.* **215** (2001), 517–557.

[Bor2]   L. Borisov, Introduction to the vertex algebra approach to mirror symmetry, preprint, arXiv:math.AG/9912195.

[Bor3]   L. Borisov, Chiral rings of vertex algebras of mirror symmetry, preprint, arXiv:math.AG/0209301.

[BL]   L. Borisov and A. Libgober, Elliptic genera of toric varieties and applications to mirror symmetry, *Invent. Math.* **140** (2000), 453–485.

[Bou1]   N. Bourbaki, Groupes et algèbres de Lie, Chaps. 7, 8, Hermann, Paris, 1968.

[Bou2]   N. Bourbaki, Groupes et algèbres de Lie, Chaps. 4, 5, 6, Hermann, Paris, 1975.

[BSc]   P. Bouwknegt and K. Schoutens, eds., *W-symmetry*, Advanced Series in Mathematical Physics 22, World Scientific, 1995.

[BFST]   P. Bowcock, B. Feigin, A. Semikhatov and A. Taormina, $\widehat{sl}(2|1)$ and $\widehat{D}(2|1; \alpha)$ as vertex operator extensions of dual affine sl(2) algebras, *Commun. Math. Phys.* **214** (2000), 495–545.

[Bre1]   P. Bressler, Vertex algebroids I preprint, arXiv:math.AG/0202185.

[Bre2]   P. Bressler, Vertex algebroids II preprint, arXiv:math.AG/0304115.

292   References

[Br]   R. C. Brower, Spectrum-generating algebra and no-ghost theorem for the dual model, *Phys. Rev.*
          D6 (1972), 1655–1662.
[BG1]   R. C. Brower and P. Goddard, Collinear algebra for the dual model, *Nucl. Phys.* **B40** (1972),
          437–444.
[BG2]   R. C. Brower and P. Goddard, Physical states in dual resonance model, in: *Proceedings of
          the International School of Physics "Enrico Fermi" Course LIV,* Academic Press, New York,
          London, 1973, 98–110.
[BN]   D. Brungs and W. Nahm, The associative algebras of conformal field theory, *Lett. Math. Phys.*
          47 (1999), 379–383.
[Bu1]   G. Buhl, A spanning set for VOA modules, *J. Algebra* **254** (2002), 125–151.
[Bu2]   G. Buhl, Rationality and C_2-cofiniteness is regularity, Ph.D. thesis, University of California,
          Santa Cruz, 2003.
[CHSW]   P. Candelas, G. Horowitz, A. Strominger and E. Witten, Vacuum configurations for superstrings,
          Nucl. Phys. **B258** (1985), 46.
[C1]   S. Capparelli, Vertex operator relations for affine algebras and combinatorial identities, Ph.D.
          thesis, Rutgers University, 1988.
[C2]   S. Capparelli, On some representations of twisted affine Lie algebras and combinatorial identities,
          J. Algebra **154** (1993), 335–355.
[CLM]   S. Capparelli, J. Lepowsky and A. Milas, The Rogers–Ramanujan recursion and intertwining
          operators, *Commun. Contemp. Math.,* to appear.
[Ch]   C. Chevalley, Sur certains groupes simples, *Tohoku Math. J.* **7** (1955), 14–66.
[Co1]   J. H. Conway, A group of order 8, 315, 553, 613, 086, 720, 000, *Bull. London Math. Soc.* **1**
          (1969), 79–88.
[Co2]   J. H. Conway, A characterization of Leech's lattice, *Invent. Math.* **7** (1969), 137–142.
[Co3]   J. H. Conway, Monsters and moonshine, *Math. Intelligencer* **2** (1980), 165–172.
[Co4]   J. H. Conway, A simple construction for the Fischer–Griess monster group, *Invent. Math.* **79**
          (1985), 513–540.
[Co5]   J. H. Conway, The monster group and its 196884-dimensional space, Chap. 29 of: J. H. Conway
          and N. J. A. Sloane, *Sphere Packings, Lattices and Groups,* Springer-Verlag, 1988, 555–567.
[CN]   J. H. Conway and S. P. Norton, Monstrous moonshine, *Bull. London Math. Soc.* **11** (1979),
          308–339.
[CS]   J. H. Conway and N. J. A. Sloane, *Sphere Packings, Lattices and Groups*, Springer-Verlag, 1988.
[CF]   E. F. Corrigan and D. B. Fairlie, Off-shell states in dual resonance theory, *Nucl. Phys.* **B91**
          (1975), 527–545.
[CH1]   E. F. Corrigan and T. J. Hollowood, A bosonic representation of the twisted string emission
          vertex, *Nucl. Phys.* **B303** (1988), 135–148.
[CH2]   E. F. Corrigan and T. J. Hollowood, Comments on the algebra of straight, twisted and intertwining
          vertex operators, *Nucl. Phys.* **B304** (1988), 77–107.
[CO]   E. F. Corrigan and D. Olive, Fermion-meson vertices in dual theories, *Nuovo Cim.* **11A** (1972),
          749–773.
[CGH]   B. Craps, M. R. Gaberdiel and J. A. Harvey, Monstrous branes, *Commun. Math. Phys.* **234**
          (2003), 229–251.
[CG]   C. J. Cummins and T. Gannon, Modular equations and the genus zero property for moonshine
          functions, *Invent. Math.* **129** (1997), 413–443.
[DJKM1]   E. Date, M. Jimbo, M. Kashiwara, and T. Miwa, Operator approach to the Kadomtsev-
          Petviashvili equation—transformation groups for soliton equations III, *J. Phys. Soc. Japan* **50**
          (1981), 3806–3812.
[DJKM2]   E. Date, M. Jimbo, M. Kashiwara, and T. Miwa, Transformation groups for soliton equations—
          Euclidean Lie algebras and reduction of the KP hierarchy, Publ. Research Inst. for Math. Sciences,
          Kyoto Univ., 18 (1982), 1077–1110.
[DJKM3]   E. Date, M. Jimbo, M. Kashiwara, and T. Miwa, Transformation groups for soliton equations,
          in: *Nonlinear Integrable Systems—Classical Theory and Quantum Theory,* edited by M. Jimbo
          and T. Miwa, World Scientific, 1983, 39–120.

[DJMM] E. Date, M. Jimbo, A. Matsuo, and T. Miwa, Hypergeometric type integrals and the $SL(2, \mathbb{C})$ Knizhnik–Zamolodchikov equations, *Internat. J. Modern Phys.* **B4** (1990), 1049–1057.

[DKM]   E. Date, M. Kashiwara and T. Miwa, Vertex operators and τ functions—transformation groups for soliton equations II, *Proc. Japan Acad. Ser. A Math. Sci.* **57** (1981), 387–392.

[De]   P. Deligne et al., eds., *Quantum Fields and Strings, A Course for Mathematicians I, II*, Amer. Math. Soc., Providence, 1999.

[DVVV] R. Dijkgraaf, C. Vafa, E. Verlinde, and H. Verlinde, Operator algebras of orbifold models, *Commun. Math. Phys.* **123** (1989), 485–526.

[DGZ]   X.-M. Ding, M. D. Gould and Y.-Z. Zhang, Twisted parafermions, *Phys. Lett.* **B530** (2002), 197–201.

[DFMS] L. Dixon, D. Friedan, E. Martinec and S. Shenker, The conformal field theory of orbifolds, *Nucl. Phys.* **B282** (1987), 13–73.

[DGHa] L. Dixon, P. Ginsparg and J. A. Harvey, Beauty and the Beast: Superconformal symmetry in a Monster module, *Commun. Math. Phys.* **119** (1989), 203–206.

[DHVW] L. Dixon, J. A. Harvey, C. Vafa and E. Witten, Strings on orbifolds, I. *Nucl. Phys.* **B261** (1985), 678–686; II. *Nucl. Phys.* **B274** (1986), 285–314.

[Do]   L. Dolan, Superstring twisted conformal field theory, in: *Moonshine, the Monster and Related Topics, Proc. Joint Summer Research Conference, Mount Holyoke*, 1994, ed. by C. Dong and G. Mason, Contemporary Math. **193**, Amer. Math. Soc., Providence, 1996, 9–24.

[DGM1] L. Dolan, P. Goddard and P. Montague, Conformal field theory, triality and the Monster group, *Phys. Lett.* **B236** (1990), 165–172.

[DGM2] L. Dolan, P. Goddard and P. Montague, Conformal field theory of twisted vertex operators, *Nucl. Phys.* **B338** (1990), 529–601.

[DGM3] L. Dolan, P. Goddard and P. Montague, Conformal field theories, representations and lattice constructions, *Commun. Math. Phys.* **179** (1996), 61–120.

[D1]   C. Dong, Structure of some nonstandard modules for $C_n^{(1)}$, *J. Algebra* **120** (1989), 301–338.

[D2]   C. Dong, Vertex algebras associated with even lattices, *J. Algebra* **160** (1993), 245–265.

[D3]   C. Dong, Twisted modules for vertex algebras associated with even lattices, *J. Algebra* **165** (1994), 91–112.

[D4]   C. Dong, Representations of the moonshine module vertex operator algebra, in: *Mathematical Aspects of Conformal and Topological Field Theories and Quantum Groups, Proc. Joint Summer Research Conference, Mount Holyoke*, 1992, ed. by P. J. Sally Jr., M. Flato, J. Lepowsky, N. Reshetikhin and G. J. Zuckerman, Contemporary Math. **175**, Amer. Math. Soc., Providence, 1994, 27–36.

[D5]   C. Dong, Introduction to vertex operator algebras I, in: *Moonshine and Vertex Operator Algebra, Proc. of 1994 Workshop*, ed. by M. Miyamoto, Publ. Research Inst. for Math. Sciences, Kyoto Univ., 904 (1995), 1–26.

[DG1]   C. Dong and R. L. Griess Jr., Rank one lattice type vertex operator algebras and their automorphism groups, *J. Algebra* **208** (1998), 262–275.

[DG2]   C. Dong and R. L. Griess Jr., Automorphism groups and derivation algebras of finitely generated vertex operator algebras, *Michigan Math. J.* **50** (2002), 227–239.

[DGH]   C. Dong, R. L. Griess Jr., and G. Höhn, Framed vertex operator algebras, codes and the Moonshine module, *Commun. Math. Phys.* **193** (1998), 407–448.

[DGR]   C. Dong, R. L. Griess Jr. and A. Ryba, Rank one lattice type vertex operator algebras and their automorphism groups. II. E-series, *J. Algebra* **217** (1999), 701–710.

[DLTYY] C. Dong, C. H. Lam, K. Tanabe, H. Yamada and K. Yokoyama, Z_3 symmetry and W_3 algebra in lattice vertex operator algebras, arXiv:math.QA/0302314.

[DLY1] C. Dong, C. H. Lam and H. Yamada, Decomposition of the vertex operator algebra $V_{\sqrt{2}A_3}$, *J. Algebra* **222** (1999), 500–510.

[DLY2] C. Dong, C. H. Lam and H. Yamada, Decomposition of the vertex operator algebra $V_{\sqrt{2}D_l}$, *Commun. Contemp. Math.* **3** (2001), 137–151.

[DL1]   C. Dong and J. Lepowsky, A Jacobi identity for relative vertex operators and the equivalence of Z-algebras and parafermion algebras, in: *Proc. XVIIth Intl. Colloq. on Group Theoretical*

294   References

Methods in Physics, Ste-Adèle, June, 1988, ed. by Y. Saint-Aubin and L. Vinet, World Scientific, Singapore, 1989, 235–238.

[DL2]   C. Dong and J. Lepowsky, Abelian intertwining algebras—a generalization of vertex operator algebras, in: *Algebraic Groups and Their Generalizations, Proc. 1991 American Math. Soc. Summer Research Institute,* ed. by W. Haboush and B. Parshall, Proc. Symp. Pure Math. **56**, Part 2, Amer. Math. Soc., Providence, 1994, 261–293.

[DL3]   C. Dong and J. Lepowsky, *Generalized Vertex Algebras and Relative Vertex Operators,* Progress in Math., Vol. 112, Birkhäuser, Boston, 1993.

[DL4]   C. Dong and J. Lepowsky, The algebraic structure of relative twisted vertex operators, *J. Pure Applied Algebra* **110** (1996), 259–295.

[DLM1]   C. Dong, H.-S. Li and G. Mason, Some twisted sectors for the Moonshine module, in: *Moonshine, the Monster and Related Topics, Proc. Joint Summer Research Conference, Mount Holyoke,* 1994, ed. by C. Dong and G. Mason, Contemporary Math. **193**, Amer. Math. Soc., Providence, 1996, 25–43.

[DLM2]   C. Dong, H.-S. Li and G. Mason, Simple currents and extensions of vertex operator algebras, *Commun. Math. Phys.* **180** (1996), 671–707.

[DLM3]   C. Dong, H.-S. Li and G. Mason, Compact automorphism groups of vertex operator algebras, *Internat. Math. Res. Notices* **18** (1996), 913–921.

[DLM4]   C. Dong, H.-S. Li and G. Mason, Regularity of rational vertex operator algebras, *Advances in Math.* **132** (1997), 148–166.

[DLM5]   C. Dong, H.-S. Li and G. Mason, Vertex operator algebras associated to admissible representations of $\widehat{sl_2}$, *Commun. Math. Phys.* **184** (1997), 65–93.

[DLM6]   C. Dong, H.-S. Li and G. Mason, Certain associative algebras similar to $U(sl_2)$ and Zhu's algebra $A(V_L)$, *J. Algebra* **196** (1997), 532–551.

[DLM7]   C. Dong, H.-S. Li and G. Mason, Twisted representations of vertex operator algebras, *Math. Annalen* **310** (1998), 571–600.

[DLM8]   C. Dong, H.-S. Li and G. Mason, Vertex operator algebras and associative algebras, *J. Algebra* **206** (1998), 67–96.

[DLM9]   C. Dong, H.-S. Li and G. Mason, Twisted representations of vertex operator algebras and associative algebras, *Internat. Math. Res. Notices* **8** (1998), 389–397.

[DLM10]   C. Dong, H.-S. Li and G. Mason, Modular invariance of trace functions in orbifold theory, *Commun. Math. Phys.* **214** (2000), 1–56.

[DLM11]   C. Dong, H.-S. Li and G. Mason, Vertex Lie algebra, vertex Poisson algebras and vertex algebras, in: *Recent Developments in Infinite-Dimensional Lie Algebras and Conformal Field Theory (Charlottesville, VA, 2000),* Contemporary Math. **297**, Amer. Math. Soc., Providence, 2002, 69–96.

[DLMM]   C. Dong, H.-S. Li, G. Mason and P. Montague, The radical of a vertex operator algebra, in: *The Monster and Lie Algebras (Columbus, OH, 1996),* Ohio State Univ. Math. Res. Inst. Publ., 7, de Gruyter, Berlin, 1998, 17–25.

[DLMN]   C. Dong, H.-S. Li, G. Mason and S. Norton, Associative subalgebras of the Griess algebra and Related Topics, in: *The Monster and Lie Algebras (Columbus, OH, 1996),* Ohio State Univ. Math. Res. Inst. Publ., 7, de Gruyter, Berlin, 1998, 27–42.

[DLin]   C. Dong and Z. Lin, Induced modules for vertex operator algebras, *Commun. Math. Phys.* **179** (1996), 157–183.

[DLinM]   C. Dong, Z. Lin and G. Mason, On vertex operator algebras as sl_2-modules, in: *Groups, Difference Sets, and the Monster (Columbus, OH, 1993),* Ohio State Univ. Math. Res. Inst. Publ. **4**, de Gruyter, Berlin, 1996, 349–362.

[DLiuM1] C. Dong, K. Liu and X. Ma, On orbifold elliptic genus, in: *Orbifolds in Mathematics and Physics (Madison, WI, 2001),* Contemporary Math. **310**, Amer. Math. Soc., Providence, 2002, 87–105.

[DLiuM2] C. Dong, K. Liu and X. Ma, Elliptic genus and vertex operator algebras, preprint, arXiv:math.QA/0201135.

[DM1]   C. Dong and G. Mason, On the construction of the moonshine module as a $\mathbb{Z}_p$-orbifold, in: *Mathematical Aspects of Conformal and Topological Field Theories and Quantum Groups, Proc. Joint Summer Research Conference, Mount Holyoke, 1992*, ed. by P. J. Sally Jr., M. Flato, J. Lepowsky, N. Reshetikhin and G. J. Zuckerman, Contemporary Math. **175**, Amer. Math. Soc., Providence, 1994, 37–52.

[DM2]   C. Dong and G. Mason, Nonabelian orbifolds and the boson-fermion correspondence, *Commun. Math. Phys.* **163** (1994), 523–559.

[DM3]   C. Dong and G. Mason, Orbifold theory of genus zero associated to the sporadic group M_{24}, *Commun. Math. Phys.* **164** (1994), 87–104.

[DM4]   C. Dong and G. Mason, Vertex operator algebras and Moonshine: a survey, in: *Progress in Algebraic Combinatorics*, Adv. Study Pure Math. 24, Math. Soc. Japan, Tokyo, 1996, 101–136.

[DM5]   C. Dong and G. Mason, On quantum Galois theory, *Duke Math. J.* **86** (1997), 305–321.

[DM6]   C. Dong and G. Mason, Vertex operator algebras and their automorphism groups, in: *Representations and Quantizations (Shanghai, 1998)*, China High. Educ. Press, Beijing, 2000, 145–166.

[DM7]   C. Dong and G. Mason, Quantum Galois theory for compact Lie groups, *J. Algebra* **214** (1999), 92–102.

[DM8]   C. Dong and G. Mason, Monstrous moonshine of higher weight, *Acta Math.* **185** (2000), 101–121.

[DM9]   C. Dong and G. Mason, The radical of a vertex operator algebra associated to a module, *J. Algebra* **242** (2001), 360–373.

[DM10] C. Dong and G. Mason, Transformation laws for theta functions, in: *Proceedings on Moonshine and Related Topics (Montreal, QC, 1999)*, CRM Proc. Lecture Notes **30**, Amer. Math. Soc., Providence, RI, 2001, 15–26.

[DM11] C. Dong and G. Mason, Rational vertex operator algebras and the effective central charge, preprint, arXiv:math.QA/0201318.

[DM12] C. Dong and G. Mason, Holomorphic vertex operator algebras of small central charges, preprint, arXiv:math.QA/0203005.

[DMN]   C. Dong, G. Mason and K. Nagatomo, Quasi-modular forms and trace functions associated to free boson and lattice vertex operator algebras, *Internat. Math. Res. Notices* **8** (2001), 409–427.

[DMZ]   C. Dong, G. Mason and Y.-C. Zhu, Discrete series of the Virasoro algebra and the moonshine module, in: *Algebraic Groups and Their Generalizations: Quantum and Infinite-dimensional Methods, Proc. 1991 American Math. Soc. Summer Research Institute*, ed. by W. J. Haboush and B. J. Parshall, Proc. Symp. Pure Math. **56**, Part 2, Amer. Math. Soc., Providence, 1994, 295–316.

[DN1]   C. Dong and K, Nagatomo, Representations of vertex operator algebra V_L^+ for rank one lattice L, *Commun. Math. Phys.* **202** (1999), 169–195.

[DN2]   C. Dong and K, Nagatomo, Classification of irreducible modules for the vertex operator algebra $M(1)^+$, *J. Algebra* **216** (1999), 384–404.

[DN3]   C. Dong and K, Nagatomo, Automorphism groups and twisted modules for lattice vertex operator algebras, in: *Recent Developments in Quantum Affine Algebras and Related Topics*, ed. by N. Jing and K. C. Misra, Contemporary Math. **248**, Amer. Math. Soc., Providence, 1999, 117–133.

[DN4]   C. Dong and K, Nagatomo, Classification of irreducible modules for the vertex operator algebra $M(1)^+$. II. Higher rank, *J. Algebra* **240** (2001), 289–325.

[DY]   C. Dong and G. Yamskulna, Vertex operator algebras, generalized doubles and dual pairs, *Math. Z.* **241** (2002), 397–423.

[DLMi] B. Doyon, J. Lepowsky and A. Milas, Twisted modules for vertex operator algebras and Bernoulli polynomials, *Internat. Math. Res. Notices* (2003), to appear.

[D]   F. J. Dyson, Missed opportunities, *Bull. Amer. Math. Soc.* **78** (1972), 635–652.

[Eb]   H. Eberle, Twistfield perturbations of vertex operators in the Z_2-orbifold model, *J. High Energy Phys.* **6** (2002), 19.

[EG]   W. Eholzer and M. Gaberdiel, Unitarity of rational N=2 superconformal theories, *Commun. Math. Phys.* **186** (1997), 61–85.

[EM]   S. Eswara Rao and R. V. Moody, Vertex representations for n-toroidal Lie algebras and a generalization of the Virasoro algebra, *Commun. Math. Phys.* **159** (1994), 239–264.

[EFK]   P. Etingof, I. Frenkel and A. Kirillov Jr., *Lectures on Representation Theory and Knizhnik–Zamolodchikov Equations*, University Lecture Series, Vol. 21, Amer. Math. Soc., Providence, 1998.

[EK]   P. Etingof and D. Kazhdan, Quantization of Lie bialgebras V, Quantum vertex operator algebras, *Selecta Mathematica*, New Series **6** (2000), 105–130.

[Fe1]   B. Feigin, The semi-infinite cohomology of Kac–Moody and Virasoro Lie algebras, *Usp. Mat. Nauk* **39** (1984), 195–196; English transl., *Russian Math. Surveys* **39** (1984), 155–156.

[Fe2]   B. Feigin, Conformal field theory and cohomologies of the Lie algebra of holomorphic vector fields on a complex curve, in: *Proc. Internat. Congr. Math., Kyoto, 1990*, Vol. I, Japan Math. Soc., 1991, 71–85.

[FF1]   B. Feigin and E. Frenkel, A family of representations of affine Lie algebras, *Russian Math. Surveys* **43** (1988), 221–222.

[FF2]   B. Feigin and E. Frenkel, Affine Kac–Moody algebras and semi-infinite flag manifolds, *Commun. Math. Phys.* **128** (1990), 161–189.

[FF3]   B. Feigin and E. Frenkel, Representations of affine Kac–Moody Lie algebras and bosonization, in: *Physics and Mathematics of Strings*, World Scientific, 1990, 271–316.

[FF4]   B. Feigin and E. Frenkel, Representations of affine Kac–Moody Lie algebras, bosonization and resolutions, *Lett. Math. Phys.* **19** (1990), 307–317.

[FF5]   B. Feigin and E. Frenkel, Quantization of the Drinfeld–Sokolov reduction, *Phys. Lett.* **B246** (1990), 75–81.

[FF6]   B. Feigin and E. Frenkel, Semi-infinite Weil complex and the Virasoro algebra, *Commun. Math. Phys.* **137** (1991), 617–639.

[FF7]   B. Feigin and E. Frenkel, Duality in W-algebras, *Internat. Math. Res. Notices* (in *Duke Math. J.*) **6** (1991), 75–82.

[FF8]   B. Feigin and E. Frenkel, Affine Kac–Moody algebras at the critical level and Gelfand–Dikii algebras, *Internat. J. Modern Phys.* **A7**, Supplement 1A (1992), 197–215.

[FF9]   B. Feigin and E. Frenkel, Kac–Moody groups and integrability of soliton equations, *Invent. Math.* **120** (1995), 379–408.

[FF10]   B. Feigin and E. Frenkel, Integrals of motion and quantum groups, in: *Integrable Systems and Quantum Groups*, Lecture Notes in Math. **1620**, Springer-Verlag, 1996, 349–418.

[FF11]   B. Feigin and E. Frenkel, Integrable hierarchies and Wakimoto modules, in: *Differential Topology, Infinite-dimensional Lie Algebras, and Applications, D. B. Fuchs 60th Anniversary Collection*, AMS Translations (2) **194**, 1999, 27–60.

[FFRe]   B. Feigin, E. Frenkel and N. Reshetikhin, Gaudin model, Bethe Ansatz and critical level, *Commun. Math. Phys.* **166** (1994), 27–62.

[FFu1]   B. Feigin and D. B. Fuchs, Skew-symmetric invariant differential operators on the line and Verma modules over the Virasoro algebra, Funkts. Anal. Prilozh. 16, No. 2, 47–63; English transl., *Funct. Anal. Appl.* **16** (1982), 114–126.

[FFu2]   B. Feigin and D. B. Fuchs, Verma modules over the Virasoro algebra, in: *Topology*, Lecture Notes in Math. **1060**, Springer-Verlag, 1984, 230–245.

[FFu3]   B. Feigin and D. B. Fuchs, Cohomology of some nilpotent subalgebras of Virasoro algebra and affine Kac–Moody Lie algebras, *J. Geom. Phys.* **5** (1988), 209.

[FFu4]   B. Feigin and D. B. Fuchs, Representations of the Virasoro algebra, in: *Representations of Lie Groups and Related Topics*, ed. by A. Vershik and D. Zhelobenko, Gordon and Breach, New York, 1990, 465–554.

[FJM]   B. Feigin, M. Jimbo and T. Miwa, Vertex operator algebra arising from the minimal series $M(3, p)$ and monomial basis, in: *Math. Phys. Odyssey, 2001*, Prog. Math. Phys., 23, Birkhäuser, Boston, 2002, 179–204.

[FLT]   B. Feigin, S. Loktev and I. Tipunin, Coinvariants for lattice VOAs and q-supernomial coefficients, *Commun. Math. Phys.* **229** (2002), 271–292.

[FM1]   B. Feigin and F. Malikov, Fusion algebra at a rational level and cohomology of nilpotent subalgebras of $\widehat{sl}(2)$, *Lett. Math. Phys.* **31** (1994), 315–325.

[FM2]   B. Feigin and F. Malikov, Modular functor and representation theory of $\widehat{sl}_2$ at a rational level, in: *Operads: Proceedings of Renaissance Conferences*, ed. by J.-L. Loday, J. Stasheff and A. A. Voronov, Contemporary Math. **202**, Amer. Math. Soc., Providence, 1997, 357–405.

[FM]   B. Feigin and T. Miwa, Extended vertex operator algebras and monomial bases, in: *Statistical Physics on the Eve of the 21st Century*, Ser. Adv. Statist. Mech., 14, World Sci. Publishing, River Edge, NJ, 1999, 366–390.

[FSV]   B. Feigin, V. Schechtman and A. Varchenko, On algebraic equations satisfied by correlators in Wess–Zumino–Witten models, *Lett. Math. Phys.* **20** (1990), 291–297.

[FSe]   B. Feigin and A. Semikhatov, The $\widehat{sl}(2) \oplus \widehat{sl}(2)/\widehat{sl}(2)$ coset theory as a Hamiltonian reduction of $\widehat{D}(2|1; \alpha)$, *Nucl. Phys.* **B610** (2001), 489–530.

[Fei1]   A. J. Feingold, Some applications of vertex operators to Kac–Moody algebras, in: *Vertex Operators in Mathematics and Physics, Proc. 1983 M.S.R.I. Conference*, ed. by J. Lepowsky, S. Mandelstam and I. M. Singer, Publ. Math. Sciences Res. Inst. #3, Springer-Verlag, 1985, 185–206.

[Fei2]   A. J. Feingold, Fusion rules for affine Kac–Moody algebras, preprint, arXiv:math.QA/0212387.

[FeF]   A. J. Feingold and I. B. Frenkel, Classical affine Lie algebras, *Advances in Math.* **56** (1985), 117–172.

[FFR]   A. J. Feingold, I. B. Frenkel and J. F. X. Ries, *Spinor Construction of Vertex Operator Algebras, Triality, and $E_8^{(1)}$*, Contemporary Math. **121**, 1991.

[FL]   A. J. Feingold and J. Lepowsky, The Weyl–Kac character formula and power series identities, *Advances in Math.* **29** (1978), 271–309.

[FRW]   A. J. Feingold, J. F. Ries and M. Weiner, Spinor construction of the $c = 1/2$ minimal model, in: *Moonshine, the Monster and Related Topics, Proc. Joint Summer Research Conference, Mount Holyoke*, 1994, ed. by C. Dong and G. Mason, Contemporary Math. **193**, Amer. Math. Soc., Providence, 1996, 45–92.

[Fi1]   L. Figueiredo, Calculus of principally twisted vertex operators, Ph.D. thesis, Rutgers University, 1986.

[Fi2]   L. Figueiredo, Calculus of principally twisted vertex operators, *Memoirs Amer. Math. Soc.* **69**, 1987.

[FMS]   P. Di Francesco, P. Mathieu and D. Sénéchal, *Conformal Field Theory*, Springer-Verlag, 1996.

[F1]   E. Frenkel, Affine Kac–Moody algebras and quantum Drinfeld–Sokolov reduction, Ph.D. thesis, Harvard University, 1991.

[F2]   E. Frenkel, W-algebras and Langlands–Drinfeld correspondence, in: *New symmetries in quantum field theory*, ed. by J. Fröhlich, et al., Plenum Press, New York, 1992, 433–447.

[F3]   E. Frenkel, Free field realizations in representation theory and conformal field theory, in: *Proc. Internat. Congr. Math., Zürich, 1994*, Birkhäuser, Boston, 1995, 1256–1269.

[F4]   E. Frenkel, Affine algebras, Langlands duality and Bethe Ansatz, in: *Proc. Internat. Congr. Math. Phys., Paris, 1994*, International Press, 1995, 606–642.

[F5]   E. Frenkel, Vertex algebras and algebraic curves, Séminaire Bourbaki, exposé no 276, 1999/2000, *Astérisque* (2002), 299–339.

[F6]   E. Frenkel, Lectures on Wakimoto modules, opers and the center at the critical level, preprint, arXiv:math.QA/0210029.

[FB]   E. Frenkel and D. Ben-Zvi, *Vertex Algebras and Algebraic Curves*, Mathematical Surveys and Monographs, Vol. 88, Amer. Math. Soc., Providence, 2001.

[FKRW]   E. Frenkel, V. Kac, A. Ratiu and W. Wang, $W_{1+\infty}$ and $W(gl_\infty)$ with central charge N, *Commun. Math. Phys.* **170** (1995), 337–357.

[FKW]   E. Frenkel, V. Kac and M. Wakimoto, Characters and fusion rules for W-algebras via quantized Drinfeld–Sokolov reduction, *Commun. Math. Phys.* **147** (1992), 295–328.

[FrR]   E. Frenkel and N. Reshetikhin, Towards deformed chiral algebras, in: *Quantum Group Symposium, Proc. of 1996 Goslar conference*, ed. by H.-D. Doebner and V. K. Dobrev, Heron Press, Sofia, 1997, 27–42.

[FrSz1]   E. Frenkel and M. Szczesny, Twisted modules over vertex algebras on algebraic curves, preprint, arXiv:math.AG/0112211.

[FrSz2]   E. Frenkel and M. Szczesny, Chiral de Rham complex and orbifolds, preprint, arXiv:math.AG/0307181.

[Fr1]   I. B. Frenkel, Spinor representations of affine Lie algebras, *Proc. Natl. Acad. Sci. USA* **77** (1980), 6303–6306.

[Fr2]   I. B. Frenkel, Two constructions of affine Lie algebra representations and boson-fermion correspondence in quantum field theory, *J. Funct. Analysis* **44** (1981), 259–327.

[Fr3]   I. B. Frenkel, Representations of affine Lie algebras, Hecke modular forms and Korteweg–DeVries type equations, in: *Proc. 1981 Rutgers Conf. on Lie Algebras and Related Topics*, ed. by D. Winter, Lecture Notes in Math. **933**, Springer-Verlag, 1982, 71–110.

[Fr4]   I. B. Frenkel, Representations of Kac–Moody algebras and dual resonance models, in: *Applications of Group Theory in Physics and Mathematical Physics, Proc. of 1982 Chicago Conference*, Lectures in Appl. Math. **21**, Amer. Math. Soc., Providence, 1985, 325–353.

[Fr5]   I. B. Frenkel, Beyond affine Lie algebras, in: *Proc. Internat. Congr. Math., Berkeley, 1986*, Vol. 1, Amer. Math. Soc., Providence, 1987, 821–839.

[Fr6]   I. B. Frenkel, Talk presented at The Institute for Advanced Study, 1988.

[FGZ]   I. B. Frenkel, H. Garland and G. Zuckerman, Semi-infinite cohomology and string theory, *Proc. Natl. Acad. Sci. USA* **83** (1986), 8442–8446.

[FHL]   I. B. Frenkel, Y.-Z. Huang and J. Lepowsky, On axiomatic approaches to vertex operator algebras and modules, preprint, 1989; *Memoirs Amer. Math. Soc.* **104**, 1993.

[FJ]   I. B. Frenkel and N.-H. Jing, Vertex representations of quantum affine algebras, *Proc. Natl. Acad. Sci. USA* **85** (1988), 9373–9377.

[FJW1]   I. B. Frenkel, N.-H. Jing and W.-Q. Wang, Vertex representations via finite groups and the McKay correspondence, *Internat. Math. Res. Notices* (2000), 195–222

[FJW2]   I. B. Frenkel, N.-H. Jing and W.-Q. Wang, Quantum vertex representations via finite groups and the McKay correspondence, *Commun. Math. Phys.* **211** (2000), 365–393.

[FJW3]   I. B. Frenkel, N.-H. Jing and W.-Q. Wang, Twisted vertex representations via spin groups and the McKay correspondence, *Duke Math. J.* **111** (2002), 51–96.

[FK]   I. B. Frenkel and V. Kac, Basic representations of affine Lie algebras and dual resonance models, *Invent. Math.* **62** (1980), 23–66.

[FLM1]   I. B. Frenkel, J. Lepowsky and A. Meurman, An E_8-approach to F_1, in: *Finite Groups—Coming of Age, Proc. 1982 Montreal Conference*, ed. by J. McKay, Contemporary Math. **45**, Amer. Math. Soc., Providence, 1985, 99–120.

[FLM2]   I. B. Frenkel, J. Lepowsky and A. Meurman, A natural representation of the Fischer–Griess Monster with the modular function J as character, *Proc. Natl. Acad. Sci. USA* **81** (1984), 3256–3260.

[FLM3]   I. B. Frenkel, J. Lepowsky and A. Meurman, A moonshine module for the Monster, in: *Vertex Operators in Mathematics and Physics, Proc. 1983 M.S.R.I. Conference*, ed. by J. Lepowsky, S. Mandelstam and I. M. Singer, Publ. Math. Sciences Res. Inst. # 3, Springer-Verlag, 1985, 231–273.

[FLM4]   I. B. Frenkel, J. Lepowsky and A. Meurman, An introduction to the Monster, in: *Unified String Theories, Proc. 1985 Inst. for Theoretical Physics Workshop*, ed. by M. Green and D. Gross, World Scientific, Singapore, 1986, 533–546.

[FLM5]   I. B. Frenkel, J. Lepowsky and A. Meurman, Vertex operator calculus, in: *Mathematical Aspects of String Theory, Proc. 1986 Conference, San Diego*, ed. by S.-T. Yau, World Scientific, Singapore, 1987, 150–188.

[FLM6]   I. B. Frenkel, J. Lepowsky and A. Meurman, *Vertex Operator Algebras and the Monster*, Pure and Applied Math., Vol. 134, Academic Press, Boston, 1988.

[FR]   I. B. Frenkel and N. Reshetikhin, Quantum affine algebras and holonomic difference equations, *Commun. Math. Phys.* **146** (1992), 1–60.

[FW]   I. B. Frenkel and W.-Q. Wang, Virasoro algebra and wreath product convolution, *J. Algebra* **242** (2001), 656–671.

[FZ]   I. B. Frenkel and Y.-C. Zhu, Vertex operator algebras associated to representations of affine and Virasoro algebras, *Duke Math. J.* **66** (1992), 123–168.

[FQS1]   D. Friedan, Z. Qiu and S. Shenker, Conformal invariance, unitarity, and critical exponents in two dimensions, *Phys. Rev. Lett.* **52** (1984), 1575.

[FQS2]   D. Friedan, Z. Qiu and S. Shenker, Conformal invariance, unitarity, and two dimensional critical exponents, in: *Vertex Operators in Mathematics and Physics, Proc. 1983 M.S.R.I. Conference*, ed. by J. Lepowsky, S. Mandelstam and I. M. Singer, Publ. Math. Sciences Res. Inst. # 3, Springer-Verlag, 1985, 419–449.

[FQS3]   D. Friedan, Z. Qiu and S. Shenker, Superconformal invariance in two dimensions and the tricritical Ising model, *Phys. Lett.* **B151** (1985), 37–43.

[FQS4]   D. Friedan, Z. Qiu and S. Shenker, Details of the non-unitarity proof, *Commun. Math. Phys.* **107** (1986), 535–542.

[FS]   D. Friedan and S. Shenker, The analytic geometry of two-dimensional conformal field theory, *Nucl. Phys.* **B281** (1987), 509–545.

[FV1]   S. Fubini and G. Veneziano, Duality in operator formalism, *Nuovo Cim.* **67A** (1970), 29–47.

[FV2]   S. Fubini and G. Veneziano, Algebraic treatment of subsidiary conditions in dual resonance models, *Ann. Phys.* **63** (1971), 12–27.

[Fu1]   D. B. Fuchs, *Cohomology of infinite-dimensional Lie algebras*, Consultants Bureau, New York, 1986.

[Fu2]   D. B. Fuchs, Two projections of singular vectors of Verma modules over the affine Lie algebra $A_1^{(1)}$, *Funct. Anal. Appl.* **23** (1989), 81–83.

[Fu3]   D. B. Fuchs, *Affine Lie Algebras and Quantum Groups*, Cambridge University Press, 1995.

[FSc]   J. Fuchs and C. Schweigert, Category theory for conformal boundary conditions, in: *Vertex Operator Algebras in Mathematics and Physics, Proc. of Workshop at Fields Institute for Research in Mathematical Sciences, 2000*, ed. by S. Berman, Y. Billig, Y.-Z. Huang and J. Lepowsky, Fields Institute Communications **39**, Amer. Math. Soc., 2003.

[Gab1]   M. Gaberdiel, A general transformation formula for conformal fields, *Phys. Lett.* **B325** (1994), 366–370.

[Gab2]   M. Gaberdiel, An explicit construction of the quantum group in chiral WZW-models, *Commun. Math. Phys.* **173** (1995), 357–378.

[Gab3]   M. Gaberdiel, Fusion of twisted representations, *Internat. J. Modern Phys.* **A12** (1997), 5183–5208.

[Gab4]   M. Gaberdiel, An introduction to conformal field theory, *Rept. Prog. Phys.* **63** (2000), 607–667.

[GG]   M. Gaberdiel and P. Goddard, Axiomatic conformal field theory, *Commun. Math. Phys.* **209** (2000), 549–594.

[GK1]   M. Gaberdiel and H. Kausch, Indecomposable fusion products, *Nucl. Phys.* **B477** (1996), 293–318.

[GK2]   M. Gaberdiel and H. Kausch, A rational logarithmic conformal field theory, *Phys. Lett.* **B386** (1996), 131–137.

[GN]   M. Gaberdiel and A. Neitzke, Rationality, quasirationality and finite W-algebras, *Commun. Math. Phys.* **238** (2003), 305–331.

[Gai]   D. Gaitsgory, Notes on $2D$ conformal field theory and string theory, in: *Quantum fields and strings: a course for mathematicians*, Vol. 2, Amer. Math. Soc., Providence, 1999, 1017–1089.

[Ga1]   T. Gannon, Monstrous moonshine and the classification of CFT, preprint, arXiv:math.QA/9906167.

[Ga2]   T. Gannon, Postcards from the edge, or snapshots of the theory of generalised moonshine, *Canad. Math. Bull.* **45** (2002), 606–622.

[Gao1]   Y. Gao, Vertex operators arising from the homogeneous realization for $\hat{gl}_N$, *Commun. Math. Phys.* **211** (2000), 745–777.

[Gao2]   Y. Gao, Representations of extended affine Lie algebras coordinated by certain quantum tori, *Compositio Mathematica* **123** (2000), 1–25.

[Gao3]   Y. Gao, Fermionic and bosonic representations of the extended affine Lie algebra $\widetilde{gl_N(\mathbb{C}_q)}$, *Canad. Math. Bull.* **45** (2002), 623–633.

[GaoL]   Y.-C. Gao and H.-S. Li, Generalized vertex algebras generated by parafermion-like vertex operators, *J. Algebra* **240** (2001), 771–807.

[Gar1]   H. Garland, The arithmetic theory of loop algebras, *J. Algebra* **53** (1978), 480–551.

[Gar2]   H. Garland, The arithmetic theory of loop groups, *Inst. Hautes Etudes Sci. Publ. Math.* **52** (1980), 5–136.

[Gar3]   H. Garland, Lectures on loop algebras and the Leech lattice, Yale University, Spring 1980.

[GarL]   H. Garland and J. Lepowsky, Lie algebra homology and the Macdonald–Kac formulas, *Invent. Math.* **34** (1976), 37–76.

[Gaw]   K. Gawedzki, Conformal field theory, Séminaire Bourbaki, exposé no 704, 1988/89, *Astérisque* **177–178** (1989), 95–126.

[Geb]   R. W. Gebert, Introduction to vertex algebras, Borcherds algebras and the monster Lie algebra, *Internat. J. Modern Phys.* **A8** (1993), 5441–5503.

[GF]   I. M. Gelfand and D. B. Fuchs, The cohomology of the Lie algebra of vector fields on a circle, *Funk. Anal. i Prilozhen.* **2** (1968), 92–93; English transl. *Funct. Anal. Appl.* **2** (1968), 342–343.

[GS]   I. M. Gelfand and G. E. Shilov, *Generalized Functions*, Vol. 1, Academic Press, New York and London, 1964.

[Ge1]   G. Georgiev, Combinatorial constructions of modules for infinite-dimensional Lie algebras, Ph.D. thesis, Rutgers University, 1996.

[Ge2]   G. Georgiev, Combinatorial constructions of modules for infinite-dimensional Lie algebras, I. Principal subspace, *J. Pure Applied Algebra* **112** (1996), 247–286.

[Ge3]   G. Georgiev, Combinatorial constructions of modules for infinite-dimensional Lie algebras, II. Parafermionic space, preprint, arXiv:q-alg/9504024.

[G]   D. Gepner, New conformal field theories associated with Lie algebras and their partition functions, *Nucl. Phys.* **B290** (1987), 10–24.

[GW]   D. Gepner and E. Witten, String theory on group manifolds, *Nucl. Phys.* **B278** (1986), 493–549.

[Gl]   J. Glass, Rank 24 holomorphic vertex operator algebras, *Commun. Algebra* **24** (1996), 1857–1896.

[GVF]   E. Del Giudice, P. Di Vecchia and S. Fubini, General properties of the dual resonance model, *Ann. Phys.* **70** (1972), 378.

[Go1]   P. Goddard, Meromorphic conformal field theory, in: *Infinite Dimensional Lie Algebras and Groups*, Proceedings of the conference held at CIRM, Luminy, edited by V. G. Kac, World Scientific Singapore, 1989, 556–587.

[Go2]   P. Goddard, The work of R. E. Borcherds, in: *Proc. Internat. Congr. Math., Vol. I, Berlin, 1998*, Doc. Math. 1998, Extra Vol. I, 99–108.

[GKO1]   P. Goddard, A. Kent and D. Olive, Virasoro algebras and coset space models, *Phys. Lett.* **B152** (1985), 88–92.

[GKO2]   P. Goddard, A. Kent and D. Olive, Unitary representations of the Virasoro and super-Virasoro algebra, *Commun. Math. Phys.* **103** (1986), 105–119.

[GNORS]   P. Goddard, W. Nahm, D. Olive, H. Ruegg and A. Schwimmer, Fermions and octonions, *Commun. Math. Phys.* **112** (1987), 385–408.

[GNOS]   P. Goddard, W. Nahm, D. Olive and A. Schwimmer, Vertex operators for non-simply-laced algebras, *Commun. Math. Phys.* **107** (1986), 179–212.

[GO1]   P. Goddard and D. Olive, Algebras, lattices and strings, in: *Vertex Operators in Mathematics and Physics, Proc. 1983 M.S.R.I. Conference*, ed. by J. Lepowsky, S. Mandelstam and I. M. Singer, Publ. Math. Sciences Res. Inst. # 3, Springer-Verlag, 1985, 51–96.

[GO2]   P. Goddard and D. Olive, Kac–Moody and Virasoro algebras in relation to quantum physics, *Internat. J. Modern Phys.* **A1** (1986), 303–414.

[GO3]   P. Goddard and D. Olive, Algebras, lattices and strings 1986, *Physica Scripta* **T15** (1987), 19–25.

[GT]   P. Goddard and C. B. Thorn, Compatibility of the dual Pomeron with unitarity and the absence of ghosts in the dual resonance model, *Phys. Lett.* **40B** (1972), 235.

[GW1]   R. Goodman and N. R. Wallach, Structure of unitary cocycle representations of loop groups and the group of diffeomorphisms of the circle, *J. Reine Angew. Math.* **347** (1984), 69–133.

[GW2]   R. Goodman and N. R. Wallach, Projective unitary positive-energy representations of $Diff(S^1)$, *J. Funct. Anal.* **63** (1985), 199–321.

[GMS1]   V. Gorbounov, F. Malikov and V. Schechtman, Gerbes of chiral differential operators, I, *Math. Res. Lett.* **7** (2000), 55–66.

[GMS2]   V. Gorbounov, F. Malikov and V. Schechtman, Gerbes of chiral differential operators, II, preprint, arXiv:math.AG/0003170.

[GMS3]   V. Gorbounov, F. Malikov and V. Schechtman, Gerbes of chiral differential operators, III, preprint, arXiv:math.AG/0005201.

[GMS4]   V. Gorbounov, F. Malikov and V. Schechtman, On chiral differential operators over homogeneous spaces, *Internat. J. Math. Math. Sci.* **26** (2001), 83–106.

[Gor]   D. Gorenstein, Classifying the finite simple groups, Colloquium Lectures, Anaheim, January 1985, Amer. Math. Soc., *Bull. Amer. Math. Soc. (New Series)* **14** (1986), 1–98.

[GSc]   M. B. Green and J. H. Schwarz, Anomaly cancellations in supersymmetric $D = 10$ gauge theory and superstring theory, *Phys. Lett.* **149B** (1984), 117.

[GSW]   M. B. Green, J. H. Schwarz and E. Witten, *Superstring Theory,* **1** and **2**, Cambridge Univ. Press, Cambridge, 1987.

[Gre]   B. Greene, *The Elegant Universe,* W. W. Norton and Co., New York, 1999.

[G1]   R. L. Griess Jr., The structure of the "Monster" simple group, in: *Proc. of the Conference on Finite Groups,* ed. by W. R. Scott and R. Gross, Academic Press, New York, 1976, 113–118.

[G2]   R. L. Griess Jr., A construction of F_1 as automorphisms of a 196, 883 dimensional algebra, *Proc. Natl. Acad. Sci. USA* **78** (1981), 689–691.

[G3]   R. L. Griess Jr., The Friendly Giant, *Invent. Math.* **69** (1982), 1–102.

[G4]   R. L. Griess Jr., The sporadic simple groups and construction of the Monster, in: *Proc. Internat. Congr. Math., Warsaw, 1983,* North-Holland, Amsterdam, 1984, 369–384.

[G5]   R. L. Griess Jr., The Monster and its nonassociative algebra, in: *Finite Groups—Coming of Age, Proc. 1982 Montreal Conference,* ed. by J. McKay, Contemporary Math. **45**, Amer. Math. Soc., Providence, 1985, 121–157.

[G6]   R. L. Griess Jr., A brief introduction to the finite simple groups, in: *Vertex Operators in Mathematics and Physics, Proc. 1983 M.S.R.I. Conference,* ed. by J. Lepowsky, S. Mandelstam and I. M. Singer, Publ. Math. Sciences Res. Inst. # 3, Springer-Verlag, 1985, 217–229.

[G7]   R. L. Griess Jr., A vertex operator algebra related to E_8 with automorphism group $O^+(10, 2)$, in: *The Monster and Lie Algebras (Columbus, OH, 1996),* Ohio State Univ. Math. Res. Inst. Publ., 7, de Gruyter, Berlin, 1998, 43–58.

[G8]   R. L. Griess Jr., GNAVOA, I. Studies in groups, nonassociative algebras and vertex operator algebras, in: *Vertex Operator Algebras in Mathematics and Physics, Proc. of Workshop at Fields Institute for Research in Mathematical Sciences, 2000,* ed. by S. Berman, Y. Billig, Y.-Z. Huang and J. Lepowsky, Fields Institute Communications **39**, Amer. Math. Soc., 2003.

[GH]   R. L. Griess Jr. and G. Höhn, Virasoro frames and their stablizers for the E_8 lattice type vertex operator algebra, preprint, arXiv:math.QA/0101054.

[GHMR]   D. J. Gross, J. A. Harvey, E. Martinec and R. Rohm, Heterotic string theory (I). The free heterotic string, *Nucl. Phys.* **B256** (1985), 253–284; (II). The interacting heterotic string, *Nucl. Phys.* **B267** (1986), 74–124.

[Gu]   H. Guo, On abelian intertwining algebras and modules, Ph.D. thesis, Rutgers University, 1993.

[Ha]   M. B. Halpern, Quantum "solitons" which are $SU(N)$ fermions, *Phys. Rev.* **D12** (1975), 1684–1699.

[HV]   S. Hamidi and C. Vafa, Interactions on orbifolds, *Nucl. Phys.* **B279** (1987), 465–513.

[HMT]   A. Hanaki, M. Miyamoto and D. Tambara, Quantum Galois theory for finite groups, *Duke Math. J.* **97** (1999), 541–544.

[HJKOS]   Y. Hara, M. Jimbo, H. Konno, S. Odake, and J. Shiraishi, On Lepowsky–Wilson's Z-algebra, in: *Recent Developments in Infinite-Dimensional Lie Algebras and Conformal Field Theory (Charlottesville, VA, 2000),* Contemporary Math. **297**, Amer. Math. Soc., Providence, 2002, 143–149.

[HLa]   K. Harada and C. H. Lam, "Inverse" of Virasoro operators, *Commun. Algebra* **23** (1995), 4405–4413.

[HLan]   K. Harada and M. L. Lang, Modular forms associated with the Monster module, in: *The Monster and Lie Algebras (Columbus, OH, 1996),* Ohio State Univ. Math. Res. Inst. Publ., 7, de Gruyter, Berlin, 1998, 59–83.

302   References

[Har]   J. A. Harvey, Twisting the heterotic string, in: *Unified String Theories, Proc. 1985 Inst. for Theoretical Physics Workshop*, ed. by M. Green and D. Gross, World Scientific, Singapore, 1986, 704–718.

[H1]   G. Höhn, Selbstduale Vertexoperatorsuperalgebren und das Babymonster, Ph.D. thesis, Universität Bonn, 1995; Bonner Mathematische Schriften **286**.

[H2]   G. Höhn, Self-dual vertex operator superalgebras with shadows of large minimal weight, *Internat. Math. Res. Notices* **13** (1997), 613–621.

[H3]   G. Höhn, Genera of vertex operator algebras and three-dimensional topological quantum field theories, in: *Vertex Operator Algebras in Mathematics and Physics, Proc. of Workshop at Fields Institute for Research in Mathematical Sciences, 2000*, ed. by S. Berman, Y. Billig, Y.-Z. Huang and J. Lepowsky, Fields Institute Communications **39**, Amer. Math. Soc., 2003.

[H4]   G. Höhn, The group of symmetries of the shorter moonshine module, preprint, arXiv:math.QA/0210076.

[HuK]   P. Hu and I. Kriz, Vertex operator algebras and elliptic cohomology, preprint.

[Hua1]   Y.-Z. Huang, On the geometric interpretation of vertex operator algebras, Ph.D. thesis, Rutgers University, 1990.

[Hua2]   Y.-Z. Huang, Geometric interpretation of vertex operator algebras, *Proc. Natl. Acad. Sci. USA* **88** (1991), 9964–9968.

[Hua3]   Y.-Z. Huang, Applications of the geometric interpretation of vertex operator algebras, in: *Proc. 20th International Conference on Differential Geometric Methods in Theoretical Physics, New York, 1991*, ed. by S. Catto and A. Rocha, World Scientific, Singapore, 1992, Vol. 1, 333–343.

[Hua4]   Y.-Z. Huang, Vertex operator algebras and conformal field theory, *Internat. J. Modern Phys.* **A7** (1992), 2109–2151.

[Hua5]   Y.-Z. Huang, Binary trees and Lie algebras, in: *Algebraic Groups and Their Generalizations: Quantum and Infinite-dimensional Methods, Proc. 1991 American Math. Soc. Summer Research Institute*, ed. by W. J. Haboush and B. J. Parshall, Proc. Symp. Pure Math. **56**, Part 2, Amer. Math. Soc., Providence, 1994, 337–348.

[Hua6]   Y.-Z. Huang, Operadic formulation of topological vertex algebras and Gerstenhaber or Batalin–Vilkovisky algebras, *Commun. Math. Phys.* **164** (1994), 105–144.

[Hua7]   Y.-Z. Huang, Introduction to vertex operator algebras III, in: *Moonshine and Vertex Operator Algebra, Proc. of 1994 Workshop*, ed. by M. Miyamoto, Publ. Research Inst. for Math. Sciences, Kyoto Univ., 904 (1995), 51–77.

[Hua8]   Y.-Z. Huang, A theory of tensor products for module categories for a vertex operator algebra, IV, *J. Pure Applied Algebra* **100** (1995), Special Volume on Algebra and Physics, guest ed., J. D. Stasheff, 173–216.

[Hua9]   Y.-Z. Huang, A nonmeromorphic extension of the moonshine module vertex operator algebra, in: *Moonshine, the Monster and Related Topics, Proc. Joint Summer Research Conference, Mount Holyoke, 1994*, ed. by C. Dong and G. Mason, Contemporary Math. **193**, Amer. Math. Soc., Providence, 1996, 123–148.

[Hua10]   Y.-Z. Huang, Virasoro vertex operator algebras, (nonmeromorphic) operator product expansion and the tensor product theory, *J. Algebra* **182** (1996), 201–234.

[Hua11]   Y.-Z. Huang, Intertwining operator algebras, genus-zero modular functors and genus-zero conformal field theories, in: *Operads: Proceedings of Renaissance Conferences*, ed. by J.-L. Loday, J. Stasheff, and A. A. Voronov, Contemporary Math. **202**, Amer. Math. Soc., Providence, 1997, 335–355.

[Hua12]   Y.-Z. Huang, Genus-zero modular functors and intertwining operator algebras, *Internat. J. Math.* **9** (1998), 845–863.

[Hua13]   Y.-Z. Huang, *Two-Dimensional Conformal Field Theory and Vertex Operator Algebras*, Progress in Math., Vol. 148, Birkhäuser, Boston, 1998.

[Hua14]   Y.-Z. Huang, A functional-analytic theory of vertex (operator) algebras, I, *Commun. Math. Phys.* **204** (1999), 61–84.

[Hua15]   Y.-Z. Huang, A functional-analytic theory of vertex (operator) algebras, II, *Commun. Math. Phys.*, to appear.

[Hua16] Y.-Z. Huang, Generalized rationality and a "Jacobi identity" for intertwining operator algebras, *Selecta Mathematica*, New Series 6 (2000), 225–267.

[Hua17] Y.-Z. Huang, Review of "Vertex Algebras and Algebraic Curves" by E. Frenkel and D. Ben-Zvi, *Bull. Amer. Math. Soc.* 39 (2002), 585–591.

[Hua18] Y.-Z. Huang, Riemann surfaces with boundaries and the theory of vertex operator algebras, in: *Vertex Operator Algebras in Mathematics and Physics, Proc. of Workshop at Fields Institute for Research in Mathematical Sciences, 2000*, ed. by S. Berman, Y. Billig, Y.-Z. Huang and J. Lepowsky, Fields Institute Communications 39, Amer. Math. Soc., 2003.

[Hua19] Y.-Z. Huang, Differential equations and intertwining operators, preprint, arXiv:math.QA/0206206.

[Hua20] Y.-Z. Huang, Conformal-field-theoretic analogues of codes and lattices, in: *Proceedings of Ramanujan International Symposium on Kac–Moody Lie Algebras and Applications*, ed. by N. Sthanumoorthy, Contemporary Math., Amer. Math. Soc., to appear.

[Hua21] Y.-Z. Huang, Differential equations, duality and modular invariance, arXiv:math.QA/0303049.

[Hua22] Y.-Z. Huang, Differential equations and conformal field theories, arXiv:math.QA/0303269.

[HK] Y.-Z. Huang and L. Kong, Open-string vertex algebras, tensor categories and operads, preprint.

[HL1] Y.-Z. Huang and J. Lepowsky, Toward a theory of tensor products for representations of a vertex operator algebra, in *Proc. 20th International Conference on Differential Geometric Methods in Theoretical Physics, New York, 1991*, ed. by S. Catto and A. Rocha, World Scientific, Singapore, 1992, Vol. 1, 344–354.

[HL2] Y.-Z. Huang and J. Lepowsky, Vertex operator algebras and operads, in: *The Gelfand Mathematical Seminars, 1990–1992*, ed. by L. Corwin, I. Gelfand and J. Lepowsky, Birkhäuser, Boston, 1993, 145–161.

[HL3] Y.-Z. Huang and J. Lepowsky, Operadic formulation of the notion of vertex operator algebra, in: *Mathematical Aspects of Conformal and Topological Field Theories and Quantum Groups, Proc. Joint Summer Research Conference, Mount Holyoke, 1992*, ed. by P. Sally, M. Flato, J. Lepowsky, N. Reshetikhin and G. Zuckerman, Contemporary Math. 175, Amer. Math. Soc., Providence, 1994, 131–148.

[HL4] Y.-Z. Huang and J. Lepowsky, A theory of tensor products for module categories for a vertex operator algebra, I, *Selecta Mathematica*, New Series 1 (1995), 699–756.

[HL5] Y.-Z. Huang and J. Lepowsky, A theory of tensor products for module categories for a vertex operator algebra, II, *Selecta Mathematica*, New Series 1 (1995), 757–786.

[HL6] Y.-Z. Huang and J. Lepowsky, Tensor products of modules and vertex tensor categories, in: *Lie Theory and Geometry, in Honor of Bertram Kostant*, Progress in Math., Vol. 123, Birkhäuser, Boston, 1994, 349–383.

[HL7] Y.-Z. Huang and J. Lepowsky, A theory of tensor products for module categories for a vertex operator algebra, III, *J. Pure Applied Algebra* 100 (1995), Special Volume on Algebra and Physics, guest ed., J. D. Stasheff, 141–171.

[HL8] Y.-Z. Huang and J. Lepowsky, On the $\mathcal{D}$-module and formal-variable approaches to vertex algebras, in: *Topics in Geometry: In Memory of Joseph D'Atri*, ed. by S. Gindikin, Progress in Nonlinear Differential Equations, Vol. 20, Birkhäuser, Boston, 1996, 175–202.

[HL9] Y.-Z. Huang and J. Lepowsky, Intertwining operator algebras and vertex tensor categories for affine Lie algebras, *Duke Math. J.* 99 (1999), 113–134.

[HL10] Y.-Z. Huang and J. Lepowsky, Vertex operator, article in: *Encyclopaedia of Mathematics, Supplement III*, ed. by M. Hazewinkel, Kluwer Academic Publishers, 2002, 420–421.

[HL11] Y.-Z. Huang and J. Lepowsky, Vertex operator algebra, article in: *Encyclopaedia of Mathematics, Supplement III*, ed. by M. Hazewinkel, Kluwer Academic Publishers, 2002, 421–424.

[HM1] Y.-Z. Huang and A. Milas, Intertwining operator superalgebras and vertex tensor categories for superconformal algebras, I, *Commun. Contemp. Math.* 4 (2002), 327–355.

[HM2] Y.-Z. Huang and A. Milas, Intertwining operator superalgebras and vertex tensor categories for superconformal algebras, II, *Trans. Amer. Math. Soc.* 354 (2002), 363–385.

[HZ] Y.-Z. Huang and W. Zhao, Semi-infinite forms and topological vertex operator algebras, *Commun. Contemp. Math.* 2 (2000), 191–241.

[Hum]   J. E. Humphreys, *Introduction to Lie Algebras and Representation Theory*, Graduate Texts in Mathematics, Vol. 9, Springer-Verlag, 1972.

[Hur1]   K. L. Hurley, Strongly holomorphic $c = 24$ vertex operator algebras and modular forms, Ph.D. thesis, University of California, Santa Cruz, 2002.

[Hur2]   K. L. Hurley, Highest-weight vectors of the moonshine module with non-zero graded trace, *J. Algebra* **261** (2003), 411–433.

[Hus1]   C. Husu, Extensions of the Jacobi identity for vertex operators, and standard $A_1^{(1)}$-modules, Ph.D. thesis, Rutgers University, 1990.

[Hus2]   C. Husu, Extensions of the Jacobi identity for vertex operators, and standard $A_1^{(1)}$-modules, *Memoirs Amer. Math. Soc.* **106**, 1993.

[Hus3]   C. Husu, Generating function identities for untwisted standard modules of affine Lie algebras, *Commun. Algebra* **28** (2000), 4411–4432.

[Hus4]   C. Husu, Extensions of the Jacobi identity for relative untwisted vertex operators, and generating function identities for untwisted standard modules: the $A_1^{(1)}$-case, *J. Pure Applied Algebra* **98** (1995), 163–187.

[Hus5]   C. Husu, Extensions of the Jacobi identity for generalized vertex algebras, *J. Pure Applied Algebra* **106** (1996), 127–139.

[IFZ]   C. Itzykson, H. Saleur and J.-B. Zuber, eds., Conformal invariance and applications to statistical mechanics, World Scientific, 1988.

[JM1]   P. Jacob and P. Mathieu, Parafermionic character formulae, *Nucl. Phys.* **B587** (2000), 514–542.

[JM2]   P. Jacob and P. Mathieu, Parafermionic quasi-particle basis and fermionic-type characters, *Nucl. Phys.* **B620** (2002), 351–379.

[JM3]   P. Jacob and P. Mathieu, Graded parafermions: standard and quasi-particle bases, *Nucl. Phys.* **B630** (2002), 433–452.

[J]   N. Jacobson, *Lie Algebras*, New York–London, Wiley Interscience, 1962.

[JW]   N.-H. Jing and W.-Q. Wang, Twisted vertex representations and spin characters, *Math. Z.* **239** (2002), 715–746.

[Ju1]   E. Jurisich, Generalized Kac–Moody Lie algebras and their relation to free Lie algebras, Ph.D. thesis, Rutgers University, 1994.

[Ju2]   E. Jurisich, An exposition of generalized Kac–Moody algebras, in: *Lie Algebras and Their Representations (Seoul, 1995)*, Contemporary Math. **194**, Amer. Math. Soc., Providence, 1996, 121–159.

[Ju3]   E. Jurisich, Generalized Kac–Moody Lie algebras, free Lie algebras and the structure of the monster Lie algebra, *J. Pure Applied Algebra* **126** (1998), 233–266.

[JLW]   E. Jurisich, J. Lepowsky and R. L. Wilson, Realizations of the Monster Lie algebra, *Selecta Mathematica*, New Series **1** (1995), 129–161.

[K1]   V. G. Kac, Simple irreducible graded Lie algebras of finite growth, *Izv. Akad. Nauk SSSR* **32** (1968), 1323–1367. English transl., *Math. USSR Izv.* **2** (1968), 1271–1311.

[K2]   V. G. Kac, Infinite-dimensional Lie algebras and Dedekind's η-function, *Funk. Anal. i Prilozhen.* **8** (1974), 77–78; English transl. *Funct. Anal. Appl.* **8** (1974), 68–70.

[K3]   V. G. Kac, Lie superalgebras, *Advances in Math.* **26** (1977), 8–96.

[K4]   V. G. Kac, Highest weight representations of infinite-dimensional Lie algebras, in: *Proc. Internat. Congr. Math., Helsinki, 1978*, Acad. Scientiarum Fennica, Helsinki, 1980, 299–304.

[K5]   V. G. Kac, Contravariant form for infinite-dimensional Lie algebras and superalgebras, Lecture Notes in Physics **94**, Springer-Verlag, 1979, 441–445.

[K6]   V. G. Kac, *Infinite-dimensional Lie Algebras*, 3rd ed., Cambridge Univ. Press, Cambridge, 1990.

[K7]   V. G. Kac, *Vertex Algebras for Beginners*, University Lecture Series, Vol. 10, Amer. Math. Soc., 1997.

[K8]   V. G. Kac, Formal distribution algebras and conformal algebras, in: *Proc. XIIth Internat. Congr. Math. Phys., Brisbane, 1997*, International Press, Cambridge, 1999, 80–97.

[KKLW]   V. G. Kac, D. A. Kazhdan, J. Lepowsky and R. L. Wilson, Realization of the basic representations of the Euclidean Lie algebras, *Advances in Math.* **42** (1981), 83–112.

[KP1]   V. G. Kac and D. H. Peterson, Spin and Wedge representations of infinite-dimensional Lie algebras and groups, *Proc. Natl. Acad. Sci. USA* **78** (1981), 3308–3312.

[KP2]   V. G. Kac and D. H. Peterson, Infinite-dimensional Lie algebras, theta functions and modular forms, *Advances in Math.* **53** (1984), 125–264.

[KP3]   V. G. Kac and D. H. Peterson, 112 constructions of the basic representation of the loop group of E_8, in: *Symposium on Anomalies, Geometry, Topology, 1985, Proceedings,* ed. by W. A. Bardeen and A. R. White, World Scientific, Singapore, 1985, 276–298.

[KRad]   V. G. Kac and A. D. Radul, Representation theory of the vertex algebra $W_{1+\infty}$, *Transformation Groups* **1** (1996), 41–70.

[KR]   V. G. Kac and A. K. Raina, Bombay Lectures on Highest Weight Representations of Infinite-dimensional Lie Algebras, *Advanced Ser. in Math. Phys.*, Vol. 2, World Scientific, 1987.

[KT]   V. G. Kac and I. Todorov, Superconformal current algberas and their unitary representations, *Commun. Math. Phys.* **102** (1985), 337–347.

[KW1]   V. G. Kac and M. Wakimoto, Modular and conformal invariance constraints in representation theory of affine algebras, *Advances in Math.* **70** (1988), 156–234.

[KW2]   V. G. Kac and M. Wakimoto, Modular invariant representations of infinite-dimensional Lie algebras and superalgebras, *Proc. Natl. Acad. Sci. USA* **85** (1988), 4956–4960.

[KWa]   V. G. Kac and W.-Q. Wang, Vertex operator superalgebras and their representations, in: *Mathematical Aspects of Conformal and Topological Field Theories and Quantum Groups, Proc. Joint Summer Research Conference, Mount Holyoke, 1992,* ed., by P. J. Sally Jr., M. Flato, J. Lepowsky, N. Reshetikhin and G. J. Zuckerman, Contemporary Math. **175**, Amer. Math. Soc., Providence, 1994, 161–191.

[Ka]   L. Kadanoff, Operator algebra and the determination of critical indices, *Phys. Rev. Lett.* **23** (1969), 1430–1433.

[K]   I. L. Kantor, Graded Lie algebras, *Trudy Sem. Vect. Tens. Anal., Moscow State University* **15** (1970), 227–266 (in Russian).

[KO]   A. Kapustin and D. Orlov, Vertex algebras, mirror symmetry, and D-branes: The case of complex tori, *Commun. Math. Phys.* **233** (2003), 79–136.

[KLi]   M. Karel and H.-S. Li, Certain generating subspaces for vertex operator algebras, *J. Algebra* **217** (1999), 393–421.

[KSU1]   T. Katsura, Y. Shimizu and K. Ueno, Formal groups and conformal field theory over $\mathbb{Z}$, in: *Integrable Systems in Quantum Field Theory and Statistical Mechanics,* Adv. Stud. Pure Math. **19**, Academic Press, Boston, 1989, 347–366.

[KSU2]   T. Katsura, Y. Shimizu and K. Ueno, Complex cobordism ring and conformal field theory over $\mathbb{Z}$, *Math. Annalen* **291** (1991), 551–571.

[KLu0]   D. Kazhdan and G. Lusztig, Affine Lie algebras and quantum groups, *Duke Math. J.* **2** (1991), 21–29.

[KLu1]   D. Kazhdan and G. Lusztig, Tensor structures arising from affine Lie algebras, I, *J. Amer. Math. Soc.* **64** (1993), 905–9047

[KLu2]   D. Kazhdan and G. Lusztig, Tensor structures arising from affine Lie algebras, II, *J. Amer. Math. Soc.* **64** (1993), 9047–1011.

[Ki]   T. Kimura, Operads of moduli spaces and algebraic structures in conformal field theory, in: *Moonshine, the Monster and Related Topics, Proc. Joint Summer Research Conference, Mount Holyoke, 1994,* ed. by C. Dong and G. Mason, Contemporary Math. **193**, Amer. Math. Soc., Providence, 1996, 159–189.

[KSV]   T. Kimura, J. D. Stasheff and A. A. Voronov, On operad structures of moduli spaces and string theory, *Commun. Math. Phys.* **171** (1995), 1–25.

[KV]   T. Kimura and A. A. Voronov, The cohomology of algebras over moduli spaces, in: Progress in Math., Vol. 129, Birkhäuser, Boston, 1995, 305–334.

[KVZ]   T. Kimura, A. A. Voronov and G. Zuckerman, Homotopy Gerstenhaber algebras and topological field theory, in: *Operads: Proceedings of Renaissance Conferences,* Contemporary Math. **202**, Amer. Math. Soc., Providence, 1997, 305–333.

[KMY]   M. Kitazume, M. Miyamoto and H. Yamada, Ternary codes and vertex operator algebras, *J. Algebra* **223** (2000), 379–395.

[Kl]   A. Kleinschmidt, Lattice vertex algebras on general even, self-dual lattices, arXiv:math.QA/0210451.

[Klev]   S. Klevtsov, On vertex operator construction of quantum affine algebras, preprint, arXiv:hep-th/0110148.

[KZ]   V. G. Knizhnik and A. B. Zamolodchikov, Current algebra and Wess–Zumino model in two dimensions, *Nucl. Phys.* **B247** (1984), 83–103.

[Ko]   M. L. Kontsevich, The Virasoro algebra and Teichmüller spaces (in Russian), *Funk. Anal. i Prilozhen* **21** (1987), 78–79. English translation, *Func. Anal. Appl.* **21** (1987), 156–157.

[La1]   C. H. Lam, Construction of vertex operator algebras from commutative associative algebras, *Commun. Algebra* **24** (1996), 4339–4360.

[La2]   C. H. Lam, On the structure of vertex operator algebras and their weight two subspaces, Ph.D. thesis, The Ohio State University, 1996.

[La3]   C. H. Lam, Induced modules for orbifold vertex operator algebras, *J. Math. Soc. Japan* **53** (2001), 541–557.

[LLY]   C. Lam, N. Lam and H. Yamauchi, Extension of unitary Virasoro vertex operator algebra by a simple module, *Internat. Math. Res. Notices* **11** (2003), 577–611.

[L1]   J. Leech, Some sphere packings in higher space, *Can. J. Math.* **16** (1964), 657–682.

[L2]   J. Leech, Notes on sphere packings, *Can. J. Math.* **19** (1967), 251–267.

[Le1]   J. Lepowsky, Existence of conical vectors in induced modules, *Annals of Math.* **102** (1975), 17–40.

[Le2]   J. Lepowsky, Lectures on Kac–Moody Lie algebras, Univ. Paris VI, 1978.

[Le3]   J. Lepowsky, Lie algebras and combinatorics, in: *Proc. Internat. Congr. Math., Helsinki, 1978*, Acad. Scientiarum Fennica, Helsinki, 1980, 579–584.

[Le4]   J. Lepowsky, Generalized Verma modules, loop space cohomology and Macdonald-type identities, *Ann. Sci. Ecole Norm. Sup.* **12** (1979), 169–234.

[Le5]   J. Lepowsky, Euclidean Lie algebras and the modular function j, in: *Santa Cruz Conference on Finite Groups, 1979*, Proc. Symp. Pure Math. **37**, Amer. Math. Soc., Providence, 1980, 567–570.

[Le6]   J. Lepowsky, Affine Lie algebras and combinatorial identities, in: *Proc. 1981 Rutgers Conf. On Lie Algebras and Related Topics*, ed. by D. Winter, Lecture Notes in Math. **933**, Springer-Verlag, 1982, 130–156.

[Le7]   J. Lepowsky, Some constructions of the affine Lie algebra $A_1^{(1)}$, in: *Proc. 1982 Summer Seminar on Applications of Group Theory in Physics and Mathematical Physics*, Lectures in Applied Math. **21**, Amer. Math. Soc., Providence, 1985, 375–397.

[Le8]   J. Lepowsky, Calculus of twisted vertex operators, *Proc. Natl. Acad. Sci. USA* **82** (1985), 8295–8299.

[Le9]   J. Lepowsky, Perspectives on vertex operators and the Monster, in: *Proceedings of the Symposium on the Mathematical Heritage of Hermann Weyl, Duke University, May, 1987*, Proc. Symp. Pure Math. **48**, Amer. Math. Soc., Providence, 1988, 181–197.

[Le10]   J. Lepowsky, Remarks on vertex operator algebras and moonshine, in: *Proc. 20th International Conference on Differential Geometric Methods in Theoretic Physics, New York, 1991*, ed. by S. Catto and A. Rocha, World Scientific, Singapore, 1992, 361–370.

[Le11]   J. Lepowsky, The work of Richard E. Borcherds, a segment of: The mathematical work of the 1998 Fields Medalists (J. Lepowsky, J. Lindenstrauss, Y. Manin and J. Milnor), feature article, *Notices of Amer. Math. Soc.* **46** (1999), 17–19.

[Le12]   J. Lepowsky, Vertex operator algebras and the zeta function, in: *Recent Developments in Quantum Affine Algebras and Related Topics*, ed. by N. Jing and K. C. Misra, Contemporary Math. **248**, Amer. Math. Soc., Providence, 1999, 327–340.

[Le13]   J. Lepowsky, Application of a "Jacobi identity" for vertex operator algebras to zeta values and differential operators, *Lett. Math. Phys.* **53** (2000), 87–103.

[LMS]   J. Lepowsky, S. Mandelstam and I. M. Singer, eds., *Vertex Operators in Mathematics and Physics, Proc. 1983 M.S.R.I. Conference*, Publ. Math. Sciences Res. Inst. # 3, Springer-Verlag, 1985.

[LM]   J. Lepowsky and A. Meurman, An E_8-approach to the Leech lattice and the Conway group, *J. Algebra* **77** (1982), 484–504.

[LP1]   J. Lepowsky and M. Primc, Standard modules for type one affine Lie algebras, in: *Number Theory, New York, 1982*, Lecture Notes in Math. **1052**, Springer-Verlag, 1984, 194–251.

[LP2]   J. Lepowsky and M. Primc, Structure of the standard modules for the affine Lie algebra $A_1^{(1)}$, *Contemporary Math.* **46**, 1985.

[LP3]   J. Lepowsky and M. Primc, Structure of the standard $A_1^{(1)}$-modules in the homogeneous picture, in: *Vertex Operators in Mathematics and Physics, Proc. 1983 M.S.R.I. Conference*, ed. by J. Lepowsky, S. Mandelstam and I. M. Singer, Publ. Math. Sciences Res. Inst. # 3, Springer-Verlag, 1985, 143–162.

[LW1]   J. Lepowsky and R. L. Wilson, Construction of the affine Lie algebra $A_1^{(1)}$, *Commun. Math. Phys.* **62** (1978), 43–53.

[LW2]   J. Lepowsky and R. L. Wilson, The Rogers–Ramanujan identities: Lie theoretic interpretation and proof, *Proc. Natl. Acad. Sci. USA* **78** (1981), 699–701.

[LW3]   J. Lepowsky and R. L. Wilson, A Lie theoretic interpretation and proof of the Rogers–Ramanujan identities, *Advances in Math.* **45** (1982), 21–72.

[LW4]   J. Lepowsky and R. L. Wilson, A new family of algebras underlying the Rogers–Ramanujan identities and generalizations, *Proc. Natl. Acad. Sci. USA* **78** (1981), 7254–7258.

[LW5]   J. Lepowsky and R. L. Wilson, The structure of standard modules, I: Universal algebras and the Rogers–Ramanujan identities, *Invent. Math.* **77** (1984), 199–290.

[LW6]   J. Lepowsky and R. L. Wilson, The structure of standard modules, II: The case $A_1^{(1)}$, principal gradation, *Invent. Math.* **79** (1985), 417–442.

[LW7]   J. Lepowsky and R. L. Wilson, Z-algebras and the Rogers–Ramanujan identities, in: *Vertex Operators in Mathematics and Physics, Proc. 1983 M.S.R.I. Conference*, ed. by J. Lepowsky, S. Mandelstam and I. M. Singer, Publ. Math. Sciences Res. Inst. # 3, Springer-Verlag, 1985, 97–142.

[Li1]   H.-S. Li, Symmetric invariant bilinear forms on vertex operator algebras, *J. Pure Applied Algebra* **96** (1994), 279–297.

[Li2]   H.-S. Li, Introduction to vertex operator algebras II, in: *Moonshine and Vertex Operator Algebra, Proc. of 1994 Workshop*, ed. by M. Miyamoto, Publ. Research Inst. for Math. Sciences, Kyoto Univ., 904 (1995), 26–50.

[Li3]   H.-S. Li, Local systems of vertex operators, vertex superalgebras and modules, preprint, 1993; *J. Pure Applied Algebra* **109** (1996), 143–195.

[Li4]   H.-S. Li, Local systems of twisted vertex operators, vertex superalgebras and twisted modules, in: *Moonshine, the Monster and Related Topics, Proc. Joint Summer Research Conference, Mount Holyoke*, 1994, ed. by C. Dong and G. Mason, Contemporary Math. **193**, Amer. Math. Soc., Providence, 1996, 203 236.

[Li5]   H.-S. Li, Representation theory and tensor product theory for modules for a vertex operator algebra, Ph.D. thesis, Rutgers University, 1994.

[Li6]   H.-S. Li, The physics superselection principle in vertex operator algebra theory, preprint, 1995; *J. Algebra* **196** (1997), 436–457.

[Li7]   H.-S. Li, Extension of vertex operator algebras by a self-dual simple module, *J. Algebra* **187** (1997), 236–267.

[Li8]   H.-S. Li, An analogue of the Hom functor and a generalized nuclear democracy theorem, *Duke Math. J.* **93** (1998), 73–114.

[Li9]   H.-S. Li, Some finiteness properties of regular vertex operator algebras, *J. Algebra* **212** (1999), 495–514.

[Li10]   H.-S. Li, Determining fusion rules by $A(V)$-modules and bimodules, *J. Algebra* **212** (1999), 515–556.

[Li11]   H.-S. Li, On genus-zero correlation functions of vertex operator algebras, in: *Representations and Quantizations (Shanghai, 1998)*, China High. Educ. Press, Beijing, 2000, 297–316.

[Li12]   H.-S. Li, On abelian coset generalized vertex algebras, *Commun. Contemp. Math.* **3** (2001), 287–340.

[Li13]   H.-S. Li, Certain extensions of vertex operator algebras of affine type, *Commun. Math. Phys.* **217** (2001), 653–696.

[Li14]   H.-S. Li, Regular representations of vertex operator algebras, *Commun. Contemp. Math.* **4** (2002), 639–683.

[Li15]   H.-S. Li, The regular representation, Zhu's $A(V)$-theory and induced modules, *J. Algebra* **238** (2001), 159–193.

[Li16]   H.-S. Li, The regular representations and the $A_n(V)$-algebras, in: *Proceedings on Moonshine and Related Topics (Montreal, QC, 1999)*, CRM Proc. Lecture Notes **30**, Amer. Math. Soc., Providence, RI, 2001, 99–116.

[Li17]   H.-S. Li, Regular representations and Huang–Lepowsky tensor functors for vertex operator algebras, *J. Algebra* **255** (2002), 422–462.

[Li18]   H.-S. Li, Axiomatic G_1-vertex algebras, *Commun. Contemp. Math.* **5** (2003), 281–327.

[Li19]   H.-S. Li, A higher-dimensional generalization of the notion of vertex algebra, *J. Algebra* **262** (2003), 1–41.

[Li20]   H.-S. Li, Vertex (operator) algebras are "algebras" of vertex operators, in: *Vertex Operator Algebras in Mathematics and Physics, Proc. of Workshop at Fields Institute for Research in Mathematical Sciences, 2000*, ed. by S. Berman, Y. Billig, Y.-Z. Huang and J. Lepowsky, Fields Institute Communications **39**, Amer. Math. Soc., 2003, 111–122.

[Li21]   H.-S. Li, Simple vertex operator algebras are nondegenerate, *J. Algebra* **267** (2003), 199–211.

[Li22]   H.-S. Li, Vertex algebras and vertex Poisson algebras, *Commun. Contemp. Math.*, to appear.

[Li23]   H.-S. Li, On certain categories of modules for affine Lie algebras, preprint, arXiv:math.QA/0306293.

[LiW]   W. Li, Representations of vertex operator superalgebras and abelian intertwining algebras, Ph.D. thesis, Rutgers University, 1997.

[LQ1]   W.-P. Li and Z.-B. Qin, Vertex operator algebras and the blowup formula for the S-duality conjecture of Vafa and Witten, *Math. Res. Lett.* **5** (1998), 791–798.

[LQ2]   W.-P. Li and Z.-B. Qin, Vector bundles, the S-duality conjecture of Vafa and Witten, and vertex algebras, *First International Congress of Chinese Mathematicians (Beijing, 1998)*, AMS/IP Stud. Adv. Math., 20, Amer. Math. Soc., Providence, 2001, 235–240.

[LQ3]   W.-P. Li and Z.-B. Qin, On blowup formulae for the S-duality conjecture of Vafa and Witten. III. Relations with vertex operator algebras, *J. Reine Angew. Math.* **542** (2002), 173–217.

[LQW1]   W.-P. Li, Z.-B. Qin and W.-Q. Wang, Vertex algebras and the cohomology ring structure of Hilbert schemes of points on surfaces, *Math. Annalen* **324** (2002), 105–133.

[LQW2]   W.-P. Li, Z.-B. Qin and W.-Q. Wang, Generators for the cohomology ring of Hilbert schemes of points on surfaces, *Internat. Math. Res. Notices* **20** (2001), 1057–1074.

[LQW3]   W.-P. Li, Z.-B. Qin and W.-Q. Wang, Hilbert schemes and W algebras, *Internat. Math. Res. Notices* **27** (2002) 1427–1456.

[LX]   H.-S. Li and X.-P. Xu, A characterization of vertex algebras associated to even lattices, *J. Algebra* **173** (1995), 253–270.

[Lia]   B.-H. Lian, On the classification of simple vertex operator algebras, *Commun. Math. Phys.* **163** (1994), 307–357.

[LZ1]   B.-H. Lian and G. Zuckerman, BRST cohomology and noncompact coset models, in: *Proceedings of the XXth International Conference on Differential Geometric Methods in Theoretical Physics, Vol. 1, 2 (New York, 1991)*, World Sci. Publishing, River Edge, NJ, 1992, 849–865.

[LZ2]   B.-H. Lian and G. Zuckerman, New perspectives on the BRST-algebraic structure of string theory, *Commun. Math. Phys.* **154** (1993), 613–646.

[LZ3]   B.-H. Lian and G. Zuckerman, Some classical and quantum algebras, in: *Lie theory and Geometry*, ed. by J.-L. Brylinski, R. Brylinski, V. Guillemin and V. Kac, Progress in Math., Vol. 123, Birkhäuser, Boston, 1994, 509–529.

[LZ4]   B.-H. Lian and G. Zuckerman, Commutative quantum operator algebras, *J. Pure Applied Algebra* **100** (1995), Special Volume on Algebra and Physics, guest ed., J. D. Stasheff, 117–139.

[LZ5]   B.-H. Lian and G. Zuckerman, Moonshine cohomology, *Moonshine and Vertex Operator Algebra, Proc. of 1994 Workshop*, ed. by M. Miyamoto, Publ. Research Inst. for Math. Sciences, Kyoto Univ., 904 (1995), 87–115.

[Li]   K. J. Linde, Towards vertex algebras of Krichever–Novikov type, Part I, arXiv:math.QA/0305428.

[Mac]   I. G. Macdonald, Affine root systems and Dedekind's η-function, *Invent. Math.* **15** (1972), 91–143.

[MFF]   A. Malikov, B. L. Feigin and D. B. Fuchs, Singular vectors in Verma modules over Kac–Moody algebras, *J. Funct. Anal. Appl.* **20** (1986), 25–37.

[MSc]   A. Malikov and V. Schechtman, Chiral de Rham complex, II, in: *Differential topology, infinite-dimensional Lie algebras, and applications,* D.B. Fuchs's 60th Anniversary Collection, AMS Translations (2) **194** (1999), 149–188.

[MSV]   A. Malikov, V. Schechtman and A. Vaintrob, Chiral de Rham complex, *Commun. Math. Phys.* **204** (1999), 439–473.

[M]   S. Mandelstam, Dual resonance models, *Phys. Reports* **13** (1974), 259–353.

[Man1]   M. Mandia, Structure of the level one standard modules for the affine Lie algebras $B_l^{(1)}$, $F_4^{(1)}$ and $G_2^{(1)}$, Ph.D. thesis, Rutgers University, 1986.

[Man2]   M. Mandia, Structure of the level one standard modules for the affine Lie algebras $B_l^{(1)}$, $F_4^{(1)}$ and $G_2^{(1)}$, *Memoirs Amer. Math. Soc.* **65**, 1987.

[Mas1]   G. Mason, Modular forms and the theory of Thompson series, in: *Proc. Rutgers Group Theory Year, 1983–1984,* ed. by M. Aschbacher et al., Cambridge Univ. Press, Cambridge, 1984, 391–407.

[Mas2]   G. Mason, Finite groups and modular functions (with an appendix by S. P. Norton), in: *Representations of Finite Groups, Proc. 1986 Arcata Summer Research Institute,* ed. by P. Fong, Proc. Symp. Pure Math. **47**, Amer. Math. Soc., Providence, 1987, 181–210.

[MT]   G. Mason and M. P. Tuite, Torus chiral n-point functions for free boson and lattice vertex operator algebras, *Commun. Math. Phys.* **235** (2003), 47–68.

[MN1]   A. Matsuo and K. Nagatomo, A note on free bosonic vertex algebra and its conformal vectors, *J. Algebra* **212** (1999), 395–418.

[MN2]   A. Matsuo and K. Nagatomo, Axioms for a Vertex Algebra and the Locality of Quantum Fields, Memoirs Japan Math. Soc. **4**, 1999.

[MP1]   A. Meurman and M. Primc, Annihilating ideals of standard modules of $\widetilde{\mathfrak{sl}(2,\mathbb{C})}$ and combinatorial identities, *Advances in Math.* **64** (1987), 177–240.

[MP2]   A. Meurman and M. Primc, Vertex operator algebras and representations of affine Lie algebras, *Acta Applicandae Math.* **44** (1996), 207–215.

[MP3]   A. Meurman and M. Primc A basis of the basic $\tilde{\mathfrak{sl}}(3, C)$-module, *Commun. Contemp. Math.* **3** (2001), 593–614.

[MP4]   A. Meurman and M. Primc, Annihilating Fields of Standard Modules of $\widetilde{\mathfrak{sl}(2,\mathbb{C})}$ and Combinatorial Identities, preprint, 1994; Memoirs Amer. Math. Soc. **652**, 1999.

[Mi1]   A. Milas, Correlation functions, vertex operator algebras and ζ-functions, Ph.D. thesis, Rutgers University, 2001.

[Mi2]   A. Milas, Formal differential operators, vertex operator algebras and zeta-values, I, *J. Pure Applied Algebra,* to appear.

[Mi3]   A. Milas, Formal differential operators, vertex operator algebras and zeta-values, II, *J. Pure Applied Algebra,* to appear.

[Mi4]   A. Milas, Weak modules and logarithmic intertwining operators for vertex operator algebras, in: *Recent Developments in Infinite-Dimensional Lie Algebras and Conformal Field Theory (Charlottesville, VA, 2000),* Contemporary Math. **297**, Amer. Math. Soc., Providence, 2002, 201–225.

[Mi5]   A. Milas, Fusion rings for degenerate minimal models, *J. Algebra* **254** (2002), 300–335.

[Mi6]   A. Milas, Correlation functions, differential operators and vertex operator algebras, in: *Vertex Operator Algebras in Mathematics and Physics, Proc. of Workshop at Fields Institute for Research in Mathematical Sciences, 2000,* ed. by S. Berman, Y. Billig, Y.-Z. Huang and J. Lepowsky, Fields Institute Communications **39**, Amer. Math. Soc., 2003.

[Mis1]   K. C. Misra, Structure of the standard modules for $A_n^{(1)}$ and $C_n^{(1)}$, Ph.D. thesis, Rutgers University, 1982.

[Mis2]   K. C. Misra, Structure of certain standard modules for $A_n^{(1)}$ and the Rogers–Ramanujan identities, *J. Algebra* **88** (1984), 196–227.

[Mis3]   K. C. Misra, Structure of some standard modules for $C_n^{(1)}$, *J. Algebra* **90** (1984), 385–409.

[Mis4]   K. C. Misra, Standard representations of some affine Lie algebras, in: *Vertex Operators in Mathematics and Physics, Proc. 1983 M.S.R.I. Conference,* ed. by J. Lepowsky, S. Mandelstam and I. M. Singer, Publ. Math. Sciences Res. Inst. # 3, Springer-Verlag, 1985, 163–183.

[Mis5]   K. C. Misra, Realization of the level two standard $\mathfrak{sl}(2k+1, \mathbb{C})$-modules, *Trans. Amer. Math. Soc.* **316** (1989), 295–309.

[Mis6]   K. C. Misra, Realization of the level one standard $\widetilde{C}_{2k+1}$-modules, *Trans. Amer. Math. Soc.* **321** (1990), 483–504.

[Mis7]   K. C. Misra, Level one standard modules for affine symplectic Lie algebras, *Math. Annalen* **287** (1990), 287–302.

[Mis8]   K. C. Misra, Level two standard $\tilde{A}_n$-modules, *J. Algebra* **137** (1991), 56–76.

[Miy1]   M. Miyamoto, 21 involutions acting on the Moonshine module, *J. Algebra* **175** (1995), 941–965.

[Miy2]   M. Miyamoto, Griess algebras and conformal vectors in vertex operator algebras, *J. Algebra* **179** (1996), 523–548.

[Miy3]   M. Miyamoto, Binary codes and vertex operator (super)algebras, *J. Algebra* **181** (1996), 207–222.

[Miy4]   M. Miyamoto, Representation theory of code vertex operator algebra, *J. Algebra* **201** (1998), 115–150.

[Miy5]   M. Miyamoto, A Hamming code vertex operator algebra and construction of vertex operator algebras, *J. Algebra* **215** (1999), 509–530.

[Miy6]   M. Miyamoto, A modular invariance on the theta functions defined on vertex operator algebras, *Duke Math. J.* **101** (2000), 221–236.

[Miy7]   M. Miyamoto, 3-state Potts model and automorphisms of vertex operator algebras of order 3, *J. Algebra* **239** (2001), 56–76.

[Miy8]   M. Miyamoto, Intertwining operators and modular invariance, preprint, arXiv:math.QA/0010180.

[Miy9]   M. Miyamoto, Modular invariance of vertex operator algebras satisfying C_2-cofiniteness, preprint, arXiv:math.QA/0209101

[Mon1]   P. Montague, Orbifold constructions and the classification of self-dual $c = 24$ conformal field theories, *Nucl. Phys.* **B428** (1994), 233–258.

[Mon2]   P. Montague, Third and higher order NFPA twisted constructions of conformal field theories from lattices, *Nucl. Phys.* **B441** (1995), 337–382.

[Mon3]   P. Montague, On representations of conformal field theories and the construction of orbifolds, *Lett. Math. Phys.* **38** (1996), 1–11.

[Mo]   R. V. Moody, A new class of Lie algebras, *J. Algebra* **10** (1968), 211–230.

[MoP]   R. V. Moody and A. Pianzola, *Lie Algebras with Triangular Decomposition*, Canadian Mathematical Society Series of Monographs and Advanced Texts, Wiley, New York, 1995.

[MRY]   R. V. Moody, S. Eswara Rao and T. Yokonuma, Toroidal Lie algebras and vertex representations, *Geom. Dedicata* **35** (1990), 283–307.

[MSe]   G. Moore and N. Seiberg, Classical and quantum conformal field theory, *Commun. Math. Phys.* **123** (1989), 177–254.

[Mos1]   G. Mossberg, A generalized Jacobi identity in axiomatic vertex operator algebras, licentiate thesis, Univ. of Lund, 1992, Report Univ. of Lund, Dept. of Math., 1992:10.

[Mos2]   G. Mossberg, Axiomatic vertex algebras and the Jacobi identity, *J. Algebra* **170** (1994), 956–1010.

[NT]   K. Nagatomo and A. Tsuchiya, Conformal field theories associated to regular chiral vertex operator algebras I: theories over the projective line, preprint, arXiv:math.QA/0206223.

[Na]   W. Nahm, Quantum field theories in one and two dimensions, *Duke Math. J.* **54** (1987), 579–613.

[NS]   A. Neveu and J. H. Schwarz, Factorizable dual model of pions, *Nucl. Phys.* **B31** (1971), 86–112.

[Ni]   N. M. Nikolov, Vertex algebras in higher dimensions and global conformally invariant quantum field theory, preprint, arXiv:hep-th/0307235.

[N]   S. P. Norton, More on moonshine, in: *Computational Group Theory,* ed. by M. D. Atkinson, Academic Press, New York, 1984, 185–193.

[O]   A. P. Ogg, Automorphismes des courbes modulaires, Séminaire Delange–Pisot–Poitou, 16e année (1974/75), no 7.

[Pf]   R. Pfister, Spin representations of $A_1^{(1)}$, Ph.D. thesis, Rutgers University, 1984.

[Pl]   V. Pless, On the uniqueness of the Golay codes, *J. Combinatorial Theory* **5** (1968), 215–228.

[Po]   J. Polchinski, *String Theory*, **1** and **2**, Cambridge University Press, Cambridge, 1998.

[PS]   A. Pressley and G. Segal, *Loop Groups*, Oxford Univ. Press, Oxford, 1986.

[P1]   M. Primc, Standard representations of $A_n^{(1)}$, in: *Infinite Dimensional Lie Algebras and Groups, Advanced Series in Math. Physics*, Vol. 7, ed. by V. Kac, World Scientific, Singapore, 1989, 273–282.

[P2]   M. Primc, Vertex operator construction of standard modules for $A_n^{(1)}$, *Pacific J. Math.* **162** (1994), 143–187.

[P3]   M. Primc, Introduction to vertex algebras, in: *Functional Analysis IV, Dubrovnik Conference Proceedings*, Various Publication Series, No. 43, Aarhus University, 1994, 221–244.

[P4]   M. Primc, Some crystal Rogers–Ramanujan type identities, *Glas. Mat. Ser. III* **34** (1999), 73–86.

[P5]   M. Primc, Vertex algebras generated by Lie algebras, *J. Pure Applied Algebra* **135** (1999), 253–293.

[P6]   M. Primc, Basic representations for classical affine Lie algebras, *J. Algebra* **228** (2000), 1–50.

[P7]   M. Primc, Vertex algebras and combinatorial identities. The 2000 Twente Conference on Lie Groups (Enschede), *Acta Appl. Math.* **73** (2002), 221–238.

[P8]   M. Primc, Relations for annihilating fields of standard modules for affine Lie algebras, in: *Vertex Operator Algebras in Mathematics and Physics, Proc. of Workshop at Fields Institute for Research in Mathematical Sciences, 2000*, ed. by S. Berman, Y. Billig, Y.-Z. Huang and J. Lepowsky, Fields Institute Communications **39**, Amer. Math. Soc., 2003.

[P9]   M. Primc, Generators of relations for annihilating fields, preprint, arXiv:math.QA/0204283.

[QW]   Z.-B. Qin and W.-Q. Wang, Hilbert schemes and symmetric products: a dictionary, in: *Orbifolds in Mathematics and Physics (Madison, WI, 2001)*, Contemp. Math., 310, Amer. Math. Soc., Providence, 2002, 233–257.

[Rad]   D. Radnell, Schiffer variation in Teichmüller space, determinant line bundles and modular functors, Ph.D. thesis, Rutgers University, 2003.

[R]   P. Ramond, Dual theory for free fermions, *Phys. Rev.* **D3** (1971), 2415–2418.

[Ray]   U. Ray, Generalized Kac–Moody algebras and some related topics, *Bull. Amer. Math. Soc. (New Series)* **38** (2001), 1–42.

[Re]   A. Recknagel, From branes to boundary conformal field theory: draft of a dictionary, in: *Vertex Operator Algebras in Mathematics and Physics, Proc. of Workshop at Fields Institute for Research in Mathematical Sciences, 2000*, ed. by S. Berman, Y. Billig, Y.-Z. Huang and J. Lepowsky, Fields Institute Communications **39**, Amer. Math. Soc., 2003.

[Ret]   A. Retakh, Associative conformal algebras of linear growth, *J. Algebra* **237** (2001), 769–788.

[Roc]   A. Rocha-Caridi, Vacuum vector representations of the Virasoro algebra, in: *Vertex Operators in Mathematics and Physics, Proc. 1983 M.S.R.I. Conference*, ed. by J. Lepowsky, S. Mandelstam and I. M. Singer, Publ. Math. Sciences Res. Inst. # 3, Springer-Verlag, 1985, 451–473.

[Roi1]   M. Roitman, On free conformal and vertex algebras, *J. Algebra* **217** (1999), 496–527.

[Roi2]   M. Roitman, On conformal and vertex algebras, Ph.D. thesis, Yale University, 2000.

[Roi3]   M. Roitman, Universal enveloping conformal algebras, *Selecta Mathematica*, New Series **6** (2000), 319–345.

[Roi4]   M. Roitman, Conformal subalgebras of lattice vertex algebras, preprint, arXiv:math.QA/0011243.

[Roi5]   M. Roitman, On twisted representations of vertex algebras, *Advances in Math.* **176** (2003), 53–88.

[Roi6]   M. Roitman, Combinatorics of free vertex algebras, *J. Algebra* **255** (2002), 297–323.

[Roi7]   M. Roitman, Invariant bilinear forms on a vertex algebra, preprint, arXiv:math.QA/0210432.

[Ros]   M. Rosellen, OPE-algebras, Ph.D. thesis, Max-Planck Institute, Bonn, Germany, 2002.

[SYa]   S. Sakuma and H. Yamauchi, Vertex operator algebra with two Miyamoto involutions generating S_3, preprint, arXiv:math.QA/0207117.

[Sc1]   A. Schellekens, Classification of ten-dimensional heterotic strings, *Phys. Lett.* **B277** (1992), 277–284.

312   References

[Sc2]   A. Schellekens, Meromorphic $c = 24$ conformal field theories, *Commun. Math. Phys.* **153** (1993), 159–185.

[SY]   A. Schellekens and S. Yankielowicz, Extended chiral algebras and modular invariant partition functions, *Nucl. Phys.* **B327** (1989), 673–703.

[Sch]   V. Schomerus, Open strings and non-commutative geometry, in: *Vertex Operator Algebras in Mathematics and Physics, Proc. of Workshop at Fields Institute for Research in Mathematical Sciences, 2000,* ed. by S. Berman, Y. Billig, Y.-Z. Huang and J. Lepowsky, Fields Institute Communications **39**, Amer. Math. Soc., 2003.

[SF]   C. Schweigert and J. Fuchs, The world sheet revisited, in: *Vertex Operator Algebras in Mathematics and Physics, Proc. of Workshop at Fields Institute for Research in Mathematical Sciences, 2000,* ed. by S. Berman, Y. Billig, Y.-Z. Huang and J. Lepowsky, Fields Institute Communications **39**, Amer. Math. Soc., 2003.

[Se1]   G. Segal, Unitary representations of some infinite-dimensional groups, *Commun. Math. Phys.* **80** (1981), 301–342.

[Se2]   G. Segal, The definition of conformal field theory, preprint.

[Se3]   G. Segal, Two-dimensional conformal field theories and modular functors, in: *Proc. IXth Internat. Congr. Math. Phys., Swansea, 1988,* Hilger, Bristol, 1989, 22–37.

[Sh]   H. Shimakura, Decompositions of the Moonshine Module with respect to subVOAs associated to codes over $\mathbb{Z}_{2k}$, *J. Algebra* **251** (2002), 308–322.

[SU]   Y. Shimizu and K. Ueno *Advances in Moduli Theory,* Translations of Mathematical Monographs, Vol. 206, American Math. Soc., 1999.

[St]   J. Stillwell, Modular miracles, *American Math. Monthly 108* (2001), 70–76.

[Su]   H. Sugawara, A field theory of currents, *Phys. Rev.* **170** (1968), 1659–1662.

[Sz]   M. Szczesny, Wakimoto modules for twisted affine Lie algebras, *Math. Res. Lett.* **9** (2002), 433–448.

[T1]   S. Tan, TKK algebras and vertex operator representations, *J. Algebra* **211** (1999), 298–342.

[T2]   S. Tan, Vertex operator representations for toroidal Lie algebras of type B_l, *Commun. Algebra* **27** (1999), 3593–3618.

[T3]   S. Tan, Principal construction of the toroidal Lie algebra of type A_1, *Math. Z.* **230** (1999), 621–657.

[Th1]   J. G. Thompson, Finite Groups and modular functions, *Bull. London Math. Soc.* **11** (1979), 347–351.

[Th2]   J. G. Thompson, Some numerology between the Fischer–Griess Monster and elliptic modular functions, *Bull. London Math. Soc.* **11** (1979), 352–353.

[Ti1]   J. Tits, Le Monstre (d'après R. Griess, B. Fischer et al.), Séminaire Bourbaki, exposé no 620, 1983/84, *Astérisque* **121–122** (1985), 105–122.

[Ti2]   J. Tits, On R. Griess' "Friendly Giant," *Invent. Math.* **78** (1984), 491–499.

[Ti3]   J. Tits, Le module du "moonshine" (d'après I. Frenkel, J. Lepowsky et A. Meurman), Séminaire Bourbaki, exposé no 684, 1986/87, *Astérisque* **152–153** (1987), 5, 285–303 (1988).

[T]   A. Tsuchiya, Moduli of stable curves, conformal field theory and affine Lie algebras, in: *Proc. Internat. Congr. Math., Kyoto, 1990,* 1409–1491.

[TK1]   A. Tsuchiya and Y. Kanie, Fock space representations of the Virasoro algebra—intertwining operators, Publ. Research Inst. for Math. Sciences, Kyoto Univ., Vol. 22 (1986), 259–327.

[TK2]   A. Tsuchiya and Y. Kanie, Vertex operators in conformal field theory on $\mathbf{P}^1$ and monodromy representations of braid group, *Lett. Math. Phys.* **13** (1987), 303.

[TK3]   A. Tsuchiya and Y. Kanie, Vertex operators in conformal field theory on $\mathbf{P}^1$ and monodromy representations of braid group, in: *Conformal Field Theory and Solvable Lattice Models, Advanced Studies in Pure Math.,* Vol. 16, Kinokuniya Company Ltd., Tokyo, 1988, 297–372.

[TUY]   A. Tsuchiya, K. Ueno and Y. Yamada, Conformal field theory on moduli family of stable curves with gauge symmetries, in: *Conformal Field Theory and Solvable Lattice Models, Advanced Studies in Pure Math.,* Vol. 19, Kinokuniya Company Ltd., Tokyo, 1989, 459–566.

[Ts1]   H. Tsukada, String path integral realization of vertex operator algebras, Ph.D. thesis, Rutgers University, 1988.

[Ts2]   H. Tsukada, Vertex operator superalgebras, *Commun. Algebra* **18** (1990), 2249–2274.

[Ts3]   H. Tsukada, String path integral realization of vertex operator algebras, *Memoirs Amer. Math. Soc.* **91**, 1991.

[Tu1]   M. Tuite, Monstrous moonshine from orbifolds, *Commun. Math. Phys.* **146** (1992), 277–309.

[Tu2]   M. Tuite, Generalized moonshine and abelian orbifold constructions, in: *Moonshine, the Monster and Related Topics, Proc. Joint Summer Research Conference, Mount Holyoke*, 1994, ed. by C. Dong and G. Mason, Contemporary Math. **193**, Amer. Math. Soc., Providence, 1996, 353–368.

[Tu3]   M. Tuite, On the relationship between monstrous moonshine and the uniqueness of the moonshine module, *Commun. Math. Phys.* **166** (1995), 495–532.

[Va]   C. Vafa, Conformal theories and punctured surfaces, *Phys. Lett.* **B199** (1987), 195–202.

[Ve]   G. Veneziano, Construction of a crossing-symmetric, Regge-behaved amplitude for linearly rising trajectories, *Nuovo Cim.* **57A** (1968), 190–197.

[Ve]   E. Verlinde, Fusion rules and modular transformations in 2D conformal field theory, *Nucl. Phys.* **B300** (1988), 360–376.

[V]   M. A. Virasoro, Subsidiary conditions and ghosts in dual-resonance models, *Phys. Rev.* **D1** (1970), 2933–2936.

[W]   M. Wakimoto, Fock representations of affine Lie algebra $A_1^{(1)}$, *Commun. Math. Phys.* **104** (1986), 605–609.

[Wa1]   W.-Q. Wang, Rationality of Virasoro vertex operator algebras, *Internat. Math. Res. Notices* (in *Duke Math. J.*) **7** (1993), 197–211.

[Wa2]   W.-Q. Wang, $W_{1+\infty}$ algebra, W_3 algebra, and Friedan–Martinec–Shenker bosonization, *Commun. Math. Phys.* **195** (1998), 95–111.

[Wa3]   W.-Q. Wang, Duality in infinite dimensional Fock representations, *Commun. Contemp. Math.* **1** (1999), 155–199.

[Wa4]   W.-Q. Wang, Equivariant K-theory, wreath products, and Heisenberg algebra, *Duke Math. J.* **103** (2000), 1–23.

[Wa5]   W.-Q. Wang, Algebraic structures behind Hilbert schemes and wreath products, in: *Recent Developments in Infinite-dimensional Lie Algebras and Conformal Field Theory (Charlottesville, VA, 2000)*, Contemp. Math., 297, Amer. Math. Soc., Providence, 2002, 271–295.

[Wa6]   W.-Q. Wang, Vertex algebras and the class algebras of wreath products, preprint, arXiv:math.QA/0203004; *Proc. London Math. Soc.*, to appear.

[Wa7]   W.-Q. Wang, Equivariant K-theory, generalized symmetric products, and twisted Heisenberg algebra, *Commun. Math. Phys.* **234** (2003), 101–127.

[Wi]   A. S. Wightman, Quantum field theory in terms of vacuum expectation values, *Phys. Rev.* **101** (1956), 860–866.

[W1]   K. Wilson, Non-Lagrangian models of current algebra, *Phys. Rev.* **179** (1969), 1499–1512.

[W2]   K. Wilson, OPE and anomalous dimensions in the Thirring model, *Phys. Rev.* **D2** (1970), 1473–1477.

[Wi1]   E. Witten, Non-abelian bosonization in two dimensions, *Commun. Math. Phys.* **92** (1984), 455–472.

[Wi2]   E. Witten, Physics and Geometry, *Proc. Internat. Congr. Math., Berkeley, 1986*, Vol. 1, American Math. Soc., 1987, 267–303.

[Xi1]   C.-F. Xie, On the relations between standard modules and vertex operator algebras, Ph.D. thesis, Rutgers University, 1994.

[Xi2]   C.-F. Xie, On the relations between standard modules and vertex operator algebras, *J. Algebra* **176** (1995), 591–620.

[Xi3]   C.-F. Xie, The equivalence between two Jacobi identities for twisted vertex operators, *Commun. Algebra* **23** (1995), 2453–2467.

[Xu1]   X.-P. Xu, Untwisted and twisted gluing techniques for constructing self-dual lattices, Ph.D. thesis, Rutgers University, 1992.

[Xu2]   X.-P. Xu, Characteristics of spinor vertex algebras and their modules, Hong Kong University of Sciences and Technology, 1992, Technical report 92-1-2.

[Xu3]   X.-P. Xu, Lattice coset of vertex operators, preprint 1991; *Alg. Colloq.* **2** (1995), 241–268.

[Xu4]   X.-P. Xu, Analogue of the vertex operator triality for ternary codes, preprint 1991; *J. Pure Applied Algebra* **99** (1995), 53–111.

[Xu5]   X.-P. Xu, K-graded algebras and vertex operator algebras, *Commun. Algebra* **23** (1995), 291–311.

[Xu6]   X.-P. Xu, Intertwining operators for twisted modules of a colored vertex operator superalgebra, *J. Algebra* **175** (1995), 241–273.

[Xu7]   X.-P. Xu, Twisted modules of colored lattice vertex operator superalgebras, *Quart. J. Math. Oxford* **47** (1996), 233–259.

[Xu8]   X.-P. Xu, On vertex operator algebras, Lecture Notes Series 32, Research Institute of Mathematics/Global Analysis Research Center, Seoul National University, 1996.

[Xu9]   X.-P. Xu, On spinor vertex operator algebras and their modules, preprint 1992; *J. Algebra* **191** (1997), 427–460.

[Xu10]   X.-P. Xu, Braid group actions, intertwining operators and ternary moonshine spaces, *Far East J. Math. Sci.* **5** (1997), 337–380.

[Xu11]   X.-P. Xu, Vertex operator algebras stemming from graded associative algebras, preprint 1996.

[Xu12]   X.-P. Xu, Introduction to Vertex Operator Superalgebras and Their Modules, Kluwer Academic, Dordrecht/Boston/London, 1998.

[Ya1]   H. Yamauchi, Orbifold Zhu theory associated to intertwining operators, preprint, arXiv:math.QA/0201054.

[Ya2]   H. Yamauchi, Modularity on vertex operator algebras arising from semisimple primary vectors, preprint, arXiv:math.QA/0209026.

[Ya3]   H. Yamauchi, Module categories of simple current extensions of vertex operator algebras, preprint, arxiv:math.QA/0211255.

[Y1]   G. Yamskulna, Vertex operator algebras and dual pairs, Ph.D. thesis, University of California, Santa Cruz, 2001.

[Y2]   G. Yamskulna, The relationship between skew group algebras and orbifold theory, *J. Algebra* **256** (2002), 502–517.

[Y3]   G. Yamskulna, C_2-cofiniteness of vertex operator algebra V_L^+ when L is a rank one lattice, *Commun. Algebra*, to appear.

[Zak]   I. Zahkarevich, Geometric vertex operators, preprint, arXiv:math.AG/9805096.

[Za]   A. B. Zamolodchikov, Infinite additional symmetries in two dimensional conformal quantum field theory, *Theor. Math. Phys.* **65** (1985), 1205–1213.

[ZF1]   A. B. Zamolodchikov and V. A. Fateev, Nonlocal (parafermion) currents in two-dimensional conformal quantum field theory and self-dual critical points in Z_N-symmetric statistical systems, *Sov. Phys. JETP* **62** (1985), 215–225.

[ZF2]   A. B. Zamolodchikov and V. A. Fateev, Disorder fields in two-dimensional conformal quantum field theory and $N = 2$ extended supersymmetry, *Sov. Phys. JETP* **63** (1986), 913–919.

[Ze]   E. Zelmanov, On the structure of conformal algebras, in: *Combinatorial and Computational Algebra (Hong Kong, 1999)*, Contemp. Math. **264**, Amer. Math. Soc., Providence, RI, 2000, 139–153.

[Zh]   W. Zhao, Some generalizations of genus zero two-dimensional conformal field theory, in: *Recent Developments in Infinite-Dimensional Lie Algebras and Conformal Field Theory (Charlottesville, VA, 2000)*, Contemporary Math. **297**, Amer. Math. Soc., Providence, 2002, 297–334.

[Z1]   Y.-C. Zhu, Vertex operator algebras, elliptic functions and modular forms, Ph.D. thesis, Yale University, 1990.

[Z2]   Y.-C. Zhu, Modular invariance of characters of vertex operator algebras, *J. Amer. Math. Soc.* **9** (1996), 237–302.

[Z3]   Y.-C. Zhu, Global vertex operators on Riemann surfaces, *Commun. Math. Phys.* **165** (1994), 485–531.

Index

Progress in Mathematics

Edited by:

Hyman Bass
Dept. of Mathematics
University of Michigan
Ann Arbor, MI 48109
USA
hybass@umich.edu

Joseph Oesterlé
Equipe de théorie des nombres
Université Paris 6
175 rue du Chevaleret
75013 Paris
FRANCE
oesterle@math.jussieu.fr

Alan Weinstein
Dept. of Mathematics
University of California
Berkeley, CA 94720
USA
alanw@math.berkeley.edu

Progress in Mathematics is a series of books intended for professional mathematicians and scientists, encompassing all areas of pure mathematics. This distinguished series, which began in 1979, includes authored monographs and edited collections of papers on important research developments as well as expositions of particular subject areas.

We encourage preparation of manuscripts in some form of T$_E$X for delivery in camera-ready copy which leads to rapid publication, or in electronic form for interfacing with laser printers or typesetters.

Proposals should be sent directly to the editors or to: Birkhäuser Boston, 675 Massachusetts Avenue, Cambridge, MA 02139, USA or to Birkhauser Verlag, 40-44 Viadukstrasse, CH-4051 Basel, Switzerland.